T0132700

CHAPMAN & HALL/CRC
Monographs and Surveys in
Pure and Applied Mathematics   114

FUNDAMENTALS OF

INFINITE DIMENSIONAL

REPRESENTATION

THEORY

CHAPMAN & HALL/CRC
Monographs and Surveys in
Pure and Applied Mathematics   114

FUNDAMENTALS OF

INFINITE DIMENSIONAL

REPRESENTATION

THEORY

RAYMOND C. FABEC

CHAPMAN & HALL/CRC

Boca Raton   London   New York   Washington, D.C.

Library of Congress Cataloging-in-Publication Data

Fabec, Raymond C.
　　Fundamentals of infinite dimensional representation theory / Raymond C. Fabec.
　　　　p.　cm. — (Monographs and surveys in pure and applied mathematics ; 114)
　　Includes bibliographical references and index.
　　ISBN 1-58488-212-3 (alk. paper)
　　1. Representations of groups. 2. Infinite groups I. Title. II. Chapman & Hall/CRC
monographs and surveys in pure and applied mathematics ; 114.
QA178 .F33 2000
512′.2—dc21
　　　　　　　　　　　　　　　　　　　　　　　　　　　　　00-038422
　　　　　　　　　　　　　　　　　　　　　　　　　　　　　CIP

No claim to original U.S. Government works
International Standard Book Number 1-58488-212-3
Library of Congress Card Number 00-038422
Printed in the United States of America 1 2 3 4 5 6 7 8 9 0
Printed on acid-free paper

Dedication

To Sandra, Kevin, and Brian

Contents

Raymond C. Fabec received his Bachelor of Arts degree in Mathematics in 1968. He worked at the National Center for Atmospheric Research in Boulder, Colorado in the early 70's as a scientific programmer and received his Ph. D. degree in Mathematics from the University of Colorado–Boulder in 1976. After graduate school, he became Assistant Professor of Mathematics at Louisiana State University in Baton Rouge where he currently is Professor of Mathematics. His publications have focused on abstract group representation theory, the structure of ergodic group actions, and the use of Fourier analysis in analyzing operators and functions space on groups. He is a member of the American Mathematical Society.

PREFACE

Infinite dimensional representation theory has blossomed in the later half of the 20th century. It started in earnest with the development of quantum mechanics and the importance of symmetry groups in particle mechanics. Motivation came from the success of the theory of characters developed by Frobenius and Schur toward the end of the 19th century, the results of Weyl on representation theory of compact Lie groups, the result of Haar on the existence of invariant measures on locally compact groups, and the generalization of Fourier analysis to arbitrary abelian locally compact groups. The correspondence between observables in quantum mechanics with self adjoint operators in operator algebras in the logical foundations of quantum mechanics spurred further developments, particularly the works of Murray and von Neumann on operator rings. These were used by Mackey and others to develop a general abstract model for unitary representation theory and to the development of the notion of induced representation that now plays such an important role in the subject. This notion has been refined and continues to be central in nonunitary and algebraic infinite dimensional representation theory.

Our intent is to present many of the central results of analytic abstract representation theory. In this development, we hope to present Mackey's methods in an accessible manner, emphasizing the notion of inducing actions and viewing unitary representations as actions by unitary operators. We present the technical substructure of the subject to justify the methods produced and to make them accessible for other applications. In this presentation, the subject is developed from an abstract point of view, where the existence and properties of concepts introduced must be established from abstract principles in functional analysis and measure theory. Some measure theoretic difficulties which never occur in practice if resolved could enhance the presentation. For instance, whether every locally simple unitary representation of a locally compact group is supported by a standard measure on the dual remains unknown. This leads to statements assuming both hypotheses.

The correspondence between unitary representations of groups and *-representations of their group algebras intimately links the representation theory of groups and the representation theory of C* and von Neumann algebras. Consequently, both are covered in the text.

The basic theory of operator algebra theory and the structure of Hilbert algebras are presented along with a relatively readable account of Tomita–Takesaki theory on modular Hilbert algebras.

Chapters II, III, and IV can be read independently. Chapters III and IV do depend in part on Chapter I, but material there can be referred to according to need. Chapter V can be read after IV and most of Chapter VII after Chapters II and III. Chapter VIII depends on nearly every earlier chapter except possibly Chapter V.

The book is self contained in the sense it depends only on a firm background in real analysis. The spectral theorem for bounded self adjoint operators is derived as part of representation theory. We do use the spectral theorem for unbounded operators in a few instances in Chapter VII.

Topics selected were the choice of the author. They were chosen in part for their relevance to representation theory and in part because of their familiarity to the author. We have made use of many classic texts on representation theory and operator algebras as well as classes and papers from the developers of the subject.

We apologize for the book's shortcomings, including any errors for which the author takes responsibility. We hope, however, that it will serve both as a general reference and as a text to the student needing access to basic group-operator algebra representation theory. Furthermore, to the careful reader, some of the beauty of the subject may be revealed.

CHAPTER I

BOREL SPACES AND SELECTION THEOREMS

Complete separable metric spaces and measures on these spaces are fundamental in many areas of analysis. This is especially true for both the unitary representation theory of locally compact groups and the structure of separable von Neumann algebras.

1. Borel Structures on Polish Spaces

Let X be a topological space. X is said to be a **Polish space** if it is second countable and has a complete metric which generates the topology. The Borel algebra $\mathcal{B}$ of sets is the sigma-algebra generated by the open sets in X. If $\mathcal{M}$ is a σ-algebra of subsets of X, $\mathcal{M}$ is called a **standard Borel structure** on X if $\mathcal{M}$ is the Borel algebra of a Polish topology on X; then X together with $\mathcal{M}$ is called a **standard Borel space**.

A σ-algebra $\mathcal{B}$ on a set X is said to be **countably generated** if it has a countable subcollection which separates points and for which the smallest σ-algebra containing it is $\mathcal{B}$.

THEOREM I.1. *A space X is countably generated iff it is Borel isomorphic to a subset of the Cantor set with the relative Borel structure.*

PROOF. Let A_i be a countable separating family of Borel sets. Define $f(x) = (\chi_i(x))$ where χ_i is the characteristic function of the set A_i. Then f is a one-to-one correspondence between X and $f(X)$ which maps the algebra generated by the A_i to the restriction of the algebra of cylinder sets in the Cantor set to the image of f. $\square$

REMARK. This proof shows that a Borel space X is countably separated iff there is a one-to-one Borel mapping of X into the Cantor set.

A space is **zero dimensional** if the topology has a base consisting of open and closed sets.

PROPOSITION I.1. *A subset of a Polish space with the relative topology is Polish iff it is a G_δ.*

PROOF. Let X be a Polish space with metric ρ, and let A be a subspace with complete separable metric ρ_0. For each $n > 0$ and each $a \in A$ choose an open set $U(n, a)$ containing a which has ρ diameter less than $\frac{1}{2^n}$ and whose intersection with A has ρ_0 diameter less than $\frac{1}{2^n}$. Set $G_n = \cup_a U(n, a)$. Suppose $x \in G_n \forall n$. Then $x \in U(n, a_n)$ for some a_n. Clearly a_n converges to x in X. But $U(n, a_n) \cap U(m, a_m) \neq \emptyset$ implies $\rho_0(a_n, a_m) < \frac{1}{2^n} + \frac{1}{2^m}$ and hence a_n converges in A since ρ_0 is complete. Thus $x \in A$ and $A = \cap_n G_n$ is a G_δ.

Suppose $A = \cap_n G_n$ is a G_δ in X. Define $f : A \to X \times \mathbb{R}^\mathbb{N}$ by $f(a) = (a, (\frac{1}{\rho(a, X - G_n)}))$. Then f is a homeomorphism from A onto a closed set of the Polish space $X \times \mathbb{R}^\mathbb{N}$. Hence A is Polish. □

Define η to be the space of all sequences $a = \{a_n\}$ of natural numbers with metric $\rho(a, b) = \sum \frac{|a_i - b_i|}{2^i(1 + |a_i - b_i|)}$. This complete separable space is known to be topologically isomorphic to the space of positive irrational numbers with the relative topology from the reals. Note that η is a zero dimensional perfect space.

LEMMA I.1. *Let X be a perfect Polish space. Then every nonempty open subset of X is uncountable.*

PROOF. Let U be a nonempty open subset of X. Then U with the relative topology is both perfect and Polish. Hence one only needs to show perfect Polish spaces X are uncountable. Suppose $X = \{x_i\}$ is countable. By the Baire category theorem, there exists i such that $\overline{\{x_i\}}$ has interior. Thus $\{x_i\}$ is an open set and x_i is an isolated point. □

THEOREM I.2. *Let X be zero dimensional perfect Polish space. Then if D is any countable infinite dense subset of X, $X - D$ is homeomorphic to η.*

PROOF. We note that if U is any open neighborhood of x and $\epsilon > 0$, then $U - \{x\}$ can be written as a union of a sequence of disjoint nonempty open closed sets of diameter less than ϵ. Indeed, write U as a union of open closed sets V_i of diameter $< \epsilon$. Assume $x \in V_1$ and take W_i a strictly decreasing sequence of open closed sets whose intersection is $\{x\}$. Then $V_1 - W_1, W_1 - W_2, W_2 - W_3, \ldots$ along with $V_2 - V_1, V_3 - V_2 - V_1, V_4 - V_3 - V_2 - V_1, \ldots$ form a countable collection of open closed sets infinitely many of which are nonempty and thus form the desired sequence.

Let $D = \{a_i\}$. Since X is zero dimensional, there is a collection U_i of pairwise disjoint nonempty open sets of ρ diameter < 1 whose union is X. Let b_i be the first element in D which is in U_i. By the note above, each $U_i - \{b_i\}$ can be written as a union of a sequence of nonempty open closed sets $U_{i,j}$ of diameter $< \frac{1}{2}$. Let $b_{i,j}$ be the first element in D not amongst the b_i that is in $U_{i,j}$. Inductively there is a cover $\cup_{i_{n+1}} U_{i_1,i_2,\ldots,i_{n+1}}$ of $U_{i_1,i_2,\ldots,i_n} - \{b_{i_1,i_1,\ldots,i_n}\}$ by disjoint nonempty open closed sets of diameter $< \frac{1}{n+1}$. Thus if $c = (c_i) \in \eta$, then $\cap_k U_{c_1,c_2,\ldots,c_k} = \{f(c)\}$. Clearly f is one-to-one and onto $X - D$; and since $f(\{c_1\} \times \cdots \times \{c_k\} \times \eta) = U_{c_1,\ldots,c_k}$ and $f^{-1}(U_{c_1,\ldots,c_k}) = \{c_1\} \times \cdots \times \{c_k\} \times \eta$, f maps a base for the topology on η to a base for the topology on $X - D$. Thus X is homeomorphic with η. $\square$

THEOREM I.3. *Let X be a Polish space. Then if B is a Borel subset of X, there is a finer zero dimensional Polish topology for X for which B is an open and closed set and which has the same Borel sets as the original topology.*

PROOF. Let $\mathcal{T}$ be the topology for X and let $\mathcal{B}_X$ be the corresponding collection of Borel sets. Let $\mathcal{B}'$ be the collection of sets $A \in \mathcal{B}_X$ for which there exists a Polish topology $\mathcal{T}_A$ with $\mathcal{T} \subset \mathcal{T}_A \subset \mathcal{B}_X$ such that A is an open and closed set in $\mathcal{T}_A$. Suppose A and $X - A$ are G_δ subsets with complete separable metrics ρ_1 and ρ_2. We may assume $\rho_1 < 1$ and $\rho_2 < 1$. Define a metric ρ by $\rho = \rho_1$ on $A \times A$, $\rho = \rho_2$ on $(X - A) \times (X - A)$, and $\rho = 1$ on the rest of $X \times X$. Then ρ is a complete separable metric and A is open and closed in the topology of ρ. Hence $A \in \mathcal{B}'$. In particular, the collection $\mathcal{B}'$ contains the open sets and is closed under complements.

To see that $\mathcal{B}'$ is a σ-algebra, it suffices to show it is closed under countable unions. Let $A_n \in \mathcal{B}'$ for all n. Set $\mathcal{T}_1$ to be the topology generated by $\cup_n \mathcal{T}_{A_n}$. Then $\mathcal{T}_1$ is Polish for if ρ_n is a complete separable metric for $\mathcal{T}_{A_n}$, then $\sum \frac{1}{2^n} \frac{\rho_n}{1+\rho_n}$ is a complete separable metric for $\mathcal{T}_1$. Since $A = \cup_n A_n$ is open in $\mathcal{T}_1$, one can enlarge $\mathcal{T}_1$ as earlier to a Polish topology $\mathcal{T}_A$ containing the G_δ sets A and $X - A$. Thus $\mathcal{B}' = \mathcal{B}_X$.

Let B be a Borel set. We may assume that B is open and closed. Let A_i be a base for the topology $\mathcal{T}$. Let $\mathcal{T}_i$ be a finer Polish topology so that $\mathcal{T}_i|_{A_i} = \mathcal{T}|_{A_i}$ and $\mathcal{T}_i|_{X-A_i} = \mathcal{T}|_{X-A_i}$ and A_i is both open and closed in $\mathcal{T}_i$. The topology $\mathcal{T}'$ generated by all the $\mathcal{T}_i$ is Polish and contains B as an open and closed subset. Moreover, using $\mathcal{T}_i|_{A_i} = \mathcal{T}|_{A_i}$ and $\mathcal{T}_i|_{X-A_i} = \mathcal{T}|_{X-A_i}$, one can show sets of the form $B_{i_1} \cap B_{i_2} \cap \cdots \cap B_{i_n}$ where each B_{i_j} or its complement belongs to the collection $\{A_i\}_{i=1}^{\infty}$ form a base for the topology $\mathcal{T}'$. Thus $\mathcal{T}'$ is zero dimensional. $\square$

COROLLARY I.1. *Let f be a Borel function from a Polish space X into a Polish space Y. Then there is a finer zero dimensional Polish topology on X under which f becomes a continuous function.*

PROPOSITION I.2. *Let X be a zero dimensional Polish space. Then $X = C \cup P$ where C is a countable open set and P is a disjoint closed perfect subset.*

PROOF. Let C be the union of all countable open subsets of X. Then C is countable since X is second countable, and clearly every nonempty open subset of P, the complement of C, is uncountable. $\square$

THEOREM I.4. *Let $\mathcal{B}$ be the Borel sets for a Polish space X. Then X is either countable or $(X, \mathcal{B})$ is Borel isomorphic to $[0, 1]$ with its usual topology and Borel sets.*

PROOF. Assume X is uncountable. We have seen that we may assume both X and $[0, 1]$ are zero dimensional, and that one can remove countable infinite sets from each so that both resulting spaces are homeomorphic to η. $\square$

We have seen that if X is a standard Borel space and if B is a Borel subset of X, then B with the relative Borel structure is a standard Borel space.

2. Analytic Sets, Analytic Spaces and Kuratowski's Theorem

Let X be a standard Borel space. A subset A of X is an **analytic set** if it is the Borel image of a standard Borel space. More precisely, A is an analytic set if there is a standard Borel space Y, a Borel function $f : Y \to X$ such that $f(B) = A$.

PROPOSITION I.3. *Let X be a Polish space. Then there is a continuous function mapping η onto X.*

PROOF. By refining the topology of X we may assume X is zero dimensional. Let U_i be a cover of X by open-closed sets of diameter < 1. Inductively obtain sets $U_{i_1,\ldots,i_{n+1}}$ which are open and closed and have diameter $< \frac{1}{n+1}$ and which cover $U_{i_1,\ldots,i_n}$. Define f by $\{f(a)\} = \cap_n U_{a_1,a_2,\ldots,a_n}$. Then f is continuous and onto. □

THEOREM I.5. *A subset A of a Polish space X is analytic iff it is the image of η under a continuous map.*

PROOF. Let A be analytic. Then $A = f(Y)$ where Y is a standard Borel space and f is Borel. Put a Polish topology on Y which generates the Borel structure. Let $\{C_n\}$ be a base for the topology of X. By further refining the topology of Y we may assume each $f^{-1}(C_n)$ is open. Let h be a continuous mapping of η onto Y. Then $f \circ h$ is continuous and $f \circ h(\eta) = A$. □

THEOREM I.6. *Countable unions and intersections of analytic sets are analytic.*

PROOF. Let $\{A_k\}$ be a sequence of analytic sets in a Polish space X. Let g_k be a continuous map from η into X with range A_k. Then the graph G_k of g_k is a closed set in $\eta \times X$, and thus $\cup G_k$ is a Borel set. The projection of $\eta \times X$ onto X is Borel and carries this union onto $\cup A_k$. Hence $\cup A_k$ is analytic.

To see $\cap A_k$ is analytic, note $B = \{(\{x_k\}, a) : \{x_k\}_{k=1}^\infty \in \eta^{\mathbb{N}}, a \in X, g_k(x_k) = a \,\forall k\}$ is a closed set in $\eta^{\mathbb{N}} \times X$ whose projection onto X is $\cap A_k$. □

THEOREM I.7. *If A_1 and A_2 are disjoint analytic sets in a Polish space X, then there exist disjoint Borel sets B_1 and B_2 such that $A_1 \subset B_1$ and $A_2 \subset B_2$.*

PROOF. Two sets E_1 and E_2 are Borel separable if there is a pair of disjoint Borel sets B_1 and B_2 with $E_1 \subseteq B_1$ and $E_2 \subseteq B_2$. Note if E_i and F_j are sequences of pairwise Borel separable sets, then $\cup E_i$ and $\cup F_j$ are Borel separable.

Now let $f_i : \eta \to A_i$ be continuous onto maps. Set

$$\eta_{k_1,k_2,\ldots,k_n} = \{(k_1, k_2, \ldots, k_n)\} \times \eta.$$

Then if A_1 is not Borel separable from A_2, there exist sequences a and b in η such that $f_1(\eta_{a_1,\ldots,a_k})$ is not Borel separable from $f_2(\eta_{b_1,\ldots,b_k})$ for all k. But by continuity of f_1 and f_2, $f_1(\eta_{a_1,\ldots,a_k})$ converges to $f_1(a)$ and $f_2(\eta_{b_1,\ldots,b_k})$ converges to $f_2(b)$. Since $f_1(a) \neq f_2(b)$, these sets are eventually separable by open sets. □

COROLLARY I.2. *If A is both analytic and coanalytic, then A is Borel.*

COROLLARY I.3. *If A_k are a pairwise disjoint sequence of analytic sets, then they can be covered by a pairwise disjoint sequence of Borel sets B_k.*

THEOREM I.8 (KURATOWSKI). *Let f be a Borel function on a standard space X with values in a countably generated Borel space Y. If f is one-to-one, then $f(X)$ is a Borel subset of Y; and f is a Borel isomorphism of X onto $f(X)$ with the relative Borel structure.*

PROOF. Since Y is countably generated, Y is Borel isomorphic to a subset of the Cantor space with the relative Borel structure. Thus we may assume Y is Polish. We first show f is a Borel isomorphism. To see this note that if A is a Borel subset of X, then since f is one-to-one, $f(A)$ and $f(X - A)$ are disjoint analytic sets, and thus can be covered by disjoint Borel sets E_1 and E_2. Thus $f(A) = f(X) \cap E_1$ is Borel in the relative Borel structure.

To show $f(X)$ is a Borel set in Y, we start by putting a compatible Polish topology on X. By refining it if necessary, we may assume f is continuous and the topology is zero dimensional. By Theorem I.2, there exists a countable set C such that $P = X - C$ is homeomorphic to η. It thus suffices to show $g(\eta)$ is a Borel subset of Y if g is a continuous one-to-one function from η into Y. For $(n_1, \dots, n_k) \in \mathbb{N}^k$, set $F_{n_1, \dots, n_k} = \{(n_1, \dots, n_k)\} \times \eta$. Since g is one-to-one, $g(F_{n_1, \dots, n_k})$ are disjoint analytic sets and thus can be covered by disjoint Borel sets $E_{n_1, \dots, n_k}$. Define $\hat{E}_{n_1, \dots, n_k} = E_{n_1} \cap E_{n_1, n_2} \cap \dots \cap E_{n_1, n_2, \dots, n_k} \cap \overline{g(F_{n_1, \dots, n_k})}$. $\hat{E}_{n_1, \dots, n_k}$ are Borel sets and $g(F_{n_1, \dots, n_k}) \subset \hat{E}_{n_1, \dots, n_k} \subset \overline{g(F_{n_1, \dots, n_k})}$. Suppose

$$y \in \cap_{k=1}^{\infty} \cup_{(n_1, \dots, n_k)} \hat{E}_{n_1, \dots, n_k}.$$

Then there exists a unique $a \in \eta$ such that $y \in \hat{E}_{a_1, \dots, a_k} \forall k$. But since g is continuous $\cap \overline{g(F_{a_1, \dots, a_k})} = \{g(a)\}$. Thus $y = g(a)$. Hence $g(\eta) = \cap_{k=1}^{\infty} \cup_{(n_1, \dots, n_k)} \hat{E}_{n_1, \dots, n_k}$ is a Borel set. $\square$

DEFINITION. An analytic Borel space is a countably generated Borel space which is the image of a standard Borel space under a Borel mapping.

COROLLARY I.4. *Y is an analytic Borel space iff it is Borel isomorphic to an analytic set in a Polish space with the relative Borel structure.*

COROLLARY I.5. *Let E be a subset of a standard Borel space X with the relative Borel structure. If E is a standard Borel space, then E is a Borel subset of X. If E is an analytic Borel space, then E is an analytic subset of X.*

Recall by Theorem I.1 that every countably generated Borel space is isomorphic to a subspace of a Polish space with the relative Borel structure.

COROLLARY I.6. *If X is a standard Borel space, then any countable separating family of Borel sets generates the Borel structure on X.*

THEOREM I.9. *Let f be a one-to-one Borel function from an analytic space X onto a countably generated Borel space Y. Then f is a Borel isomorphism.*

PROOF. Since Y is countably generated, we may assume that Y is a subspace of a Polish space Z with the relative Borel structure. Let E be a Borel subset of X. Since f is Borel and one-to-one, $f(E)$ and $f(X - E)$ are disjoint analytic subsets of Z. Thus, by Theorem I.7, there exist disjoint Borel sets E_1 and E_2 in Z which cover these two sets, respectively. Therefore, $f(E) = Y \cap E_1$ is a Borel subset of Y. Hence f^{-1} is Borel. $\square$

COROLLARY I.7. *Let X be an analytic Borel space. Then any countable separating family of Borel sets generates the Borel structure on X.*

PROPOSITION I.4. *Let Y be an analytic Borel space. For each natural number n, let F_n be a Borel function from Y into a countably separated Borel space Z_n. If the functions F_n separate the points of Y, then a function f from a Borel space X into Y is Borel if and only if $F_n \circ f$ is a Borel function for each n.*

PROOF. Note that if f is Borel, then clearly $F_n \circ f$ is Borel for each n.

Suppose next that $F_n \circ f$ is Borel for each n. Let $\mathcal{C}_n$ be a countable separating family of Borel sets for the space Z_n. Then since the Borel functions F_n separate the points of Y, the collection $\mathcal{A} = \cup_n F_n^{-1}(\mathcal{C}_n)$ is a countable separating family for the Borel sets in Y. By Corollary I.7, the smallest σ-algebra on Y containing the collection $\mathcal{A}$ is the σ-algebra of Borel sets.

Let $\mathcal{B}_0$ be the collection of Borel subsets E of Y such that $f^{-1}(E)$ is a Borel subset of X. This collection is a σ-algebra. Moreover, if $E \in \mathcal{A}$, then $E = F_n^{-1}(W)$ for some n and some Borel subset W of Z_n. Therefore $f^{-1}(E) = f^{-1}(F_n^{-1}(W)) = (F_n \circ f)^{-1}(W)$ is a Borel subset of X. It follows that $\mathcal{B}_0$ contains $\mathcal{A}$ and thus is the σ-algebra of Borel subsets of Y. $\square$

Let f be a mapping from a Borel space X into a set Y. Then the **quotient Borel structure** on Y is the largest σ-algebra on Y for which f is Borel. It consists of all subsets E of Y such that $f^{-1}(E)$ is a Borel subset of X.

THEOREM I.10. *Let ϕ be a mapping from an analytic Borel space X onto a set Y. If the quotient Borel structure on Y is countably separated, then it is analytic.*

PROOF. Let $\mathcal{C}$ be a countable separating family for the quotient Borel structure on Y. Let $\mathcal{B}'$ be the σ-algebra generated by $\mathcal{C}$. Let E be any Borel subset of Y and set $\mathcal{B}''$ to be the σ-algebra generated by $\mathcal{C} \cup \{E\}$. Both resulting Borel spaces on Y are analytic for both are countably generated and thus can be considered subspaces of a Polish space with the relative Borel structure. The identity map is a Borel isomorphism by Theorem I.9. $\square$

LEMMA I.2. *Let A be an analytic set in a compact Polish space X. Then there exist a compact Polish space F, a continuous function f from F into X, and an $F_{\sigma\delta}$ set B such that $f(B) = A$.*

PROOF. Let $g : \eta \to X$ be continuous with $g(\eta) = A$. Set $F = X \times \mathbb{R}_p^{\mathbb{N}}$ where $\mathbb{R}_p$ is the one point compactification of $\mathbb{R}$. Set $B = \{(g(a), a) : a \in \eta\}$. Then B is a closed set in $X \times \eta$ whose topology is the relative topology from F. Thus $B = (X \times \eta) \cap \overline{B}$. Hence B is an $F_{\sigma\delta}$ if $X \times \eta$ is a G_δ set, for open sets in separable metric spaces are F_σ sets. But $X \times \eta = \cap G_k$ where G_k is the open set consisting of those (x, a) in F whose first k coordinates for a are within $\frac{1}{k}$ of a natural number. Finally note $A = f(B)$ where f is the coordinate projection of F onto X. $\square$

3. Measures and Selection Theorems on Polish Spaces

Our discussion now will involve measures on Polish spaces and measurable selection theorems. We first note measures on Polish spaces have to be regular in the following sense.

THEOREM I.11. *Let ν be a σ-finite measure on a Polish space X. Let E be a Borel set. Then there exist a G_δ set U and a σ-compact set W with $W \subset E \subset U$ and $\nu(U - W) = 0$.*

PROOF. We may assume ν is finite. Let $\mathcal{B}$ be the collection of Borel sets E such that for each $\epsilon > 0$ there exist an open set U and a closed set F such that $F \subset E \subset U$ and $\nu(U - F) < \epsilon$. It is clear that $\mathcal{B}$ is closed under complements. Moreover, it contains the open sets since they are F_σ's. It is also closed under countable unions for if $A_i \in \overline{\mathcal{B}}$ for each i, then there exist open sets U_i and closed sets F_i satisfying $F_i \subset A_i \subset U_i$ and $\nu(U_i - F_i) < \frac{\epsilon}{2^i}$. Then $\lim_{n\to\infty} \nu(\cup_{i=1}^{\infty} U_i - \cup_{i=1}^{n} F_i) < \epsilon$. It follows that every Borel set belongs to $\mathcal{B}$; and thus, if E is a Borel set, there exist a G_δ set U and an F_σ set W such that $W \subset E \subset U$ and $\nu(U - W) = 0$.

To finish the proof it suffices to show that X contains a conull σ-compact set. To see this, let $\epsilon > 0$. For each n, let $\{S_{n,k}\}_{k=1}^{\infty}$ be a sequence of closed neighborhoods of radius $\frac{1}{n}$ that cover X. Choose k_n so that $\nu(X - \cup_{k=1}^{k_n} S_{n,k}) < \frac{\epsilon}{2^n}$. Set $K_\epsilon = \cap_n \cup_{k=1}^{k_n} S_{k,n}$. Then $\nu(X - K_\epsilon) < \epsilon$. To see that K_ϵ is compact, let $\{y_j\}_{j=1}^{\infty}$ be a sequence in K_ϵ. Then there exists a $j_1 \le k_1$ and a subsequence $\{y_{1,j}\}_{j=1}^{\infty}$ of $\{y_j\}_{j=1}^{\infty}$ such that each $y_{1,j}$ belongs to S_{1,j_1}. Repeating the process and using Cantor diagonalization, one sees that there is a sequence j_n of natural numbers and a subsequence $\{y_{j,j}\}_{j=1}^{\infty}$ such that $j_n \le k_n$ and $y_{j,j} \in S_{n,j_n}$ for $j \ge n$. It follows that the subsequence $\{y_{j,j}\}_{j=1}^{\infty}$ converges, and its limit is in K_ϵ. Thus K_ϵ is compact. Finally take $K = \cup_n K_{\frac{1}{n}}$. $\square$

THEOREM I.12. *Let A be an analytic set in a standard Borel space X. If ν is a σ-finite Borel measure on X, then there exist Borel sets E_1 and E_2 satisfying $E_1 \subset A \subset E_2$ and $\nu(E_2 - E_1) = 0$.*

PROOF. We may assume X is compact and ν is finite. First note if ν^* is the outer measure associated with ν, then $\nu^*(\cup C_n) = \lim_{n\to\infty} \nu^*(C_n)$ whenever C_n is an increasing sequence of sets.

Now since $\overline{A}$ is compact, there is a compact metric space F, a double sequence $B_{m,n}$ of closed subsets of F, and a continuous function f from F into X such that $A = f(B)$ where $B = \cap_m \cup_n B_{m,n}$. We clearly may assume $B_{m,n}$ is increasing in n for each m. Let $\epsilon > 0$. Then $f(B \cap B_{1,k})$ increases to $A = f(B)$. Hence there exists a n_1 such that $\nu^*(f(B \cap B_{1,n_1})) > \nu^*(A) - \epsilon$. Now $f(B \cap B_{1,n_1} \cap B_{2,k})$ increases to $f(B \cap B_{1,n_1})$. Hence there is an n_2 with $\nu^*(f(B \cap B_{1,n_1} \cap$

B_{2,n_2})) $> \nu^*(A) - \epsilon$. Repeating inductively one sees there is a sequence $n_1, n_2, n_3, \ldots$ such that $\nu^*(f(B \cap B_{1,n_1} \cap B_{2,n_2} \cap \cdots \cap B_{k,n_k})) > \nu^*(A) - \epsilon$. Let $C = \cap_k B_{k,n_k}$. Note C and $f(C)$ are compact. Moreover, $\nu(f(C)) = \lim_k \nu(f(B_{1,n_1} \cap \cdots \cap B_{k,n_k})) \geq \nu^*(A) - \epsilon$. Hence $f(C) \subset A$ and $\nu^*(A) - \nu(f(C)) \leq \epsilon$. The result now follows easily. $\square$

THEOREM I.13 (VON NEUMANN SELECTION LEMMA).
Let A be an analytic set in $X \times Y$ where X and Y are Polish spaces. Assume the projection of A into X is onto. Then there exists a function $\phi : X \to Y$ such that $(x, \phi(x)) \in A$ for all x and for each Borel set E of Y, $\phi^{-1}(E)$ is in the σ-algebra generated by the analytic subsets of X.

PROOF. Let $f : \eta \to X \times Y$ be a continuous function with $f(\eta) = A$. Then $f(a) = (g(a), h(a))$ where g and h are continuous functions from η into X and Y with g onto. Linearly order η lexicographically. Thus $(a_i) < (b_i)$ iff $\exists k$ such that $a_1 = b_1, \ldots, a_k = b_k$ but $a_{k+1} < b_{k+1}$. Then any nonempty subset S of η has a greatest lower bound $\inf S$ which belongs to the closure of S. Set $a(x) = \inf g^{-1}(x)$. Since $g^{-1}(x)$ is closed, $a(x)$ is the smallest element in $g^{-1}(x)$. Define $\phi(x) = h(a(x)) = h \circ a(x)$. Clearly $(x, \phi(x)) = (g(a(x)), h(a(x))) \in A$.

Let U be an open subset of Y. Since $h^{-1}(U)$ is an open set, to see $\phi^{-1}(U) = a^{-1}(h^{-1}(U))$ is in the σ-algebra generated by the analytic sets, it suffices to see $A_{n_1,\ldots,n_k} = a^{-1}(\{(n_1, \ldots, n_k)\} \times \eta)$ is in this algebra for all $n_1, \ldots, n_k$. But $A_{n_1,\ldots,n_k}$ is the set of points x in $g(\{(n_1, \ldots, n_k)\} \times \eta)$ where x is not in $g(\{(m_1, \ldots, m_k)\} \times \eta)$ if $(m_1, \ldots, m_k, 1, 1, \ldots) < (n_1, \ldots, n_k, 1, 1, \ldots)$. Thus $A_{n_1,\ldots,n_k}$ is in the algebra because the set $g(\{(m_1, \ldots, m_k)\} \times \eta)$ is analytic for all $m_1, \ldots, m_k$. $\square$

The following theorem is an easy consequence of this lemma and plays an important role for obtaining Borel selections.

THEOREM I.14. *Let X and Y be standard Borel spaces and let A be an analytic subset of $X \times Y$ whose projection into X is onto. Let μ be a σ-finite measure on X. Then there is a conull Borel subset X_0 in X and a Borel function $\phi : X_0 \to Y$ such that $(x, \phi(x)) \in A$ for all $x \in X$.*

PROOF. Choose ϕ as in the von Neumann selection lemma. Let U_i be a countable basis for the open sets in Y. By Theorem I.12, each $\phi^{-1}(U_i)$ is measurable relative to the completion of μ. Hence there exist Borel sets E_i and F_i with $E_i \subset \phi^{-1}(U_i) \subset F_i$ and $\mu(F_i - E_i) = 0$. Set $X_0 = X - \cup(F_i - E_i)$ and restrict ϕ to X_0. $\square$

Our next topic deals with cross sections for equivalence relations. Namely suppose X is a standard Borel space and R is an equivalence relation on X. Let X/R be the space of equivalence classes. Set $[x]$ to be the equivalence class of x, and let $[F]$ be $\cup_{x \in F}[x]$. $[F]$ is called the **saturation** of F. Give X/R the quotient Borel structure defined by the mapping $x \mapsto [x]$.

PROPOSITION I.5. *Suppose E is a Borel set that meets each equivalence class of R in exactly one point. If, in addition, the saturation $[F]$ of each Borel subset F of E is Borel, then the mapping $x \mapsto [x]$ is a Borel isomorphism of E onto X/R. In particular, if X is standard, then the quotient Borel structure on X/R is standard.*

PROOF. By definition of the quotient Borel structure, the mapping $x \mapsto [x]$ is Borel. Moreover, by hypothesis, the image of a Borel set under this map is Borel in the quotient Borel structure. $\square$

The following theorem shows the existence of Borel cross sections for equivalence relations under appropriate conditions.

THEOREM I.15. *Let X be a separable metric space with an equivalence relation R whose equivalence classes are complete. Then if the saturation of every open set is a Borel set, there exists a Borel set meeting each equivalence class in exactly one point.*

PROOF. Let Π be the set of all finite sequences of natural numbers. Set Π_k to be the subset consisting of sequences with k elements. Since X is separable, there is a sequence of open sets U_i which cover X and have diameter < 1. Again by separability, each U_i can be covered by open sets $U_{i,j}$ of diameter $< \frac{1}{2}$ satisfying $\overline{U}_{i,j} \subset U_i$. Repeating one sees by induction that there is a family $\{U_a : a \in \Pi\}$ of open sets satisfying:

(1) $U_a = \cup_i U_{a,i}$ for each a
(2) $\mathrm{diam} U_a < \frac{1}{k}$ if $a \in \Pi_k$
(3) $\overline{U}_{a,i} \subset U_a$ for all i and $\cup_i U_i = X$.

The above family is called a sifting of the metric space X. Note there is a natural well ordering on Π. Indeed, let every element in Π_k be less than every element in Π_j if $k < j$, and order each Π_k lexicographically. For each $a \in \Pi$ we define a Borel subset H_a inductively. Set $H_1 = U_1, H_2 = U_2 - [U_1], \ldots, H_i = U_i - \cup_{j<i}[U_j], \ldots$.

Next define $H_{a,1} = U_{a,1} \cap H_a$ and $H_{a,i} = U_{a,i} \cap H_a - \cup_{j<i}[U_{a,j}]$. We note that if W is an equivalence class and k is given, then there is a unique a in Π_k with $W \cap H_a \neq \emptyset$ and in this case $W \cap H_a = W \cap U_a$. Let $B = \cap_k \cup_{a \in \Pi_k} H_a$. One thus has for each equivalence class W, a unique sequence $\{n_k\}$ of natural numbers such that if $a_k = n_1, \ldots, n_k$, then $W \cap H_{a_k} \neq \emptyset$. Hence $W \cap B = \cap_k (W \cap H_{a_k}) = \cap_k (W \cap U_{a_k}) \supset \cap_k (W \cap \overline{U}_{a_{k+1}})$ which is nonempty since W is complete. Moreover the diameter of $W \cap B$ is less than $\frac{1}{k}$ for all k. Thus $W \cap B$ consists of exactly one point. $\square$

THEOREM I.16 (FEDERER AND MORSE SELECTION LEMMA).
Let f be a continuous function on a compact metric space X which maps onto a metric space Y. Then there is a Borel set $E \subset X$ such that f restricted to E is one-to-one and onto. In particular, f is a Borel isomorphism from E onto Y.

PROOF. Define an equivalence relation $\mathcal{R}$ on X by $x_1 \sim x_2$ iff $f(x_1) = f(x_2)$. The equivalence classes are complete since they are closed. Moreover, if U is an open set in X, then it is σ-compact. Hence the saturation $[U] = f^{-1}(f(U))$ is an F_σ set and thus is Borel. By Theorem I.15, there is a Borel set E which meets each set $f^{-1}(y)$ in exactly one point. Since f restricted to E is one-to-one, the result follows by Theorem I.8. $\square$

A measure on a Borel space is said to be **standard** if there is a conull Borel subset whose relative Borel structure is standard.

THEOREM I.17 (VON NEUMANN SELECTION THEOREM).
Let μ be a standard finite measure on a Borel space X. Suppose f is a Borel function on X with values in a countably generated Borel space Y. Then there is a Borel subset E of X on which f is one-to-one, $\mu(X - f^{-1}(f(E))) = 0$, and which is standard with the relative Borel structure.

PROOF. By Theorem I.1 and Corollary I.1 we may assume X is Polish, Y is Polish, and f is continuous. By Theorem I.11, there is a conull σ-compact subset. Replacing X by this subset, we may assume f is a continuous function from a σ-compact metric space X into a Polish space Y. Hence suppose $X = \cup_n K_n$ where each K_n is compact. By the Federer–Morse Lemma, there is a Borel set E_n in K_n such that f restricted to E_n is a Borel isomorphism onto the compact Borel space $f(K_n)$. Set $E = \cup_n (E_n - f^{-1}(f(K_1 \cup \cdots \cup K_{n-1})))$. Then f restricted to E is one-to-one and $f^{-1}(f(E)) = X$.

Moreover, since each set $f^{-1}(f(K_1 \cup \cdots \cup K_n))$ is closed, E is a Borel set. $\square$

COROLLARY I.8. *Let f be a continuous function from a σ-compact metric space X onto a metric space Y. Then there is a Borel function g from Y into X such that $f(g(y)) = y$.*

4. Uniformization of Borel Subsets with σ-Compact Sections

The final selection theorem we present depends on obtaining some fine separation theorems for analytic sets. We begin by showing an analytic set is describable in terms of a family of closed sets.

Let Π be the set of all finite sequences f of natural numbers. If $f \in \Pi$, let $l(f)$ denote the length of the sequence f. Let Γ be the set of all infinite sequences of 0's and 1's, with all but finitely many of the terms 0. Lexicographic orderings make both Γ and Π well ordered. If $a \in \eta$, define $a(n)$ in Π by $a(n) = (a_1, \dots, a_n)$.

PROPOSITION I.6. *Let A be an analytic set in a Polish space X. Then there is a collection of closed sets $\{A_\gamma\}_{\gamma \in \Gamma}$ such that A is the collection of all points x for which there exists an increasing sequence $\gamma_1 < \gamma_2 < \cdots$ with $x \in \cap A_{\gamma_i}$.*

PROOF. Suppose A is an analytic set. Let ϕ be a continuous function from η onto A. Let f belong to Π and set N_f to be the finite collection $\{f_1, f_1 + f_2, \dots, f_1 + f_2 + \cdots + f_{l(f)}\}$. Define γ_f to be the characteristic function of N_f. Then the mapping $f \mapsto \gamma_f$ is a one-to-one mapping between Π and Γ.

Define A_{γ_f} to be the closure of the set $\{\phi(b) : b(l(f)) = f\}$. Suppose $x = \phi(a)$. Then note $\gamma_{a(1)} < \gamma_{a(2)} < \cdots$ and x is in the closure of $\{\phi(b) : b(n) = a(n)\}$ for each n. Conversely, suppose $x \in \cap A_{\gamma_i}$ where $\gamma_1 < \gamma_2 < \cdots$. By choosing a subsequence of this sequence, we may assume $\gamma_i(j) = \gamma_k(j)$ if $j \le i < k$. Hence there is a mapping $\gamma : \mathbb{N} \to \{0, 1\}$ such that $\gamma(j) = \gamma_i(j)$ if $j \le i$. If $\{j : \gamma(j) = 1\} = \{f_1, f_1 + f_2, \cdots\}$, then $a = (f_1, f_2, \cdots) \in \eta$. Moreover, since $x \in \cap A_{\gamma_i}$ and $\gamma(j) = \gamma_i(j)$ for $j \le i$, we see for each n, there is an m such that $\{j \le f_1 + f_2 + \cdots + f_n : \gamma_i(j) = 1\} = \{f_1, f_1 + f_2, \cdots, f_1 + f_2 + \cdots + f_n\}$ for each $i \ge m$. The continuity of ϕ and $x \in A_{\gamma_i} \subset \overline{\{\phi(b) : b(n) = (f_1, \cdots, f_n)\}}$ for $i \ge m$ imply $\phi(a) = x \in A$. $\square$

Suppose A_γ, $\gamma \in \Gamma$ is a collection of closed subsets in the Polish X. For each $x \in X$, the set $\mathrm{ord}_A(x) \equiv \{\gamma : x \in A_\gamma\}$ is well ordered and thus defines an ordinal number which we will also denote by $\mathrm{ord}_A(x)$. Recall that if α and β are two ordinal numbers; that is two well ordered sets, then $\alpha \leq \beta$ if there exists a one-to-one mapping from α into β which preserves order. If α is countable, this relation is equivalent to the existence of sequences ζ_k in α and η_k in β satisfying $\alpha = \{\zeta_k : k \in \mathbb{N}\}$ and $\zeta_i < \zeta_j$ implies $\eta_i < \eta_j$. Moreover, if m and n are natural numbers, $m\alpha + n$ is the ordinal number obtained by ordering the set $\{1,\ldots,n\} \cup \{1,\ldots,m\} \times \alpha$ by $1 < 2 < 3 \cdots < n < (k,\zeta)$ for all (k,ζ) and $(k,\zeta) < (l,\zeta')$ if $k < l$ or $k = l$ and $\zeta < \zeta'$.

PROPOSITION I.7. *Let A_γ and B_γ, $\gamma \in \Gamma$, be families of closed subsets of Polish spaces X and Y, respectively. Then the set $\{(x,y) : m\,\mathrm{ord}_A(x) + n \leq m'\,\mathrm{ord}_B(y) + n'\}$ is an analytic subset of $X \times Y$.*

PROOF. Let α be the well ordered set giving the ordinal $m\Gamma + n$ and let β be the well ordered set defining the ordinal $m'\Gamma + n'$. These sets are countable and hence Polish with the discrete topology. Thus the spaces $\alpha^\mathbb{N}$ and $\beta^\mathbb{N}$ are Polish spaces with the infinite product topology.

Let A_1 be the collection of all (x,y,ζ,ξ) with $x \in X$, $y \in Y$, $\zeta \in \alpha^\mathbb{N}$ with $\zeta_i \in m\,\mathrm{ord}_A(x) + n$, and $\xi \in \beta^\mathbb{N}$. The condition that $\zeta_i \in m\,\mathrm{ord}_A(x) + n$ is equivalent to either $\zeta_i \in \{1,\ldots,n\}$ or $\zeta_i = (k,\gamma)$ where $k \in \{1,\ldots,m\}$ and $x \in A_\gamma$. Thus

$$A_1 = \bigcap_i \Big(\bigcup_{r=1}^m \bigcup_\gamma A_\gamma \times Y \times \{\zeta_i = (r,\gamma)\} \times \beta^\mathbb{N} \cup \bigcup_{r=1}^n X \times Y \times \{\zeta : \zeta_i = r\} \times \beta^\mathbb{N} \Big).$$

This shows that A_1 is Borel. A similar argument shows the set A_2 consisting of all tuples (x,y,ζ,ξ) where $\xi_i \in m'\,\mathrm{ord}_B(y) + n'$ is Borel.

Next, let A_3 be the set of all (x,y,ζ,ξ) where for each ξ' in $m\,\mathrm{ord}_A(x) + n$, there is a k with $\zeta_k = \xi'$. Thus $A_3 = B_1 \cap B_2$ where $B_1 = \cap_{r=1}^m \{(x,y,\zeta,\xi) : \gamma \in \mathrm{ord}_A(x) \text{ implies } \exists k \ni \zeta_k = (r,\gamma)\}$ and $B_2 = \cap_{r=1}^n \{(x,y,\zeta,\xi) : \exists k \ni \zeta_k = r\}$. But each set $\{(x,y,\zeta,\xi) : \gamma \in \mathrm{ord}_A(x) \text{ implies } \exists k \ni \zeta_k = (r,\gamma)\}$ is Borel, for it equals $\cap_\gamma (\cup_k (A_\gamma \times Y \times \{\zeta : \zeta_k = (r,\gamma)\} \times \beta^\mathbb{N} \cup A'_\gamma \times Y \times \alpha^\mathbb{N} \times \beta^\mathbb{N})$. Thus B_1 is Borel. That B_2 is Borel follows from the fact that each set $\{(x,y,\zeta,\xi) : \exists k \ni \zeta_k = r\}$ equals $\cup_{k=1}^\infty X \times Y \times \{\zeta_k = r\} \times \beta^\mathbb{N}$ which is Borel. Thus A_3 is Borel.

Finally let A_4 be the set consisting of all (x,y,ζ,ξ) such that $\xi_i < \xi_j$ whenever $\zeta_i < \zeta_j$. Then $A_4 = X \times Y \times \cap_{i,j}(\{(\zeta,\xi) : \zeta_i < \zeta_j \text{ and } \xi_i < \xi_j\} \cup \{'(\zeta,\xi) : \zeta_i \geq \zeta_j\})$. Thus A_4 is Borel.

It follows that the coordinate projection of the set $A = A_1 \cap A_2 \cap A_3 \cap A_4$ into $X \times Y$ is analytic. Since this set is $\{(x,y) : m\,\mathrm{ord}_A(x) + n \leq m'\,\mathrm{ord}_B(y) + m'\}$, the result follows. $\square$

Let A be an analytic set. Choose a family $\{A_\gamma\}_{\gamma \in \Gamma}$ of closed sets satisfying $x \in A$ iff there is an increasing sequence $\gamma_k \in \Gamma$ with $x \in \cap_k A_{\gamma_k}$. Again set $\mathrm{ord}_A(x) = \{\gamma : x \in A_\gamma\}$. Then if Ω is the ordinal number defined by the natural numbers, then $x \in A$ iff $\Omega \leq \mathrm{ord}_A(x)$.

THEOREM I.18 (THE REDUCTION THEOREM). *Let* $U_1, U_2, \ldots$ *be a sequence of coanalytic subsets of a Polish space* Y. *Then there exists a sequence* $V_1, V_2, \cdots$ *of pairwise disjoint coanalytic sets such that* $V_i \subset U_i \, \forall i$ *and* $\cup V_i = \cup_i U_i$. *If* $\cup U_i$ *is a Borel set, then each* V_i *is a Borel set.*

PROOF. Since U_n is coanalytic, Proposition I.6 implies the existence of a family U_γ^n, $\gamma \in \Gamma$, of closed sets with $x \in Y - U_n$ iff $\mathrm{ord}_n(x) = \{\gamma : x \in U_\gamma^n\}$ defines an ordinal number with $\Omega \leq \mathrm{ord}_n(x)$. Next note that if α and β are ordinal numbers with $(2\alpha + 1)2^n = (2\beta+1)2^k$, then one has $k = n$ and $\alpha = \beta$.

Define

$$V_n = U_n \cap \bigcap_{k \neq n} \{x : (2\,\mathrm{ord}_k(x) + 1)2^k \leq (2\,\mathrm{ord}_n(x) + 1)2^n\}'.$$

By Proposition I.7, the sets V_n are coanalytic. Note that if $x \in V_k \cap V_n$, then $(2\,\mathrm{ord}_k(x)+1)2^k > (2\,\mathrm{ord}_n(x)+1)2^n > (2\,\mathrm{ord}_k(x)+1)2^k$ which is impossible. Thus the sets V_n are pairwise disjoint.

Next suppose $x \in \cup U_n$. Then $\mathrm{ord}_n(x) < \Omega$ for some n. Hence $\mathrm{ord}_n(x)$ is finite. Choose n such that $(2\,\mathrm{ord}_n(x)+1)2^n$ is a minimum. Then note if $k \neq n$, then $(2\,\mathrm{ord}_k(x) + 1)2^k \geq (2\,\mathrm{ord}_n(x) + 1)2^n$. But one can't have equality here for otherwise $k = n$. Thus $x \in V_n$. So $\cup V_n = \cup U_n$.

If we assume $\cup V_n$ is Borel, then $(\cup V_n) - V_k = \cup_{n \neq k} V_n$ is analytic and coanalytic and hence must be Borel. It follows that each V_k is a Borel set. $\square$

THEOREM I.19 (FIRST GENERALIZED SEPARATION THEOREM). *Let* $\{A_n\}_{n=1}^\infty$ *be a sequence of analytic subsets of a Polish space* Y. *If* $\cap_n A_n = \emptyset$, *then there exist Borel sets* $B_1, B_2, \cdots$ *with* $B_n \supset A_n$ *and* $\cap_n B_n = \emptyset$.

PROOF. Apply the reduction theorem to the coanalytic sets A'_n. □

THEOREM I.20 (SECOND GENERALIZED SEPARATION THEOREM).
Let A_n be a sequence of analytic sets in a Polish space. Then there exists a sequence B_n of coanalytic sets with $A_n - \cap_k A_k \subset B_n$ and $\cap B_n = \emptyset$.

PROOF. By the reduction theorem, there exist disjoint coanalytic sets V_n with $V_n \subset A'_n$ and $\cup V_n = \cup A'_n$. Again using the reduction theorem, for each n there exist two disjoint coanalytic sets, B_n and C_n, with $B_n \cup C_n = \cup_{k \neq n} V_k \cup A'_n$, $B_n \subset \cup_{k \neq n} V_k$, and $C_n \subset A'_n$. Note $\cap B_n \subset \cap_n (\cup_{k \neq n} V_k) = \cap_n (\cup A'_k - V_n) = (\cup V_n)' \cap (\cup A'_n) = \emptyset$. Moreover, if $x \in A_n - \cap_k A_k$, then since $V_n \subset A'_n$, we have $x \notin V_n$ and $x \in \cup_{k \neq n} A'_k$. Thus $x \in \cup_{k \neq n} V_k$ and $x \notin A'_n$. Thus $x \in B_n$. □

LEMMA I.3 (KUNUGUI). *Let Y be a Polish space and suppose ϕ is a continuous mapping of a closed nonempty subset η_0 of η whose range is contained in an F_σ set E. Then there is an f in Π such that $\phi(f) \equiv \{\phi(a) : a(l(f)) = f\} \neq \emptyset$ and $\overline{\phi(f)} \subset E$.*

PROOF. Let $\{G_n\}$ be a decreasing sequence of open sets such that $Y - E = \cap G_n$. Suppose $\phi(f) = \emptyset$ or $\overline{\phi(f)} \cap (Y - E) \neq \emptyset$ for all f. Since $\phi(\emptyset)$ is nonempty, $\overline{\phi(\emptyset)} \cap G_1 \neq \emptyset$. Thus $\phi(\emptyset) \cap G_1 \neq \emptyset$. Hence there is a $b_1 \in \eta_0$ with $\phi(b_1) \in G_1$. Since ϕ is continuous, there is an n_1 such that $\phi(a) \in G_1$ whenever $a(n_1) = b_1(n_1)$. Set $f_1 = b_1(n_1)$. Then $\phi(f_1) \neq \emptyset$. Hence $\overline{\phi(f_1)} \cap G_2 \neq \emptyset$. Thus there exists a $b_2 \in \eta_0$ with $b_2(l(f_1)) = f_1$ and $\phi(b_2) \in G_2$. By continuity there is an n_2 with $n_2 > n_1$ such that $\phi(a) \in G_2$ whenever $a(n_2) = b_2(n_2)$. Set $f_2 = b_2(n_2)$. Repeating inductively and using the fact that η_0 is closed, we obtain a $b \in \eta_0$ and a sequence $n_1 < n_2 < n_3 \cdots$ satisfying $\phi(a) \in G_k$ if $a(n_k) = b(n_k)$. Thus $\phi(b)$ is in $\cap G_n = Y - E$. This is a contradiction. □

Let X and Y be Polish spaces. If M is a subset of $X \times Y$, then the vertical section M^x is the set $(\{x\} \times Y) \cap M$. The **Y-closure** of a subset M of $X \times Y$ is the union over all x of the closures of the sets M^x.

DEFINITION. Suppose X is a Polish space and Y is a **compact** metric space. A subset A of $X \times Y$ is an elementary coanalytic set if there exist a system $B_{m,i}$, $C_{m,i}$, $i = 1, \cdots, p(m)$ of closed subsets of Y and a sequence of coanalytic sets D_m of $X \times Y$ such that $A =$

$\cap_{m=1}^{\infty} \cup_{i=1}^{p(m)} A_{m,i}$ where $A_{m,i} = \{(x, y) : y \in C_{m,i}$ and $\{x\} \times B_{m,i} \subset D_m\}$.

Note that if A is an elementary coanalytic set and E is a Borel subset of X and B is a closed subset of Y, then $A \cap (E \times B)$ is an elementary coanalytic set.

Let p_X be the natural coordinate projection $(x, y) \rightarrow x$.

LEMMA I.4. *If A is an elementary coanalytic subset of $X \times Y$, then A is a Y-closed coanalytic set and $p_X(A)$ is a coanalytic subset of X.*

PROOF. Clearly A is Y-closed. Note that $A_{m,i} = (X - p_X[(X \times B_{m,i}) \cap (X \times Y - D_m)] \times C_{m,i}$ is coanalytic. Thus A is coanalytic.

To show $p_X(A)$ is coanalytic, first set $A_m = \cup_{i=1}^{p(m)} A_{m,i}$. Then $x \notin p_X(A)$ iff $\cap_{m=1}^{\infty} A_m^x = \emptyset$. Since each A_m^x is compact, this occurs iff there exists an n with $\cap_{m=1}^{n} A_m^x = \emptyset$. But since $A_m^x = \cup_{i=1}^{p(m)} A_{m,i}^x$, this intersection is empty if and only if whenever $\{x\} \times B_{1,i_1} \subset D_1, \cdots, \{x\} \times B_{n,i_n} \subset D_n$ where $1 \leq i_m \leq p(m)$, then one has $\cap_{m=1}^{n} C_{m,i_m} = \emptyset$. Hence $x \notin p_X(A)$ iff there exists an n such that for every finite sequence $i_1, \ldots, i_n$ with $\cap_{m=1}^{n} C_{m,i_m} \neq \emptyset$, one has $\{x\} \times B_{m,i_m} \not\subset D_m$ for some $1 \leq m \leq n$. Thus

$$X - p_X(A) = \bigcup_n \bigcap_{C_{1,i_1} \cap \cdots \cap C_{n,i_n} \neq \emptyset} \bigcup_{m=1}^{n} (X - p_X(A_{m,i_m})).$$

But each $X - p_X(A_{m,i}) = p_X(X \times B_{m,i} - D_m)$ is analytic. Thus $p_X(A)$ is coanalytic. $\square$

PROPOSITION I.8. *Let A be an elementary coanalytic subset of $X \times Y$. Suppose $p_X(A)$ is a Borel set. Then there exists a Borel subset E of $X \times Y$ which meets each nonempty vertical section of A in exactly one point.*

PROOF. Let $B_{m,i}$, $i = 1, \ldots, q(m)$ be closed balls in Y with diameters $< 1/m$ whose union is Y. We construct a decreasing sequence E_k of Y-closed Borel subsets of $X \times Y$ such that

(1) $E_k \cap A$ is a finite union of elementary coanalytic sets,
(2) $p_X(E_k) = p_X(A) = p_k(E_k \cap A)$ and
(3) diam $E_k^x \leq 1/k$ for all x.

Since each A^x is compact, it would then follow that $E = \cap E_k$ is the desired Borel set.

Let $k = 1$. The sets $A \cap (X \times B_{1,i})$, $i = 1, \dots, q(1)$, are elementary coanalytic with union A. Thus, by Lemma I.4, the sets $p_X(A \cap (X \times B_{1,i}))$ are coanalytic and have union equal to the Borel set $p_X(A)$. By the reduction theorem, there exist disjoint Borel sets $G_1, \dots, G_{p(1)}$ in X with $p_X(A) = \cup G_i$ and $G_i \subset p_X(A \cap (X \times B_{1,i}))$. Set $E_1 = \cup(G_i \times B_{1,i})$. Then E_1 satisfies (1), (2) and (3).

Suppose E_k has been defined. Then $E_k \cap A \cap (X \times B_{k+1,i})$ is a finite union of elementary coanalytic sets. Again by Lemma I.4, the sets $p_X(E_k \cap A \cap (X \times B_{k+1,i}))$, $i = 1, \dots, q(k+1)$, are coanalytic sets whose union is $p_X(A)$. Thus by the reduction theorem, there exist disjoint Borel sets $G_1, \cdots, G_{p(k+1)}$ with $\cup G_i = p_X(A)$ and $G_i \subset p_X(E_k \cap A \cap (X \times B_{k+1,i}))$. Set $E_{k+1} = \cup_{i=1}^{q(k+1)} E_k \cap (G_i \times B_{k+1,i})$. Then E_{k+1} is a finite union of elementary coanalytic sets. E_{k+1} clearly satisfies (2) and (3). $\square$

THEOREM I.21. *Let X and Y be Polish spaces. Suppose M is a Borel subset of $X \times Y$ such that each vertical section $M^x \equiv \{(x,y) : (x,y) \in M\}$ is a σ-compact set. Then if p_X is the canonical projection from $X \times Y$ onto X, then $p_X(M)$ is a Borel set. Moreover, there is a Borel set which meets each nonempty vertical section of M in exactly one point.*

PROOF. Let Y^* be the one point compactification of Y. Then Y^* is a Polish space, and M is a Borel subset of $X \times Y^*$ with σ-compact vertical sections. Hence we may assume Y is compact.

Let ϕ be a continuous function of η into $X \times Y$ with $\phi(\eta) = M$. Define $\eta_x = \{a : \phi(a) \in M^x\}$. Then η_x is a closed subset of η. Again for $f \in \Pi$, set $\phi(f) = \{\phi(a) : a(l(f)) = f\}$. Let $\phi_x(f) = M^x \cap \phi(f)$, and define $\phi^*(f)$ to be the set of all points $(x,y) \in \phi(f)$ with $\overline{\phi_x(f)} \subset M^x$. By Kunugui's Lemma, for each x with $M^x \neq \emptyset$ there is an f with $\emptyset \neq \overline{\phi_x(f)} \subset M^x$. Thus $p_X(M) = \cup_{f \in \Pi} p_X(\phi^*(f))$. This implies that $p_X(M)$ is coanalytic if for each f there is a coanalytic set $A(f)$ with $p_X(\phi^*(f)) \subset A(f) \subset p_X(M)$.

Fix f in Π. For each m, choose closed balls $B_{m,1}, \dots, B_{m,p(m)}$ of Y each with diameter smaller than $1/m$ which cover Y. Set $C_{m,i} = p_X(\phi(f) \cap (X \times B_{m,i})) \times B_{m,i}$ for $i = 1, \dots, p(m)$. Define $C_m = \cup C_{m,i}$ and set $C = \cap_m C_m$. Note that C is the Y closure of $\phi(f)$. Each C_m is an analytic set. By the Second Separation Theorem, there exist coanalytic sets D_m with $C_m - C \subset D_m$ and $\cap D_m = \emptyset$.

Fix (x, y) in $\phi^*(f)$. Then $C^x = \overline{\phi(f)^x} = \overline{\phi_x(f)}$ is a nonempty subset of M^x. Hence for each m, there is a $q(m)$ with $(x, y) \in \{x\} \times B_{m,q(m)} \subset C_m^x \subset ((C_m - C) \cup M)^x \subset (D_m \cup M)^x$. Hence if $(x, y) \in \phi^*(f)$, then $x \notin p_X((X \times B_{m,q(m)}) \cap M' \cap D_m')$. Let $A_{m,i} = (X - p_X(X \times B_{m,i} \cap M' \cap D_m')) \times B_{m,i}$. Then $(x, y) \in A_{m,q(m)}$. Note each set $A_{m,i}$ is the union of all sets $\{x\} \times B_{m,i}$ with $\{x\} \times B_{m,i} \subset D_m \cup M$. Hence if we define $A_m = \cup A_{m,i}$ and $A_f = \cap A_m$, then A_f is an elementary coanalytic set and $\phi^*(f) \subset A_f \subset \cap(D_n \cup M) = M$.

Set $A(f) = p_X(A_f)$. By Lemma I.4, $A(f)$ is a coanalytic subset of X. Since $p_X(\phi^*(f)) \subset A(f) \subset p_X(M)$, we have $p_X(M) = \cup_{f \in \Pi} A(f)$ is coanalytic. Since it is also analytic, we see from Corollary I.2 that $p_X(M)$ is Borel.

Since $p_X(M) = \cup_f p_X(A_f)$ is Borel and each $p_X(A_f)$ is coanalytic, it follows by the reduction theorem that there exist pairwise disjoint Borel sets X_f with $X_f \subset p_X(A_f)$ and $\cup X_f = p_X(M)$. Thus there exists a Borel subset of M which meets each nonempty vertical section of M in exactly one point if and only if such a set exists for each elementary coanalytic set $(X_f \times Y) \cap A_f$. Since each $p_X((X_f \times Y) \cap A_f) = X_f$ is Borel, such sets exist by Proposition I.8. $\square$

THEOREM I.22. *Let f be a Borel function defined on a Borel subset E of a Polish space X with values in a countably separated Borel space Y. If each $f^{-1}(y)$ is a σ-compact set, then $f(E)$ is a Borel set, and there is a Borel function $g : f(E) \to E$ such that $f(g(y)) = y$ for all $y \in f(E)$.*

PROOF. Apply Theorem I.21 to the Borel set $\{(x, y) : f(x) = y\}$. $\square$

5. Borel Sets and the Baire Property

A subset of a metric space X is **nowhere dense** if its closure has no interior. A set which is a countable union of nowhere dense subsets is said to be **meager** in X or to be **first category** in X. A subset A of X has the **Baire property** if there is an open set G in X such that $A \triangle G = (A - G) \cup (G - A)$ is meager in X.

PROPOSITION I.9. *Every Borel set has the Baire property.*

PROOF. Note $E_1 \sim E_2$ if $E_1 \triangle E_2$ is a meager set defines an equivalence relation with the property that if E_i and F_i are sequences of sets with $E_i \sim F_i \, \forall i$, then $\cup_i E_i \sim \cup_i F_i$.

Let $\mathcal{B}'$ be the collection of Borel sets equivalent to an open set. Then $\mathcal{B}'$ contains the open sets and is closed under countable unions. To see that $\mathcal{B}'$ is closed under complements, it suffices to show that closed sets are in $\mathcal{B}'$. Indeed, $A \sim G$ where G is open implies $X - A \sim X - G$. But if F is a closed set, then $F = F^\circ \cup \partial F$. Hence F is equivalent to the interior of F since the boundary is nowhere dense. $\square$

PROPOSITION I.10. *Let $f : X \to Y$ be a Borel function from a metric space X into a second countable topological space Y. Then there is a meager subset P in X such that f restricted to $X - P$ is continuous.*

PROOF. Let $\{V_n\}$ be a countable base for the topology on Y. Each $f^{-1}(V_n)$ is a Borel set and thus has the Baire property. Hence $f^{-1}(V_n) = (G_n - R_n) \cup S_n$ where G_n is an open set and R_n and S_n are meager sets. Set $P = \bigcup_n (R_n \cup S_n)$. Then P is meager and if g is the restriction of f to $X - P$, then $g^{-1}(V_n) = G_n \cap (X - P)$. Thus g is continuous. $\square$

6. Measure Algebras and Their Homomorphisms

Let μ be a finite measure on a Borel σ-algebra on a set X. The **measure algebra** associated with μ is obtained by identifying Borel sets whose symmetric differences have measure 0. It will be denoted by $B(X, \mu)$. There is a complete metric ρ on $B(X, \mu)$ defined by $\rho(E, F) = \mu(E \bigtriangleup F)$.

PROPOSITION I.11. *Suppose the measure algebra $B(X, \mu)$ is separable. Then there is a standard Borel space Y and a finite measure ν on Y such that $B(X, \mu)$ is isomorphic to $B(Y, \nu)$ as measure algebras.*

PROOF. Let $\mathcal{C}$ be a countable dense algebra of Borel sets in X. Define an equivalence relation $\sim$ on X by $x \sim x'$ iff x and x' are not separated by the sets in $\mathcal{C}$. Let $\mathcal{E}$ be the space of equivalence classes. Then each set $E \in \mathcal{C}$ is a union of equivalence classes and thus can be considered to be a subset $\Phi(E)$ of $\mathcal{E}$. Let $\mathcal{M}$ be the σ-algebra on $\mathcal{E}$ generated by the sets $\Phi(E)$. $\mathcal{M}$ is countably generated and the map Φ is an algebra homomorphism of $\mathcal{C}$ into the algebra of subsets of $\mathcal{E}$. Let ϕ be the quotient map from X to $\mathcal{E}$. The σ-algebra $\mathcal{M}$ is a countably generated subalgebra of the quotient algebra consisting of those sets whose preimages under ϕ are Borel subsets of X. Thus $\nu(E) = \mu(\phi^{-1}(E))$ is a measure on $\mathcal{M}$. The map ϕ^* from the measure algebra defined by ν on $\mathcal{M}$ to the measure algebra of μ defined by

$\phi^*(E) = \phi^{-1}(E)$ is a σ-algebra homomorphism which is an isometry of the dense subalgebra $\Phi(\mathcal{C})$ of $\mathcal{M}$ onto the dense subalgebra $\mathcal{C}$ of the measure algebra for μ. By completeness, this map extends to an onto isometry.

Since every countably generated Borel space is Borel isomorphic to a subset of a Polish space with the relative Borel structure, the result now follows. $\square$

THEOREM I.23. *Let (X, μ) and (Y, ν) be standard atom free probability spaces. Then there is a Borel isomorphism ϕ of X onto Y such that $\mu(\phi^{-1}(E)) = \nu(E)$ for all Borel subsets E of Y.*

PROOF. We may assume $X = Y = [0, 1]$, and μ is Lebesgue measure m on X. Define $\psi(r) = \nu([0, r))$. Since ν has no atoms, ψ is a continuous increasing function on $[0, 1]$ satisfying $\psi(0) = 0$ and $\psi(1) = 1$. Thus $\psi^{-1}(s)$ is a closed interval for each number s. It follows that the set C of those $s \in [0, 1]$ for which these intervals are not a point form a countable set. Moreover, since $m([0, \psi(r)) = \nu[0, r)$ for all r, it follows that $m[0, s) = \nu(\psi^{-1}[0, s))$ for all s. Thus $m(E) = \nu(\psi^{-1}(E))$ for all Borel sets E. In particular $D = \psi^{-1}(C)$ has ν measure 0. Thus ψ is strictly increasing on $[0, 1] - D$ and has range $[0, 1] - C$, and the restriction ϕ of ψ to $[0, 1] - D$ is a Borel isomorphism onto $[0, 1] - C$ satisfying $m(E) = \nu(\phi^{-1}(E))$ for all Borel sets E.

To finish the proof, let U be an uncountable Borel subset of $[0, 1] - C$ with Lebesgue measure 0. Then $V = \phi^{-1}(U)$ is an uncountable Borel subset of $[0, 1] - D$ with ν measure 0. Since both spaces $V \cup D$ and $U \cup C$ are standard, redefine and extend ϕ to be a Borel isomorphism between these two measure 0 sets. $\square$

The above result can easily be extended to σ-Boolean algebras having a measure. To do this we use Stone's Theorem characterizing Boolean algebras as an algebra of sets. Recall that a **Boolean algebra** is a set with two commutative, associative, binary operations $(a, b) \mapsto a \wedge b$ and $(a, b) \mapsto a \vee b$, and a unary operation $a \mapsto a'$ along with two special distinct elements 0 and 1 that satisfy the following conditions:

(1) $a \wedge (b \vee c) = (a \wedge b) \vee (a \wedge c)$
(2) $(a')' = a$
(3) $(a \wedge b)' = a' \vee b'$
(4) $a' \wedge a = 0$

(5) $a \wedge 0 = 0$ and $a \vee 0 = a$

The element 1 defined by $1 = 0'$ satisfies $1 \wedge a = a$ and $1 \vee a = 1$ for all a.

A Boolean algebra $\mathcal{A}$ has a partial order; namely $a \leq b$ iff $a \wedge b = a$. In this ordering, any two elements a and b have a least upper bound $a \vee b$ and a greatest lower bound $a \wedge b$. If every countable subset of $\mathcal{A}$ has a least upperbound, the algebra is said to be a σ-Boolean algebra, and the least upper bound of the sequence $\{a_i\}$ is denoted by $\vee_i a_i$. A measure on a σ-Boolean algebra $\mathcal{A}$ is a nonnegative function μ on $\mathcal{A}$ satisfying:

(1) $\mu(1) = 1$
(2) $\mu(\vee_i a_i) = \sum_i \mu(a_i)$ for any sequence $\{a_i\}$ satisfying $a_i \wedge a_j = 0$ for $i \neq j$
(3) $\mu(a) = 0$ iff $a = 0$.

If μ is a measure on $\mathcal{A}$, then $\mathcal{A}$ is a complete metric space with metric $\rho(a, b) = \mu((a \wedge b') \vee (a' \wedge b))$.

An **ideal** in a Boolean algebra is a subset $\mathcal{I}$ which contains 0 but does not contain 1, is closed under the operation $\vee$, and has the additional property that $a \wedge b \in \mathcal{I}$ whenever $b \in \mathcal{I}$ and $a \in \mathcal{A}$. The simplest Boolean algebra is $\{0, 1\}$. Any Boolean algebra homomorphism of $\mathcal{A}$ into this algebra has a kernel $\mathcal{I}$ that is an ideal having the property that for any $a \in \mathcal{A}$, either a or a' is in $\mathcal{I}$.

LEMMA I.5. *An ideal $\mathcal{I}$ in a Boolean algebra $\mathcal{A}$ is maximal if and only if it has the property that for any element $a \in \mathcal{A}$, either $a \in \mathcal{I}$ or $a' \in \mathcal{I}$. Moreover, if $a \in \mathcal{A}$ and $a \neq 1$, then there is a maximal ideal containing a.*

PROOF. Suppose a or a' is in $\mathcal{I}$ for all a and $b \notin \mathcal{I}$. Then b cannot be in any ideal containing $\mathcal{I}$ for then $1 = b \vee b'$ would be in that ideal. Therefore $\mathcal{I}$ is maximal.

Suppose $\mathcal{I}$ is maximal with $a' \notin \mathcal{I}$. Set $\mathcal{J} = \{c \vee b : b \in \mathcal{I} \ c < a'\}$. This cannot be an ideal which means it must contain 1. Therefore $a \in \mathcal{I}$.

Finally note if $a \neq 1$, then the set $\{a \wedge b : b \in \mathcal{A}\}$ is an ideal containing a which by Zorn's Lemma is contained in a maximal ideal. $\square$

THEOREM I.24 (STONE). *Any Boolean algebra is isomorphic to the algebra of open closed sets in some compact Hausdorff space.*

PROOF. Let H be the set of all homomorphisms from $\mathcal{A}$ onto the Boolean algebra $\{0, 1\}$. H is a closed subset of the Cantor space $2^{\mathcal{A}}$

and hence is a compact Hausdorff space. Define $\Phi(a) = \{h \in H : h(a) = 1\}$. The map Φ satisfies $\Phi(a \vee b) = \Phi(a) \cup \Phi(b)$, $\Phi(a \wedge b) = \Phi(a) \cap \Phi(b)$, and $\Phi(a') = H - \Phi(a)$. Moreover, the sets $\Phi(a), a \in \mathcal{A}$, are open and closed sets in the relative topology on H; they form a basis for this topology. By compactness, every open and closed subset of H is a finite union of such sets and thus is in the range of Φ. That Φ is one-to-one is an easy consequence of Lemma I.5. $\square$

COROLLARY I.9. *If $\mathcal{A}$ is a σ-Boolean algebra, then there exists a σ-algebra of sets $\mathcal{M}$ containing a σ-ideal $\mathcal{N}$ such that $\mathcal{A}$ is isomorphic to $\mathcal{M}/\mathcal{N}$ as σ-algebras.*

PROOF. Let Φ be an isomorphism of $\mathcal{A}$ onto the open and closed subsets of a compact Hausdorff space H. By the proof of Stone's Theorem, we may assume H has a base of open closed sets. Let $\mathcal{M}$ be the σ-algebra generated by the open closed sets. Let $\mathcal{N}$ be the collection of all meager sets in $\mathcal{M}$. The map $\Psi(a) = \Phi(a) \mod \mathcal{N}$ is an into algebra isomorphism since, by the Baire category theorem, no compact and open subset is meager. Next note that $\Psi(\vee a_i) = \vee \Psi(a_i)$. Indeed note $\Phi(\vee a_i) \geq \vee \Phi(a_i)$. We show the difference is a closed nowhere dense set. Indeed, it is closed; and if it had interior, it would contain an open closed set $\Phi(b)$ where $b \in \mathcal{A}$. Thus $\Phi(b) \wedge \Phi(a_i) = 0$ and $\Phi(b) \leq \Phi(\vee a_i)$. Hence $b' \geq a_i$ for all i, and we see $b' \geq \vee a_i$. But $b \leq \vee a_i$. Hence $b = 0$. The map is onto since the image is closed under countable unions. $\square$

THEOREM I.25. *Let $\mathcal{A}$ be a σ-Boolean algebra with a separable atom free measure μ. Then $\mathcal{A}$ is isomorphic to $B([0,1], m)$ where m is Lebesgue measure on $[0,1]$.*

PROOF. Using Corollary I.9, one sees that one may lift the measure μ from the quotient algebra $\mathcal{M}/\mathcal{N}$ to a probability measure on the σ-algebra $\mathcal{M}$ of subsets of some set X. Thus $\mathcal{A}$ is isomorphic to the separable measure algebra $B(X, \mu)$. The result now follows from Proposition I.11 and Theorem I.23. $\square$

Let φ be a Borel map from X into Y. Suppose μ is a measure on X. Then the image $\varphi_* \mu$ of μ under φ is the Borel measure on Y defined by $\varphi_* \mu(E) = \mu(\varphi^{-1}(E))$. If φ is a Borel mapping from a standard measure space (X, μ) into a standard measure space (Y, ν) such that $\varphi_* \mu \prec \nu$, then the set mapping φ^{-1} induces a σ-homomorphism, also denoted by φ^{-1}, from the measure algebra $B(Y, \nu)$ into the measure

algebra $B(X,\mu)$. The following theorem shows all such homomorphisms arise this way.

THEOREM I.26. *Let Φ be a σ-homomorphism from the σ-Boolean
algebra $B(Y,\nu)$ into the σ-Boolean algebra $B(X,\mu)$ where μ is a finite
measure on a standard Borel space X and ν is a finite measure on a
standard Borel space Y. Then there exists an a.e. uniquely defined
Borel function φ from X into Y such that $\varphi_*\mu \prec \nu$ and $\Phi = \varphi^{-1}$.
The map Φ is one-to-one iff $\varphi_*\mu \sim \nu$. Moreover, suppose X and
Y are uncountable. Then, if Φ is onto, φ may be taken to be one-
to-one; and if Φ is an isomorphism, φ may be taken to be a Borel
isomorphism of X onto Y.*

PROOF. The case when X or Y are countable is easy and will be
omitted. In the uncountable cases, we may assume $X = [0,1]$ and
$Y = [0,1]$. Let C be the set of rationals in $[0,1]$. For each $r \in C$, let
E_r be a Borel set representing $\Phi([0,r])$. Define $U_r = \cap_{s>r}E_s$. Then
$U_r \subset U_t$ for $r < t$ and $\Phi[0,r]$ is represented by U_r. Define $\varphi(x) =$
$\inf\{r : x \in U_r\}$. Note $\Phi([0,r]) = \Phi(\cap_{s>r}[0,s]) = \cap_{s>r}\Phi([0,s])$ is
represented by U_r. But $\varphi^{-1}([0,r]) = \cap_{s>r}U_s$. Hence $\Phi = \varphi^{-1}$ on the
generating family $\{[0,r] : r \in C\}$. Hence $\Phi = \varphi^{-1}$. Since Φ takes
the 0 element to the 0 element, $\varphi_*\mu \prec \nu$. Moreover, note that Φ is
one-to-one iff the kernel of Φ consists only of 0. This is equivalent
to the property $\varphi^{-1}(E)$ is null iff E is null. Hence Φ is one-to-one iff
$\varphi_*\mu \sim \nu$. That φ is unique a.e. depends on the fact that any other
φ which induces Φ must have the property that $\varphi(x) \leq r$ a.e. on U_r.

Next suppose Φ is onto. Then there exist Borel sets F_r in Y such
that $[0,r]$ represents the set $\Phi(F_r)$ for $r \in C$. Set $V_r = \cap_{s>r}F_s$.
Then $V_r \subset V_t$ for $r < t$ and $[0,r]$ represents $\Phi(V_r)$. Hence the set
$X_0 = \cup_r\varphi^{-1}(V_r) \triangle [0,r]$ is μ null, and if $x \neq y$ and x and y both
do not belong to X_0, then we may assume $x < r < y$ and hence
$\varphi(x) \in V_r$ and $\varphi(y) \notin V_r$. Thus φ is one-to-one off the null set X_0.

Finally, note that if Φ is an isomorphism, then $\varphi(X - X_0)$ is a
conull Borel subset of Y. Choose a Borel subset X_1 of $X - X_0$ such
that $X_0 \cup X_1$ is null and uncountable and $Y - \varphi(X - X_0 - X_1)$ is
uncountable. Extend the mapping $\varphi|_{X-X_0-X_1}$ to X so that it is a
Borel one-to-one isomorphism of $X_0 \cup X_1$ onto $Y - \varphi(X-X_0-X_1)$. $\square$

REMARK. If Φ is an isomorphism but X and Y do not have the
same cardinality, then there are conull Borel subsets X_0 and Y_0 of X
and Y and a Borel isomorphism $\phi : X_0 \to Y_0$ such that $\phi^{-1} = \Phi$. In
particular, we say a Borel mapping ϕ between standard Borel spaces

X and Y is **essentially one-to-one** if ϕ is one-to-one on a conull Borel subset; we say ϕ is an **essential Borel isomorphism** if ϕ is essentially one-to-one and $\phi_* \mu \sim \nu$.

7. The Weak Borel Structure on the Space of Finite Measures

Let X be a standard Borel space, and let $M(X)$ be the set of all finite measures on X. Equip $M(X)$ with the smallest Borel structure for which $\mu \mapsto \mu(E)$ is a Borel function for each Borel subset E of X. We call this the **weak Borel** structure on $M(X)$.

PROPOSITION I.12. *Let X be a standard Borel space. Then the weak Borel structure on $M(X)$ is standard.*

PROOF. We may assume $X = [0, 1]$ with the usual topology. By the Riesz representation theorem, $M(X)$ is isomorphic to the space of positive continuous linear functionals on $C[0, 1]$. We can thus equip $M(X)$ with the relative weak $*$ topology. This topology will be a Polish topology on $M(X)$ which generates the weak Borel structure.

We use the notation $\mu(f) = \int f(x)\, d\mu(x)$ for $f \in C[0, 1]$ and $\mu \in M([0, 1])$. By the Stone-Weierstrass Theorem, the space of polynomials functions with rational coefficients is uniformly dense in $C[0, 1]$. Let $f_1 = 1, f_2, f_3, \cdots$ enumerate these functions. Define a metric on $M(X)$ by $\rho(\mu_1, \mu_2) = \sum_n \frac{|\mu_1(f_n) - \mu_2(f_n)|}{2^n(1 + |\mu_1(f_n) - \mu_2(f_n)|)}$. Then by uniform density of the functions $f_1, f_2, f_3, \cdots$, convergence in $M(X)$ relative to this metric defines the relative weak $*$ topology on $M([0, 1])$. Moreover, μ_n is Cauchy iff $\mu_n(f_k)$ is Cauchy for each k iff $\mu_n(f)$ is Cauchy for each $f \in C[0, 1]$. Indeed, if $\mu_n(f_k)$ is Cauchy for every k, $\mu_n(X) = \mu_n(1)$ converges and thus is bounded by some constant M. Hence

$$|\mu_n(f) - \mu_m(f)| \leq |\mu_n(f - f_k)| + |\mu_n(f_k) - \mu_m(f_k)| + |\mu_m(f_k - f)|$$
$$\leq 2M\|f - f_k\| + |\mu_n(f_k) - \mu_m(f_k)|$$
$$< \epsilon$$

if f_k is chosen with $\|f - f_k\|_\infty < \frac{\epsilon}{3M}$ and n and m are large. The Riesz representation theorem now implies ρ is a complete metric on $M([0, 1])$. To see that $M([0, 1])$ is separable, one can easily check using uniform continuity of functions in $C([0, 1])$ that the set of finite nonnegative rational linear combinations of point mass measures at rational points in $[0, 1]$ form a countable dense subspace.

We next show the open sets in $M([0,1])$ relative to the metric ρ generate the weak Borel subsets of $M([0,1])$. To see all open sets are Borel, it suffices to show $\{\mu : |\mu(f_1) - \mu_0(f_1)| < \epsilon, \cdots, |\mu(f_n) - \mu_0(f_n)| < \epsilon\}$ is a weak Borel set for each μ_0 and each n. This follows easily for $\mu \mapsto \mu(f) = \int f \, d\mu$ is Borel on $M([0,1])$ for each $f \in C[0,1]$. Hence the σ-algebra generated by the weak $*$ open subsets are a subalgebra of the σ-algebra of weak Borel subsets of $M([0,1])$.

Hence to complete the proof one need only show that $\mu \mapsto \mu(E)$ is Borel in the weak $*$ topology sense for every Borel subset $E \subseteq [0,1]$. But $\mu \mapsto \mu(f)$ is weak $*$ continuous for every $f \in C([0,1])$. By taking a sequence of continuous functions f_n converging pointwise to 1_E, this implies $\mu \mapsto \mu(E)$ is weak $*$ Borel for open intervals E. Since the set of Borel subsets E of $[0,1]$ for which $\mu \mapsto \mu(E)$ is Borel in the weak $*$ sense is a σ-algebra, we see $\mu \mapsto \mu(E)$ is Borel in the weak $*$ sense for all Borel subsets of $[0,1]$. Since the weak Borel structure on $M([0,1])$ is defined to be the smallest for which these functions are Borel, we see the weak Borel structure on $M([0,1])$ is generated by the weak $*$ open subsets of $M[0,1]$. Hence $M(X)$ is standard. $\square$

COROLLARY I.10. *The set of discrete measures is weakly dense in* $M(X)$.

PROPOSITION I.13. *The subsets* $\{(\mu,\lambda) : \mu \prec \lambda\}$, $\{(\mu,\lambda) : \mu \succ \lambda\}$ *and* $\{(\mu,\lambda) : \mu \sim \lambda\}$ *are Borel subsets of* $M(X) \times M(X)$. *Thus, if* $\lambda \in M(X)$, *then* $\{\mu : \mu \prec \lambda\}$, $\{\mu : \lambda \prec \mu\}$ *and* $\{\mu : \mu \sim \lambda\}$ *are Borel subsets of* $M(X)$.

PROOF. Let $\mathcal{C}$ be a countable generating algebra for the Borel σ-algebra on X. Then since this algebra is dense in the measure algebra defined by any finite measure, one has $\mu \prec \lambda$ iff for each $\epsilon > 0$, there is a $\delta > 0$ such that $\mu(E) \le \epsilon$ whenever $\lambda(E) < \delta$ and $E \in \mathcal{C}$. Thus

$$\{(\mu,\lambda) : \mu \prec \lambda\} = \bigcap_{k=1}^{\infty} \bigcup_{n=1}^{\infty} \bigcap_{E \in \mathcal{C}} \left(\left\{ (\mu,\lambda) : \mu(E) \le \tfrac{1}{k} \right\} \cup \left\{ (\mu,\lambda) : \lambda(E) > \tfrac{1}{n} \right\} \right)$$

is a Borel set in $M(X) \times M(X)$. The other conclusions follow from this. $\square$

PROPOSITION I.14. *Let* $n \in \{\infty, 0, 1, 2, \dots\}$. *Then the set of measures in* $M(X)$ *having* n *atoms is a Borel subset of* $M(X)$. *Moreover, the collection of purely atomic measures is a Borel subset of* $M(X)$.

PROOF. Let $\mathcal{C}$ be a countable subalgebra of Borel subsets of X which generate the Borel structure on X. Set

$$M_n = \cup_{k>0} \cup_{A_1, A_2, \ldots, A_n \in \mathcal{C}} \{\mu : \mu(A_i) \geq \frac{1}{k}, \mu(A_i \cap A_j) < \frac{1}{2k} \text{ for } i \neq j,$$

$$\text{and } \mu(A \cap A_i) \geq \frac{1}{k} \text{ or } \mu(A \cap A_i) < \frac{1}{2k} \text{ for all } A \in \mathcal{C}\}.$$

Then M_n is the subset of $M(X)$ of all measures having at least n atoms. Note each M_n is a Borel set. Also $M_\infty = \cap_n M_n$ is the set of measures having infinitely many atoms. Set

$$P = \cap_{k>0} \cup_{n>0} \cup_{m>kn} \cup_{A_1, A_2, \ldots, A_n \in \mathcal{A}_0} \{\mu : |\mu(\cup A_i) - \mu(X)| < \frac{1}{k},$$

$$\mu(\cup_{i \neq j} A_i \cap A_j) < \frac{1}{2m}, \text{ and } \mu(A \cap A_i) < \frac{1}{2m} \text{ or } \mu(A \cap A_i) \geq \frac{1}{m} \forall A \in \mathcal{C}\}.$$

Then P is clearly Borel, and a standard argument shows P is the set of all μ such that μ is purely atomic. $\square$

For weakly Borel finite measure valued functions we have the following Fubini type result.

PROPOSITION I.15. *Suppose h is a nonnegative Borel function on the Borel space $X \times Y$. Then $(\mu, y) \mapsto \int h(x, y) \, d\mu(x)$ is a Borel function on $M(X) \times Y$.*

PROOF. By using simple functions, it suffices to show this for characteristic functions $h = 1_E$. But the set of all Borel sets E for which the resulting function is Borel is a monotone class containing the algebra of all measurable rectangles. But the smallest monotone class containing an algebra is the σ-algebra it generates. The result follows. $\square$

COROLLARY I.11. *Suppose f is a positive Borel function on $X \times Y$ and $\mu \in M(X)$. Then the mapping $(E, y) \mapsto \int_E f(x, y) \, d\mu(x)$ is Borel on $B(X, \mu) \times Y$.*

PROOF. Define $\mu_E(F) = \mu(E \cap F)$. Then $E \mapsto \mu_E$ is a Borel mapping of $B(X, \nu)$ into $M(X)$. Hence $(E, y) \mapsto (\mu_E, y)$ is Borel from $B(X, \mu) \times Y$ into $M(X) \times Y$ and the result follows. $\square$

8. Disintegration on Standard Borel Spaces

The following decomposition of the measure μ is called a **disintegration** of μ over the fibers of φ.

THEOREM I.27. *Let φ be a Borel mapping from a standard Borel space X into a Borel space Y. Suppose μ is a σ-finite measure on X and ν is a σ-finite measure on Y equivalent to the image $\varphi_* \mu$. Then there exists a mapping $y \mapsto \mu_y$ from Y into the set of σ-finite measures on X such that*

(1) *$y \mapsto \mu_y(E)$ is a Borel function for each Borel set E*
(2) *$\mu(E) = \int \mu_y(E)\, d\nu(y)$ for each Borel set E*
(3) *the function $y \mapsto \mu_y$ is unique ν a.e y and*
(4) *if Y is countably separated, then $\mu_y(X - \varphi^{-1}(y)) = 0\,\nu$ a.e. y.*

PROOF. The case when X is countable is trivial. We thus suppose X is the Cantor space $2^{\mathbb{N}}$. Let $\mathcal{A}$ be the countable algebra of open and closed subsets of X. For each A in $\mathcal{A}$, define a measure ν_A on Y by $\nu_A(F) = \mu(A \cap \varphi^{-1}(F))$. The measure ν_A is absolutely continuous with respect to ν. Since both measures are σ-finite, there is a Radon-Nikodym derivative $y \mapsto \mu_y(A)$. Moreover, by the a.e. uniqueness of Radon-Nikodym derivatives and the countability of $\mathcal{A}$, there is a subset Y_0 of Y whose complement has ν measure 0 and on which the following are true:

(a) $\mu_y(\emptyset) = 0$
(b) $\mu_y(A_1 \cup \cdots \cup A_n) = \mu_y(A_1) + \cdots + \mu_y(A_n)$ for any finite sequence $A_1, \ldots, A_n$ of disjoint sets in $\mathcal{A}$.
(c) $\mu_y(A) < \infty$ if $\mu(A) < \infty$.

It follows that if $y \in Y_0$, then μ_y is countably additive. Indeed, if $A \in \mathcal{A}$ is a disjoint union $\cup_{i=1}^{\infty} A_i$ of sets $A_i \in \mathcal{A}$, then by compactness of A, all but finitely many A_i are empty. Thus $\mu_y(A) = \sum_{i=1}^{\infty} \mu_y(A_i)$. By the Caratheódory extension theorem, for each $y \in Y_0$ there is a unique measure μ_y defined on all the Borel sets of X which extends the measure μ_y on $\mathcal{A}$. Define $\mu_y = 0$ for $y \in Y - Y_0$. The set of all Borel subsets E of X for which (1) and (2) hold is a monotone class containing the algebra $\mathcal{A}$ and thus contains the σ-algebra generated by $\mathcal{A}$. Thus (1) and (2) hold for every Borel set. That each measure μ_y is σ-finite follows from the finiteness of the numbers $\mu_y(A)$ when $A \in \mathcal{A}$ has μ finite measure. Essential uniqueness of the measures μ_y follows from the a.e. uniqueness of the Radon-Nikodym derivatives $y \mapsto \mu_y(A)$ for $A \in \mathcal{A}$.

Next suppose Y is countably separated. Then since $y \mapsto \mu_y(A)$ is the Radon-Nikodym derivative of ν_A, one has $\mu(A \cap \varphi^{-1}(F) \cap \varphi^{-1}(E)) = \int_E \chi_F(y)\mu_y(A)\,d\nu(y)$ for each Borel subset F of Y. But by (2) one has $\mu(A \cap \varphi^{-1}(F) \cap \varphi^{-1}(E)) = \int_E \mu_y(A \cap \varphi^{-1}(F))\,d\nu(y)$. By uniqueness of Radon-Nikodym derivatives, one has $\chi_F(y)\mu_y(A) = \mu_y(A \cap \varphi^{-1}(F))$ ν a.e. y for each $A \in \mathcal{A}$ and each Borel subset F of Y. Let $\{F_i\}$ be a countable separating family of Borel subsets of Y. Since $\mathcal{A}$ is countable, there is a Borel subset Y_1 in Y whose complement in Y has ν measure 0 and for which $\chi_{F_i}(y)\mu_y(A) = \mu_y(A \cap \varphi^{-1}(F_i))$ for all i and all $A \in \mathcal{A}$. Let $y \in Y_1$. Then $\mu_y(A \cap \varphi^{-1}(y)) = \mu_y(A \cap (\bigcap_{y \in F_i} \varphi^{-1}(F_i))) = \mu_y(A)$ for any A with $\mu_y(A)$ finite. Suppose E is a Borel set having finite μ_y measure. By Theorem I.11, for each $\epsilon > 0$ there is a compact set W and an open set U with $W \subset E \subset U$ and $\mu_y(U - W) < \epsilon$. Since W is compact and $\mathcal{A}$ is a base for the topology, there exist sets $A_1, A_2, \ldots, A_k$ in $\mathcal{A}$ with $W \subset A \subset U$ where $A = A_1 \cup A_2 \cup A_3 \cup \cdots \cup A_k$. It follows that $\mu_y(A \triangle E) < \epsilon$ and $\mu_y((A \cap \varphi^{-1}(y)) \triangle (E \cap \varphi^{-1}(y))) < \epsilon$. Since $\mu_y(A) = \mu_y(A \cap \varphi^{-1}(y))$, these combine to yield $|\mu_y(E) - \mu_y(E \cap \varphi^{-1}(y))| < 2\epsilon$. This gives $\mu_y(E \cap \varphi^{-1}(y)) = \mu_y(E)$ for all Borel sets E of μ finite measure. Since the μ_y are σ-finite, (3) holds a.e. y. $\quad\square$

REMARK. Using Theorem I.26 one can rewrite Theorem I.27 in terms of homomorphisms of measure algebras. More specifically, let X and Y be standard Borel spaces and $\mu \in M(X)$ and $\nu \in M(Y)$. Suppose $\epsilon : B(Y, \nu) \to B(X, \mu)$ is a σ-homomorphism. Then there is a Borel mapping $y \mapsto \mu_y \in M(X)$ such that $\mu(\epsilon(E) \cap F) = \int_E \mu_y(F)\,d\nu(y)$ for Borel subsets $E \subseteq Y$ and Borel subsets $F \subseteq X$. Furthermore, μ_y is uniquely determined ν a.e. y; and if $p^{-1} = \epsilon$, then μ_y is concentrated on $p^{-1}(y)$ for ν a.e. y.

PROOF. Everything is immediate except the finiteness of the measures μ_y. But $\mu(X) = \int \mu_y(X)\,d\nu(y) < \infty$. Thus $\mu_y(X) < \infty$ ν a.e. y. Redefine $\mu_y = 0$ if $\mu_y(X) = \infty$. $\quad\square$

We write $\mu = \int^\epsilon \mu_y\,d\nu(y)$ to denote such a decomposition of a finite measure μ on X relative to a σ-homomorphism $\epsilon : B(Y, \nu) \to B(X, \mu)$.

PROPOSITION I.16. Suppose $B(Z, \lambda) \xrightarrow{\epsilon} B(Y, \nu) \xrightarrow{\delta} B(X, \mu)$ are σ-homomorphisms and $\nu = \int^\epsilon \nu_z\,d\lambda(z)$ and $\mu = \int^\delta \mu_y\,d\nu(y)$. Set

$\mu_z = \int \mu_y \, d\nu_z(y)$. *Then*

$$\mu = \int^{\delta \circ \epsilon} \mu_z \, d\lambda(z).$$

PROOF. Immediate by calculation. $\square$

COROLLARY I.12. *Suppose* $\mu^1 \prec \mu^2$. *Let* $I : B(X, \mu^2) \to B(X, \mu^1)$ *be the mapping* $E \mapsto E$. *Let* $\epsilon : B(Y, \nu) \to B(X, \mu^2)$, *and let* $\mu^2 = \int^{\epsilon} \mu_y^2 \, d\nu(y)$ *and* $\mu^1 = \int^{I \circ \epsilon} \mu_y^1 \, d\nu(y)$. *Then* $\mu_y^1 \prec \mu_y^2 \ \nu$ *a.e.* y.

PROOF. Use Proposition I.16 and $\mu^1 = \int^I f(x) \epsilon_x \, d\mu^2(x)$ where $f = \frac{d\mu^1}{d\mu^2}$. $\square$

9. Standard Borel Structures on Function Spaces

We now show how to put a natural standard Borel structure on the space of measurable functions on a standard measure space. Let X and Y be Polish spaces and let μ be a σ-finite measure on X. Let $\mathcal{M}(X, \mu, Y)$ be the space of all Borel functions f from X into Y identified when they are equal a.e. μ. Let δ be a bounded complete separable metric which defines the topology on Y. Let w be a strictly positive L^1 function on X. Define a metric ρ on $\mathcal{M}(X, \mu, Y)$ by

$$\rho(f_1, f_2) = \int \delta(f_1(x), f_2(x)) w(x) \, d\mu(x).$$

The topology defined by the metric ρ is the topology of convergence in measure on sets of finite measure.

THEOREM I.28. ρ *is a complete separable metric for* $\mathcal{M}(X, \mu, Y)$.

PROOF. By Corollary I.7, there is a countable algebra $\mathcal{A}$ that generates the Borel sets on X. Let D be a countable dense set in Y. Then the simple functions whose level sets are in $\mathcal{A}$ and whose values are in D form a countable dense subset of $\mathcal{M}(X, \mu, Y)$. To see that ρ is complete, note that if f_n is Cauchy in measure on each subset of X of finite measure, then by the completeness of Y, there is a measurable function f such that f_n converges in measure to f on all sets of finite measure. $\square$

PROPOSITION I.17. *Let $\{g_i\}$ be a countable dense subset of $L^1(X)$ and let $\{h_j\}$ be a countable separating family of bounded Borel functions on Y. Then a mapping $z \mapsto f_z$ from a Borel space Z into $\mathcal{M}(X,\mu,Y)$ is Borel iff*

$$z \mapsto \int g_i(x)h_j(f_z(x))\,d\mu(x)$$

is Borel for all i and j.

PROOF. By Proposition I.4, it suffices to show that the functions

$$f \mapsto \int g_i(x)h_j(f(x))\,d\mu(x)$$

are Borel and separate points.

To see that these functions separate points, note that if

$$\int g_i(x)h_j(f_1(x))\,d\mu(x) = \int g_i(x)h_j(f_2(x))\,d\mu(x)$$

for all i and j, then by the L^1 density of the functions g_i, one has $h_j(f_1(x)) = h_j(f_2(x))$ for all j for a.e. x. Since the h_j separate the points of Y, $f_1(x) = f_2(x)$ a.e. x.

We now show they are Borel. First note that if h is continuous and bounded, then $f \mapsto \int g_i(x)h(f(x))\,d\mu(x)$ is continuous by the dominated convergence theorem. Hence, if the collection of functions $\{h_j\}$ consists entirely of bounded continuous functions, the result would follow. Now note that for any complete metric space Y, there is a sequence of continuous bounded functions on Y which separate points. In particular, if the h_j were such a sequence, the standard Borel structure on $\mathcal{M}(X,\mu,Y)$ would be the smallest for which the functions $f \mapsto \int g_i(x)h_j(f(x))\,d\mu(x)$ are Borel. Consequentially, if we refine the topology, the Borel structure on $\mathcal{M}(X,\mu,Y)$ would remain the same. In particular, by Theorem I.3, we may assume the topology on Y has a base consisting of sets which are both open and closed.

Now note that for each closed subset W of Y, there is a continuous function h such that $h = 1$ on W and $0 \le h < 1$ off W. Thus $f \mapsto \int g_i(x)1_W(f(x))\,d\mu(x) = \lim \int g_i(x)h^n(f(x))\,d\mu(x)$ is Borel. Hence the collection of all Borel sets E in Y such that $f \mapsto \int g_i(x)1_E(f(x))\,d\mu(x)$ is a Borel function contains the algebra of open and closed subsets of Y. But the set of all such E is a monotone class. It therefore contains the σ-algebra generated by the algebra of open closed sets. We conclude $f \mapsto \int g_i(x)1_E(f(x))\,d\mu(x)$

is Borel on $\mathcal{M}(X, \mu, Y)$ for every Borel subset E of Y. Each h_j being the limit of a bounded sequence of simple functions implies that $f \mapsto \int g_i(x) h_j(f(x)) \, d\mu(x)$ is Borel for all i and j. $\square$

COROLLARY I.13. *Suppose X and Y are standard Borel spaces. Let $\mu \in M(X)$. Then $\mathcal{M}(X, \mu, Y)$ is a standard Borel space under the smallest Borel structure for which $f \mapsto \int g(x) h(f(x)) \, d\mu(x)$ is Borel for all bounded Borel functions g on X and h on Y.*

COROLLARY I.14. *Suppose X, Y and Z are standard Borel spaces and $f : X \times Y \to Z$ is a Borel function. Let $\mu \in M(X)$. Then $y \mapsto f_y$ where $f_y(x) = f(x, y)$ is a Borel function from Y into $\mathcal{M}(X, \mu, Z)$.*

The set of constant functions in $\mathcal{M}(X, \mu, Z)$ is a closed set. Hence if $f : X \times Y \to Z$ is Borel, the set of y with $x \mapsto f(x, y)$ constant μ a.e. x is a Borel subset of Y. This can be generalized somewhat.

PROPOSITION I.18. *Let $f : X \times Y \to Z$ be a Borel function. If Y and Z are countably generated and ν is a σ-finite measure on Y, then*

$$\{x : f(x, y) \text{ is a constant } \nu \text{ a.e. } y\}$$

is a Borel subset of X.

PROOF. By Theorem I.1 and Theorem I.4, we may assume $Z = [0, 1]$. We may also replace ν by an equivalent finite measure. Let $\mathcal{A}$ be a countable generating family for the Borel sets of Y. Then if $f(x, y)$ is constant a.e. y, we have

$$\nu(F) \int_E f(x, y) \, d\nu(y) = \nu(E) \int_F f(x, y) \, d\nu(y)$$

for all E and F in $\mathcal{A}$.

Conversely if $\nu(F) \int_E f(x, y) \, d\nu(y) = \nu(E) \int_F f(x, y) \, d\nu(y)$ for all E and F in $\mathcal{A}$, then since $\mathcal{A}$ is a generating family for the Borel sets in Y and since both sides are measures in the variables E and F, we see the equality would then hold for all Borel subsets E and F of Y. Hence, if $\mu(E) \equiv \int_E f(x, y) \, d\nu(y)$, then $\mu \prec \nu$ and both $y \mapsto f(x, y)$ and the constant $\nu(F)^{-1} \int_F f(x, y) \, d\nu(y)$ are the Radon-Nikodym derivative. Hence $f(x, y)$ is constant a.e. in y. But by Fubini's Theorem, the set consisting of those x satisfying $\nu(F) \int_E f(x, y) \, d\nu(y) = \nu(E) \int_F f(x, y) \, d\nu(y)$ is Borel. This together with the countability of $\mathcal{A}$ implies $\{x : f(x, y) \text{ is constant a.e. in } y\}$ is a Borel set. $\square$

The following is useful for showing functions are Borel relative to the product Borel structure.

PROPOSITION I.19. *Let X be a separable metric space. Suppose Y is a Borel space and Z is a Hausdorff topological space. Then if $f : X \times Y \to Z$ is a function such that $y \mapsto f(x, y)$ is Borel for each x and $x \mapsto f(x, y)$ is continuous for each y, then f is a Borel function on the product Borel space $X \times Y$.*

PROOF. Let ρ be a metric for the topology on X, and let $\{x_n\}_{n=1}^{\infty}$ be a dense subset of X. For each n define a function f_n on $X \times Y$ by $f_n(x, y) = f(x_i, y)$ if x_i is the first member in the sequence satisfying $\rho(x_i, x) < \frac{1}{n}$. Each function f_n is Borel on the product Borel space $X \times Y$. Moreover, $f_n(x, y) = f(x_{i_n}, y)$ where $\{x_{i_n}\}_{n=1}^{\infty}$ is a sequence converging to x. The continuity of f in x for each y implies the sequence f_n converges pointwise to f. Therefore, f is a Borel function. $\square$

PROPOSITION I.20. *The set of ϕ in $\mathcal{M}(X, \mu, Y)$ which are essentially one-to-one is a Borel subset of $\mathcal{M}(X, \mu, Y)$.*

PROOF. Let $\nu = \phi_* \mu$ and $\phi^{-1} : B(Y, \nu) \to B(X, \mu)$ be the induced σ-homomorphism. By Theorem I.26, ϕ is essentially one-to-one iff ϕ^{-1} is onto.

Let $\mathcal{A}_0$ and $\mathcal{B}_0$ be countable generating algebras for the Borel subsets of X and Y. Then ϕ^{-1} is onto iff for each $\epsilon > 0$ and each $A \in \mathcal{A}_0$, there is a $B \in \mathcal{B}_0$ such that $\mu(\phi^{-1}(B) + A) < \epsilon$. But

$$\mu(\phi^{-1}(B) + A) = \int \chi_B \circ \phi(x) \chi_{X-A}(x) d\mu(x) + \int \chi_{Y-B} \circ \phi(x) \chi_A(x) \, d\mu(x)$$

is Borel in ϕ. Hence

$$\{\phi : \phi \text{ is one-to-one}\} = \cap_{k>0} \cap_{A \in \mathcal{A}_0} \cup_{B \in \mathcal{B}_0} \{\phi : \mu(\phi^{-1}(B) + A) < \frac{1}{k}\}$$

is Borel. $\square$

LEMMA I.6. *$\phi \mapsto \phi_* \mu$ is a Borel function on $\mathcal{M}(X, \mu, Y)$ into $M(Y)$.*

PROOF. By Corollary I.13, $\phi_* \mu(E) = \mu(\phi^{-1} E) = \int \chi_E \circ \phi(x) \, d\mu(x)$ is Borel in ϕ for fixed Borel subsets E of Y. $\square$

PROPOSITION I.21. *Let $\nu \in M(Y)$. Then $\{\phi \in \mathcal{M}(X, \mu, Y) : \phi_* \mu \prec \nu\}$ and $\{\phi \in \mathcal{M}(X, \mu, Y) : \phi_* \mu \sim \nu\}$ are Borel subsets of $\mathcal{M}(X, \mu, Y)$.*

PROOF. Immediate from Proposition I.13. □

LEMMA I.7. *Let* $q : Y \to Z$ *be Borel. Then the map* $f \mapsto q \circ f$
from $\mathcal{M}(X, \mu, Y)$ *to* $\mathcal{M}(X, \mu, Z)$ *is Borel.*

PROOF. We may assume $Z = [0, 1]$. Let h_i be a countable family
of bounded real valued Borel functions on X which form a dense
subset of $L^1(X, \mu)$. Then each function $f \mapsto \int q \circ f(x) h_i(x) \, d\mu(x)$ is
Borel. But the functions $\phi \mapsto \int \phi(x) h_i(x) \, d\mu(x)$ form a separating
family of Borel functions on $\mathcal{M}(X, \mu, Z)$. Hence, by Proposition I.4,
$f \mapsto q \circ f$ is Borel. □

Let $f_+ = f \vee 0 = \max(f, 0)$ and $f_- = -f \wedge 0 = \min(-f, 0)$.

LEMMA I.8. *The maps* $f \mapsto f_+$, $f \mapsto f_-$ *and* $f \mapsto |f|$ *are Borel on*
$\mathcal{M}(X, \lambda, \mathbb{R})$.

PROOF. $f \mapsto \int \chi_{[0,\infty)} \circ f(x) h(x) \, d\lambda(x)$ is a Borel function for every
bounded positive Borel function h. Since countably many of these
functions separate the points in $\mathcal{M}(X, \lambda, [0, \infty))$, $f \mapsto \chi_{[0,\infty)} \circ f$ is
Borel from $\mathcal{M}(X, \lambda, \mathbb{R})$ into $\mathcal{M}(X, \lambda, [0, \infty))$. The rest are immediate. □

REMARK. The above maps are continuous relative to the topology
of convergence in measure. Moreover, pointwise addition and multiplication of functions are continuous and thus are Borel mappings.

COROLLARY I.15. $(f, g) \mapsto f \wedge g$ *and* $(f, g) \mapsto f \vee g$ *are Borel*
functions from $\mathcal{M}(X, \lambda, \mathbb{R}) \times \mathcal{M}(X, \lambda, \mathbb{R}) \to \mathcal{M}(X, \lambda, \mathbb{R})$.

Let $\phi \in \mathcal{M}(X, \mu, Y)$ and assume $\phi_* \mu \prec \nu$. For each nonnegative Borel function f on X, define a measure ν_f on Y by $\nu_f(E) = \int_{\phi^{-1}(E)} f(x) \, d\mu(x)$. Then $\nu_f \prec \nu$ and hence there is a positive Borel
function f_ϕ on Y such that

$$\int g \circ \phi(x) \, d\mu(x) = \int g(y) f_\phi(y) \, d\nu(y)$$

for all bounded Borel functions g on Y. We thus obtain a map $f \mapsto f_\phi$ from $\mathcal{M}(X, \mu, [0, \infty))$ to $\mathcal{M}(Y, \nu, [0, \infty))$. The mapping $f_\phi \circ \phi$ is
the "conditional expectation" of f relative to the subalgebra $\phi^{-1}(\mathcal{B})$
where $\mathcal{B}$ is the σ-algebra of Borel subsets of Y.

PROPOSITION I.22. *The mapping* $f \mapsto f_\phi$ *is Borel.*

PROOF. $f \mapsto \int g(y) f_\phi(y) \, d\nu(y) = \int g \circ \phi(x) f(x) \, d\mu(x)$ is Borel for
all bounded Borel functions g on Y. □

Let $\mathcal{M}(X,\mu,Y,\nu) = \{\phi \in \mathcal{M}(X,\mu,Y) : \phi_*\mu \prec \nu\}$. Then by Lemma I.6 and Proposition I.13, $\mathcal{M}(X,\mu,Y,\nu)$ is a standard Borel space.

PROPOSITION I.23. *The mapping* $(\phi,\psi) \mapsto \psi \circ \phi$ *is a Borel function from* $\mathcal{M}(X,\mu,Y,\nu) \times \mathcal{M}(Y,\nu,Z)$ *into* $\mathcal{M}(X,\mu,Z)$.

PROOF. Let g be a bounded real valued Borel function on Z and let f be a positive bounded Borel function on X. It suffices to show $(\phi,\psi) \mapsto \int g \circ \psi \circ \phi(x)f(x)\,d\mu(x)$ is Borel. But

$$\int g \circ \psi \circ \phi(x)f(x)\,d\mu(x) = \int g \circ \psi(y)f_\phi(y)\,d\nu(y).$$

But then by a limit argument, it suffices to show

$$(\phi,\psi) \mapsto \int g \circ \psi(y)(f_\phi \wedge n)(y)\,d\nu(y)$$

is Borel for each n. Let K_i be an o.n. basis of $L^2(Y,\nu)$. Then

$$\int g \circ \psi(y)(f_\phi \wedge n)(y)\,d\nu(y) =$$
$$\sum_i (\int g \circ \psi(y)\bar{K}_i(y)\,d\nu(y))(\int f_\phi \wedge n(y)\bar{K}_i(y)\,d\nu(y))$$

which is Borel in ψ and ϕ for $\phi \mapsto f_\phi \wedge n$ is Borel. $\square$

Let μ and ν be finite measures on X and Y, respectively. Let $\mathcal{I}(X,\mu,Y,\nu)$ be the standard Borel space consisting of those $\phi \in \mathcal{M}(X,\mu,Y)$ such that ϕ is essentially one-to-one and $\phi_*\mu \sim \nu$.

LEMMA I.9. $\phi \mapsto \phi^{-1}$ *is a Borel isomorphism of* $\mathcal{I}(X,\mu,Y,\nu)$ *onto* $\mathcal{I}(Y,\nu,X,\mu)$.

PROOF. We show $\phi \mapsto \phi^{-1}$ is Borel. Let W be a Borel subset of $\mathcal{I}(Y,\nu,X,\mu)$. Then $\{\phi : \phi^{-1} \in W\} = \pi_2\{(\psi,\phi) : \psi \in W$ and $\psi\phi = I\}$ and hence is analytic. But the complement of $\{\phi : \phi^{-1} \in W\}$ is $\{\phi : \phi^{-1} \notin W\}$ which is again analytic. Hence $\{\phi : \phi^{-1} \in W\}$ is Borel. $\square$

PROPOSITION I.24. *Let* $\mathcal{I}(X,\mu) = \mathcal{I}(X,\mu,X,\mu)$. *Then* $\mathcal{I}(X,\mu)$ *is a standard Borel group; i.e. it is a standard Borel space and composition and inversion are Borel functions.*

10. Borel Fields of Functions

We discuss maps from Borel spaces to function spaces, define when they are Borel and define a Borel structure for spaces of such maps.

DEFINITION. Let S be a standard Borel space and let $s \mapsto \mu_s$ be a mapping from S into $M(X)$ such that $s \mapsto \mu_s(X)$ is a bounded function. Then a mapping $s \mapsto \phi_s \in \mathcal{M}(X, \mu_s, Y)$ is said to be a Borel field of functions if $s \mapsto \int g \circ \phi_s(x) h(x) \, d\mu_s(x)$ is Borel for all bounded Borel functions g on Y and h on X.

REMARK. $s \mapsto \phi_s$ is Borel iff $s \mapsto \mu_s(\phi_s^{-1}(E) \cap F)$ is Borel for arbitrary Borel subsets $E \subseteq Y$ and $F \subseteq X$. In particular $s \mapsto \mu_s \in M(X)$ is Borel.

DEFINITION. Suppose $m \in M(S)$ and $s \mapsto \mu_s \in M(X)$ is Borel. Define $\mathcal{F}(S, m, \mu, Y)$ to be the space of all Borel fields of functions $s \mapsto \phi_s \in \mathcal{M}(X, \mu_s, Y)$, identifying any two equal m a.e. s. Give $\mathcal{F}(S, m, \mu, Y)$ the smallest Borel structure such that $\phi \mapsto \int f(s) \int g \circ \phi_s(x) h(x) \, d\mu_s(x) \, dm(s)$ is Borel for bounded Borel functions f on S, g on Y and h on X.

Define a finite measure $m * \mu$ on $S \times X$ by

$$m * \mu = \int^{\pi_1^{-1}} \epsilon_s \times \mu_s \, dm(s).$$

Note this is a measure on the product space $S \times X$ having the property that the disintegration of $m * \mu$ over the fibers of the mapping $\pi_1(s, x) = s$ is $s \mapsto \epsilon_s \times \mu_s$.

PROPOSITION I.25. *The map* $\phi \mapsto \phi(s, \cdot) \in \mathcal{M}(X, \mu_s, Y)$ *is a Borel isomorphism of* $\mathcal{M}(S \times X, m * \mu, Y)$ *onto* $\mathcal{F}(S, m, \mu, Y)$. *In particular,* $\mathcal{F}(S, m, \mu, Y)$ *is a standard Borel space.*

PROOF. The map is clearly well defined and one-to-one. We show it is onto. We may assume $Y = [0, 1]$. Let $s \mapsto \phi_s$ be a Borel field of functions. We first remark that $s \mapsto \int \phi_s(x) h(s, x) \, d\mu_s(x)$ is Borel for bounded Borel functions h on $S \times X$. Note this is clearly true for characteristic functions of finite disjoint unions of Borel rectangles in $S \times X$. Moreover, the collection of all Borel subsets W of $S \times X$ for which $s \mapsto \int \phi_s(x) 1_W(s, x) \, d\mu_s(x)$ is Borel is a monotone class; it thus contains every element in the σ-algebra generated by the algebra of finite disjoint unions of Borel rectangles. Hence $s \mapsto \int \phi_s(x) h(s, x) \, d\mu_s(x)$ is Borel for all simple functions h and by a limit argument for all bounded Borel functions h.

Define a measure μ' on $S \times X$ by

$$\mu'(E) = \iint \phi_s(x)\chi_E(s,x)\,d\mu_s(s)\,dm(s).$$

Then $\mu' \prec m * \mu$. Hence, the Radon-Nikodym derivative ϕ is a Borel function on $S \times X$ satisfying $\mu'(E) = \iint \chi_E(s,x)\phi(s,x)\,d\mu_s(x)\,dm(s)$. Furthermore,

$$\mu'(\pi_1^{-1}(F) \cap E) = \iint \chi_F \circ \pi_1(s,x)\chi_E(s,x)\phi(s,x)\,d\mu_s(s)\,dm(s)$$

$$= \int_F \int \chi_E(s,x)\phi(s,x)\,d\mu_s(x)\,dm(s)$$

$$= \int_F \int \chi_E(s,x)\phi_s(x)\,d\mu_s(x)\,dm(s).$$

Hence, for each Borel subset $E \subseteq S \times X$, one has

$$\int \chi_E(s,x)\phi(s,x)\,d\mu_s(x) = \int \chi_E(s,x)\phi_s(x)\,d\mu_s(x)$$

for m a.e. s. Taking the E's from a countable generating algebra and using the fact that as functions of E, both sides of the equality are measures, we may conclude that for m a.e. s that

$$\int \chi_E(s,x)\phi(s,x)\,d\mu_s(x) = \int \chi_E(s,x)\phi_s(x)\,d\mu_s(x)$$

for all Borel subsets E of $S \times X$. This implies $\phi(s,x) = \phi_s(x)$ for μ_s a.e. x for m a.e. s.

The map $\phi \mapsto (s \mapsto \phi(s,\cdot))$ is a Borel isomorphism for the mappings $\phi \mapsto \iint f \circ \phi_s(x)g(s,x)\,d\mu_s(x)\,dm(s)$ for bounded Borel functions f on Y and g on $S \times X$ are Borel on $\mathcal{F}(S,m,\mu,Y)$, and the Borel structure on $M(S \times X, m * \mu, Y)$ is the smallest for which $\phi \mapsto \int \int f \circ \phi(s,x)g(s,x)\,d(m*\mu)$ is Borel for bounded Borel functions f on Y and g on $S \times X$. $\square$

COROLLARY I.16. *Let $\lambda \in M(X)$ and $\nu \in M(Y)$. The mapping $\phi \mapsto (x \mapsto \phi(x,\cdot))$ is a Borel isomorphism from $\mathcal{M}(X \times Y, \lambda \times \nu, Z)$ onto $\mathcal{M}(X, \lambda, \mathcal{M}(Y, \nu, Z))$.*

PROOF. Take $\mu_x = \nu$ for $x \in X$ and note

$$\mathcal{F}(X, \lambda, \mu, Z) = \mathcal{M}(X, \lambda, \mathcal{M}(Y, \nu, Z)).$$

$\square$

DEFINITION. Let $(s \mapsto \phi_s) \in \mathcal{F}(S, m, \mu, Y)$. This field is said to be **realizable** if there exists a $\phi \in \mathcal{M}(X, \mu, Y)$ such that $\phi_s(x) = \phi(x)$ for μ_s a.e. x for m a.e. s. In this case we write $\phi = \int \phi_s \, dm(s)$ and call ϕ a realization of the Borel field $s \mapsto \phi_s$.

PROPOSITION I.26. *Every Borel field $s \mapsto \phi_s$ is realizable iff there exists a $\tau : B(S, m) \to B(X, \mu)$ such that $\mu = \int^\tau \mu_s \, dm(s)$.*

PROOF. We assume $Y = [0, 1]$. Suppose $\mu = \int^\tau \mu_s \, dm(s)$ and $s \mapsto \phi_s \in \mathcal{M}(X, \mu_s, Y)$ is Borel. Define $\mu'(E) = \int \int_E \phi_s(x) \, d\mu_s(x) \, dm(s)$. Then $\mu' \prec \mu$, and hence the Radon-Nikodym derivative ϕ is a Borel function on X satisfying

$$\mu'(E) = \int_E \phi(x) \, d\mu(x) = \int \int \chi_E(x) \phi(x) \, d\mu_s(x) \, dm(s).$$

An easy argument shows $\mu' = \int^\tau \mu'_s \, dm(s) = \int^\tau \mu''_s \, dm(s)$ where $\mu'_s(E) = \int_E \phi(x) \, d\mu_s(x)$ and $\mu''_s(E) = \int_E \phi_s(x) \, d\mu_s(x)$. Hence $\mu'_s = \mu''_s$ m a.e. s, and consequently $\phi_s(x) = \phi(x)$ for μ_s a.e. x for m a.e. s.

Conversely, suppose every $(s \mapsto \phi_s) \in \mathcal{F}(S, m, \mu, Y)$ has a realization. Let E be a Borel subset of S. Define $\phi_s(x) = \chi_E(s)$. Then $s \mapsto \phi_s$ is Borel, and hence has a realization ϕ_E. Define $\tau(E) = \{x : \phi_E(x) = 1\}$. τ is easily shown to be a σ-homomorphism from $B(S, m)$ into $B(X, \mu)$. Moreover,

$$\mu(\tau(E) \cap F) = \int \mu_s(\{x : \phi_E(x) = 1\} \cap F) \, dm(s)$$

$$= \int \mu_s(\{x : \phi_s(x) = 1\} \cap F) \, dm(s)$$

$$= \int_E \mu_s(F) \, dm(s).$$

Thus $\mu = \int^\tau \mu_s \, dm(s)$. $\square$

COROLLARY I.17. *Suppose $\mu \in M(X)$ and $\mu = \int^\tau \mu_s \, dm(s)$ where $\tau : B(S, m) \to B(X, \mu)$. Then $\mathcal{F}(S, m, \mu, Y)$ is Borel isomorphic to $\mathcal{M}(X, \mu, Y)$.*

THEOREM I.29. *Let $m \in M(S)$ and $n \in M(T)$. Let $t \mapsto \mu_t \in M(X)$ be Borel. Then $\mathcal{M}(S, m, \mathcal{F}(T, n, (t \mapsto \mu_t), Y))$ is Borel isomorphic to $\mathcal{F}(S \times T, m \times n, ((s, t) \mapsto \mu_t), Y)$ under the mapping*

$$\phi \mapsto ((s, t) \mapsto \phi(s)_t).$$

PROOF. By an application of Proposition I.25 and its corollary,

$$
\begin{aligned}
\mathcal{M}(S, m, \mathcal{F}(T, n, (t \mapsto \mu_t), Y)) &\cong \mathcal{M}(S, m, \mathcal{M}(T \times X, n * \mu, Y)) \\
&\cong \mathcal{M}(S \times T \times X, m \times (n * \mu), Y) \\
&\cong \mathcal{M}(S \times T \times X, (m \times n) * \mu, Y) \\
&\cong \mathcal{F}(S \times T, m \times n, ((s, t) \mapsto \mu_t), Y).
\end{aligned}
$$

The compositions of these isomorphisms is the isomorphism in the statement of the theorem. □

COROLLARY I.18. *Suppose* $\mu = \int^\tau \mu_t \, dn(t)$ *where* $\tau : B(T, n) \to B(X, \mu)$. *Then* $\mathcal{M}(S, m, \mathcal{M}(X, \mu, Y))$ *is Borel isomorphic to* $\mathcal{F}(S \times T, m \times n, ((s, t) \mapsto \mu_t), Y)$.

PROOF. $\mathcal{F}(T, n, (t \mapsto \mu_t), Y)$ is Borel isomorphic to $\mathcal{M}(X, \mu, Y)$. □

LEMMA I.10. *Let* $m \in M(S)$, *and suppose* $s \mapsto \phi_s \in M(X, \lambda_s, Y)$ *and* $s \mapsto \psi_s \in M(Y, \nu_s, Z)$ *are Borel fields of functions. Then if* $\phi_* \lambda_s \prec \nu_s$ *for* m *a.e.* s, *then* $s \mapsto \psi_s \circ \phi_s$ *is a Borel field on a conull Borel subset of* S.

PROOF. Proposition I.25 implies there are Borel functions $\Phi : S \times X \to Y$ and $\Psi : S \times Y \to Z$ such that $\phi_s(x) = \Phi(s, x)$ for λ_s a.e. x and $\psi_s(x) = \Psi(s, x)$ for ν_s a.e. x for m a.e. s. Since $\phi_{s*} \lambda_s \prec \nu_s$ for m a.e. s, one has $\Psi(s, \phi_s(x)) = \psi_s(\phi_s(x))$ for λ_s a.e. x for m a.e. s. Let g_i and h_j be bounded sequences of Borel functions on Z and X which separate points. Proposition I.15 implies $s \mapsto \int g_i(\Psi(s, \Phi(s, x)))h_j(x) \, d\lambda_s(x)$ is Borel for each i and j. Hence, there is a conull Borel subset S_0 of S such that

$$
s \mapsto \int g_i(\psi_s(\phi_s(x)))h_j(x) \, d\lambda_s(x)
$$

is Borel on S_0 for all i and j. Since the functions

$$
\xi \mapsto \int g_i(\xi(s, x))h_j(x) \, d\lambda_s(x) \, dm(s)
$$

are Borel on $\mathcal{M}(S_0 \times X, m * \lambda, Z)$ and separate points, Proposition I.25 and Proposition I.4 imply $s \mapsto \psi_s(\phi_s(\cdot))$ is a Borel field on some conull subset of S. □

PROPOSITION I.27. *There is a canonical Borel isomorphism between $\mathcal{F}(S, m, \lambda, Y)$ and $\{\Phi \in \mathcal{M}(S \times X, m * \lambda, S \times Y) : \pi_1 \circ \Phi = \pi_1\}$ determined by $(s \mapsto \phi_s) \leftrightarrow \Phi$ iff $\Phi(s, x) = \phi_s(x)$ for λ_s a.e. x for m a.e. s. Furthermore, Φ is essentially one-to-one iff ϕ_s is essentially one-to-one for m a.e. s, and if $s \mapsto \nu_s$ is Borel into $M(Y)$, then $\Phi_* m * \lambda \prec m * \nu$ iff $\phi_{s*} \lambda_s \prec \nu_s$ m a.e. s and $\Phi_* m * \lambda \sim m * \nu$ iff $\phi_{s*} \lambda_s \sim \nu_s$ for m a.e. s.*

PROOF. Let $\mathcal{M}_0$ be the space of all $\Phi \in \mathcal{M}(S \times X, m * \lambda, S \times Y)$ such that $\pi_1 \circ \Phi = \pi_1$. By Lemma I.7, $\mathcal{M}_0$ is a Borel subset of $\mathcal{M}(S \times X, m * \lambda, S \times Y)$ and thus is a standard Borel space. For $\Phi \in \mathcal{M}_0$, define $\phi_s(x) = \pi_2 \circ \Phi(s, x)$. Then $s \mapsto \phi_s$ is in $\mathcal{F}(S, m, \lambda, Y)$; and since $\int h(s) \int f(x) g \circ \phi_s(x) \, d\lambda_s(x) \, dm(s) = \int \int h(s) f(x) g(\pi_2 \circ \Phi(s, x)) \, d\lambda_s(x) \, dm(s)$ for bounded Borel functions h on S, f on X and g on Y, we see the mapping $\Phi \mapsto (s \mapsto \phi_s)$ is a Borel mapping from $\mathcal{M}_0$ into $\mathcal{F}(S, m, \lambda, Y)$.

Suppose $s \mapsto \phi_s \in \mathcal{F}(S, m, \lambda, Y)$. Then $s \mapsto \phi_s$ is a Borel field of functions. By Proposition I.25, there is a $\Psi \in \mathcal{M}(S \times X, m * \lambda, Y)$ such that $\Psi(s, x) = \phi_s(x)$ for λ_s a.e. x for m a.e. s. Define $\Phi(s, x) = (s, \Psi(s, x))$. Then $\pi_1 \circ \Phi = \pi_1$ and $\phi_s = \pi_2 \circ \Phi(s, \cdot)$ for m a.e. s. Hence the mapping $\Phi \mapsto (s \mapsto \phi_s)$ is onto. Thus by Kuratowski's Theorem, $\mathcal{M}(S, m, \lambda, Y)$ is Borel isomorphic to $\mathcal{F}(S, m, \lambda, Y)$.

Suppose $s \mapsto \nu_s$ is a Borel mapping of S into $M(Y)$ and $s \mapsto \nu_s(Y)$ is bounded. Suppose $\Phi_* m * \lambda \prec m * \nu$. Since $m * \lambda = \int^{\pi_1^{-1}} \epsilon_s \times \lambda_s \, dm(s)$,

$$\Phi_*(m * \lambda) = \int^{\Phi^{-1} \circ \pi_1^{-1}} \Phi_*(\epsilon_s \times \lambda_s) \, dm(s)$$

$$= \int^{\pi_1^{-1}} \Phi_*(\epsilon_s \times \lambda_s) \, dm(s) \prec \int^{\pi_1^{-1}} \epsilon_s \times \nu_s \, dm(s).$$

Corollary I.12 implies $\Phi_*(\epsilon_s \times \lambda_s) \prec \epsilon_s \times \nu_s$ for m a.e. s. Conversely, an easy argument shows $\Phi_* m * \lambda \prec m * \nu$ if $\lambda_s \prec \nu_s$ for m a.e. s.

Now, using Corollary I.12, note that

$$\Phi_* m * \lambda \sim m * \nu$$

$$\text{iff } \int^{\pi_1^{-1}} \Phi_* \epsilon_s \times \lambda_s \, dm(s) \sim \int^{\pi_1^{-1}} \epsilon_s \times \nu_s \, dm(s)$$

$$\text{iff } \Phi_*(\epsilon_s \times \lambda_s) \sim \epsilon_s \times \nu_s \text{ for } m \text{ a.e. } s$$

$$\text{iff } (\pi_2 \circ \Phi)_*(\epsilon_s \times \lambda_s) \sim \nu_s \text{ for } m \text{ a.e. } s.$$

Suppose Φ is essentially one-to-one. Then there is a conull Borel subset $W \subseteq S \times X$ such that $\Phi|_W$ is one-to-one. Since $\pi_1 \circ \Phi = \pi_1$ a.e. $m * \lambda$, we may assume $\pi_1 \circ \Phi(s, x) = s$ for $(s, x) \in W$. Hence W is $\epsilon_s \times \lambda_s$ conull for m a.e. s, and $x \mapsto \pi_2 \circ \Phi(s, x)$ is ν_s essentially one-to-one on $W_s = \pi_2(\{s\} \times X \cap W)$ for m a.e. s.

We remark that the theorem is proved if we can show Φ is $m * \lambda$ essentially one-to-one if ϕ_s is λ_s essentially one-to-one for m a.e. s. Set $\nu_s = \phi_{s*}\lambda_s$. By Theorem I.26, it suffices to show the σ-homomorphism Φ^{-1} from $B(S \times Y, m * \nu)$ into $B(S \times X, m * \lambda)$ is onto.

Let $W \subseteq S \times X$ be Borel. Then $\Phi(W) \subseteq S \times Y$ is an analytic and hence measurable set. Therefore, there is a Borel subset $E \supseteq \Phi(W)$ such that $(m * \nu)(E - \Phi(W)) = 0$. We claim $\Phi^{-1}(E) = W$ in $B(S \times X, m * \lambda)$. Clearly $\Phi^{-1}(E) \supseteq W$. But since $\epsilon_s \times \nu_s(E - \Phi(W)) = 0$ for m a.e. s, $\epsilon_s \times (\pi_2 \circ \Phi)_*(\epsilon_s \times \lambda_s)(E - \Phi(W)) = 0$ for m a.e. s. Hence $\lambda_s\{x : \Phi(s, x) \in E, \Phi(s, x) \notin \Phi(W)\} = 0$ for m a.e. s. But then $(\epsilon_s \times \lambda_s)(\Phi^{-1}(E) - W) = 0$ for m a.e. s. Thus $\Phi^{-1}(E) = W$. $\square$

Let Y be a standard Borel space. For $n = 1, 2, 3, \ldots$, let J_n be the Borel subset of $M(Y)$ consisting of those measures which are purely atomic with n atoms. For $n = 0, -1, -2, \ldots$, let J_n be the Borel subset of $M(Y)$ consisting of those measures having continuous part and $-n$ atoms. For $\nu \in M(Y)$, we say $\nu \in M(Y)$ has type n precisely if $\nu \in J_n$.

COROLLARY I.19. *Suppose $n \in \mathbb{Z}$ and $s \mapsto \nu_s \in J_n$ is Borel. Let T be a standard Borel space, and let τ be a fixed measure of type n on T. Then there exists a Borel isomorphism $\phi \in \mathcal{I}(S \times T, m \times \tau, S \times Y, m * \nu)$ such that $\pi_1 \circ \phi = \pi_1$ and $(t \mapsto \pi_2 \circ \phi(s, t)) \in \mathcal{I}(T, \tau, S, \nu_s)$ for m a.e. s. If, in addition, $n = 0$ and τ and all ν_s with $s \in S$ are probability measures, then ϕ may be chosen to preserve measure; i.e. $\phi_* m \times \tau = m * \nu$.*

PROOF. Let $T = [0, 1]$ and let $\tau \in M(T)$ be a measure of type n. Take A to be the set of all pairs (s, ψ) where ψ is in $\mathcal{M}(T, \tau, Y)$, $\psi_* \tau \sim \nu_s$, and ψ is essentially one-to-one. Then by Lemma I.6, Proposition I.13, and Proposition I.20, A is a Borel subset of $S \times \mathcal{M}(T, \tau, Y)$. Furthermore, $\pi_1 A = S$. Using von Neumann's selection theorem, Theorem I.14, there is a Borel function $s \mapsto \psi_s \in \mathcal{M}(T, \tau, Y)$ with $(s, \psi_s) \in A$ for m a.e. s. Therefore $s \mapsto \psi_s$ is a Borel field of functions for which ψ_s is essentially one-to-one and $\psi_{s*}\tau \sim \nu_s$ for m a.e. s.

Hence, there exists a $\Psi \in \mathcal{I}(S \times T, m \times \tau, S \times Y, m*\nu)$ with $\pi_1 \circ \Psi = \pi_1$ and $(t \mapsto \pi_2 \circ \Psi(s,t)) \in \mathcal{I}(T, \tau, S, \nu_s)$ for m a.e. s.

To obtain the last statement of the corollary, change the condition $\psi_* \tau \sim \nu_s$ in the definition of A to $\psi_* \tau = \nu_s$. □

COROLLARY I.20. *Let* $\nu = \int \nu_s \, dm(s) \in M(Y)$. *Then there is a Borel isomorphism* $\Phi \in \mathcal{I}(Y, \nu, S \times Y, m*\nu)$ *such that* $\Phi \in \mathcal{I}(Y, \nu_s, S \times Y, \epsilon_s \times \nu_s)$ *for* m *a.e.* s *iff there is a* σ-*homomorphism* $\delta : M(S, m) \to M(Y, \nu)$ *such that* $\nu = \int^{\delta} \nu_s \, dm(s)$. *In this case* $\delta = \Phi^{-1} \circ \pi_1^{-1}$.

PROOF. Suppose $\nu = \int^{\delta} \nu_s \, dm(s)$. Let $\phi_s : (Y, \nu_s) \to (S \times Y, \epsilon_s \times \nu_s)$ and $\phi_s^{-1} : (S \times Y, \epsilon_s \times \nu_s) \to (Y, \nu_s)$ be the natural mappings $y \mapsto (s, y)$ and $(s, y) \mapsto y$. Using Proposition I.15, $s \mapsto \phi_s$ and $s \mapsto \phi_s^{-1}$ are Borel. By Proposition I.26, they each have Borel realizations Φ and Φ^{-1}. Then Φ is the desired mapping. One may easily show $\nu = \int^{\Phi^{-1} \circ \pi_1^{-1}} \nu_s \, dm(s)$, and hence $\delta = \phi^{-1} \circ \pi_1^{-1}$.

Suppose such a Φ exists. By Proposition I.27 every Borel field $s \mapsto \psi_s \in \mathcal{M}(Y, \nu_s, Z)$ may be integrated to obtain a $\Psi : S \times Y \to Z$ such that $\Psi(s, y) = \psi_s(y)$ for ν_s a.e. y for m a.e. s. Hence every Borel field $s \mapsto \psi_s$ has a realization, namely $\Psi \circ \Phi$. The result follows from Proposition I.26. □

REMARK. Note the above Φ is measure preserving, and $\Phi_* \nu_s = \epsilon_s \times \nu_s$ m a.e. s.

COROLLARY I.21. *Let* $\lambda = \int^{\epsilon} \lambda_s \, dm(s)$ *and* $\nu = \int^{\delta} \nu_s \, dm(s)$. *Then there is a Borel isomorphism from Borel fields* $s \mapsto \phi_s \in \mathcal{M}(X, \lambda_s, Y, \nu_s)$ *to* $\Phi \in \mathcal{M}(X, \lambda, Y, \nu)$ *satisfying* $\Phi^{-1} \circ \delta = \epsilon$. *The field* $(s \mapsto \phi_s)$ *corresponds to* Φ *iff* $\Phi(x) = \phi_s(x)$ *for* λ_s *a.e.* x *for* m *a.e.* s. *Furthermore,* $\Phi_* \lambda \sim \nu$ *iff* $\phi_{s*} \lambda_s \sim \nu_s$ *for* m *a.e.* s, *and* Φ *is essentially one-to-one iff* ϕ_s *is essentially one-to-one for* m *a.e.* s.

PROOF. Apply Corollary I.20 and Proposition I.27. □

COROLLARY I.22. *Suppose* $\nu = \int^{\delta} \nu_s \, dm(s)$ *and* ν_s *is a type* n *measure for* m *a.e.* s. *If* T *is a standard space and* $\tau \in M(T)$ *is type* n, *then there exists a* $\Phi \in \mathcal{I}(Y, \nu, S \times T, m \times \tau)$ *such that* $\Phi \in \mathcal{I}(Y, \nu_s, S \times T, \epsilon_s \times T)$ *for* m *a.e.* s. *Furthermore, if* $n = 0$ *and* $\nu_s \in M_1(Y)$ *for* m *a.e.* s *and* $\tau \in M_1(T)$, *then* Φ *may be taken to be measure preserving. Also* $\delta = \Phi^{-1} \pi_1^{-1}$.

PROOF. This follows from Corollaries I.20 and I.19. □

COROLLARY I.23. *Let* $\epsilon : B(S, m) \to B(X, \lambda)$. *Suppose a is in* $\mathcal{I}(S, m)$. *Let* $\lambda = \int^{\epsilon} \lambda_s \, dm(s)$. *Then the mapping* $(s \mapsto \phi(s)) \mapsto \Phi = \int \phi(s) \, dm(s)$ *is a Borel isomorphism of the Borel subset of* $\mathcal{F}(S, m, \lambda_{a(\cdot)}, X)$ *consisting of those Borel fields* $s \mapsto \phi_s$ *such that* $\phi_{s*}\lambda_{a(s)} \prec \lambda_s$ *for m a.e. s onto* $\{\Phi \in \mathcal{M}(X, \lambda, X, \lambda) : \epsilon \circ a = \Phi^{-1} \circ \epsilon\}$. *Furthermore,* $\Phi_*\lambda \sim \lambda$ *iff* $\phi_{s*}\lambda_{a(s)} \sim \lambda_s$ *for m a.e. s, and* Φ *is essentially one-to-one iff* ϕ_s *is essentially one-to-one for m a.e. s.*

PROOF. Let $\nu = \int \lambda_{a(s)} \, dm(s)$. Note $\nu = \int^{\epsilon \circ a} \lambda_{a(s)} \, dm(s)$. Indeed, if E is a Borel subset of S and F is a Borel subset of X, then

$$\nu(\epsilon \circ a(E) \cap F) = \int \lambda_{a(s)}(\epsilon \circ a(E) \cap F) \, dm(s)$$

$$= \int \lambda_s(\epsilon \circ a(E) \cap F) \, da_* m(s)$$

$$= \int \lambda_s(\epsilon(a(E)) \cap F) \frac{da_* m}{dm}(s) \, dm(s)$$

$$= \int_{a(E)} \lambda_s(F) \frac{da_* m}{dm}(s) \, dm(s)$$

$$= \int \chi_{a(E)}(s) \lambda_s(F) \, da_* m(s)$$

$$= \int \chi_{a(E)}(a(s)) \lambda_{a(s)}(F) \, dm(s)$$

$$= \int \chi_E(s) \lambda_{a(s)}(F) \, dm(s)$$

$$= \int_E \lambda_{a(s)}(F) \, dm(s).$$

Clearly $\nu \sim \lambda$. Hence $B(X, \lambda) = B(X, \nu)$. By Corollary I.21, the mapping that associates Borel fields $s \mapsto \phi_s \in \mathcal{M}(X, \lambda_{a(s)}, X, \lambda_s)$ to Borel functions $\Phi \in \mathcal{M}(X, \nu, X, \lambda)$ where $\Phi^{-1} \circ \epsilon = \epsilon \circ a$ and $\Phi(x) = \phi_s(x)$ for $\lambda_{a(s)}$ a.e. x for m a.e. s is a Borel isomorphism. That $\Phi_*\lambda \sim \lambda$ iff $\phi_{s*}\lambda_{a(s)} \sim \lambda_s$ for m a.e. s and that Φ is essentially one-to-one iff ϕ_s is essentially one-to-one for m a.e. s follow from the same corollary. $\square$

CHAPTER II

PRELIMINARIES ON C* ALGEBRAS

In this chapter, C* algebras and their representations are defined, and their basic properties are developed. We begin with Gelfand's results on commutative C* algebras. Next representations of commutative C* algebras are defined in terms of projection valued measures. This description is used to obtain the spectral theorem. Then for noncommutative C* algebras the Gelfand-Naimark-Segal construction is described along with the relation between positive linear functionals and cyclic representations. The chapter ends with Gelfand's characterization of C* algebras as normed closed ∗ algebras of bounded operators on a Hilbert space.

1. Basic Definitions and the Gelfand Theorem

DEFINITION. A complex Banach algebra is an algebra $\mathcal{B}$ with a complete norm $|| \cdot ||$ satisfying the property $||xy|| \leq ||x|| \, ||y||$ for all x and y in $\mathcal{B}$. If, in addition, there is an involution $x \mapsto x^*$ on $\mathcal{B}$ satisfying

    (a) $(x + y)^* = x^* + y^*$
    (b) $(cx)^* = \bar{c}\, x^*$
    (c) $(xy)^* = y^* x^*$ and
    (d) $||x^*|| = ||x||$

for all $c \in \mathbb{H}$ and x and y in $\mathcal{B}$, then $\mathcal{B}$ is said to be a Banach ∗ algebra.

Elements in a Banach ∗ algebra satisfying $x^* = x$ are said to be self adjoint.

DEFINITION. A C* algebra is a nonzero Banach $*$ algebra $\mathcal{A}$ for which the norm satisfies $||x^*x|| = ||x||^2$.

An example of a C* algebra is a norm closed $*$ subalgebra of $B(\mathbb{H})$ where $\mathbb{H}$ is a Hilbert space and $B(\mathbb{H})$ is the set of all bounded linear operators on $\mathbb{H}$. The adjoint is the usual adjoint operation defined by

$$\langle Av, w \rangle = \langle v, A^*w \rangle$$

if $A \in B(\mathbb{H})$ and $v, w \in \mathbb{H}$.

Commutative C* algebras may be obtained from locally compact spaces. More precisely, let X be a locally compact Hausdorff space, and let $C_0(X)$ be the space of all continuous complex valued functions f on X vanishing at ∞ (i.e., for each $\epsilon > 0$ the set $\{x : |f(x)| \geq \epsilon\}$ is compact). Set $||f|| = \max |f(x)|$, and define $f^*(x) = \overline{f(x)}$. Let addition, multiplication and scalar multiplication be defined pointwise.

A Banach algebra $\mathcal{B}$ has an identity if there is an element e satisfying $||e|| = 1$ and

$$ae = ea = a \text{ for all } a \in \mathcal{A}.$$

If $\mathcal{B}$ is a Banach $*$ algebra with an identity e, then $e^* = e$ and $||e|| = 1$. Indeed, $e^* = e^*e = (e^*e)^* = (e^*)^* = e$.

Suppose $\mathcal{B}$ has an identity e. Let $x \in \mathcal{B}$. Then $\sigma(x) = \{\lambda \in \mathbb{C} : \lambda e - x$ has a multiplicative inverse in $\mathcal{B}\}$ is called the **spectrum** of x. The complement of $\sigma(x)$ is the **resolvent** set for x and is denoted by $\rho(x)$.

THEOREM II.1. *Let $\mathcal{B}$ be a Banach algebra with an identity. Elements in $\{x : ||x - e|| < 1\}$ are invertible. Indeed, if $||x|| < 1$, then $(e - x)^{-1} = \sum_{n=0}^{\infty} x^n$. For each $x \in \mathcal{B}$, the spectrum $\sigma(x)$ is a closed bounded nonempty set; and the spectral radius, $||x||_\sigma$, defined by*

$$||x||_\sigma = \max\{|\lambda| : \lambda \in \sigma(x)\}$$

satisfies

$$||x||_\sigma = \lim ||x^n||^{\frac{1}{n}} \leq ||x||.$$

If $\mathcal{B}$ is a C algebra and x is self adjoint, then $||x|| = ||x||_\sigma$.*

PROOF. Suppose $||x|| < 1$. Then $(e - x)^{-1} = \sum x^n$. Indeed, $(e - x)\sum_{n=0}^{N} x^n = e - x^N$. Thus $(e - x)\sum x^n = e$. Similarly $\sum x^n(e - x) = e$. Hence if $|\lambda| > ||x||$, $(\lambda e - x)^{-1} = \lambda^{-1}(e - x/\lambda)^{-1}$ exists. Hence $\sigma(x) \subset \{z : |z| \leq ||x||\}$. Moreover, $\rho(x)$ is an open set, for if $z_0 \in \rho(x)$, then $ze - x = z_0 e - x + (z - z_0)e = (z_0 e - x)(e + (z - z_0)(z_0 e - x)^{-1})$

which is invertible if $z - z_0$ is near 0. Moreover, one sees the function $ze - x$ is analytic in z near any z_0 in $\rho(x)$. In fact

$$(ze - x)^{-1} = \sum (-1)^n (z - z_0)^n (z_0 e - x)^{-n} (z_0 e - x)^{-1}$$

for z near z_0. Hence if $\rho(x) = \mathbb{C}$, $g(z) = (ze - x)^{-1}$ is analytic on $\mathbb{C}$. We claim it is bounded. Indeed, it is bounded on $|z| \leq 2||x||$. Moreover for $|z| > 2||x||$, we have

$$||(ze - x)^{-1}|| = |z|^{-1} \left|\left|(e - \frac{x}{z})^{-1}\right|\right| \leq \sum \frac{||x||^n}{|z|^{n+1}} < \sum \frac{1}{2^{n+1}||x||}.$$

Thus $(ze-x)^{-1}$ is a bounded entire function. By Liouville's Theorem, it must be constant. This gives a contradiction.

Note $f(z) = (\frac{1}{z}e - x)^{-1} = z(e - zx)^{-1}$ is analytic for $|z| < \frac{1}{||x||}$, for it is $\sum z^{n+1}x^n$. The radius of convergence for this power series is $R = \frac{1}{\overline{\lim}||x^n||^{1/n}}$. Thus z is in the resolvent if $|z| > \overline{\lim}||x^n||^{\frac{1}{n}}$ and $\max\{|\lambda| : \lambda \in \sigma(x)\} = \overline{\lim}||x^n||^{1/n}$. To see that $\lim ||x^n||^{1/n}$ exists, it suffices to note $||x^m||^{1/m} \geq \overline{\lim}||x^n||^{1/n}$. Indeed, $||x^n||^{\frac{1}{n}} = ||x^{lm+r}||^{\frac{1}{lm+r}}$ where $n = lm + r$ and $0 \leq r < m$. Hence

$$||x^n||^{1/n} \leq ||x^m||^{\frac{l}{lm+r}}||x||^{\frac{r}{lm+r}} = (||x^m||^{1/m})^{\frac{lm}{lm+r}}||x||^{\frac{r}{n}}.$$

Let $n \to \infty$. Then $\overline{\lim}||x^n||^{1/n} \leq ||x^m||^{\frac{1}{m}}$.

Suppose $\mathcal{B}$ is a C* algebra and x is self adjoint. Then

$$||x^{2^n}|| = ||x^{2^{n-1}}x^{2^{n-1}}|| = ||x^{2^{n-1}}||^2 = \cdots = ||x||^{2^n}.$$

Thus $\lim||x^n||^{\frac{1}{n}} = ||x||$. $\square$

PROPOSITION II.1. *Let $\mathcal{B}$ be a Banach algebra, and let M be a closed proper two sided ideal. Define $||x + M|| = \inf_m ||x + m||$. Then $\mathcal{B}/M$ is a Banach algebra.*

PROOF. A standard argument shows $\mathcal{B}/M$ is a normed space. To see it is complete, let $\{x_n + M\}_{n=1}^\infty$ be a Cauchy sequence. By taking a subsequence, we may assume $||x_n - x_{n+1} + M|| \leq \frac{1}{2^{n+1}}$ for all n. Set $y_1 = x_1$. Suppose $y_j = x_j + m_j$ for $j \leq n$ have been chosen such that $||y_j - y_{j+1}|| \leq \frac{1}{2^j}$ for $j = 1, 2, \cdots, n-1$. Then since $||y_n - x_{n+1} + M|| = ||x_n - x_{n+1} + M|| \leq \frac{1}{2^{n+1}}$, there is an $m_{n+1} \in M$ with $||y_n - x_{n+1} - m_{n+1}|| \leq \frac{1}{2^n}$. Set $y_{n+1} = x_{n+1} + m_{n+1}$. Hence, by induction we see there is a sequence $m_n \in M$ with $||x_n + m_n - (x_{n+1} +$

$m_{n+1})|| \leq \frac{1}{2^n}$ for all n. It follows that the sequence $\{x_n + m_n\}_{n=1}^{\infty}$ is Cauchy in $\mathcal{B}$. Let x be its limit. Then

$$||x_n - x + M|| = ||x_n + m_n - x + M||$$
$$\leq ||x_n + m_n - x|| \to 0 \text{ as } n \to \infty.$$

Thus $\lim_n(x_n + M) = x + M$. Moreover,

$$||xy + M|| = \inf_{m \in M} ||xy + m||$$
$$\leq \inf_{m_1 \in M, m_2 \in M} ||xy + m_1 y + x m_2 + m_1 m_2||$$
$$\leq \inf_{m_1 \in M, m_2 \in M} ||x + m_1|| \, ||y + m_2||$$
$$= ||x + M|| \, ||y + M||.$$

$\square$

Let $\mathcal{A}$ be a commutative C* algebra with identity. Let Δ be the set of nonzero algebraic homomorphisms h from $\mathcal{A}$ into $\mathbb{C}$. Thus $h \in \Delta$ iff $h(x + y) = h(x) + h(y)$, $h(\lambda x) = \lambda h(x)$, and $h(xy) = h(x)h(y)$ for all $x, y \in \mathcal{A}$ and $\lambda \in \mathbb{C}$. We note since $h \neq 0$, that $h(e) = 1$. The set Δ is called the **Gelfand spectrum** of $\mathcal{A}$.

Let X be a Banach space, and let X^* be its dual space. Then the smallest topology on X^* making the mappings $f \mapsto f(x)$ continuous for $x \in X$ is called the weak $*$ topology on X^*. Note $X^* \subset \prod_x \mathbb{C}$, and X^* with the weak $*$ topology is the relative product topology. The unit ball, $\{f \in X^* : ||f|| \leq 1\}$, is a subset of X^* which is compact and Hausdorff.

THEOREM II.2 (GELFAND). *Let $\mathcal{A}$ be a commutative C* algebra with an identity. Then Δ is a compact subset of $\mathcal{A}^*$ under the weak $*$ topology, and the mapping $x \mapsto \hat{x}$ defined by*

$$\hat{x}(h) = h(x)$$

is an isometric $*$-isomorphism of $\mathcal{A}$ onto $C(\Delta)$.

PROOF. The weak $*$ topology on $\mathcal{A}^*$ is the smallest topology making the functions $f \mapsto f(x)$, $x \in \mathcal{A}$, continuous. The unit ball in $\mathcal{A}^*$ is compact in this topology.

Let $h \in \Delta$. We claim $||h|| \leq 1$. Recall $h(e) = 1$. We show $|h(x)| \leq ||x||$. Suppose $|h(x)| > ||x||$, then $\frac{x}{h(x)} - e$ is invertible. Thus $h(\frac{x}{h(x)} - e) \neq 0$. This is a contradiction. Thus $||h|| \leq 1$.

Next suppose x is self adjoint. Set $w = x + ite$; then $h(w) = h(x) + it$. Thus $|h(x)|^2 - ith(x) + it\overline{h(x)} + t^2 = |h(w)|^2 \leq ||w||^2 = ||w^*w|| = ||x^2 + t^2e|| \leq ||x||^2 + t^2$. This implies

$$|h(x)|^2 + 2t\text{Im}h(x) + t^2 \leq ||x||^2 + t^2$$

for all t. Hence $\text{Im}(h(x)) = 0$, and thus $h(x)$ is real. Hence for any $x \in \mathcal{A}$,

$$\begin{aligned}
h(x^*) &= h(\frac{1}{2}(x^* + x) - \frac{i}{2}(ix^* - ix)) \\
&= \frac{1}{2}h(x^* + x) - \frac{i}{2}h(ix^* - ix) \\
&= \frac{1}{2}\overline{h(x^* + x)} + i\overline{h(ix^* - ix)} \\
&= \frac{1}{2}\overline{h(x^* + x - x^* + x)} \\
&= \overline{h(x)}
\end{aligned}$$

Δ is a closed subset of the unit ball of $\mathcal{A}^*$. Indeed, if h_η is a net in Δ converging to $f \in \mathcal{A}^*$, then $h_\eta(x)$ converges to $f(x)$ for each x. Uniqueness of limits and continuity imply that $f(xy) = f(x)f(y)$, $f(x + y) = f(x) + f(y)$, $f(cx) = cf(x)$ and $f(e) = 1$. Thus $f \in \Delta$. Hence Δ is a compact Hausdorff space.

The mapping $x \mapsto \hat{x}$ preserves pointwise operations and hence is a homomorphism of $\mathcal{A}$ into $C(\Delta)$.

Let $\lambda \in \sigma(x)$. Then $x - \lambda e$ is not invertible. By Zorn's Lemma, there is a maximal proper ideal M containing $x - \lambda e$. It is closed; for if not, then $\bar{M}$ is a proper ideal containing $x - \lambda e$ but not containing e. (There is a neighborhood of e missing M.) By Proposition II.1, $\mathcal{A}/M$ is a Banach algebra. Moreover, since M is maximal, every nonzero element in $\mathcal{A}/M$ is invertible. Thus if $y + M \in \mathcal{A}/M$, then $y = se + M$ for the unique s in its spectrum. The mapping $y \mapsto s$ is a nonzero homomorphism h from $\mathcal{A}$ into $\mathbb{C}$. Since $x = \lambda e + M$, $h(x) = \lambda$. Thus

$$\begin{aligned}
||x||^2 &= ||x^*x|| \\
&= ||x^*x||_\sigma \\
&= \max\{|\lambda| : \lambda \in \sigma(x^*x)\} \\
&= \max\{|h(x^*x)| : h \in \Delta\} \\
&= \max_{h \in \Delta} |\hat{x}(h)|^2.
\end{aligned}$$

This gives $||x|| = ||\hat{x}||_\infty$.

Finally $\hat{A}$ is a conjugate closed subalgebra of $C(\Delta)$ which separates points and contains the constants. Since it is complete, $\hat{A} = C(\Delta)$ by the Stone–Weierstrass Theorem. $\square$

COROLLARY II.1. *If x is a self adjoint element in a C^* algebra $\mathcal{A}$ having an identity, then $\sigma(x) \subset [-||x||, ||x||]$.*

PROOF. Let $\mathcal{A}_x$ be the commutative C* subalgebra of $\mathcal{A}$ obtained by taking the norm closure of the algebra generated by x and e. We first note an element $y \in \mathcal{A}_x$ has an inverse in $\mathcal{A}_x$ iff y has an inverse in $\mathcal{A}$. Indeed, suppose y has an inverse y^{-1} in $\mathcal{A}$. Let Δ be the Gelfand spectrum of $\mathcal{A}_x$, and let $\hat{y}$ be the Gelfand transform of y. Then $y^{-1} \in \mathcal{A}_x$ iff $\hat{y}$ is never 0 on Δ. Suppose there is a $h \in \Delta$ such that $\hat{y}(h) = h(y) = 0$. Since Δ is compact and Hausdorff, Δ is a Tychonoff space; and thus there is a sequence $\{f_n\}_{n=1}^{\infty}$ of continuous functions on Δ such that $f_n(h) = 1$ and $f_n\hat{y}$ converges uniformly to 0. Let x_n be the element in $\mathcal{A}_x$ with $\hat{x}_n = f_n$. Then $||x_ny|| \to 0$ as $n \to \infty$ while $||x_n|| \geq 1$ for all n. But then $x_n = (x_ny)y^{-1} \to 0$ in $\mathcal{A}$. But $\{x_n\}_{n=1}^{\infty}$ does not converge to 0 in $\mathcal{A}$. Thus $\hat{y}$ cannot vanish, and y^{-1} is in $\mathcal{A}_x$.

Hence $\lambda \in \sigma(x)$ iff $\lambda e - x$ is not invertible in $\mathcal{A}_x$ iff $\lambda \in \hat{x}(\Delta)$. But for $h \in \Delta$, $\hat{x}(h) = \widehat{(x^*)}(h) = \overline{\hat{x}(h)}$. Thus $\hat{x}(\Delta) \subseteq \mathbb{R}$. Since $||\hat{x}||_\infty = ||x||$, we see $\sigma(x) \subseteq [-||x||, ||x||]$. $\square$

COROLLARY II.2. *Let $\mathcal{A}$ be a C^* subalgebra of C^* algebra $\mathcal{B}$. Assume $\mathcal{B}$ has an identity which is also in $\mathcal{A}$. Then for $x \in \mathcal{A}$, then $\sigma_{\mathcal{A}}(x) = \sigma_{\mathcal{B}}(x)$.*

PROOF. The argument in the above corollary shows a self adjoint element x in $\mathcal{A}$ is invertible in $\mathcal{A}$ iff it is invertible in $\mathcal{B}$. A general element x in a C* algebra is invertible iff the self adjoint element x^*x is invertible. Thus we see an element in $\mathcal{A}$ is invertible in $\mathcal{A}$ iff it is invertible in $\mathcal{B}$. In particular $\mathbb{C} - \sigma_{\mathcal{A}}(x) = \mathbb{C} - \sigma_{\mathcal{B}}(x)$ for $x \in \mathcal{A}$. $\square$

PROPOSITION II.2. *Let $\mathcal{A}$ be a C^* algebra with no identity. Let $\mathcal{A}_e$ be the set consisting of all formal sums $se + x$ where $s \in \mathbb{C}$ and $x \in \mathcal{A}$. Define $||se + x|| = \sup\{||sy + xy|| : y \in \mathcal{A}, ||y|| = 1\}$. Set $(se + x)^* = \bar{s}e + x^*$ and define addition and multiplication naturally. Then $\mathcal{A}_e$ is a C^* algebra containing $\mathcal{A}$ isometrically as a subalgebra.*

PROOF. $\mathcal{A}_e$ is a * algebra with an identity. Since $||se+x+te+y|| = \sup_{||z||=1}||sz+xz+tz+yz|| \leq \sup_{||z||=1}||sz+xz|| + \sup_{||z||=1}||tz+yz||$, we see $||se + x + te + y|| \leq ||se + x|| + ||te + y||$.

We show $||se+x|| = 0$ implies $x = 0$ and $s = 0$. Suppose $||se+x|| = 0$. Then $sy + xy = 0$ for all y. If $s = 0$, then $xx^* = 0$; and since $||xx^*|| = ||x^*||^2 = ||x||^2$, one has $x = 0$. Suppose $s \neq 0$. Then $y = -s^{-1}xy$ for all y. This implies $\mathcal{A}$ has a left identity. Call it e_L. Then since $ye_L^* = (e_L y^*)^* = (y^*)^* = y$, we see e_L^* is a right identity. But $e_L = e_L e_L^* = e_L^*$. Thus $\mathcal{A}$ has an identity. This is a contradiction.

Next note $||0e + x|| \leq \sup_{||y||=1} ||xy|| \leq ||x||$. If $x \neq 0$ then $||x\frac{x^*}{||x^*||}|| = ||x^*|| = ||x||$; and thus $||0e + x|| \geq ||x||$. Hence, $||0e + xe|| = ||x||$ for all x, and the norm on $\mathcal{A}$ is extended to $\mathcal{A}_e$.

Clearly one has $||e + 0|| = 1$. Also since $||se + x|| = ||L_{se+x}||$ where L_{se+x} is the operator defined on $\mathcal{A}$ by $L_{se+x}z = (se + x)z$, we see $||(se + x)(te + y)|| = ||L_{se+x}L_{te+y}|| \leq ||L_{se+x}||||L_{te+y}|| = ||se+x||\,||te+y||$. To see $||\bar{s}e+x^*|| = ||se+x||$, note $||z(\bar{s}e + x^*)y|| = ||y^*(se + x)z^*|| \leq ||y||\,||se + x||\,||z||$. Thus $\sup_{||y||=1,||z||=1} ||z(\bar{s}e + x^*)y|| \leq ||se + x||$. But $||z(\bar{s}e + x^*)y|| \leq ||(\bar{s}e + x^*)y||$ if $||z|| = 1$. By taking $z^* = \frac{(\bar{s}e+x^*)y}{||(\bar{s}e+x)y||}$, we see $||z(\bar{s}e + x^*)y|| = ||(\bar{s}e + x)y||$. Thus $\sup_{||y||=1,||z||=1} ||z(\bar{s}e + x^*)y|| = ||\bar{s}e + x^*||$. This implies $||\bar{s}e + x^*|| = ||se + x||$.

Next we claim $\mathcal{A}_e$ is complete. Note the mapping h defined by $h(se + x) = s$ is linear and has kernel $\mathcal{A}$. Since $\mathcal{A}$ is complete, it is hence closed; and consequentially h is a bounded linear functional. Let $s_n e + x_n$ be Cauchy. Since both h and the sequence $s_n e + x_n$ are bounded, $\{s_n\}$ is a bounded sequence. By taking a subsequence, we may assume s_n converges to s. But then $||x_m - x_n|| \leq ||s_m e + x_m - (s_n e + x_n)|| + ||s_n e - s_m e|| \to 0$ as $m, n \to \infty$. Thus, there is an x such that $x_n \to x$ in $\mathcal{A}$ as $n \to \infty$. It follows that $||s_n e + x_n - (se + x)|| \leq |s_n - s| + ||x_n - x|| \to 0$. Thus $\mathcal{A}_e$ is complete. Finally to see $\mathcal{A}$ is a C* algebra, note $||(se + x)^*(se + x)|| \leq ||se + x||^2$. Let $r < ||se + x||$. Choose y with $||y|| = 1$ and $||(se + x)y|| > r$. Then $||(se + x)^*(se + x)||^2 \geq ||y^*(se + x)^*(se + x)y|| > r^2$. Since this is true for every $r < ||se + x||$, we see $||(se + x)^*(se + x)|| \geq ||se + x||^2$. Thus $||(se + x)^*(se + x)|| = ||se + x||^2$, and $\mathcal{A}_e$ is a C* algebra. $\square$

PROPOSITION II.3. *Let $\mathcal{A}$ be a C* subalgebra of a C* algebra $\mathcal{B}$. Assume $\mathcal{B}$ has an identity which is also in $\mathcal{A}$. Then for $x \in \mathcal{A}$, then $\sigma_{\mathcal{A}}(x) = \sigma_{\mathcal{B}}(x)$.*

PROOF. Suppose x is self adjoint and invertible in $\mathcal{B}$. Let $\mathcal{B}_x$ be the smallest C* subalgebra of $\mathcal{B}$ containing e, x and x^{-1}. Since $\mathcal{B}_x$ is commutative, it is isomorphic to $C(\Delta)$ where Δ is the Gelfand spectrum of $\mathcal{B}_x$. In particular x is a nowhere zero function on Δ.

The function x separates points on Δ. Indeed, if $h_1, h_2 \in \Delta$ and $h_1 \neq h_2$ and $h_1(x) = h_2(x)$, then $h_1(x^{-1}) = h_2(x^{-1})$. Thus $h_1 = h_2$ on the smallest C^* algebra containing x, x^{-1} and e. Therefore the set of all $P(x)$ where P is a complex polynomial is a $*$ subalgebra which separates points and contains the constants. Hence, its closure in $C(\Delta)$ is $C(\Delta)$. This implies the C* algebra generated by e and x is $\mathcal{B}_x$. In particular $x^{-1} \in \mathcal{A}$. Hence if x is self adjoint, then x is invertible in $\mathcal{A}$ iff x is invertible in $\mathcal{B}$. Since x is invertible iff x^*x is invertible we see an element in $\mathcal{A}$ is invertible in $\mathcal{A}$ iff it is invertible in $\mathcal{B}$. One concludes $\mathbb{C} - \sigma_{\mathcal{A}}(x) = \mathbb{C} - \sigma_{\mathcal{B}}(x)$ for $x \in \mathcal{A}$. $\square$

A $*$ homomorphism Φ between two C* algebras is said to be **faithful** if $\ker \Phi = \{0\}$.

PROPOSITION II.4. *Let $\mathcal{A}$ and $\mathcal{B}$ be C^* algebras. Assume Φ is a faithful $*$-homomorphism of $\mathcal{A}$ into $\mathcal{B}$. Then Φ is an isometry.*

PROOF. If $\mathcal{A}$ and $\mathcal{B}$ have no identities, we may extend Φ from $\mathcal{A}_e$ to $\mathcal{B}_e$ by setting $\Phi(ze + a) = ze + \Phi(a)$. If $\mathcal{B}$ has an identity e and $\Phi(a) = e$ then a is an identity on $\mathcal{A}$ for $\Phi(aa') = \Phi(a')$ and $\Phi(a'a) = \Phi(a')$. Since Φ is faithful, $aa' = a'$ and $a'a = a'$ for all a'. Hence again we see $\mathcal{A}$ and $\mathcal{B}$ have identities. If $\mathcal{B}$ has an identity e and $\mathcal{A}$ has no identity and e is not in the range of Φ, then defining $\Phi(ze + a) = ze + \Phi(a)$ extends Φ from $\mathcal{A}$ to $\mathcal{A}_e$; and the extension is still faithful. Finally we note if Φ is a homomorphism between C* algebras having identities, then $\Phi(e) = e$ (Homomorphisms are assumed to preserve the identities when they exist.)

Suppose x is self adjoint. We show $\|\Phi(x)\| = \|x\|$. By restricting to the smallest C* algebra $\mathcal{A}_x$ of $\mathcal{A}$ containing x and e and replacing $\mathcal{B}$ by the norm closure of $\Phi(\mathcal{A}_x)$, we may assume $\mathcal{A}$ and $\mathcal{B}$ are commutative and $\Phi(\mathcal{A})$ is dense in $\mathcal{B}$. Let $\Delta_{\mathcal{A}}$ be the Gelfand spectrum of $\mathcal{A}$ and $\Delta_{\mathcal{B}}$ be the Gelfand spectrum of $\mathcal{B}$. Define a mapping $J : \Delta_{\mathcal{B}} \to \Delta_{\mathcal{A}}$ by $J(h)(a) = h(\Phi(a))$. The mapping J is continuous. It is one-to-one, for $\Phi(\mathcal{A})$ is a dense subspace of $\mathcal{B}$. We claim J is onto. Indeed, if not, $J(\Delta_{\mathcal{B}})$ would be a proper compact subset of $\Delta_{\mathcal{A}}$. Hence there would be an element in $C(\Delta_{\mathcal{A}})$ which is nonzero and whose restriction to $J(\Delta_{\mathcal{B}})$ would be 0. Thus there is a nonzero element $x \in \mathcal{A}$ with $\hat{x} = 0$ on $J(\Delta_{\mathcal{B}})$. But then $\widehat{\Phi(x)}(h) = h(\Phi(x)) = J(h)(x) = \hat{x}(J(h)) = 0$ for all $h \in \Delta_{\mathcal{B}}$. Thus $\Phi(x) = 0$, and since Φ is faithful, $x = 0$, a contradiction. Thus J is onto. Hence

$$||x|| = \max_{h \in \Delta_A} |\hat{x}(h)|$$
$$= \max_{h \in \Delta_B} |\hat{x}(J(h))|$$
$$= \max_{h \in \Delta_A} |\widehat{\Phi(x)}(h)|$$
$$= ||\Phi(x)||.$$

In general for nonself adjoint elements x, one has $||\Phi(x^*x)|| = ||x^*x||$, and thus $||\Phi(x)||^2 = ||x||^2$. We conclude Φ is an isometry. $\square$

THEOREM II.3. *Let A be a commutative C* algebra without an identity. Let $\sigma(A)$ be the set of all nonzero homomorphisms from A into $\mathbb{C}$. Give $\sigma(A)$ the weak $*$ topology. Then $\sigma(A)$ is a locally compact Hausdorff space, and the mapping $a \mapsto \hat{a}$ defined by $\hat{a}(h) = h(a)$ is an isometric isomorphism of the C* algebra A onto $C_0(\sigma(A))$.*

PROOF. Let A_e be the extension of A to a C* algebra containing an identity e. Let h_∞ be the homomorphism on A_e defined by $h_\infty(\lambda e + x) = \lambda$. If $h \in \sigma(A)$, let h_e be the homomorphism on A_e defined by $h_e(\lambda e + x) = \lambda + h(x)$. Then $\sigma(A)_e \cup \{h_\infty\} = \sigma(A_e)$ for if $h \in \sigma(A_e)$ and $h \neq h_\infty$, then $h = (h|_A)_e$ for $h|_A \neq 0$.

Moreover, since the topology on $\sigma(A_e)$ is the smallest topology making the functions $h \mapsto h(\lambda e + x)$ continuous for each $x \in A$ and $\lambda \in \mathbb{C}$, it follows that the relative topology on $\sigma(A)_e$ is the smallest topology on $\sigma(A)_e$ making the functions $h_e \mapsto h_e(x) = h(x)$ continuous for all $x \in A$. Identifying $\sigma(A)$ with $\sigma(A)_e$ via the correspondence $h \leftrightarrow h_e$, we see $\sigma(A)_e$ with the relative topology of $\sigma(A_e)$ is $\sigma(A)$ with the weak $*$ topology. Since $\sigma(A_e)$ is Hausdorff, $\sigma(A)_e$ is Hausdorff. To see it is locally compact, let $h' \in \sigma(A)_e$. Choose open neighborhoods U and V of h' and h_∞ in $\sigma(A_e)$ with $U \cap V = \emptyset$. Then $\overline{U}$ does not contain h_∞ and thus is a compact neighborhood of h' in $\sigma(A)_e$. Thus $\sigma(A)_e$ and consequentially $\sigma(A)$ is a locally compact Hausdorff space. We note $\sigma(A_e)$ is the one-point compactification of $\sigma(A)_e$. Indeed, a set E is open in the one point compactification iff E is an open subset of $\sigma(A)_e$ or $h_\infty \in E$ and $\sigma(A_e) - E$ is a compact subset of $\sigma(A)_e$. Since $\sigma(A)_e$ is an open subset of $\sigma(A_e)$ and $\sigma(A_e)$ is compact, we see these are precisely the open sets in $\sigma(A_e)$.

Now the Gelfand transform maps A_e onto $C(\sigma(A_e))$. But $A_e = \mathbb{C}e + A$, and $\mathbb{C}e$ is mapped to the constant functions. Hence if $f \in$

$C(\sigma(\mathcal{A}_e))$ and $f(h_\infty) = 0$, then since $(\widehat{\lambda e + x})(h_\infty) = h_\infty(\lambda e + x) = \lambda$, we see f must have form $\hat{x}$ for some $x \in \mathcal{A}$. This shows the Gelfand transform is a C* algebra isomorphism of $\mathcal{A}$ onto the subalgebra of $C(\sigma(\mathcal{A}_e))$ consisting of those functions vanishing at ∞. But this subalgebra is precisely $C_0(\sigma(\mathcal{A})_e)$. This gives the result. By Proposition II.4 all isomorphisms of C^* algebras are isometries. $\square$

2. Approximate Units on a C* Algebra

DEFINITION. An element a in a C* algebra having an identity is said to be positive if it is self adjoint and $\sigma(a) \subset [0, \infty)$.

PROPOSITION II.5. *Let $\mathcal{A}$ be a C^* algebra with an identity. Then the following hold:*

(a) *If x is self adjoint, then $x \geq 0$ iff $||(||x||e - x)|| \leq ||x||$.*
(b) *If $x \geq 0$ and $\lambda \geq 0$, then $\lambda x \geq 0$.*
(c) *If $x \geq 0$ and $y \geq 0$, then $x + y \geq 0$.*
(d) *If x is self adjoint and $||e - x|| \leq 1$, then $x \geq 0$.*

PROOF. In (a),(b), and (d), we may assume $\mathcal{A} = C(\Delta)$ for some compact Hausdorff space Δ and x is a continuous real valued function on Δ. Then for (a), $x \geq 0$ iff $0 \leq x \leq 2||x||$ iff $-2||x|| \leq -x \leq 0$ iff $-||x|| \leq ||x|| - x \leq ||x||$ iff $||(||x||e - x)|| \leq ||x||$. For b, if $x \in C(\Delta)$ and $x \geq 0$, then $\lambda x \geq 0$ for $\lambda \geq 0$. For (d), if $-1 \leq 1 - x \leq 1$, then $x \geq 0$.

To see (c), note (a) implies $||(||x|| + ||y||)e - (x + y)|| \leq ||\ ||x||e - x|| + ||\ ||y||e - y|| \leq ||x|| + ||y||$. Hence $||e - \frac{x+y}{||x||+||y||}|| \leq 1$. By (d), $\frac{x+y}{||x||+||y||} \geq 0$. Using (b) we obtain $x + y \geq 0$. $\square$

COROLLARY II.3. *The set P of positive elements in a C^* algebra is a norm closed set.*

LEMMA II.1. *Suppose $\mathcal{A}$ is a C^* algebra with an identity. Then $\sigma(ab) - \{0\} = \sigma(ba) - \{0\}$.*

PROOF. Indeed, suppose $w = (ab - \lambda e)^{-1}$ exists where $\lambda \neq 0$. Then $(ba - \lambda e)(bwa - e) = babwa - \lambda bwa - ba + \lambda e = b(ab - \lambda e)wa - ba + \lambda = ba - ba + \lambda e = \lambda e$ and similarly $(bwa - e)(ba - \lambda e) = \lambda e$. Hence $(ba - \lambda e)^{-1}$ exists. The other direction follows by symmetry. $\square$

PROPOSITION II.6. *Let x be a self adjoint element in a C^* algebra $\mathcal{A}$ having an identity. Then the following are equivalent:*

(a) $x \geq 0$.
(b) $x = y^*y$ *for some element y.*

(c) $x = y^2$ where $y^* = y$.

PROOF. We show (b) implies (a). Suppose $x = y^*y$. Let $\mathcal{A}_x$ be the sub C* algebra generated by e and x. Then $\mathcal{A}_x = C(\Delta)$. Hence $x = x^+ - x^-$ where x^+ and x^- are the plus and minus parts of the function x. Set $u = \sqrt{x^+}$ and $v = \sqrt{x^-}$. Then $uv = 0$ and $x = u^2 - v^2$. Moreover $(yv)^*(yv) = vxv = -v^4$, and thus $\sigma((yv)^*(yv)) \subset (-\infty, 0]$. Hence by Lemma II.1, $\sigma((yv)(yv)^*) \subset (-\infty, 0]$. Let $yv = s + it$ where s and t are self adjoint. Then

$$(yv)(yv)^* = (s + it)(s - it)$$
$$= -(yv)^*(yv) + (s - it)(s + it) + (s + it)(s - it)$$
$$= -(yv)^*(yv) + 2s^2 + 2t^2$$

is positive, for it is a sum of positive elements. Thus $\sigma((yv)^*(yv)) = \{0\}$. This implies $v = 0$. Hence $x = x^+ \geq 0$.

(c) implies (b) is obvious; and (a) implies (c) because in $C(\Delta)$, a positive element is a continuous function having nonnegative real values and in particular a continuous square root. $\square$

LEMMA II.2. *Let x and y be positive elements in a C^* algebra with an identity. If $x \leq y$ (i.e., $y - x \geq 0$), then $||x|| \leq ||y||$.*

PROOF. Indeed if y is self adjoint, then by the Gelfand theorem, $y \leq ||y||$. Hence if $x \leq y$, then $x \leq ||y||$. In particular, one has $0 \leq \hat{x} \leq ||y||$. Thus $||x|| = \max_{h \in \Delta} |\hat{x}(h)| \leq ||y||$. $\square$

LEMMA II.3. *Let $\mathcal{A}$ be a C^* algebra with an identity, and suppose $a \leq b$. Then $x^*ax \leq x^*bx$ for any x.*

PROOF. Choose y with $y^*y = b - a$. Then $x^*bx - x^*ax = x^*y^*yx \geq 0$. $\square$

LEMMA II.4. *Suppose $\mathcal{A}$ is a C^* algebra with an identity. Suppose $a > 0$ and $b > 0$ are invertible and $a \leq b$. Then $b^{-1} \leq a^{-1}$.*

PROOF. Lemma II.3 implies $e = a^{-1/2}aa^{-1/2} \leq a^{-1/2}ba^{-1/2}$. But if x is self adjoint and $x \geq e$, then $x^{-1} \leq e$. Thus $a^{1/2}b^{-1}a^{1/2} \leq e$. Hence $b^{-1} = a^{-1/2}a^{1/2}b^{-1}a^{1/2}a^{-1/2} \leq a^{-1}$. $\square$

PROPOSITION II.7. *Let $\mathcal{M}$ be a dense two sided ideal in a C^* algebra $\mathcal{A}$. Then there is a net u_λ in $\mathcal{M}$ with $||u_\lambda x - x|| \to 0$ and $||xu_\lambda - x|| \to 0$ for each $x \in \mathcal{A}$. Moreover, this net can be chosen so that $0 \leq u_\lambda$, $||u_\lambda|| \leq 1$, and $u_\lambda \leq u_{\lambda'}$ for $\lambda < \lambda'$. Furthermore, if $\mathcal{A}$ is separable, the net may be taken to be a sequence.*

PROOF. If $\mathcal{A}$ has no identity, we enlarge $\mathcal{A}$ to $\mathcal{A}_e = \mathbb{C}e + \mathcal{A}$ with its C* algebra norm. Let Λ be the set of all finite subsets of $\mathcal{M}$. Order Λ by inclusion. For $\lambda = \{x_1, x_2, \cdots, x_n\}$, set $v_\lambda = x_1^* x_1 + \cdots + x_n^* x_n$. Then $v_\lambda > 0$. Set $u_\lambda = v_\lambda(\frac{1}{n}e + v_\lambda)^{-1}$. This exists by the Gelfand theory. Indeed, if $\mathcal{B}$ is the smallest C* algebra in $\mathcal{A}$ containing e and v_λ, then $\mathcal{B}$ is a commutative C* algebra with an identity. It is isomorphic to $C(\Delta)$ where Δ is its Gelfand spectrum. Moreover, if $v_\lambda \leftrightarrow f$, then $f \geq 0$ and $\frac{1}{n} + f \geq \frac{1}{n}$. Hence $\frac{1}{n}e + v_\lambda$ is invertible. Next note $u_\lambda \leftrightarrow \frac{f}{\frac{1}{n}+f} = \frac{nf}{1+nf}$; and thus $||u_\lambda|| \leq 1$; and

$$\left(1 - \frac{nf}{1+nf}\right) f \left(1 - \frac{nf}{1+nf}\right) = \frac{1}{n^2} f \left(\frac{1}{\frac{1}{n}+f}\right)^2 \leq \frac{1}{n}$$

implies $||(1-u_\lambda)^* v_\lambda (1-u_\lambda)|| \leq \frac{1}{n}$. Hence $|| \sum (1-u_\lambda)^* x_i^* x_i (1-u_\lambda)|| \leq \frac{1}{n}$. By Lemma II.3,

$$0 \leq (1 - u_\lambda)^* x_i^* x_i (1 - u_\lambda) \leq (1 - u_\lambda)^* v_\lambda (1 - u_\lambda);$$

and since $0 < x < y \implies ||x|| \leq ||y||$, we see

$$||x_i(1 - u_\lambda)|| \leq \frac{1}{n}.$$

Hence if λ_0 is chosen so that $\frac{1}{n_0} < \epsilon$ and $x \in \lambda_0$, then $||x - xu_\lambda|| < \epsilon$ for $\lambda > \lambda_0$. Thus $||x - xu_\lambda|| \to 0$ for all $x \in \mathcal{M}$. By the density of $\mathcal{M}$ in $\mathcal{A}$, one has $||x - xu_\lambda|| \to 0$ for all $x \in \mathcal{A}$. Taking adjoints gives $||x - u_\lambda x|| \to 0$ for all $x \in \mathcal{A}$.

Next suppose $\lambda < \lambda'$. Then $\frac{1}{n} + v_\lambda \leq \frac{1}{n} + v_{\lambda'}$; and by Lemma II.4, $(\frac{1}{n} + v_{\lambda'})^{-1} \leq (\frac{1}{n} + v_\lambda)^{-1}$. Since $n \leq n'$,

$$\frac{1}{n'}\left(\frac{1}{n'} + v_{\lambda'}\right)^{-1} \leq \frac{1}{n}\left(\frac{1}{n} + v_{\lambda'}\right)^{-1}.$$

Hence

$$\frac{1}{n'}\left(\frac{1}{n'} + v_{\lambda'}\right)^{-1} \leq \frac{1}{n}\left(\frac{1}{n} + v_\lambda\right)^{-1} \leq 1.$$

Thus

$$u_\lambda = v_\lambda\left(\frac{1}{n} + v_\lambda\right)^{-1} = 1 - \frac{1}{n}\left(\frac{1}{n} + v_\lambda\right)^{-1}$$

$$\leq 1 - \frac{1}{n'}\left(\frac{1}{n'} + v_{\lambda'}\right)^{-1} = v_{\lambda'}\left(\frac{1}{n'} + v_{\lambda'}\right)^{-1} = u_{\lambda'}.$$

Finally, assume $\mathcal{A}$ is separable. Let $\{x_n\}_{n=1}^\infty$ be a sequence in $\mathcal{M}$ which gives a dense subset of $\mathcal{A}$. Replace the net u_λ by the sequence u_n where $\lambda_n = \{x_1, x_2, \cdots, x_n\}$. $\square$

COROLLARY II.4. *Let $\mathcal{N}$ be a left ideal in a C^* algebra $\mathcal{A}$. Then there is an increasing net u_λ in $\mathcal{N}$ with $0 \leq u_\lambda$ and $||u_\lambda|| \leq 1$ such that*

$$\lim_\lambda ||x - xu_\lambda|| = 0 \text{ for all } x \in \mathcal{N}.$$

COROLLARY II.5. *Let $\mathcal{I}$ be a closed two sided ideal in a C^* algebra $\mathcal{A}$. Then $\mathcal{I}$ is a $*$ ideal, and $\mathcal{A}/\mathcal{I}$ with the quotient norm $||a + \mathcal{I}|| = \inf_{b \in \mathcal{I}} ||a + b||$ is a C^* algebra.*

PROOF. The proof of Proposition II.7 shows there is a net u_λ in $\mathcal{I}$ with $0 \leq u_\lambda$ and $||u_\lambda|| \leq 1$ for all λ, and $||bu_\lambda - b|| \to 0$ for all $b \in \mathcal{I}$. Hence, if $b \in \mathcal{I}$, $||u_\lambda b^* - b^*|| = ||(bu_\lambda - b)^*|| = ||bu_\lambda - b|| \to 0$ as $\lambda \to \infty$. Thus, since $\mathcal{I}$ is norm closed, $b^* \in \mathcal{I}$.

Note if $\mathcal{A}$ does not contain an identity, $\mathcal{A}$ may be extended to a C^* algebra $\mathcal{A}_e$ containing an identity e. Hence, we may suppose $e \in \mathcal{A}$. Moreover, we have $||e - u_\lambda|| \leq 1$. Note if $b \in \mathcal{I}$, then

$$
\begin{aligned}
||a + \mathcal{I}|| &\leq \underline{\lim}||a - u_\lambda a|| \\
&\leq \overline{\lim_\lambda}||a - u_\lambda a|| \\
&= \overline{\lim_\lambda}||a - u_\lambda a + b - bu_\lambda|| \\
&= \overline{\lim_\lambda}||(a + b)(e - u_\lambda)|| \\
&\leq \overline{\lim_\lambda}||a + b|| \, ||e - u_\lambda|| \\
&\leq ||a + b||
\end{aligned}
$$

for $||b - bu_\lambda|| \to 0$. This implies $||a + \mathcal{I}|| = \lim_\lambda ||a - u_\lambda a||$. Similarly, $||a + \mathcal{I}|| = \lim_\lambda ||a - au_\lambda||$. Thus $||a^* + \mathcal{I}|| = ||a + \mathcal{I}||$. Moreover, if $b \in \mathcal{I}$, one notes

$$
\begin{aligned}
||a + \mathcal{I}||^2 &= \lim_\lambda ||a - u_\lambda a||^2 \\
&= \lim_\lambda ||(a^* - u_\lambda a)^*(a - u_\lambda a)|| \\
&= \lim_\lambda ||(a - u_\lambda a)^*(a - u_\lambda a) + (b - u_\lambda b)(e - u_\lambda)|| \\
&= \lim_\lambda ||(e - u_\lambda)a^* a(e - u_\lambda) + (e - u_\lambda)b(e - u_\lambda)|| \\
&= \lim_\lambda ||(e - u_\lambda)(a^* a + b)(e - u_\lambda)|| \\
&\leq ||a^* a + b||
\end{aligned}
$$

since $||b - u_\lambda b|| \to 0$ and $||e - u_\lambda|| \leq 1$ for all λ. Thus $||a + \mathcal{I}||^2 \leq ||a^* a + \mathcal{I}||$. Proposition II.1 implies $||a^* a + \mathcal{I}|| \leq ||a^* + \mathcal{I}|| \, ||a + \mathcal{I}|| = ||a + \mathcal{I}||^2$. Hence $||a^* a + \mathcal{I}|| = ||a + \mathcal{I}||^2$, and $\mathcal{A}/\mathcal{I}$ is a C^* algebra. $\quad\square$

3. Representations of Commutative C* Algebras

A representation of a Banach $*$ algebra $\mathcal{B}$ is a $*$ homomorphism $\pi : \mathcal{B} \to B(\mathbb{H})$. It is said to be nondegenerate if $\pi(x)v = 0$ for all x implies $v = 0$.

REMARK. π is nondegenerate iff the linear span of $\{\pi(x)v : x \in \mathcal{B}, v \in \mathbb{H}\}$ is a dense subspace of $\mathbb{H}$.

PROOF. This linear span is dense iff $\{\pi(x)v : x \in \mathcal{B}, v \in \mathbb{H}\}^{\perp} = \{0\}$ iff $\langle w, \pi(x)v \rangle = 0$ for all $x \in \mathcal{B}$ and $v \in \mathbb{H}$ implies $w = 0$ iff $\langle \pi(z)w, v \rangle = 0$ for all z and v implies $w = 0$ iff $\pi(z)w = 0$ for all z implies $w = 0$. $\square$

REMARK. If $\mathcal{B}$ contains an identity e and π is nondegenerate, then $\pi(e) = I$.

PROOF. Note $\pi(e)\pi(x)v = \pi(x)v$ for all x and v. Since the span of the vectors $\pi(x)v$ form a dense subspace of $\mathbb{H}$ and $\pi(e)$ is a bounded operator, $\pi(e)w = w$ for all $w \in \mathbb{H}$. $\square$

Replacing π by the restriction of π to the closure of the linear span of $\{\pi(x)v : x \in \mathcal{B}$ and $v \in \mathbb{H}\}$ makes π nondegenerate.

DEFINITION. Let X be a locally compact Hausdorff space, and let $\mathcal{B}$ be a σ-algebra of subsets of X. Then a projection valued measure P on $(X, \mathcal{B})$ is a function $P : \mathcal{B} \to B(\mathbb{H})$ satisfying

   (1) $P(X) = I$
   (2) $P(E \cap F) = P(E)P(F)$
   (3) $P(\cup_{i=1}^{\infty} E_i) = \sum P(E_i)$ if the E_i are pairwise disjoint and
   (4) $P(E)^* = P(E)$.

PROPOSITION II.8. *Let P be a projection valued measure based on the σ-algebra of Borel subsets of a locally compact Hausdorff space X. Let f be a function in $C_0(X)$. Then there is a unique bounded operator $\int f(x)\,dP(x)$ on $\mathbb{H}$ satisfying*

$$\langle \int f(x)\,dP(x)v, w \rangle = \int f(x)\langle dP(x)v, w \rangle$$

with $\langle dP(x)v, w \rangle = d\mu_{v,w}(x)$ where $\mu_{v,w}(E) = \langle P(E)v, w \rangle$ is a complex measure. Moreover, $f \mapsto \pi(f)$ is a representation of the C^* algebra $C_0(X)$.

PROOF. Define a sesquilinear form on $\mathbb{H}$ by

$$B_f(v, w) = \int f(x) \, d\langle P(x)v, w\rangle.$$

Note $|B_f(v, w)| \leq ||f||_\infty |\mu_{v,w}|$. But

$$|\mu_{v,w}| = \sup \sum |\langle P(E_i)v, w\rangle|$$
$$= \sup \sum |\langle P(E_i)v, P(E_i)w\rangle|$$

where the supremum is over the collections $\{E_i\}$ of countable measurable partitions of X. Hence

$$|\mu_{v,w}| \leq \sum ||P(E_i)v|| \, ||P(E_i)w||$$
$$\leq (\sum ||P(E_i)v||^2)^{\frac{1}{2}} (\sum ||P(E_i)w||^2)^{\frac{1}{2}}$$
$$= ||v|| \, ||w||.$$

Thus

$$|B_f(v, w)| \leq ||f||_\infty ||v|| \, ||w||,$$

and there is a unique bounded operator $\int f(x) dP(x)$ on $\mathbb{H}$ with $||\int f(x) dP(x)|| \leq ||f||_\infty$ satisfying the desired property. Denote this operator by M_f. Clearly $f \mapsto M_f$ is linear and $M_f^* = M_{\bar{f}}$. We claim it is multiplicative. Indeed, one has $\mu_{v, M_f w} \prec \mu_{v,w}$, and the Radon-Nikodym derivative is $\bar{f}(x)$. In fact,

$$\mu_{v, M_f w}(E) = \langle P(E)v, M_f w\rangle$$
$$= \langle M_{\bar{f}} P(E)v, w\rangle$$
$$= \int \bar{f}(x) d\langle P(x) P(E)v, w\rangle$$
$$= \int_E \bar{f}(x) \, d\langle P(x)v, w\rangle,$$

for $\langle P(F) P(E)v, w\rangle = \langle P(F \cap E)v, w\rangle$. Thus

$$\langle M_{fg}v, w\rangle = \int f(x) g(x) d\langle P(x)v, w\rangle$$
$$= \int g(x) f(x) d\mu_{v,w}(x)$$
$$= \int g(x) \, d\mu_{v, M_{\bar{f}} w}(x)$$
$$= \langle M_g v, M_{\bar{f}} w\rangle$$
$$= \langle M_f M_g v, w\rangle.$$

We conclude $M_{fg} = M_f M_g$. $\square$

REMARK. The continuity and boundedness of a function f in $C_0(X)$ can be used to show that $\int f(x)\,dP(x)$ is a norm limit of a sequence of finite sums of form $\sum f(x_k)P(E_k)$. Indeed, for each $n > 0$, take a simple function s_n of form $\sum f(x_k)1_{E_k}$ with E_k pairwise disjoint satisfying $|f(x) - s_n(x)| \leq \frac{1}{n}$ for all x. Then

$$\|(\int f(x)\,dP(x) - \sum f(x_k)P(E_k))v\|^2$$

$$= \langle(\int f(x)\,dP(x) - \sum f(x_k)P(E_k))v, (\int f(x)\,dP(x) - \sum f(x_k)P(E_k))v\rangle$$

$$= \int |f(x) - s_n(x)|^2\,d\langle P(x)v, v\rangle$$

$$\leq \frac{1}{n^2}\|v\|^2$$

for all v. Hence $\|\int f(x)\,dP(x) - \sum f(x_k)P(E_k)\| \leq \frac{1}{n}$.

PROPOSITION II.9. *If a Banach $*$ algebra $\mathcal{B}$ has an identity and π is a nondegenerate representation of $\mathcal{B}$, then $\pi(1) = I$ and $\|\pi(x)\| \leq \|x\|$ for all $x \in \mathcal{B}$.*

PROOF. Note $\lambda \notin \sigma(x) \implies \lambda \notin \sigma(\pi(x))$. Hence $\sigma(\pi(x)) \subset \sigma(x)$. Hence, if x is self adjoint, then $\|\pi(x)\| = \max\{|\lambda| : \lambda \in \sigma(\pi(x)) \leq \|x\|$; for by Theorem II.1, $\sigma(x)$ is bounded by $\|x\|$. This implies $\|\pi(x)\|^2 = \|\pi(x^*x)\| \leq \|x^*x\| \leq \|x^*\|\,\|x\| = \|x\|^2$. $\square$

COROLLARY II.6. *Let $\mathcal{B}$ be a Banach $*$ algebra, and let π be a representation of $\mathcal{B}$. Then $\|\pi(x)\| \leq \|x\|$ for all x.*

PROOF. Set $\mathcal{B}_e = \mathbb{C} \times \mathcal{B}$. Define $(s, x)(t, y) = (st, tx + sy + xy)$ and $(s, x)^* = (\bar{s}, x^*)$. Equip $\mathcal{B}_e$ with norm $\|(s, x)\| = |s| + \|x\|$. Then $\mathcal{B}_e$ is a Banach $*$ algebra. Indeed, that $\mathcal{B}_e$ is a $*$ algebra is easy to check. Note, since $\|x^*\| = \|x\|$, one has $\|(s, x)^*\| = \|(s, x)\|$. Also

$$\|(s, x)(t, y)\| = \|(st, tx + sy + xy)\|$$

$$= |st| + \|tx + sy + xy\|$$

$$\leq |s|\,|t| + |t|\,\|x\| + |s|\,\|y\| + \|x\|\,\|y\|$$

$$= \|(s, x)\|\,\|(t, y)\|.$$

$\mathcal{B}_e$ is complete; for if (s_n, x_n) is Cauchy, then s_n is Cauchy in $\mathbb{C}$ and x_n is Cauchy in $\mathcal{B}$. Thus $s_n \to s$ and $x_n \to x$ for some s and x.

Now suppose π is a representation of $\mathcal{B}$. Extend π to $\mathcal{B}_e$ by $\pi(s, x) = sI + \pi(x)$. Then π is a representation of the Banach $*$

algebra $\mathcal{B}_e$. It is nondegenerate for $\pi(1,0) = I$. Hence $||\pi(s,x)|| \leq |s| + ||x||$ for all (s,x). In particular, $||\pi(x)|| \leq ||x||$ for all x. $\square$

Recall that a Borel measure μ on a locally compact Hausdorff space is regular if it is finite on compact sets and satisfies $\mu(E)$ is the supremum of $\mu(K)$ over all compact subsets K of E and the infimum of all $\mu(U)$ over open sets containing E. A projection valued measure P is said to be **regular** if the variations of the measures $\mu_{v,w}$ are regular.

THEOREM II.4. *Suppose X is a locally compact Hausdorff space and π is a nondegenerate representation of the commutative C^* algebra $C_0(X)$. Then there is a unique regular projection valued measure P based on the Borel subsets of X such that*

$$\pi(f) = \int f(x)\, dP(x).$$

PROOF. By Corollary II.6, $||\pi(f)|| \leq ||f||_\infty$. Hence for each pair of vectors v, w in $\mathbb{H}$, the mapping

$$f \mapsto \langle \pi(f)v, w \rangle$$

is a linear functional on $C_0(X)$ with norm at most $||v||\,||w||$. By the Riesz Representation Theorem, there is a unique regular complex Borel measure $\mu_{v,w}$ on X with

$$\langle \pi(f)v, w \rangle = \int f(x)\, d\mu_{v,w}(x) \text{ and } |\mu_{v,w}| \leq ||v||\,||w||.$$

Moreover, $(v, w) \mapsto \mu_{v,w}$ is a sesquilinear function on $\mathbb{H}$ into the space of complex Borel measures on X. In particular, for each Borel subset E, the mapping

$$v, w \mapsto \mu_{v,w}(E)$$

is a sesquilinear form on $\mathbb{H}$ with $|\mu_{v,w}(E)| \leq ||v||\,||w||$. By the Riesz Representation Theorem, there is an operator $P(E)$ satisfying

$$\mu_{v,w}(E) = \langle P(E)v, w \rangle \text{ and } ||P(E)|| \leq 1.$$

Moreover, since $\langle \pi(f)v, w \rangle = \overline{\langle \pi(\bar{f})w, v \rangle}$, one sees $\mu_{v,w}(E) = \overline{\mu_{w,v}(E)}$ for each E. This gives $\langle P(E)v, w \rangle = \langle v, P(E)w \rangle$ for all v and w, and thus the operators $P(E)$ are self adjoint. To see $P(E \cap F) = P(E)P(F)$, note first that $d\mu_{v,\pi(\bar{g})w}(x) = g(x)d\mu_{v,w}(x)$. Indeed

$$\langle \pi(f)v, \pi(\bar{g})w \rangle = \langle \pi(fg)v, w \rangle$$
$$= \int f(x)g(x)d\mu_{v,w}(x)$$
$$= \int f(x)\, d\mu_{v,\pi(\bar{g})w}(x).$$

Hence, if F is a Borel set,

$$\int g(x)1_F(x)\, d\mu_{v,w}(x) = \int 1_F(x)\, d\mu_{v,\pi(\bar{g})w}(x)$$
$$= \mu_{v,\pi(\bar{g})w}(F)$$
$$= \langle P(F)v, \pi(\bar{g})w \rangle$$
$$= \langle \pi(g)P(F)v, w \rangle$$
$$= \int g(x)\, d\mu_{P(F)v,w}(x).$$

This implies

$$d\mu_{P(F)v,w}(x) = 1_F(x)\, d\mu_{v,w}(x).$$

Hence

$$\mu_{P(F)v,w}(E) = \int_E 1_F(x)\, d\mu_{v,w}(x) = \int_{E \cap F} d\mu_{v,w}(x) = \mu_{v,w}(E \cap F).$$

Consequentially, $\langle P(E)P(F)v, w \rangle = \langle P(E \cap F)v, w \rangle$ for all v, w; and thus

$$P(E \cap F) = P(E)P(F).$$

To see $P(X) = I$, note

$$\langle P(X)\pi(f)v, w \rangle = \langle \pi(f)v, P(X)w \rangle$$
$$= \int f(x)d\langle P(x)v, P(X)w \rangle$$
$$= \int f(x)d\langle P(X)P(x)v, w \rangle$$
$$= \int f(x)\, d\langle P(x)v, w \rangle$$
$$= \langle \pi(f)v, w \rangle$$

implies $P(X)\pi(f)v = \pi(f)v$ for all f and v. Since the linear span of the vectors $\pi(f)v$ is dense in $\mathbb{H}$, we see $P(X) = I$.

Now if $\{E_i\}$ is a sequence of pairwise disjoint Borel sets, then the vectors $P(E_i)v$ are pairwise orthogonal; and since

$$\sum \langle P(E_i)v, w \rangle = \langle P(\cup E_i)v, w \rangle,$$

one has

$$\left\langle \sum P(E_i)v, w \right\rangle = \langle P(\cup E_i)v, w \rangle.$$

Thus $P(\cup E_i) = \sum P(E_i)$, and P is a projection valued measure. Moreover,

$$\langle \pi(f)v, w \rangle = \int f(x)\, d\mu_{v,w}(x) = \int f(x)\, d\langle P(x)v, w \rangle.$$

We conclude $\pi(f) = \int f(x)\, dP(x)$. $\square$

4. The Spectral Theorem for Self Adjoint Operators

A useful version of the spectral theorem is that if T is a bounded self adjoint operator on a Hilbert space $\mathbb{H}$, then T is in the norm closure of the linear span of all the orthogonal projections commuting with all bounded linear operators commuting with T.

THEOREM II.5 (SPECTRAL THEOREM). *Let T be a bounded self adjoint linear operator on a Hilbert space $\mathbb{H}$. Then T is in the norm closure of the linear span of all the orthogonal projections commuting with all the bounded operators commuting with T. Moreover, there is a unique projection valued measure P on the compact Hausdorff space $\sigma(T)$ with*

$$T = \int_{\sigma(T)} \lambda\, dP(\lambda).$$

REMARK. The unique projection valued measure P is called the resolution of the identity of the operator T.

PROOF. Let $\mathcal{A}$ be the smallest C* subalgebra of $B(\mathbb{H})$ containing I and T. Define a representation of the commutative C* algebra $\mathcal{A}$ by $\pi(X) = X$ for $X \in \mathcal{A}$. Using the Gelfand transform to identify $\mathcal{A}$ with $C(\sigma(\mathcal{A}))$, Theorem II.4 implies there is a projection valued measure P on $\sigma(\mathcal{A})$ such that $X = \pi(X) = \int X(t)\, dP(t)$. In view of the remark after Proposition II.8, each X in $\mathcal{A}$ including T is in the norm closure of the linear span of the projections $P(E)$, E a Borel subset of $\sigma(\mathcal{A})$. We claim the range of P consists of orthogonal projections commuting with the operators commuting with T.

Assume A commutes with T. Recall the construction of P. Namely, $\langle P(E)v, w \rangle = \mu_{v,w}(E)$ where

$$\int X(h)\, d\mu_{v,w}(h) = \langle \pi(X)v, w \rangle = \langle Xv, w \rangle.$$

But if A commutes with T, A commutes with all X in $\mathcal{A}$. Thus

$$
\begin{aligned}
\langle AXv, w \rangle &= \langle XAv, w \rangle \\
&= \int X(h)\, d\langle P(h)Av, w \rangle \\
&= \langle Xv, A^*w \rangle \\
&= \int X(h)\, d\langle P(h)v, A^*w \rangle
\end{aligned}
$$

for all $X \in C(\sigma(\mathcal{A}))$. Since the C* algebra $\mathcal{A}$ is separable, the spectrum $\sigma(\mathcal{A})$, besides being compact, is second countable. In particular, the spectrum is Polish; and by Theorem I.11, all finite measures on a Polish space are regular. Hence $\langle P(E)Av, w \rangle = \langle P(E)v, A^*w \rangle$ for all E and v and w; and we conclude $P(E)A = AP(E)$ for all E. This gives the first statement.

To obtain the second, we define a mapping $\psi : \sigma(\mathcal{A}) \to \sigma(T)$ by $h \mapsto h(T)$. Since $\mathcal{A}$ is isomorphic to $C(\sigma(\mathcal{A}))$ under the Gelfand transform, we know $\sigma(T) = \sigma(\hat{T}) = \hat{T}(\sigma(\mathcal{A})) = \{h(T) : h \in \sigma(\mathcal{A})\}$. Thus the mapping ψ is onto. We also note it is one-to-one. Indeed if $h_1(T) = h_2(T)$, then h_1 and h_2 agree on all operators $p(T)$ where p is a polynomial. These form a dense subspace of $\mathcal{A}$. By Corollary II.6, the h_i are continuous and thus $h_1 = h_2$. Moreover, ψ is continuous by the definition of the weak $*$ topology. Hence, since these spaces are compact, ψ is a homeomorphism. Identify $\sigma(\mathcal{A})$ and $\sigma(T)$ under this homeomorphism. Then $P \circ \psi^{-1}$ is a projection valued measure on $\sigma(T)$. We calculate $\pi(T)$.

$$\begin{aligned}
T &= \pi(T) \\
&= \pi(\hat{T}) \\
&= \int_{\sigma(\mathcal{A})} \hat{T}(h) \, dP(h) \\
&= \int_{\sigma(T)} \hat{T}(\psi^{-1}(\lambda)) \, dP(\psi^{-1}(\lambda)) \\
&= \int_{\sigma(T)} \psi^{-1}(\lambda)(T) \, dP(\psi^{-1}(\lambda)) \\
&= \int_{\sigma(T)} \lambda \, dP(\psi^{-1}(\lambda)).
\end{aligned}$$

Suppose Q is another projection valued measure on $\sigma(T)$ with $T = \int \lambda \, dQ(\lambda)$. Clearly, $p(T) = \int p(\lambda) \, dQ(\lambda)$ for every polynomial function $p(x)$. By the Stone–Weierstrass Theorem, each continuous function $f(x)$ on $\sigma(T)$ is a uniform limit of a sequence of polynomial functions $p_n(x)$. Thus $\int f(\lambda) \, dP(\lambda)$ and $\int f(\lambda) \, dQ(\lambda)$ are both limits of the sequence $p_n(T)$. Thus $\int f(\lambda) \, dP(\lambda) = \int f(\lambda) \, dQ(\lambda)$ for all continuous functions $f(\lambda)$ on $\sigma(T)$. Let F be a closed subset of $\sigma(T)$. Choose a continuous function $f(\lambda)$ such that $0 \leq f(\lambda) \leq 1$ and $f(\lambda) = 1$ iff $\lambda \in F$. Then

$$\begin{aligned}
\int_F d\langle P(\lambda)v, w \rangle &= \lim_n \int f(\lambda)^n \, d\langle P(\lambda)v, w \rangle \\
&= \lim_n \int f(\lambda)^n \, d\langle Q(\lambda)v, w \rangle \\
&= \int_F d\langle Q(\lambda)v, w \rangle
\end{aligned}$$

for all vectors $v, w \in \mathbb{H}$. But the collection of all Borel sets W where

$$\int_W d\langle P(\lambda)v, w \rangle = \int_W d\langle Q(\lambda)v, w \rangle$$

is a σ-algebra. Thus the measures $E \mapsto \langle P(E)v, w \rangle$ and $E \mapsto \langle Q(E)v, w \rangle$ agree on all Borel sets for all v and w. This gives $P(E) = Q(E)$ for all Borel subsets E. $\square$

DEFINITION. A bounded linear operator A on a Hilbert space $\mathbb{H}$ is said to be compact if for every bounded sequence v_n in $\mathbb{H}$, the sequence of vectors Av_n has a convergent subsequence. The set of all bounded compact operators on $\mathbb{H}$ will be denoted by $\mathcal{K}(\mathbb{H})$.

PROPOSITION II.10. $\mathcal{K}(\mathbb{H})$ *is a norm closed $*$ ideal in $B(\mathbb{H})$.*

PROOF. Suppose $\{A_n\}$ is a sequence in $\mathcal{K}(\mathbb{H})$ which converges in norm to $A \in B(\mathbb{H})$. Let $\{v_k\}_{k=1}^\infty$ be a bounded sequence of vectors. Since each sequence $\{A_n v_k\}_{k=1}^\infty$ has a convergent subsequence, by using the Cantor diagonalization process, one can find a subsequence v_{k_j} of v_k such that $A_n v_{k_j}$ converges for each n. But if $\epsilon > 0$ and n is chosen so that $\|A_n - A\| < \frac{\epsilon}{3 \sup_k \|v_k\|}$, then

$$\begin{aligned}
\|A v_{k_j} - A v_{k_{j'}}\| &\leq \|(A - A_n) v_{k_j}\| + \|A_n v_{k_j} - A_n v_{k_{j'}}\| + \|(A_n - A) v_{k_{j'}}\| \\
&< \frac{2\epsilon}{3} + \|A_n v_{k_j} - A_n v_{k_{j'}}\|.
\end{aligned}$$

Since $\{A_n v_{k_j}\}_{j=1}^\infty$ is Cauchy, this implies $\{A v_{k_j}\}_{j=1}^\infty$ is Cauchy. Thus A is compact, and $\mathcal{K}(\mathbb{H})$ is norm closed.

If A and B are in $\mathcal{K}(\mathbb{H})$ and $\{v_n\}$ is a bounded sequence, then by replacing $\{v_n\}$ by a subsequence, we may assume $A v_n$ converges. Thus if $\{v_{n_j}\}_{j=1}^\infty$ is a sub-subsequence such that $\{B v_{n_j}\}_{j=1}^\infty$ converges, then $\{(A + cB) v_{n_j}\}_{j=1}^\infty$ converges for each complex number c; and we conclude $A + cB \in \mathcal{K}(\mathbb{H})$. Thus $\mathcal{K}(\mathbb{H})$ is a linear subspace. Also note if $\{v_n\}$ is a bounded sequence in $\mathbb{H}$ and $A \in \mathcal{K}(\mathbb{H})$ and a subsequence $\{v_{n_j}\}$ is chosen so that $\{A v_{n_j}\}_{j=1}^\infty$ converges, then $BA v_{n_j}$ converges for all $B \in B(\mathbb{H})$. This shows $\mathcal{K}(\mathbb{H})$ is a left ideal. Furthermore, if $B \in B(\mathbb{H})$, then the sequence $B v_n$ is bounded, and there exists a subsequence $B v_{n_j}$ such that $AB v_{n_j}$ converges. This implies $AB \in \mathcal{K}(\mathbb{H})$ whenever $A \in \mathcal{K}(\mathbb{H})$ and $B \in B(\mathbb{H})$. The fact that $\mathcal{K}(\mathbb{H})$ is a $*$-ideal follows from Corollary II.5. $\square$

THEOREM II.6. *Let A be a positive compact operator on a Hilbert space $\mathbb{H}$. Let P be the resolution of the identity for the operator A. Then $P([\epsilon, \delta])$ has finite dimensional range whenever $0 < \epsilon \leq \delta < \infty$. In particular, each nonzero element in the spectrum of A is an eigenvalue, and the nonzero eigenvalues have finite dimensional eigenspaces. Moreover, the only possible accumulation point in $\sigma(A)$ is 0.*

PROOF. Let $0 < \epsilon \leq \delta < \infty$. Set $B = \{v \in \mathbb{H} : \|v\| \leq 1\}$. To show the range of $P([\epsilon, \delta])$ is finite dimensional, it suffices to show $P([\epsilon, \delta])B$ is compact. Let v_i be a sequence in B. Since A is compact, the sequence $A v_i$ has a convergent subsequence $A v_{i_j}$. Thus $P([\epsilon, \delta]) A v_{i_j}$ converges. Hence $A P([\epsilon, \delta]) v_{i_j}$ converges. But $\|Aw\| \geq \epsilon \|w\|$ if $w \in P([\epsilon, \delta])\mathbb{H}$. Thus $\|P([\epsilon, \delta]) v_{i_j} - P([\epsilon, \delta]) v_{i_{j'}}\| \leq \frac{1}{\epsilon} \|A P([\epsilon, \delta]) v_{i_j} - A P([\epsilon, \delta]) v_{i_{j'}}\|$. This implies $P([\epsilon, \delta]) v_{i_j}$ is Cauchy and thus converges. Hence $P([\epsilon, \delta])B$ is compact. $\square$

DEFINITION. Let $\mathbb{H}$ be a Hilbert space and let v, w be in $\mathbb{H}$. Define $v \otimes \bar{w}$ to be the rank one operator $v \otimes \bar{w}(u) = \langle u, w \rangle v$ for $u \in \mathbb{H}$.

COROLLARY II.7. *Let A be a compact linear operator on $\mathbb{H}$. Then there are orthonormal sequences $\{e_n\}$ and $\{e_n'\}$, and a sequence $\{\lambda_n\}$ such that $\lambda_n \geq \lambda_{n+1} \geq 0$ for all n and $A = \sum \lambda_n e_n \otimes \bar{e}_n'$. Moreover, if these sequences are infinite, $\lim_n \lambda_n = 0$.*

PROOF. Let $T = A^*A$. Then T is a positive compact operator on $\mathbb{H}$. By Theorem II.6, there exist sequences $\{e_n'\}$ and $\{\mu_n\}$ where $\mu_n > 0$ and $T = \sum \mu_n e_n' \otimes \bar{e}_n'$. Namely, the e_n' form a complete list of orthonormal eigenvectors having nonzero eigenvalues, and the μ_n are the corresponding eigenvalues. Since for each $\delta > 0$, the dimension of the linear span of all eigenvectors of T with eigenvalue bounded below by δ is finite, we may rearrange the sequences so that $\{\mu_n\}$ is decreasing. Moreover, since 0 is the only possible accumulation point of $\sigma(T)$, we see $\lim_n \mu_n = 0$ if this is an infinite sequence. Let $\lambda_n = \sqrt{\mu_n}$ and set $|A| = \sum \lambda_n e_n' \otimes \bar{e}_n'$. Clearly, $|A| = \sqrt{T} = \sqrt{A^*A}$.

Define a linear operator U on the range of $|A|$ by $U|A|v = Av$ for $v \in \mathbb{H}$. Then U is an isometry. Indeed, note

$$\begin{aligned}
\langle U|A|v, U|A|v \rangle &= \langle Av, Av \rangle \\
&= \langle A^*Av, v \rangle \\
&= \langle Tv, v \rangle \\
&= \langle |A|^2 v, v \rangle \\
&= \langle |A|v, |A|v \rangle.
\end{aligned}$$

This implies U is well defined and has a unique linear extension to an isometry from the closure of the range of $|A|$ onto the closure of the range of A. We continue to call the extension U. Extend U to all of $\mathbb{H}$ by defining $U(v \oplus w) = Uv$ if $v \in \overline{\mathrm{Rang}(|A|)}$ and $w \in \overline{\mathrm{Rang}(|A|)}^{\perp}$. Then $A = U|A|$, a left **polar decomposition** of A. Set $e_n = Ue_n'$. Since the e_n' are in the range of $|A|$ and U is an isometry on this range, the vectors e_n form an orthonormal sequence in $\mathbb{H}$. Lastly, note

$$\begin{aligned}
Av = U|A|v &= U\left(\sum \lambda_n e_n' \otimes \bar{e}_n'\right)(v) \\
&= U\left(\sum \lambda_n \langle v, e_n' \rangle e_n'\right) = \sum \lambda_n \langle v, e_n' \rangle e_n = \left(\sum \lambda_n e_n \otimes \bar{e}_n'\right)(v)
\end{aligned}$$

for $v \in \mathbb{H}$. $\square$

DEFINITION. Let $\mathcal{B}$ be a Banach $*$ algebra and let π be a representation of $\mathcal{B}$ on a Hilbert space $\mathbb{H}$. π is said to be irreducible if the only closed subspaces W of $\mathbb{H}$ satisfying $\pi(x)W \subset W$ for all x in $\mathcal{B}$ are $\{0\}$ and $\mathbb{H}$.

A vector ξ in $\mathbb{H}$ is said to be cyclic if $\pi(\mathcal{B})\xi$ is a dense subspace of $\mathbb{H}$.

PROPOSITION II.11 (SCHUR'S LEMMA).
The following statements are equivalent:

- (a) π *is irreducible.*
- (b) *The only bounded linear operators commuting with* $\pi(x)$ *for all* $x \in \mathcal{B}$ *are* cI *where* $c \in \mathbb{C}$.
- (c) *Either* $\dim \mathbb{H} = 1$ *or every nonzero vector* ξ *in* $\mathbb{H}$ *is cyclic.*

PROOF. Assume (a). Let T be a bounded linear operator on $\mathbb{H}$ satisfying $T\pi(x) = \pi(x)T$ for all x. Since pi is a $*$ representation, T^* also commutes with every $\pi(x)$. From $T = \frac{T+T^*}{2} + i\frac{T-T^*}{2i}$, in order to show T is scalar, it suffices to do the case when T is self adjoint. But, by the spectral theorem, T is in the strong closure of the linear span of all the orthogonal projections commuting with all the operators commuting with T. In particular, T is in the strong closure of the linear span of the set of orthogonal projections commuting with every $\pi(x)$. But the irreducibility of π implies the only such projections are 0 and I, for the ranges of such projections are closed invariant subspaces. Hence T is a scalar multiple of the identity.

Suppose (b) holds. If W is a closed invariant subspace, the orthogonal projection onto W commutes with $\pi(x)$ for all x and thus is either 0 or I. Hence π is irreducible.

We next show (a) implies (c). Suppose $\mathbb{H}$ has dimension larger than 1. Let ξ be a nonzero vector. If $\pi(x)\xi = 0$ for all x, then $\{c\xi\}$ is a closed invariant subspace of π. Since it is nonzero, it must be all of $\mathbb{H}$; and the dimension of $\mathbb{H}$ is one. This is a contradiction. Thus $\overline{\pi(\mathcal{B})\xi}$ is a closed nonzero invariant subspace. It must therefore be equal to $\mathbb{H}$. Thus each nonzero ξ is cyclic.

Finally assume (c). Let W be a closed nonzero invariant subspace. If $\dim \mathbb{H} = 1$, then clearly $W = \mathbb{H}$. Otherwise, choose a nonzero $\xi \in W$. Then $\overline{\pi(\mathcal{B})\xi} \subset W$. Since $\pi(\mathcal{B})\xi$ is dense in $\mathbb{H}$, $W = \mathbb{H}$. $\square$

5. Positive Functionals and the GNS Construction

DEFINITION. Let $\mathcal{A}$ be a $*$ algebra. Then a linear functional $\omega : \mathcal{A} \to \mathbb{C}$ is called positive if $\omega(x^*x) \geq 0$ for all x in $\mathcal{A}$.

LEMMA II.5. *Let $\omega \geq 0$. Then*

$$\omega(b^*a) = \overline{\omega(a^*b)} \text{ and}$$

$$|\omega(a^*b)| \leq \omega(a^*a)^{\frac{1}{2}}\omega(b^*b)^{\frac{1}{2}}.$$

PROOF.

$$\omega((a+tb)^*(a+tb)) = \omega(a^*a) + t\omega(b^*a) + t\omega(a^*b) + t^2\omega(b^*b) \geq 0$$

for real numbers t. In particular, $\text{Im}(\omega(b^*a) + \omega(a^*b)) = 0$. Thus $\text{Im}(w(b^*a)) = -\text{Im}(\omega(a^*b))$ for all a, b. In particular, by taking ia instead of a, we see $\text{Re}(\omega(b^*a)) = \text{Re}(\omega(a^*b))$; and thus $\omega(b^*a) = \overline{\omega(a^*b)}$. Moreover, we have

$$\omega(a^*a) + 2t Re(\omega(a^*b)) + t^2\omega(b^*b) \geq 0 \text{ for all } t.$$

This implies

$$|Re(\omega(a^*b))| \leq \omega(a^*a)^{\frac{1}{2}}\omega(b^*b)^{\frac{1}{2}}.$$

Replacing b by $e^{i\theta}b$ where $e^{i\theta}\omega(a^*b) \geq 0$ gives the result. $\square$

COROLLARY II.8. *Suppose $\mathcal{A}$ is a Banach $*$ algebra with an identity and ω is a positive linear functional on $\mathcal{A}$. Then $\omega(a^*) = \overline{\omega(a)}$ and $||\omega|| = \omega(e)$. Conversely, when $\mathcal{A}$ is a C^* algebra with unity and ω is a linear functional on $\mathcal{A}$ and $||\omega|| = \omega(e)$, then ω is positive.*

PROOF. Let x be self adjoint with $||x|| \leq 1$. The binomial series $(1+r)^{\frac{1}{2}} = 1 + \frac{1}{2}r + (\frac{1}{2})(-\frac{1}{2})\frac{1}{2!}r^2 + \cdots$ converges absolutely on $[-1,1]$ to $\sqrt{1+r}$. Hence $1 - x = y^2$ for some self adjoint y. Therefore $\omega(e) \geq \omega(x)$ for any self adjoint x with $||x|| \leq 1$. Thus $\omega(-x) \leq \omega(e)$. Therefore $|\omega(x)| \leq \omega(e)$ for any self adjoint x with $||x|| \leq 1$. More generally, $|\omega(x)|^2 \leq \omega(e)\omega(x^*x) \leq \omega(e)^2$ if $||x|| \leq 1$. Thus $||\omega|| \leq \omega(e)$. But $||\omega|| \geq \omega(e)$. Hence $\omega(e) = ||\omega||$.

Now suppose $\mathcal{A}$ is a C^* algebra and $\omega(e) = ||\omega||$. Let x be a positive element in $\mathcal{A}$. To show ω is positive we need to show $\omega(x) \geq 0$. Let $\mathcal{B}$ be the smallest C^* subalgebra of $\mathcal{A}$ containing x and e. Thus $\mathcal{B} \cong \mathcal{C}(X)$ for some compact Hausdorff space X. Hence we may assume $\mathcal{A} = \mathcal{C}(X)$. By the Riesz representation theorem, there is a regular complex Borel measure μ on X satisfying $\omega(f) = \int f(\gamma) \, d\mu(\gamma)$ for all $f \in C(X)$. Since $\omega(e) = ||\omega||$, we see $||\omega|| = \mu(X)$. The result will follow if we show μ is a positive measure. We first claim μ is real. If not, then $\mu = \mu_1 + i\mu_2$ where μ_1 and μ_2 are finite real measures. Suppose $\mu_2(E) \neq 0$ for some Borel set E. Consider the partition $E, X - E$ of X. We see $|\mu(E)| + |\mu(X - E)| > \mu_1(E) + \mu_1(X - E) = \mu_1(X) = \mu(X) = ||\omega||$.

Thus $||\omega|| = |\mu| \geq |\mu(E)| + |\mu(X - E)| > ||\omega||$ which is a contradiction. Thus $\mu_2 = 0$, and μ is real. Now by the Hahn decomposition Theorem, there is a Borel partition E_1 and E_2 of X such that $\mu|_{E_1}$ is a positive measure and $\mu|_{E_2}$ is a negative measure. Hence $|\mu| = \mu(E_1) - \mu(E_2) = ||\omega|| = \mu(X) = \mu(E_1) + \mu(E_2)$. This gives $\mu(E_2) = 0$ and $\mu = \mu|_{E_1}$. Thus μ is positive. $\square$

PROPOSITION II.12. *Let ω be a positive linear functional on a C^* algebra $\mathcal{A}$. Then ω is continuous.*

PROOF. Suppose ω is not continuous. Then by Corollary II.8, $\mathcal{A}$ does not have an identity. Moreover, one can choose x_n with $||x_n|| \leq 1$ with $|\omega(x_n)| \to \infty$. Since each $x_n = y_n + iz_n$ where $y_n = \frac{x_n + x_n^*}{2}$ and $z_n = \frac{x_n - x_n^*}{2i}$, we may assume each x_n is self adjoint. Now if x is self adjoint, then one may write $x = u^*u - v^*v$ where $u, v \in \mathcal{A}$ and $||u||^2 \leq ||x||$ and $||v||^2 \leq ||x||$. Indeed, if $\mathcal{A}_x$ is the smallest C^* subalgebra of $\mathcal{A}$ containing x, then by Theorem 1, the Gelfand transform is an isometric $*$-isomorphism of $\mathcal{A}_x$ onto $C_0(\sigma(\mathcal{A}_x))$. In particular, $\hat{x} = f - g$ where $f = \frac{\hat{x} + |\hat{x}|}{2}$ and $g = \frac{|\hat{x}| - \hat{x}}{2}$. Clearly $|\hat{x}| \geq f \geq 0$ and $|\hat{x}| \geq g \geq 0$. Define u and v by $\hat{u} = \sqrt{f}$ and $\hat{v} = \sqrt{g}$.

Hence for each n, $x_n = u_n^*u_n - v_n^*v_n$ where $||u_n|| \leq 1$ and $||v_n|| \leq 1$. Since $\omega(x_n) = \omega(u_n^*u_n) - \omega(v_n^*v_n)$ and $\omega(u_n^*u_n) \geq 0$ and $\omega(v_n^*v_n) \geq 0$, one sees $\omega(u_n^*u_n) + \omega(v_n^*v_n) \to \infty$ as $n \to \infty$. Hence, replacing some x_n by $-x_n$ if necessary, we may assume $\omega(u_n^*u_n) \to \infty$. Let $\{\lambda_n\}_{n=1}^{\infty} \in l_1$. Thus $\sum |\lambda_n| < \infty$. Hence $\sum |\lambda_n| u_n^*u_n$ converges in $\mathcal{A}$ to some element u. Moreover, for each N one has $\sum_{n=1}^{N} |\lambda_i| u_i^*u_i \leq u$. Hence $\sum_{i=1}^{N} |\lambda_i| \omega(u_i^*u_i) \leq \omega(u)$. Thus $\sum_{n=1}^{\infty} |\lambda_n| \omega(u_n^*u_n) < \infty$ for all $\{\lambda_n\} \in l_1$. This implies the sequence $\{\omega(u_n^*u_n)\}$ is bounded, giving a contradiction. $\square$

Let $\mathcal{A}$ be a Banach algebra without identity. Define $\mathcal{A}_e = \{ze + a : z \in \mathbb{C}, a \in \mathcal{A}\}$. Set $||ze + a|| = |z| + ||a||$. Define multiplication and addition naturally. Then $\mathcal{A}_e$ is a Banach $*$-algebra with an identity. Indeed, note

$$\begin{aligned}
||(ze + a)(z'e + b)|| &= |zz'| + ||zb + z'a + ab|| \\
&\leq |zz'| + |z| \, ||b|| + |z'| \, ||a|| + ||a|| \, ||b|| \\
&= ||ze + a|| \, ||z'e + b||.
\end{aligned}$$

The rest can be shown by standard arguments.

PROPOSITION II.13. *Suppose $\mathcal{A}$ is a Banach $*$ algebra with no identity. Let ω be a continuous positive linear functional on $\mathcal{A}$. Suppose $\omega(a^*) = \overline{\omega(a)}$ and $|\omega(a)|^2 \leq ||\omega||\omega(a^*a)$ for all a. Define ω on $\mathcal{A}_e$ by $\omega(ze+a) = z||w||+\omega(a)$. Then ω is a positive linear functional on $\mathcal{A}_e$. Moreover $|\omega(a^*ba)| \leq \omega(a^*a)\,||b||$ for all a and b.*

PROOF.

$$\begin{aligned}
\omega((\bar{z}e + a^*)(ze + a)) &= \omega(|z|^2 e + \bar{z}a + za^* + a^*a) \\
&= |z|^2||\omega|| + \bar{z}\omega(a) + z\omega(a^*) + \omega(a^*a) \\
&= |z|^2||\omega|| + 2Re(\bar{z}\omega(a)) + \omega(a^*a) \\
&\geq |z|^2||\omega|| - 2|z|\omega(a^*a)^{\frac{1}{2}}||\omega||^{\frac{1}{2}} + \omega(a^*a) \\
&= (|z|\,||\omega||^{\frac{1}{2}} - \omega(a^*a)^{\frac{1}{2}})^2 \geq 0
\end{aligned}$$

since $|Re(|z|\omega(a)| \leq |z|\,||\omega||^{\frac{1}{2}}\omega(a^*a)^{\frac{1}{2}}$.

An easy argument shows $\phi(x) = \omega(a^*xa)$ is a positive linear functional on $\mathcal{A}_e$. Moreover, $||\phi|| = \phi(e) = \omega(a^*a)$. Thus $|\phi(b)| = |\omega(a^*ba)| \leq ||b||\,\omega(a^*a)$. $\square$

COROLLARY II.9. *If $\mathcal{B}$ is a Banach $*$ algebra having an approximate unit and ω is a bounded positive linear form on $\mathcal{B}$, then*

$$\omega(a)^* = \overline{\omega(a)} \quad and$$
$$|\omega(a)|^2 \leq ||\omega||\,\omega(a^*a).$$

In particular, $||\omega|| = \sup_{||a|| \leq 1} \omega(a^*a)$.

PROOF. Let u_i be an approximate unit in $\mathcal{B}$. Thus $||u_i|| \leq 1$ and $u_i x \to x$ and $xu_i \to x$ for each $x \in \mathcal{B}$. Taking adjoints, we see u_i^* is also an approximate unit. Thus $\omega(a^*u_i) = \overline{\omega(u_i^*a)}$. Continuity then implies $\omega(a^*) = \overline{\omega(a)}$. Also $|\omega(u_i^*a)|^2 \leq \omega(u_i^*u_i)\omega(a^*a) \leq ||\omega||\,||u_i||^2\omega(a^*a) \leq ||\omega||\,\omega(a^*a)$. Continuity again gives $|\omega(a)|^2 \leq ||\omega||\,\omega(a^*a)$. The final statement follows immediately from this. $\square$

REMARK. Thus any result established for positive linear functionals on Banach $*$ algebras with identities holds for bounded linear positive functionals on Banach algebras having an approximate identity.

If $\mathcal{A}$ is a normed $*$ algebra, we set $\mathcal{A}'_+ = \{\omega : \omega \in \mathcal{A}'$ with $w \geq 0\}$. The following is known as the Gelfand–Naimark–Segal Theorem.

THEOREM II.7 (GNS). *Let $\mathcal{A}$ be a Banach $*$ algebra with approximate unit. Let ω be a positive bounded linear functional on $\mathcal{A}$. Then there is a representation π of $\mathcal{A}$ on a Hilbert space $\mathbb{H}$ such that*

(a) *there is a vector $\xi \in \mathbb{H}$ with $\omega(x) = \langle \pi(x)\xi, \xi \rangle$ for $x \in \mathcal{A}$,*
(b) *$\pi(\mathcal{A})\xi$ is a dense subspace of $\mathbb{H}$, and*
(c) *the representation and vector ξ are unique up to unitary equivalence.*

PROOF. Extend ω to $\mathcal{A}_e$ by $\omega(ze + a) = z\|\omega\| + \omega(a)$. By Corollary II.9 and Proposition II.13, ω is a positive linear functional on $\mathcal{A}_e$. Let $N = \{x \in \mathcal{A}_e : \omega(x^*x) = 0\}$. We claim N is a left-ideal. Indeed, if $x \in N$, then $\omega((yx)^*xy) = \omega(x^*(y^*xy)) \le \omega(x^*x)^{\frac{1}{2}}\omega((y^*xy)^*(y^*xy))^{\frac{1}{2}} = 0$. To see it is closed under sums, note $\omega((x + y)^*(x + y)) = \omega(x^*x) + \omega(y^*x) + \omega(x^*y) + \omega(y^*y) = 0$ for $|\omega(y^*x)|^2 \le \omega(y^*y)\omega(x^*x) = 0$. Consider the vector space $\mathcal{A}_e/N$. It is a pre-Hilbert space when given the inner product defined by

$$\langle x + N, y + N \rangle = \omega(y^*x).$$

For each $x \in \mathcal{A}_e$, define an operator $\pi(x)$ on $\mathcal{A}_e/N$ by $\pi(x)(y + N) = xy + N$. Since

$$\begin{aligned}
\|xy + N\|^2 &= \omega(y^*x^*xy) \\
&\le \|x^*x\|\,\omega(y^*y) \\
&\le \|x\|^2\|y + N\|^2,
\end{aligned}$$

each $\pi(x)$ is bounded. Let $\mathbb{H}$ be the Hilbert space completion of $\mathcal{A}_e/N$. Then $\pi(x)$ extends in a unique fashion to a bounded operator on $\mathbb{H}$. Clearly, $\pi(x)\pi(y) = \pi(xy)$ and $\pi(x + \lambda y) = \pi(x) + \lambda\pi(y)$. Also

$$\begin{aligned}
\langle \pi(x^*)(y + N), z + N \rangle &= \langle x^*y + N, z + N \rangle \\
&= \omega(z^*x^*y) \\
&= \overline{\omega(y^*xz)} \\
&= \overline{\langle xz + N, y + N \rangle} \\
&= \langle y + N, \pi(x)(z + N) \rangle.
\end{aligned}$$

Hence $\pi(x^*) = \pi(x)^*$. Thus π is a *-representation of $\mathcal{A}_e$. Let $\xi = e + N$. Note

$$\langle \pi(x)\xi, \xi \rangle = \langle x + N, e + N \rangle = \omega(x).$$

Next we show $\mathcal{A} + N$ is dense in $\mathcal{A}_e/N$. This will occur if $e + N$ is a limit point of $\mathcal{A}/N$. Choose $\|x_i\| \le 1$ with $\omega(x_i) \to \|\omega\|$. Then

$\omega(x_i^* x_i) \to ||\omega||$. By replacing x_i by $x_i^* x_i$, we may assume all x_i are self adjoint. Now

$$\begin{aligned}
||x_i - e + N||^2 &= \omega((x_i - e)^*(x_i - e)) \\
&= \omega(x_i^* x_i) - \omega(x_i) - \omega(x_i^*) + \omega(e) \\
&\to ||\omega|| - ||\omega|| - \overline{||\omega||} + ||\omega|| = 0,
\end{aligned}$$

and thus $x_i + N$ converges to $e + N$. Since $\pi(\mathcal{A})\xi = \mathcal{A} + N$, $\pi(\mathcal{A})\xi$ is dense in $\mathbb{H}$.

Finally for uniqueness, assume π', $\mathbb{H}'$ and ξ' are another such triple. Define a mapping U from $\mathcal{A}/N$ into $\mathbb{H}'$ by $U(a + N) = \pi'(a)\xi'$. U is defined on a dense subspace of $\mathbb{H}$. Moreover,

$$\begin{aligned}
\langle U(a + N), U(b + N) \rangle &= \langle \pi'(a)\xi', \pi'(b)\xi' \rangle \\
&= \langle \pi'(b^* a)\xi', \xi' \rangle \\
&= \omega(b^* a) \\
&= \langle a + N, b + N \rangle.
\end{aligned}$$

Thus U is an isometry of the dense subspace $\mathcal{A}/N$ of $\mathbb{H}$ onto $\pi'(\mathcal{A})\xi'$ which is a dense subspace of $\mathbb{H}'$. Hence U extends to a unitary isometry of $\mathbb{H}$ onto $\mathbb{H}'$ satisfying $U\pi(a)\pi(b)\xi = \pi'(ab)\xi' = \pi'(a)\pi'(b)\xi' = \pi'(a)U(\pi(b)\xi)$. Thus $U\pi(a) = \pi'(a)U$ on the dense subspace $\pi(\mathcal{A})\xi$. By continuity, they are equal everywhere. Moreover,

$$\begin{aligned}
\langle \pi'(a)\xi', U\xi \rangle &= \langle U\pi(a)\xi, U\xi \rangle \\
&= \langle \pi(a)\xi, \xi \rangle \\
&= \omega(a) \\
&= \langle \pi'(a)\xi', \xi' \rangle \text{ for all } \xi
\end{aligned}$$

implies $U\xi = \xi'$. □

6. States

Let $\mathcal{B}$ be a Banach $*$ algebra. Let ω be a positive linear functional on $\mathcal{B}$ with $||\omega|| = 1$. Then ω is called a **state**. It is said to be a **pure state** if whenever $0 \leq \omega_1 \leq \omega$, then $\omega_1 = \lambda\omega$ for some λ with $0 \leq \lambda \leq 1$.

PROPOSITION II.14. *Suppose ω is a state on a Banach algebra with approximate identity. Then the representation π_ω generated by the GNS construction is irreducible iff ω is a pure state.*

PROOF. Let $\pi = \pi_\omega$, and let ξ be a cyclic vector satisfying $\omega(x) = \langle \pi(x)\xi, \xi \rangle$. Suppose ω is a pure state. Let $\mathbb{H}_0$ be a closed invariant subspace of $\mathbb{H}$, and let P be the orthogonal projection of $\mathbb{H}$ onto $\mathbb{H}_0$. Then $P\pi(x) = \pi(x)P$ for all x, and $\omega_1(x) = \langle \pi(x)P\xi, P\xi \rangle$ is a positive linear functional on $\mathcal{B}$. Note $\omega(x) - \omega_1(x) = \langle \pi(x)(\xi - P\xi), \xi \rangle = \langle (I - P)\pi(x)(\xi - P\xi), \xi \rangle = \langle \pi(x)(\xi - P\xi), \xi - P\xi \rangle$ implies $\omega - \omega_1 \geq 0$. Since ω is pure, $\omega_1 = \lambda\omega$ for some λ with $0 \leq \lambda \leq 1$. Hence $\langle \pi(x)P\xi, \xi \rangle = \lambda\langle \pi(x)\xi, \xi \rangle = \langle \pi(x)\lambda\xi, \xi \rangle$ for all x. This gives $\langle P\xi - \lambda\xi, \pi(x)\xi \rangle = 0$ for all x. Since the $\pi(x)\xi$ are dense, we see $P\xi = \lambda\xi$. $P^2 = P$ implies $\lambda^2 = \lambda$, and thus $\lambda = 0$ or $\lambda = 1$. If $\lambda = 0$, $P\pi(x)\xi = 0$ for all x and thus $P = 0$. If $\lambda = 1$, then $P\pi(x)\xi = \pi(x)\xi$ for all x and thus $P = I$. Hence $\mathbb{H}_0 = \{0\}$ or $\mathbb{H}_0 = \mathbb{H}$.

Conversely, suppose π is irreducible and $0 \leq \omega_1 \leq \omega$. Define a positive semidefinite sesquilinear form S on $\pi(\mathcal{B})\xi \times \pi(\mathcal{B})\xi$ by $S(\pi(x)\xi, \pi(y)\xi) = \omega_1(y^*x)$. Since $|\omega_1(y^*x)|^2 \leq \omega_1(y^*y)\omega_1(x^*x) \leq \omega(y^*y)\omega(x^*x)$, one has

$$|S(\pi(x)\xi, \pi(y)\xi)|^2 \leq (\pi(x)\xi, \pi(x)\xi)\,(\pi(y)\xi, \pi(y)\xi),$$

and thus

$$|S(\pi(x)\xi, \pi(y)\xi)| \leq ||\pi(x)\xi||\,||\pi(y)\xi||.$$

This implies S extends to a bounded positive semidefinite sesquilinear form on $\overline{\pi(\mathcal{B})\xi} \times \overline{\pi(\mathcal{B})\xi} = \mathbb{H} \times \mathbb{H}$. By the Riesz representation theorem, there is a bounded positive operator T on $\mathbb{H}$ such that

$$S(v, w) = \langle Tv, w \rangle$$

for all v and w. Hence

$$\langle T\pi(x)\xi, \pi(y)\xi \rangle = \omega_1(y^*x)$$

and we see

$$\begin{aligned}
\langle \pi(z)T\pi(x)\xi, \pi(y)\xi \rangle &= \langle T\pi(x)\xi, \pi(z^*y)\xi \rangle \\
&= \omega_1(y^*zx) \\
&= \langle T\pi(z)\pi(x)\xi, \pi(y)\xi \rangle
\end{aligned}$$

for all x and y. Thus $\pi(z)T = T\pi(z)$; and by Proposition II.11, we see $T = \lambda I$ for some λ. Thus $\omega_1(y^*x) = \langle \lambda\pi(x)\xi, \pi(y)\xi \rangle = \lambda\omega(y^*x)$. Clearly $0 \leq \lambda \leq 1$. $\square$

COROLLARY II.10. *Let π be a continuous representation of a Banach $*$ algebra $\mathcal{B}$. Let $\xi \in \mathbb{H}$ and let $\omega(x) = \langle \pi(x)\xi, \xi \rangle$ be the corresponding continuous positive linear functional on $\mathcal{B}$. Then if ω_1 is a positive linear functional on $\mathcal{B}$ with $\omega_1 \leq \omega$, there is a* *bounded positive operator T on $\mathbb{H}$ such that $\pi(x)T = \pi(x)T$ and $\omega_1(x) = \langle \pi(x)T\xi, T\xi \rangle$ for all $x \in \mathcal{B}$.*

PROOF. The above proof shows there is a unique bounded positive operator T_0 on the Hilbert space $\overline{\pi(\mathcal{B})\xi} \subset \mathbb{H}$ with $\omega_1(x^*y) = \langle T_0\pi(y)\xi, \pi(x)\xi \rangle$. Define T by taking $T|_{\overline{\pi(\mathcal{B})\xi}} = T_0$ and $T|_{(\pi(\mathcal{B})\xi)^\perp} = 0$. Then T is a positive linear operator and $\omega_1(x^*y) = \langle T\pi(y)\xi, \pi(x)\xi \rangle$ for all x and y. We have $\langle \pi(z)T\pi(x)\xi, \pi(y)\xi \rangle = \langle T\pi(z)\pi(x)\xi, \pi(y)\xi \rangle$. Thus $\pi(z)T = T\pi(z)$ on the subspace $\overline{\pi(\mathcal{B})\xi}$. Since $\pi(z)(\pi(\mathcal{B})\xi)^\perp \subset (\pi(\mathcal{B})\xi)^\perp$, we see $\pi(z)T = 0 = T\pi(z)$ on $(\pi(\mathcal{B})\xi)^\perp$. Combining these yield $\pi(z)T = T\pi(z)$ for all $z \in \mathcal{B}$ and $\omega_1(z) = \langle \pi(z)T\xi, T\xi \rangle$. $\square$

LEMMA II.6. *Let $\mathcal{B}$ be a Banach $*$ algebra with approximate unit. Then ω is a pure state iff ω is a nonzero extreme point in the weakly $*$ closed convex set of positive functionals ω' satisfying $\|\omega'\| \leq 1$.*

PROOF. By Corollary II.9 and Proposition II.13, we may assume $\mathcal{B}$ has an identity and $\omega(e) = \|\omega\|$. Now, if ω is pure and $\omega = \lambda\omega_1 + (1-\lambda)\omega_2$ where ω_i are positive linear functionals of norm less or equal to 1 and $\lambda \in (0,1)$, then $\omega_i = a_i\omega$ for some $a_i \geq 0$. But $a_i = a_i\omega(e) = \omega_i(e) = \|\omega_i\| \leq 1$. Since $\omega(e) = \lambda a_1\omega(e) + (1-\lambda)a_2\omega(e)$, one has $\lambda a_1 + (1-\lambda)a_2 = 1$. This can only occur if $a_1 = 1$ and $a_2 = 1$. Thus ω is extreme.

Conversely, if ω is extreme and nonzero, then $\|\omega\| = 1$; for otherwise $\omega = \|\omega\|\frac{\omega}{\|\omega\|} + (1 - \|\omega\|)0$. Moreover, if $0 \leq \omega_1 \leq \omega$ and $\omega_1 \neq 0$, then $\omega = \omega_1 + (\omega - \omega_1) = \|\omega_1\|\frac{\omega_1}{\|\omega_1\|} + (1 - \|\omega_1\|)\|\frac{\omega-\omega_1}{1-\|\omega_1\|}$. Corollary II.8 implies the norm of $\frac{\omega-\omega_1}{1-\|\omega_1\|}$ is $\frac{\omega(e)-\omega_1(e)}{1-\|\omega_1\|} = \frac{1-\|\omega_1\|}{1-\|\omega_1\|} = 1$. Since ω is extreme, $\omega_1 = \lambda\omega$ for some $\lambda \in (0,1]$. $\square$

PROPOSITION II.15. *Let $\mathcal{B}$ be a Banach $*$ algebra with approximate unit. Suppose π is a continuous representation of $\mathcal{B}$. Then if $\pi(x) \neq 0$, there is a pure state ω with $\omega(x^*x) \neq 0$.*

PROOF. Define $\omega_1(y) = \langle \pi(y)\xi, \xi \rangle$ where ξ satisfies $\pi(x^*x)\xi \neq 0$ and $\|\xi\| = 1$. Then ω_1 is a state. The Krein–Milman theorem implies there exist positive linear combinations ω_i of pure states which converge to ω in the weak $*$ topology. Thus $\omega_i(x^*x) \to \omega_1(x^*x) > 0$. In particular, there must be a pure state ω with $\omega(x^*x) > 0$. $\square$

COROLLARY II.11. *If $\mathcal{B}$ is a Banach $*$ algebra with approximate unity and there is a representation π with $\pi(x) \neq 0$, then there is an irreducible representation π' with $\pi'(x) \neq 0$.*

PROOF. Choose a pure state ω with $\omega(x^*x) > 0$. Let π be the corresponding GNS representation. Then $\langle \pi(x^*x)\xi, \xi \rangle > 0$ implies $\pi(x)\xi \neq 0$. $\square$

7. The Universal Enveloping C* algebra

Let $\mathcal{B}$ be a Banach $*$ algebra. Let π be a representation of $\mathcal{B}$. Recall that π has an extension to $\mathcal{B}_e$. Namely $\pi(ze + x) = zI + \pi(x)$. By Corollary II.6, we know $||\pi(x)|| \leq ||x||$. Hence every representation of $\mathcal{B}$ is continuous.

Let R be the set of all nondegenerate representations of $\mathcal{B}$. Let R_i be the set of all irreducible representations in R. Let P be the set of all states on $\mathcal{B}$ and let P_{ext} be the set of all pure states.

PROPOSITION II.16. *Let $\mathcal{B}$ be a Banach $*$ algebra with an approximate unity. Then*

$$\sup_{\pi \in R} ||\pi(x)||^2 = \sup_{\pi \in R_i} ||\pi(x)||^2 = \sup_{\omega \in P} \omega(x^*x) = \sup_{\omega \in P_{ext}} \omega(x^*x).$$

PROOF. We show $\sup_{\pi \in R} ||\pi(x)||^2 = \sup_{\omega \in P} \omega(x^*x)$. First suppose $\omega \in P$. Do the GNS construction to obtain a representation $\pi_\omega \in R$ and a cyclic vector $\xi \in \mathbb{H}_\omega$ with $\langle \pi_\omega(x)\xi, \xi \rangle = \omega(x)$. We note ξ is a unit vector. Indeed, in the construction of π_ω, we assumed $\mathcal{B}$ has an identity and then $\omega(e) = ||\omega|| = 1 = \langle \xi, \xi \rangle$. Hence $\omega(x^*x) = \langle \pi_\omega(x)\xi, \pi_\omega(x)\xi \rangle = ||\pi_\omega(x)\xi||^2 \leq ||\pi_\omega(x)||^2$. Hence $\sup_{\omega \in P} \omega(x^*x) \leq \sup_{\pi \in R} ||\pi(x)||^2$. Conversely, we first note if π is a $*$-representation of $\mathcal{B}$ on a Hilbert space $\mathbb{H}$, then $||\pi(x)||^2 = \sup_{||\xi||=1} \langle \pi(x)\xi, \pi(x)\xi \rangle = \sup_{||\xi||=1} \langle \pi(x^*x)\xi, \xi \rangle \leq \sup_{\omega \in P} \omega(x^*x)$ since $x \mapsto \langle \pi(x)\xi, \xi \rangle$ is a state if $\xi = 1$.

The fact that $\sup_{\pi \in R_i} ||\pi(x)||^2 = \sup_{\omega \in P_{ext}} \omega(x^*x)$ follows by the same argument and Proposition II.14 and the GNS construction.

Finally to finish the proposition, we need to show

$$\sup_{\omega \in P} \omega(x^*x) = \sup_{\omega \in P_{ext}} \omega(x^*x).$$

Note $\sup_{\omega \in P_{ext}} \omega(x^*x) \leq \sup_{\omega \in P} \omega(x^*x)$. To see the converse, recall P is a weak $*$ closed convex subset of the unit ball of the dual space of $\mathcal{B}$ and P_{ext} is the subset of extreme points in this set. By the Krein–Milman Theorem, for $\omega \in P$ there is a net ω_λ of convex linear combinations of elements of P_{ext} which converge pointwise to ω. We note $\omega_\lambda(x^*x) = \sum a_i \omega_i(x^*x) \leq \sup_{\omega' \in P_{ext}} \omega'(x^*x) \sum a_i = \sup_{\omega' \in P_{ext}} \omega'(x^*x)$ if $\omega_\lambda = \sum a_i \omega_i$ where $\omega_i \in P_{ext}$ and $\sum a_i = 1$,

$a_i \geq 0$. This gives $\omega(x^*x) \leq \sup_{\omega' \in P_{ext}} \omega'(x^*x)$ for any $\omega \in P$. Hence $\sup_{\omega \in P} \omega(x^*x) \leq \sup_{\omega \in P_{ext}} \omega(x^*x)$. $\square$

DEFINITION. Let $\mathcal{B}$ be a Banach $*$ algebra with an approximate identity. Let N be the set of all x such that $\pi(x) = 0$ for all $\pi \in R$. Then N is a two sided $*$-ideal in $\mathcal{B}$. Thus $\mathcal{B}/N$ is a $*$-algebra. Define a norm on $\mathcal{B}/N$ by:

$$||x + N||^2 = \sup_{\pi \in R} ||\pi(x)||^2$$
$$= \sup_{\pi \in R_i} ||\pi(x)||^2$$
$$= \sup_{\omega \in P} \omega(x^*x)$$
$$= \sup_{\omega \in P_{ext}} \omega(x^*x).$$

THEOREM II.8. *Let $\mathcal{B}$ be a Banach $*$ algebra with an approximate unity. The completion of $\mathcal{B}/N$ is a C^* algebra called the universal enveloping C^* algebra for $\mathcal{B}$.*

PROOF. We claim $||x + N||$ is a norm. Clearly $||x + N|| \neq 0$ if and only if $x + N \neq 0$. One easily checks $||x + N + y + N|| \leq ||x+N|| + ||y+N||$ and $||c(x+N)|| = |c| \, ||x+N||$. To see $||xy+N|| \leq ||x+N|| \, ||y+N||$, note $||\pi(xy)|| \leq ||\pi(x)|| \, ||\pi(y)|| \leq ||x+N|| \, ||y+N||$ and thus $||xy + N|| \leq ||x + N|| \, ||y + N||$. Since $||\pi(x^*)|| = ||\pi(x)||$, $||x^* + N|| = ||x + N||$. Also if $\mathcal{B}$ has an identity, $||e + N|| = 1$ since $\pi(e) = I$ for any nondegenerate representation π.

Finally we need only note

$$||x^*x + N|| = \sup_{\pi \in R} ||\pi(x^*x)||$$
$$= \sup_{\pi \in R} ||\pi(x)||^2$$
$$= ||x + N||^2.$$

Thus the completion of $\mathcal{B}/N$ is a C* algebra. $\square$

Denote the completion of $\mathcal{B}/N$ with this norm as $C^*(\mathcal{B})$ and the mapping $b \mapsto b + N$ as τ.

THEOREM II.9. *Suppose π is a $*$-representation of $\mathcal{B}$ where $\mathcal{B}$ is a Banach $*$ algebra with an approximate identity. Then there is a unique $*$-representation π' of $C^*(\mathcal{B})$ such that $\pi' \circ \tau = \pi$. The mapping π to π' is a one-to-one mapping from the $*$-representations of $\mathcal{B}$*

onto the ∗-representations of $C^*(\mathcal{B})$. Moreover, π is nondegenerate iff π' is nondegenerate, and π is irreducible iff π' is irreducible.

PROOF. Let π be a ∗-representation of $\mathcal{B}$. Define $\pi'(b+N) = \pi(b)$. Then π' is a well defined ∗-representation of $\mathcal{B}/N$. We claim it is continuous. Indeed, $||\pi(b)|| \leq \sup_{\rho \in R}||\rho(b)|| = ||b + N||$. Hence, π' extends to a ∗-representation of the C* algebra $C^*(\mathcal{B})$. Moreover, since ∗-representations of C* algebras are automatically continuous, we see π' is the unique representation of $C^*(\mathcal{B})$ with $\pi' \circ \tau = \pi$. Clearly, the mapping $\tau \mapsto \tau'$ is one-to-one. Moreover, if ρ is a ∗-representation of $C^*(\mathcal{B})$, then $\pi = \rho \circ \tau$ is a ∗-representation of $\mathcal{B}$, and π' and ρ agree on the dense subalgebra $\tau(\mathcal{B})$ in $C^*(\mathcal{B})$. Thus $\pi' = \rho$ and the mapping $\pi \mapsto \pi'$ is onto. Clearly π is nondegenerate iff π' is nondegenerate, and π is irreducible iff π' is irreducible. □

COROLLARY II.12. *Suppose $\mathcal{B}$ is a Banach ∗ algebra with an approximate identity. Then there is a one-to-one correspondence between the continuous positive linear functionals on $\mathcal{B}$ and the continuous positive linear functionals on $C^*(\mathcal{B})$. This correspondence is defined by $\omega \mapsto \omega'$ where $\omega = \omega' \circ \tau$. Moreover, states correspond to states, and pure states correspond to pure states.*

PROOF. One easily notes that in the correspondence $\pi \mapsto \pi'$ between representations of $\mathcal{B}$ and $C^*(\mathcal{B})$, the correspondence carries representations π having cyclic vector ξ to representations π' having cyclic vector ξ. Hence $\omega_\xi(b) = \langle \pi(b)\xi, \xi \rangle$ corresponds to $\omega'_\xi(\tau(b)) = \langle \pi'(b + N)\xi, \xi \rangle = \omega_\xi(b)$.

Next, note that if π is a cyclic ∗-representation of a Banach ∗ algebra with an approximate identity and $\omega(b) = \langle \pi(b)\xi, \xi \rangle$ where ξ is a cyclic vector, then $||\omega|| = ||\xi||^2$. Indeed, one clearly has $||\omega|| \leq ||\xi||^2$. But if u_i is an approximate identity, then $||u_i|| \leq 1$; and since $\pi(u_i)\pi(b)\xi \to \pi(b)\xi$ for all b and the vectors $\pi(b)\xi$ form a dense subspace, we see $\pi(u_i)\xi \to \xi$. Thus $||\omega|| \geq \sup_i \langle \pi(u_i^* u_i)\xi, \xi \rangle \geq \lim_i \langle \pi(u_i)\xi, \pi(u_i)\xi \rangle = ||\xi||^2$. Thus $||\omega|| = ||\xi||^2$. In particular, $||\omega_\xi|| = ||\omega'_\xi||$, and thus states correspond to states. It is easy to check that pure states correspond to pure states, for $\omega_1 \leq \omega_2$ iff $\omega'_1 \leq \omega'_2$. □

THEOREM II.10 (GELFAND). *Every C* algebra $\mathcal{A}$ has a faithful ∗-representation π on a Hilbert space $\mathbb{H}$. Any such π is a ∗-isometric isomorphism of $\mathcal{A}$ onto a closed C* subalgebra of $B(\mathbb{H})$.*

PROOF. Let $x \neq 0$ in $\mathcal{A}$. Then $y = x^*x \geq 0$. Let $P = \{y \in \mathcal{A} : y \geq 0\}$. Then $-x^*x \notin P$ and P is a closed convex set. By the Hahn–Banach Theorem, there is a continuous linear functional ω satisfying $\omega(-x^*x) < 0$ and $\omega(y) \geq 0$ for $y \in P$. Thus $\omega \geq 0$ and $\omega(x^*x) > 0$. By the GNS construction, there is a representation π of $\mathcal{A}$ such that $\langle \pi(x^*x)\xi, \xi \rangle = \omega(x^*x) > 0$. Thus $\pi(x) \neq 0$. Hence for each nonzero $x \in \mathcal{A}$, choose a *-representation π_x of $\mathcal{A}$ with $\pi_x(x) \neq 0$. Define a representation π on $\sum_{x \neq 0} \mathbb{H}_x$ by $\pi(y) = \sum_{x \neq 0} \pi_x(y)$. Then π is a faithful *-representation of $\mathcal{A}$. Moreover, by Proposition II.4, π is a * isometric isomorphism. □

COROLLARY II.13. *There exists a family π_i of irreducible representations of $\mathcal{A}$ on $\mathbb{H}_i$ such that $\sum \pi_i$ is a faithful * isometric isomorphism of $\mathcal{A}$ into $B(\oplus \mathbb{H}_i)$.*

PROOF. This follows from the above theorem and Propositions II.16 and II.4. □

COROLLARY II.14. *Let $\mathcal{A}$ be a C^* algebra. Then for $x \in \mathcal{A}$*

$$||x||^2 = \sup_{\pi \in R} ||\pi(x)||^2 = \sup_{\pi \in R_i} ||\pi(x)||^2 = \sup_{\omega \in P} \omega(x^*x) = \sup_{\omega \in P_{ext}} \omega(x^*x).$$

PROOF. Apply Proposition II.16, Corollary II.6 and Theorem II.10. □

CHAPTER III

TYPE ONE VON NEUMANN ALGEBRAS

This chapter will obtain the structure of Type One von Neumann algebras. Of particular interest are those whose Hilbert spaces are separable. These play a pivotal role in Mackey's method of analyzing representations by considering their restrictions to normal subgroups.

1. Structure of Projection Valued Measures

In this section all Hilbert spaces are separable. Let S be a standard Borel space. Recall that a **projection valued measure** based on S is a mapping P from the Borel subsets of S into the set of orthogonal projections on a complex Hilbert space $\mathbb{H}$ which satisfy the following conditions:

(1) $P(S) = I$
(2) $P(E \cap F) = P(E)P(F)$ for all Borel sets E and F
(3) $P(\bigcup_i E_i) = \sum_i P(E_i)$ for any sequence E_i of pairwise disjoint Borel sets.

It is easy to verify that the range of a projection valued measure forms a Boolean σ-algebra under the operations $P(E) \vee P(F) = P(E \cup F)$, $P(E) \wedge P(F) = P(E \cap F)$, and $P(E)' = I - P(E)$.

A vector v is a **separating vector** for the projection valued measure P if $P(E) = 0$ whenever $P(E)v = 0$. If v is a separating vector for P, then the measure μ defined on S by $\mu(E) = (P(E)v, v)$ induces a measure on the range of P. Specifically, $P(E) = 0$ if and only if $\mu(E) = 0$.

Define an equivalence relation $\sim$ on $\mathbb{H}$ by $v \sim w$ if $P(E)v = 0$ iff $P(E)w = 0$. If $v \in \mathbb{H}$, the **cyclic subspace** $\langle v \rangle$ generated by v is the smallest closed linear subspace of $\mathbb{H}$ containing all the vectors $P(E)v$, E a Borel subset of S. The vector v is said to be **cyclic** if this subspace is $\mathbb{H}$. Two vectors v and w are said to be **cyclicly orthogonal** if the cyclic subspaces $\langle v \rangle$ and $\langle w \rangle$ are orthogonal.

DEFINITION. The projection $P(E)$ has multiplicity n if there is a family of nonzero vectors $\{v_i\}$ of cardinality n such that $v_i \sim v_j$ for all i and j and $P(E)\mathbb{H}$ is the orthogonal sum of the cyclic subspaces $\langle v_i \rangle$.

Since we are assuming our Hilbert spaces are separable, projections can have only finite multiplicity or countable multiplicity. The second case will be denoted by $n = \infty$.

PROPOSITION III.1. *A projection valued measure on a separable Hilbert space has a separating vector.*

PROOF. By Zorn's Lemma there is a family of unit vectors v_a such that the cyclic subspaces $\langle v_a \rangle$ give a pairwise orthogonal decomposition of the Hilbert space $\mathbb{H}$. Since $\mathbb{H}$ is separable, this family is countable. Thus we may assume the indices a belong to the set of natural numbers. Set $v = \sum 2^{-a} v_a$. Then $P(E)v = 0$ implies $P(E)v_a = 0$ for all a which implies $P(E)\langle v_a \rangle = 0$ for all a and hence $P(E) = 0$. $\square$

PROPOSITION III.2.

  (a) If $P(E)$ has multiplicity n and $F \subset E$ with $P(F) \neq 0$, then $P(F)$ has multiplicity n.
  (b) If $P(E_i)$ has multiplicity n for $i = 1, 2, \ldots$, then $P(\cup E_i)$ has multiplicity n.

PROOF. (a) If $P(E)$ has multiplicity n, then $P(E)\mathbb{H} = \langle v_1 \rangle + \cdots + \langle v_n \rangle$ where the v_i are cyclicly orthogonal similar vectors. Thus $P(F)\mathbb{H} = \langle P(F)v_1 \rangle + \cdots + \langle P(F)v_n \rangle$. Since $F \subset E$, $v_i \sim v_j$, and $P(F) \neq 0$, it follows that $P(F)v_i \neq 0$ for all i. Moreover, since $v_i \sim v_j$, one has $P(F)v_i \sim P(F)v_j$ for all i and j. Thus $P(F)$ has multiplicity n.

(b) By (a), we may assume the Borel sets E_i are pairwise disjoint. Now $P(E_i)\mathbb{H} = \langle v_{i,1} \rangle + \cdots + \langle v_{i,n} \rangle$ where the $v_{i,j}$ are similar cyclicly orthogonal unit vectors. Set $w_j = \sum_i \frac{1}{2^i} v_{i,j}$ for $j = 1, 2, \ldots, n$. Since the $P(E_i)$ are pairwise orthogonal projections, one has $P(E_i)w_j = \frac{1}{2^i} v_{i,j}$ and thus $P(\cup E_i)\mathbb{H} = \langle w_1 \rangle + \cdots + \langle w_n \rangle$. Moreover, $P(E)w_j = 0$ iff $P(E \cap E_i)v_{i,j} = 0$ for all i iff $P(E \cap E_i)v_{i,k} = 0$ for all i iff $P(E)w_k = 0$. Thus all the w_j are similar. $\square$

COROLLARY III.1. *For each $n \in \{\infty, 1, 2, \dots\}$ there is a largest projection $P(E_n)$ of multiplicity n.*

PROOF. Take a maximal family of pairwise orthogonal multiplicity n projections. The family is countable by separability. □

THEOREM III.1. *If P is a projection valued measure on the standard Borel space S and $P(S) = I$ has multiplicity n, then there exists a probability measure μ on S and a unitary isomorphism U of $\mathbb{H}$ onto $L^2(S, \mu, \mathbb{H}_n)$ satisfying:*

$$UP(E)U^{-1}f = 1_E f$$

where 1_E is the characteristic function of the set E and $\mathbb{H}_n$ is a Hilbert space of dimension n.

PROOF. Since I has multiplicity n, $\mathbb{H} = \langle v_1 \rangle + \cdots + \langle v_n \rangle$ where the v_i are pairwise similar unit vectors. It follows that the measures μ_i defined by $\mu_i(E) = \langle P(E)v_i, v_i \rangle$ are pairwise equivalent. Let $\mu = \mu_1$ and let g_i be the Radon-Nikodym derivative $\frac{d\mu_i}{d\mu}$.

Let $\{e_j\}_{j=1}^n$ be an orthonormal basis of $\mathbb{H}_n$. Define $U(P(E_1)v_1 + \cdots + P(E_n)v_n) = \sqrt{g_1}\, 1_{E_1} e_1 + \cdots + \sqrt{g_n}\, 1_{E_n} e_n$. One can verify that U is norm preserving on these vectors and extends to unitary operator on all of $\mathbb{H}$. Indeed, U is norm preserving and has dense range; for the image contains all functions of form $\sqrt{g_i}\, f e_j$ where f is simple and the length of this vector is $(\int \|f\|^2 \, d\mu_i)^{1/2}$. In particular, the image contains $h e_j$ for all $h \in L^2(\mu)$; and thus U must be onto.

Finally note $U(P(E)P(E_j)v_j) = 1_E U(P(E_j)v_j)$ for all j and E_j implies by linearity and continuity that $U(P(E)v) = 1_E U(v)$ for all vectors v. □

A projection valued measure is said to have **uniform multiplicity** n if the identity projection $P(S)$ has multiplicity n.

If E is a Borel subset and $P(E)$ has multiplicity n, then the projection valued measure P^E based on E and defined on the Hilbert space $\mathbb{H}_E = P(E)\mathbb{H}$ by $P(F) = P(F \cap E)|_{\mathbb{H}_E}$ has uniform multiplicity n.

The projection valued measure defined on $L^2(S, \mu, \mathbb{H}_n)$ by $P(E)f = 1_E f$ is called the **canonical projection valued measure** for this L^2 space. Note that the maximum number of similar cyclically orthogonal separating vectors for this projection valued measure is n. We thus obtain the following corollary.

COROLLARY III.2. *If a projection valued measure has uniform multiplicities k and l, then $k = l$.*

COROLLARY III.3. *The largest projections $P(E_k)$ having multiplicity k are pairwise orthogonal.*

PROOF. $P(E_j \cap E_k) = 0$ or has multiplicity k and multiplicity j. $\square$

COROLLARY III.4. *Let P be a projection valued measure on a standard Borel space S. Let $\mathcal{B}$ be the range of P. Then $\mathcal{B}$ is a maximal Boolean algebra of projections iff every separating vector for $\mathcal{B}$ is cyclic.*

PROOF. Assume $\mathcal{B}$ is maximal, and suppose v is a separating vector for $\mathcal{B}$. Let Q be the orthogonal projection onto the closure of the linear span of all vectors having form Pv where $P \in \mathcal{B}$. Then Q commutes with every projection in $\mathcal{B}'$. Indeed, $QPP_1v = PP_1v$ for all P and P_1 in $\mathcal{B}$. Thus $QPQ = PQ$. Taking adjoint yields $PQ = QP$ for all $P \in \mathcal{B}$. The maximality of $\mathcal{B}$ implies $Q \in \mathcal{B}$. But since $(I - Q)v = 0$ and v is a separating vector, $Q = I$. Thus v is cyclic.

Conversely, assume P has a cyclic vector v. Equivalently, $P(S)$ has multiplicity one. By Theorem III.1, we may assume $\mathbb{H} = L^2(S, \mu, \mathbb{C})$ where μ is a probability measure and $P(E)f = 1_Ef$ for each Borel subset E of S. Now suppose Q is an orthogonal projection commuting with every projection $P(E)$. Set $h = Q1$. Note h is an L^2 Borel function on S and $P(E)h = P(E)Q1 = QP(E)1 = Q1_E$ for every Borel subset E of S. Thus $1_Eh = Q1_E$ for Borel subsets E of S. This implies $Qf = hf$ for every simple function $f \in L^2(S, \mu)$. Since the simple functions are L^2 dense, continuity of Q implies $Qf = hf$ for all $f \in L^2$. Now $Q^2 = Q$ implies $h^2 = h$, and thus $Q = P(F)$ where $F = \{s : h(s) = 1\}$. $\square$

PROPOSITION III.3. *If P and Q are projection valued measures on standard spaces S and S', respectively, and U is a unitary operator satisfying $U\{P(E) : E \subset S\}U^{-1} = \{Q(F) : F \subset S'\}$, then $UP(E_k)U^{-1} = Q(F_k)$ where $P(E_k)$ and $Q(F_k)$ are the largest projections of multiplicity k.*

PROOF. U preserves the property of similar cyclicly orthogonal vectors. $\square$

The following theorem along with Theorem III.1 completely describes projection valued measures on separable Hilbert spaces.

THEOREM III.2. *For $k = \infty, 1, 2, 3, \ldots$, let $P(E_k)$ be the largest projection of multiplicity k for the projection valued measure P. Then $P(E_\infty) \oplus \sum_{k=1}^{\infty} P(E_k) = I$.*

PROOF. By Proposition III.2, the separability of $\mathbb{H}$, and an application of Zorn's Lemma, it suffices to show there is a k and an E such that $P(E)$ is a nonzero projection of multiplicity k.

Take a maximal collection $\{v_a\}$ of separating cyclicly orthogonal similar vectors. If $\mathbb{H}_0 \equiv \sum_a \langle v_a \rangle = \mathbb{H}$, then $I = P(S)$ has multiplicity equal to the cardinality of this collection, and the proof is complete. Otherwise, $\mathbb{H}_0^\perp \neq 0$ and P restricted to this subspace is a projection valued measure. Let w be a separating vector for this projection valued measure. Thus $P(F)w = 0$ if and only if $P(F)\mathbb{H}_0^\perp = 0$. By the separability of $\mathbb{H}$, there is a largest projection $P(E)$ such that $P(E)w = 0$. This projection has to be nonzero for otherwise w is a separating vector for P and the family $\{v_a\} \cup \{w\}$ consists of separating cyclicly orthogonal similar vectors. This contradicts the maximality of the collection $\{v_a\}$. Since $P(E)w = 0$, $P(E)\mathbb{H}_0^\perp = 0$, and thus $P(E)\mathbb{H} = P(E)\mathbb{H}_0 = \sum_a \langle P(E)v_a \rangle$. Therefore $P(E)$ is a nonzero projection whose multiplicity is the cardinality of the family $\{v_a\}$. $\square$

2. Some Topologies on the Space of Bounded Operators

We now discuss several topologies on $B(\mathbb{H})$. The topology $\mathcal{T}_n$ of $B(\mathbb{H})$ as a normed space will be called the norm topology and will be the strongest topology we consider. There are several coarser topologies which will be needed. They are the weak topology $\mathcal{T}_w$, the strong topology $\mathcal{T}_s$, the σ-weak topology $\mathcal{T}_{\sigma w}$ and the σ-strong topology $\mathcal{T}_{\sigma s}$.

The weak topology is defined by using seminorms $T \mapsto |\langle Tv, w \rangle|$ where $v, w \in \mathbb{H}$. This topology may also be defined using the smaller family of seminorms $T \mapsto |\langle Tv, v \rangle|$. Indeed, note

$$\langle Tv, w \rangle = \frac{1}{4} \sum_{j=0}^{3} i^j \langle T(v + i^j w), v + i^j w \rangle$$

where $i = \sqrt{-1}$.

The strong topology is defined by using the seminorms $T \mapsto \|Tv\|$ for $v \in \mathbb{H}$. Note the strong topology on $B(\mathbb{H})$ is the topology of pointwise convergence; i.e., $T_\alpha \to T$ strongly iff $T_\alpha v \to Tv$ for all $v \in \mathbb{H}$. It is clear that $T_\alpha \to 0$ strongly iff $T_\alpha^* T_\alpha \to 0$ weakly.

The σ-weak topology is defined by the seminorms

$$T \mapsto |\sum_{i=1}^{\infty} \langle Tv_i, w_i \rangle|$$

where $\sum ||v_i||^2 < \infty$ and $\sum ||w_i||^2 < \infty$; and the σ-strong topology is defined by the seminorms

$$T \mapsto (\sum ||Tv_i||^2)^{1/2}$$

where $\sum ||v_i||^2 < \infty$. Again one can show the σ-weak topology is defined by seminorms

$$T \mapsto |\sum \langle Tv_i, v_i \rangle| \text{ where } \sum ||v_i||^2 < \infty$$

and $T_\alpha \to 0$ in the σ-strong topology iff $T_\alpha^* T_\alpha \to 0$ in the σ-weak topology.

Note the adjoint operation $T \mapsto T^*$ is not continuous in the strong topology or the σ-strong topology but is continuous in the weak and σ-weak topologies. For example, let e_i be an orthonormal basis of a separable Hilbert space $\mathbb{H}$. Set $T_i v = \langle v, e_i \rangle e_1$. Then T_i converges strongly and σ-strongly to 0. But $\langle T_i v, w \rangle = \langle v, e_i \rangle \langle e_1, w \rangle = \langle v, \langle w, e_1 \rangle e_i \rangle$, and thus $T_i^*(w) = \langle w, e_1 \rangle e_i$. Thus T_i^* does not converge strongly to 0.

The strong $*$ topology $\mathcal{T}_{s*}$ and the σ-strong $*$ topology $\mathcal{T}_{\sigma s*}$ are defined by

$$T_\alpha \to T$$

in these topologies iff both $T_\alpha \to T$ and $T_\alpha^* \to T^*$ in the strong, or respectively, the σ-strong topologies.

REMARK. On bounded sets, the σ-strong and strong (also the σ-weak and weak) topologies are the same. Indeed, if $T_\alpha \to T$ strongly and $||T_\alpha|| \leq K$ for all α and $\sum ||v_i||^2 < \infty$ and $\epsilon > 0$, then

$$\limsup_\alpha \sum ||(T_\alpha - T)v_i||^2 \leq \lim \sum_{i=1}^{N} ||T_\alpha v_i - Tv_i||^2 + 4K^2 \sum_{i=N+1}^{\infty} ||v_i||^2$$

$$= 4K^2 \sum_{i=N+1}^{\infty} ||v_i||^2 < \epsilon^2$$

if N is chosen sufficiently large. This gives T_α converges to T σ-strongly.

Note one has

$$\mathcal{T}_n > \mathcal{T}_{\sigma s*} > \mathcal{T}_{\sigma s} > \mathcal{T}_{\sigma w} > \mathcal{T}_w \text{ and}$$

$$\mathcal{T}_{\sigma s*} > \mathcal{T}_{s*} > \mathcal{T}_s > \mathcal{T}_w.$$

We now give a definition of a von Neumann algebra. There are many equivalent definitions, indeed even one having no reference to a Hilbert space. However, for our purposes the following definition will suffice.

DEFINITION. A **von Neumann algebra** on a Hilbert space $\mathbb{H}$ is an algebra of bounded operators on $\mathbb{H}$ which contains the identity, is closed under adjoints, and is a closed set in the strong operator topology.

In a locally convex topology, a convex set's closure is determined by continuous linear functionals. Since von Neumann algebras are convex and strongly closed, a description of strongly continuous linear functionals would be helpful. Moreover, we shall eventually see that each von Neumann algebra is the dual space of the normed closed Banach space of σ-weakly continuous linear functionals defined on it.

For this and other reasons, we now focus on continuous linear functionals in the topologies just introduced. Though the topologies are different, the continuous linear functionals have fairly explicit and similar descriptions.

Let $\mathbb{H}$ be a Hilbert space. Let $\mathbb{H}^*$ be the dual space of $\mathbb{H}$. Then each element in $\mathbb{H}^*$ has form $\phi_v(w) = \langle w, v \rangle$ for some unique vector v. The mapping $v \mapsto \phi_v$ is a conjugate linear isomorphism from $\mathbb{H}$ onto $\mathbb{H}^*$. An inner product can be introduced on $\mathbb{H}^*$ by $\langle \phi_v, \phi_{v'} \rangle = \langle v', v \rangle$. With these definitions $\mathbb{H}^*$ becomes a Hilbert space. It is called the conjugate Hilbert space of $\mathbb{H}$ and is denoted by $\bar{\mathbb{H}}$. Using the correspondence $v \leftrightarrow \phi_v$, $\bar{\mathbb{H}}$ may be identified with $\mathbb{H}$. As a Hilbert space, $\bar{\mathbb{H}}$ has the same addition as $\mathbb{H}$; however in $\bar{\mathbb{H}}$ scalar multiplication is given by $\lambda v \equiv \bar{\lambda} v$ and the inner product is defined by $\langle v, v' \rangle_{\bar{\mathbb{H}}} = \langle v', v \rangle = \overline{\langle v, v' \rangle}$.

PROPOSITION III.4. *Let f be a σ-strongly $*$ continuous linear functional on $B(\mathbb{H})$. Then f is σ-weakly continuous.*

PROOF. Note if f is σ-strong $*$ continuous, there is a sequence ξ_n of vectors with $\sum ||\xi_n||^2 < \infty$ such that $|f(A)| \le \rho(A)$ where

$$\rho(A) = \left\{ \sum_{n=1}^{\infty} (||A\xi_n||^2 + ||A^*\xi_n||^2) \right\}^{\frac{1}{2}}$$

for $A \in B(\mathbb{H})$. Set

$$\mathbb{H}_n = \begin{cases} \mathbb{H} \text{ for } n = 1, 2, \cdots \\ \bar{\mathbb{H}} \text{ for } n = -1, -2, \cdots \end{cases}$$

and define $\tilde{A}$ on the direct sum $\tilde{\mathbb{H}}$ of $\mathbb{H}_n$ by

$$(\tilde{A}v)_n = \begin{cases} Av_n \text{ if } n > 0 \\ A^*v_n \text{ if } n < 0. \end{cases}$$

Let $\tilde{\xi}$ be defined by $\tilde{\xi}_n = \xi_{|n|}$. Set $\tilde{\mathbb{H}}_0$ to be the subspace $\{\tilde{A}\tilde{\xi} : A \in B(\mathbb{H})\}$ of $\oplus \mathbb{H}_n$. Define a linear map of $\tilde{\mathbb{H}}_0$ into $\mathbb{C}$ by $F(\tilde{A}\tilde{\xi}) = f(A)$. Note F is well defined; for if $\tilde{A}\tilde{\xi} = 0$, then $A\xi_n = 0 = A^*\xi_n$ for all n, and thus $f(A) = 0$.

Since $|F(\tilde{A}\tilde{\xi})|^2 = |f(A)|^2 \le \rho(A)^2 = ||\tilde{A}\tilde{\xi}||^2$, F is continuous on $\tilde{\mathbb{H}}_0$. By the Hahn–Banach Theorem, F has a continuous extension to $\tilde{\mathbb{H}}$. The Riesz representation theorem implies there is a vector $\tilde{\eta} \in \tilde{\mathbb{H}}$ with $F(\tilde{A}\tilde{\xi}) = \langle \tilde{A}\tilde{\xi}, \tilde{\eta} \rangle$ for all $A \in B(\mathbb{H})$. This gives

$$\begin{aligned} f(A) &= \sum_{n>0} \langle A\xi_n, \eta_n \rangle + \sum_{n>0} \langle A^*\xi_n, \eta_{-n} \rangle^- \\ &= \sum_n \langle A\xi_n, \eta_n \rangle + \langle A\eta_{-n}, \xi_n \rangle \\ &= \sum \langle Av_i, w_i \rangle \end{aligned}$$

where $\sum ||v_i||^2 < \infty$ and $\sum ||w_i||^2 < \infty$. Hence f is σ-weakly continuous. $\square$

Matrix coefficients are the simplest linear functionals on $B(\mathbb{H})$. For each pair of vectors $v, w \in \mathbb{H}$, the corresponding matrix coefficient $\omega_{v,w}$ is defined by $\omega_{v,w}(A) = \langle Av, w \rangle$.

COROLLARY III.5. *Let f be a σ-strongly $*$ continuous linear functional on $B(\mathbb{H})$. Then there exist sequences $\{v_i\}$ and $\{w_i\}$ with $\sum ||v_i||^2 < \infty$ and $\sum ||w_i||^2 < \infty$ with $f = \sum \omega_{v_i,w_i}$.*

COROLLARY III.6. *Let S be a convex set in $B(\mathbb{H})$. Then S is σ-weakly closed iff S is σ-strongly closed iff S is σ-strongly $*$ closed.*

PROOF. Note a linear functional on $B(\mathbb{H})$ is σ-strongly $*$ continuous iff it is σ-strongly continuous iff it is σ-weakly continuous. Since each of these topologies is locally convex, the result follows from the Hahn–Banach Theorem for locally convex spaces. $\square$

PROPOSITION III.5. *Let f be a positive linear functional on $B(\mathbb{H})$. Then the following are equivalent:*

(a) *f is σ-strongly $*$ continuous;*
(b) *f is σ-weakly continuous;*
(c) *there is a sequence v_i in $\mathbb{H}$ with $\sum ||v_i||^2 < \infty$ such that $f = \sum_i \omega_{v_i,v_i}$.*

PROOF. The equivalence of (a) and (b) follows from Proposition III.4. Also (c) clearly implies (b).

We establish (b) implies (c). By Corollary III.5, there exist w_i and w_i' such that $f(T) = \sum_i \langle Tw_i, w_i' \rangle$ where $\sum ||w_i||^2 < \infty$ and $\sum ||w_i'||^2 < \infty$. Hence $\omega(T) \equiv \sum \langle T(w_i + w_i'), w_i + w_i' \rangle$ is a positive linear functional on $B(\mathbb{H})$ and $0 \le f \le \omega$. Consider the representation π of $B(\mathbb{H})$ on $l_2 \otimes \mathbb{H}$ defined by $\pi(T)(\xi_i)_{i=1}^\infty = (T\xi_i)_{i=1}^\infty$. Then $\omega(T) = \langle \pi(T)(w + w'), w + w' \rangle$. Corollary II.10 implies there is a positive linear operator S on $l_2 \otimes \mathbb{H}$ such that $f(T) = \langle \pi(T)S(w + w'), S(w + w') \rangle$ for $T \in B(\mathbb{H})$. Set $v_i = S(w + w')_i$. Then $\sum ||v_i||^2 < \infty$ and $f(T) = \sum_i \langle Tv_i, v_i \rangle$. Hence $f = \sum_i \omega_{v_i,v_i}$. $\square$

PROPOSITION III.6. *Let f be a strongly $*$ continuous linear functional on $B(\mathbb{H})$. Then there exist v_i, w_i $i = 1, 2, \cdots, n$ with $f = \sum_{i=1}^n \omega_{v_i,w_i}$. In particular f is weakly continuous.*

PROOF. Note if f is strongly $*$ continuous, there is a finite sequence $\xi_1, \xi_2, \ldots, \xi_n$ of vectors with $|f(A)| \le \rho(A)$ where

$$\rho(A) = \left(\sum_{i=1}^n (||A\xi_i||^2 + ||A^*\xi_i||^2) \right)^{\frac{1}{2}}$$

for $A \in B(\mathbb{H})$. Set

$$\mathbb{H}_i = \begin{cases} \mathbb{H} \text{ for } i = 1, 2, \cdots, n \\ \bar{\mathbb{H}} \text{ for } i = -1, -2, \cdots, -n \end{cases}$$

and define $\tilde{A}$ on the direct sum $\widetilde{\mathbb{H}}$ of the Hilbert spaces $\mathbb{H}_i$ by

$$(\tilde{A}v)_i = \begin{cases} Av_i \text{ if } i > 0 \\ A^*v_i \text{ if } i < 0. \end{cases}$$

Let $\tilde{\xi}$ be defined by $\tilde{\xi}_i = \xi_{|i|}$. Set $\widetilde{\mathbb{H}}_0$ to be the subspace $\{\tilde{A}\tilde{\xi} : A \in B(\mathbb{H})\}$. Define a linear map of $\widetilde{\mathbb{H}}_0$ into $\mathbb{C}$ by $F(\tilde{A}\tilde{\xi}) = f(A)$. Note F is well defined; indeed if $\tilde{A}\tilde{\xi} = 0$, then $A\xi_i = 0 = A^*\xi_i$ for all n implies $f(A) = 0$.

Since $|F(\tilde{A}\tilde{\xi})| = |f(A)| \leq \rho(A) = ||\tilde{A}\tilde{\xi}||$, F is continuous on $\widetilde{\mathbb{H}}_0$. By the Hahn–Banach Theorem, F has a continuous extension to $\widetilde{\mathbb{H}}$. By the Riesz representation theorem, there is a vector $\tilde{\eta} \in \widetilde{\mathbb{H}}$ with $F(\tilde{A}\tilde{\xi}) = \langle \tilde{A}\tilde{\xi}, \tilde{\eta} \rangle$ for all $A \in B(\mathbb{H})$. This gives

$$f(A) = \sum_{i=1}^{n} \langle A\xi_i, \eta_i \rangle + \sum_{n>0} \langle A^*\xi_i, \eta_{-i} \rangle^-$$
$$= \sum_{i} \langle A\xi_i, \eta_i \rangle + \langle A\eta_{-i}, \xi_i \rangle$$
$$= \sum_{i=1}^{n'} \langle Av_i, w_i \rangle.$$

Hence f is weakly continuous. $\square$

Hence a linear functional f is strongly $*$ continuous iff f is weakly continuous. In particular a linear subspace of $B(\mathbb{H})$ is strongly $*$ closed iff it is weakly closed.

COROLLARY III.7. *A convex subset of bounded linear operators is weakly operator closed if and only if it is $*$-strongly operator closed.*

PROOF. Both the weak operator and *-strong operator topologies are locally convex. They have the same continuous linear functionals. The result follows by the Hahn–Banach Theorem for locally convex spaces. $\square$

$B(\mathbb{H})^*$ is the space of norm continuous linear functionals on $B(\mathbb{H})$. Note $B(\mathbb{H})^*$ contains all the linear functionals continuous in any weaker locally convex topology on $B(\mathbb{H})$.

Let $B_\sim(\mathbb{H})$ be the collection of all weakly continuous linear functionals on $B(\mathbb{H})$ and let $B_*(\mathbb{H})$ be the collection of all σ-weakly linear continuous functionals on $B(\mathbb{H})$. Then $B_\sim(\mathbb{H}) \subset B_*(\mathbb{H}) \subset B(\mathbb{H})^*$.

DEFINITION. A continuous linear transformation T from $\mathbb{H}$ to $\mathbb{K}$ is said to be finite rank if the range of T is a finite dimensional subspace of $\mathbb{K}$.

LEMMA III.1. *The finite rank operators are dense in $B(\mathbb{H})$ with the σ-weak operator topology. They are also dense in $B(\mathbb{H})$ in the weak operator topology.*

PROOF. Consider a seminorm with form $||T||' = |\sum_{i=1}^{\infty} \langle Tv_i, v_i \rangle|$ where $\sum_{i=1}^{\infty} ||v_i||^2$ is finite. Balls using these seminorms define a base for the σ-weak topology. Let T_0 be a bounded operator. Let P_n be the orthogonal projection onto the linear span of the vectors $v_1, v_2, \ldots, v_n$. Then $P_n T_0$ is a finite rank operator, and

$$||P_n T_0 - T_0||' = |\sum_{i=n+1}^{\infty} \langle (P_n T_0 - T_0)v_i, v_i \rangle| \leq \sum_{i=n+1}^{\infty} 2||T_0|| \, ||v_i||^2 \to 0$$

as $n \to \infty$.

The last statement is immediate because the weak operator topology is weaker than the σ-weak topology. $\square$

Let $\mathbb{H}$ be a Hilbert space and let v, w be in $\mathbb{H}$. Recall $v \otimes \bar{w}$ denotes the rank one operator defined by

$$v \otimes \bar{w}(u) = \langle u, w \rangle v.$$

for $u \in \mathbb{H}$.

LEMMA III.2. *Let $f \in B_{\sim}(\mathbb{H})$. Then there exist orthonormal sets $\{e_i\}_{i=1}^{n}$ and $\{e_i'\}_{i=1}^{n}$ and $\lambda_i \geq 0$ with $f = \sum \lambda_i \omega_{e_i, e_i'}$. Moreover $||f|| = \sum \lambda_i$.*

PROOF. We claim $\sum \omega_{v_i, w_i} = \sum \omega_{v_j', w_j'}$ where these are finite sums iff the operators $\sum w_i \otimes \bar{v}_i$ and $\sum w_j' \otimes \bar{v}_j'$ are equal. Indeed,

$$\sum \omega_{v_i, w_i}(v \otimes \bar{w}) = \sum \langle (v \otimes \bar{w})v_i, w_i \rangle$$
$$= \sum \langle v_i, w \rangle \langle v, w_i \rangle$$
$$= \sum \langle v, \langle w, v_i \rangle w_i \rangle$$
$$= \langle v, \sum w_i \otimes \bar{v}_i(w) \rangle$$

Since every finite rank operator is a sum of simple tensors and the finite rank operators are dense in the weak topology, $\sum \omega_{v_i, w_i} = \sum \omega_{v_i', w_i'}$ iff $\sum w_i \otimes \bar{v}_i = \sum w_j' \otimes \bar{v}_j'$.

Now let f be weakly continuous. Proposition III.6 implies $f = \sum_{i=1}^{n} \omega_{v_i, w_i}$ for some vectors v_i, w_i. Let $T = \sum w_i \otimes \bar{v}_i$. Then T is zero on the orthogonal complement of the linear span W of the vectors $\{v_i, w_i : i = 1, \cdots, n\}$ and has range in this span. Hence we may assume we are in a finite dimensional vector space. The matrix T^*T is positive and self adjoint. Hence there is a positive self adjoint operator P with $P^2 = T^*T$. Choose an orthonormal basis $\{e_i\}$ of W with $Pe_i = \lambda_i e_i$ for $i = 1, 2, \ldots, \dim W$. Define U on the range of P by $UPw = Tw$. Note U is an isometry on the range of P (this implies U is well defined). Indeed, $\langle UPw_1, UPw_2 \rangle = \langle Tw_1, Tw_2 \rangle = \langle T^*Tw_1, w_2 \rangle = \langle P^2w_1, w_2 \rangle = \langle Pw_1, Pw_2 \rangle$. Extend U so that U becomes a unitary operator on all of W. Clearly $UP = T$. Set $e'_i = Ue_i$. Then $T = \sum \lambda_i e'_i \otimes \bar{e}_i$. Indeed $Te_i = UPe_i = \lambda_i Ue_i = \lambda_i e'_i = \sum \lambda_j e'_j \otimes \bar{e}_j(e_i)$. In particular, we see $f = \sum \lambda_i w_{e_i, e'_i}$. Clearly $\|f\| \leq \sum \lambda_i$. Take $S = \sum e'_i \otimes \bar{e}_i$. Then $\|S\| = 1$ and $f(S) = \sum_{i,j} \lambda_j \omega_{e_j, e'_j}(e'_i \otimes \bar{e}_i) = \sum_{i,j} \lambda_j \langle (e'_i \otimes \bar{e}_i)(e_j), e'_j \rangle$. Thus $f(S) = \sum_i \lambda_i \langle e'_i, e'_i \rangle = \sum \lambda_i$. Hence $\|f\| = \sum \lambda_i$. $\square$

3. A predual of $B(\mathbb{H})$

THEOREM III.3. *The norm closure of $B_\sim(\mathbb{H})$ in $B^*(\mathbb{H})$ is $B_*(\mathbb{H})$ and $B_\sim(\mathbb{H})^* = B_*(\mathbb{H})^* = B(\mathbb{H})$. This is an isometric isomorphism.*

PROOF. Let $f \in B_*(\mathbb{H})$. By Proposition III.6, $f = \sum w_{v_i, w_i}$ where $\sum \|v_i\|^2 < \infty$ and $\sum \|w_i\|^2 < \infty$. Since

$$\|\sum_{i=n+1}^{\infty} w_{v_i, w_i}\| \leq \sum_{i=n+1}^{\infty} \|v_i\| \|w_i\|$$

$$\leq (\sum_{i=n+1}^{\infty} \|v_i\|^2)^{1/2} (\sum_{i=n+1}^{\infty} \|w_i\|^2)^{1/2},$$

we see the norm closure of $B_\sim(\mathbb{H})$ contains $B_*(\mathbb{H})$.

Suppose f is a norm limit of elements in $B_\sim(\mathbb{H})$. We show $f \in B_*(\mathbb{H})$. Indeed, we may assume $f = \sum f_n$ where $f_n \in B_\sim(\mathbb{H})$ and $\|f_n\| \leq \frac{1}{2^n}$. By Lemma III.2, for each n there are finite orthonormal sets $e_{n,i}$ and $e'_{n,i}$ and positive numbers $\lambda_{n,i}$ such that $f_n = \sum \lambda_{n,i} \omega_{e_{n,i}, e'_{n,i}}$. Moreover $\|f\|_n = \sum \lambda_{n,i} \leq \frac{1}{2^n}$. Let $f = \sum_{n,i} \omega_{f_{n,i}, f'_{n,i}}$ where $f_{n,i} = \sqrt{\lambda_{n,i}} e_{n,i}$ and $f'_{n,i} = \sqrt{\lambda_{n,i}} e'_{n,i}$. Then f is a σ-weakly continuous linear functional, for

$$\sum_{n,i} \|f_{n,i}\|^2 = \sum_{n,i} \|f'_{n,i}\|^2 = \sum_{n,i} \lambda_{n,i} \leq \sum_n \frac{1}{2^n} < \infty.$$

We now show the dual of the space $B_*(\mathbb{H})$ is $B(\mathbb{H})$. First note if $A \in B(\mathbb{H})$, then the linear functional l_A defined on $B_*(\mathbb{H})$ by

$$l_A(f) = f(A) = \langle A, f \rangle$$

is continuous on $B_*(\mathbb{H})$ with the norm topology. Namely

$$||l_A(f)|| \leq ||f|| \, ||A||.$$

Now suppose W is a continuous linear functional on $B_*(\mathbb{H})$. For $v, w \in \mathbb{H}$ set

$$S(v, w) = W(\omega_{v,w}).$$

Then S is a sesquilinear form on $\mathbb{H}$ and

$$|S(v, w)| \leq ||W|| \, ||\omega_{v,w}|| \leq ||W|| \, ||v|| \, ||w||.$$

The Riesz representation theorem implies there is a bounded linear operator A on $\mathbb{H}$ such that

$$W(\omega_{v,w}) = \langle Av, w \rangle = \omega_{v,w}(A) = l_A(\omega_{v,w}).$$

By Proposition III.6, $W(\omega) = l_A(\omega)$ for any weakly continuous linear functional ω on $B(\mathbb{H})$. Since these form a norm dense linear subspace of $B_*(\mathbb{H})$, we see $l_A = W$. Also $|l_A(\omega)| = |\omega(A)| \leq ||\omega|| \, ||A||$ for all ω implies $||l_A|| \leq ||A||$. But $||\omega_{v,w}|| \leq 1$ if $||v|| \leq 1$ and $||w|| \leq 1$. Thus

$$\begin{aligned}
||l_A|| &\geq \sup_{||v|| \leq 1, ||w|| \leq 1} |l_A(\omega_{v,w})| \\
&= \sup_{||v|| \leq 1, ||w|| \leq 1} |\langle Av, w \rangle| \\
&= ||A||.
\end{aligned}$$

This completes the proof. $\square$

LEMMA III.3. *Assume $\mathbb{H}$ is a separable Hilbert space. Then the strong operator topology and the strong-* topology are Polish on the unit ball in $B(\mathbb{H})$.*

PROOF. Let $\{e_n\}$ be a countable orthonormal basis of $\mathbb{H}$. Let B_1 be the set of all bounded linear operators on $\mathbb{H}$ of norm less or equal to 1. Define

$$\rho(A, B) = \sum 2^{-n} ||(A - B)e_n||.$$

This is a metric defining the strong operator topology on bounded sets. This metric is complete on B_1. Indeed, note that a sequence $\{A_k\}$ in B_1 is Cauchy iff $\{A_k(v)\}_{k=1}^{\infty}$ is Cauchy for every vector v. The principle of uniform boundedness implies there is an operator $A \in B_1$ such that $A_k(v) \to A(v)$ for all $v \in \mathbb{H}$.

To see that B_1 is separable, let $\mathcal{C}$ be the collection of all finite rank operators in B_1 which vanish on the orthogonal complement of the linear span of $\{e_1, e_2, \ldots, e_N\}$ for some N and whose range is in the linear span of $\{e_1, e_2, \ldots, e_M\}$ for some M and which has a rational matrix relative to the basis $\{e_n\}$. Then C is a countable dense subset of B_1.

For the strong-$*$ topology, use the metric

$$\rho'(A, B) = \sum 2^{-n} \left(\|(A - B)e_n\| + \|(A^* - B^*)e_n\| \right)$$

and a similar argument. $\square$

LEMMA III.4. *In a separable Hilbert space, the weak operator topology has the same Borel sets as the strong operator topology.*

PROOF. Since the norm closed balls are clearly closed in both the strong and weak operator topologies, it suffices to show the two topologies have the same Borel sets on the normed closed ball of radius N. Since every weakly open set is strongly open, all weakly Borel sets are strongly Borel. Now by Lemma III.3, the strongly Borel sets make the closed ball a standard Borel space. Corollary I.6 implies it suffices to show the ball of radius N contains a countable separating family of weakly Borel sets. The linear functionals $f_{v,w}(A) = \langle Av, w \rangle$ are weakly continuous on this ball. Let $\{v_i\}_{i=1}^{\infty}$ be a countable dense subset of $\mathbb{H}$, and let $\{E_k\}_{k=1}^{\infty}$ be a countable separating family of Borel subsets in the space $\mathbb{C}$. Then the weakly Borel sets $A_{i,j,k} = f_{v_i,v_j}^{-1}(E_k)$ are a countable separating family for the space of bounded linear operators of norm N or less. $\square$

THEOREM III.4. *The unit ball of bounded linear operators on a separable Hilbert space with the weak operator topology is a compact Polish space.*

PROOF. Let $\{v_i\}_{i=1}^{\infty}$ be a countable dense subset of the unit ball of $\mathbb{H}$. Let D be the unit disk in the complex plane. Then the mapping of the unit ball of bounded operators into the Hilbert cube $\prod_{i,j} D$ defined by $A \mapsto \{(Av_i, v_j)\}_{i,j}$ is a homeomorphism of the unit ball onto a closed subset of the Hilbert cube. $\square$

LEMMA III.5. *Let S be a standard Borel space and $\mathbb{H}$ be a separable Hilbert space. Let p be the coordinate projection of $S \times \mathbb{H}$ onto S and let q be the projection onto $\mathbb{H}$. Then the product Borel structure on $S \times \mathbb{H}$ is the σ-algebra for which a function f from a Borel space X into $S \times \mathbb{H}$ is Borel whenever the functions $p \circ f$ and $x \mapsto \langle q \circ f(x), v \rangle$ for v in a countable dense subset of $\mathbb{H}$ are Borel.*

PROOF. Give $S \times \mathbb{H}$ the product Borel structure. Then this space is standard; and if f is a Borel function from X into this space, then the functions $p \circ f$ and $x \mapsto \langle f(x), v \rangle$ are Borel. To finish the proof, the condition would force the identity map on $S \times \mathbb{H}$ to be Borel and therefore the Borel structure induced by the condition is weaker than the product Borel structure. But it easy to see this induced Borel structure is countably separated. Corollary I.6 implies the result. □

4. Trace Class Operators

LEMMA III.6. *Suppose $\{v_i\}_{i=1}^{\infty}$ and $\{w_i\}_{i=1}^{\infty}$ are sequences of vectors in $\mathbb{H}$ with $\sum \|v_i\|^2 < \infty$ and $\sum \|w_i\|^2 < \infty$. Then*

$$\sum_{i=1}^{\infty} v_i \otimes \bar{w}_i$$

is a bounded operator on $\mathbb{H}$ and

$$\left\| \sum_{i=1}^{\infty} v_i \otimes \bar{w}_i \right\| \leq \left(\sum_{i=1}^{\infty} \|v_i\|^2 \right)^{1/2} \left(\sum_{i=1}^{\infty} \|w_i\|^2 \right)^{1/2}.$$

PROOF. Let $v \in \mathbb{H}$. Then using the Cauchy-Schwarz inequality for both $\mathbb{H}$ and l_2, one has

$$\sum \|v_i \otimes \bar{w}_i(v)\| = \sum |\langle v, w_i \rangle| \, \|v_i\|$$

$$\leq \sum \|v\| \, \|w_i\| \, \|v_i\|$$

$$\leq \|v\| \left(\sum \|w_i\|^2 \right)^{1/2} \left(\sum \|v_i\|^2 \right)^{1/2}$$

$$< \infty.$$

□

DEFINITION. A bounded operator T on a Hilbert space $\mathbb{H}$ is said to be trace class if it has form $\sum v_i \otimes \bar{w}_i$ where $\sum \|v_i\|^2 < \infty$ and $\sum \|w_i\|^2 < \infty$. $B_1(\mathbb{H})$ will denote the linear subspace of $B(\mathbb{H})$ consisting of the trace class operators on $\mathbb{H}$.

PROPOSITION III.7. *There is a natural linear one-to-one onto correspondence between the set of trace class operators T on $\mathbb{H}$ and the collection of σ-weakly continuous linear functionals f on $B(\mathbb{H})$. This correspondence is determined by the condition*

$$f(v \otimes \bar{w}) = \langle Tv, w \rangle$$

for all $v, w \in \mathbb{H}$. Moreover, if $T = \sum v_i \otimes \bar{w}_i$ where $\sum \|v_i\|^2 < \infty$ and $\sum \|w_i\|^2 < \infty$, then

$$f(A) = \sum \langle Av_i, w_i \rangle$$

for $A \in B(\mathbb{H})$.

PROOF. Suppose T is trace class. Then $T = \sum v_i \otimes \bar{w}_i$ where $\sum ||v_i||^2 < \infty$ and $\sum ||w_i||^2 < \infty$. Define f by $f(A) = \sum \langle Av_i, w_i \rangle$ for $A \in B(\mathbb{H})$. Then f is σ-weakly continuous. Moreover, for $v, w \in \mathbb{H}$, one has

$$
\begin{aligned}
f(v \otimes \bar{w}) &= \sum_i \langle (v \otimes \bar{w})v_i, w_i \rangle \\
&= \sum_i \langle v, w_i \rangle \langle v_i, w \rangle \\
&= \langle \sum_i \langle v, w_i \rangle v_i, w \rangle \\
&= \langle \sum_i (v_i \otimes \bar{w}_i)v, w \rangle \\
&= \langle Tv, w \rangle.
\end{aligned}
$$

By Lemma III.1, the linear span of the rank one operators $v \otimes \bar{w}$ is dense in $B(\mathbb{H})$ in the σ-weak topology. This implies f is uniquely determined by the condition $f(v \otimes \bar{w}) = \langle Tv, w \rangle$, and the linear functional f is independent of the particular decomposition of T into a sum of form $\sum v_i \otimes \bar{w}_i$. Clearly, the correspondence $T \mapsto f$ is linear, and since $f = 0$ when $T = 0$, we see this correspondence is one-to-one. Corollary III.5 implies this correspondence is onto. $\square$

DEFINITION. Let T be a trace class operator on $\mathbb{H}$. Then the **trace** of T is defined by $\mathrm{Tr}(T) = f(I)$ where f is the unique σ-weakly continuous linear functional on $B(\mathbb{H})$ satisfying $f(v \otimes \bar{w}) = \langle Tv, w \rangle$ for $v, w \in \mathbb{H}$. In particular, if $T = \sum v_i \otimes \bar{w}_i$ where $\sum ||v_i||^2 < \infty$ and $\sum ||w_i||^2 < \infty$, then

$$
\mathrm{Tr}(T) = \sum_i \langle v_i, w_i \rangle.
$$

PROPOSITION III.8. *Let T be a trace class operator and let $\{e_\alpha\}$ be a complete orthonormal basis of $\mathbb{H}$. Then $\sum_\alpha |\langle Te_\alpha, e_\alpha \rangle| < \infty$ and*

$$
Tr(T) = \sum_\alpha \langle Te_\alpha, e_\alpha \rangle.
$$

PROOF. Let $T = \sum v_i \otimes \bar{w}_i$ where $\sum ||v_i||^2 < \infty$ and $\sum ||w_i||^2 < \infty$. Then

$$
\langle Te_\alpha, e_\alpha \rangle = \sum_i \langle e_\alpha, w_i \rangle \langle v_i, e_\alpha \rangle.
$$

The Cauchy-Schwarz inequality implies

$$\sum_{\alpha} |\langle Te_\alpha, e_\alpha\rangle| \leq \sum_{i,\alpha} |\langle e_\alpha, w_i\rangle\langle v_i, e_\alpha\rangle|$$

$$\leq \left(\sum_{i,\alpha} |\langle w_i, e_\alpha\rangle|^2\right)^{1/2} \left(\sum_{i,\alpha} |\langle v_i, e_\alpha\rangle|^2\right)^{1/2}$$

$$= \left(\sum_{i} ||w_i||^2\right)^{1/2} \left(\sum_{i} ||v_i||^2\right)^{1/2} < \infty.$$

Moreover, since $\sum_{i,\alpha} |\langle e_\alpha, w_i\rangle\langle v_i, e_\alpha\rangle| < \infty$,

$$\sum_{\alpha} \langle Te_\alpha, e_\alpha\rangle = \sum_{\alpha}\sum_{i} \langle e_\alpha, w_i\rangle\langle v_i, e_\alpha\rangle$$

$$= \sum_{i}\langle \sum_{\alpha} \langle v_i, e_\alpha\rangle e_\alpha, w_i\rangle$$

$$= \sum_{i}\langle v_i, w_i\rangle$$

$$= \mathrm{Tr}(T).$$

□

PROPOSITION III.9. *Let T be a trace class operator on $\mathbb{H}$.*

(a) *If $A \in B(\mathbb{H})$, then TA and AT are trace class operators.*

(b) *There exist orthonormal sequences $\{e_i\}$ and $\{e_i'\}$ in $\mathbb{H}$ and a sequence $\{\lambda_i\}$ with $\lambda_i > 0$ such that $\sum \lambda_i < \infty$ and $T = \sum \lambda_i e_i \otimes \bar{e}_i'$.*

(c) *If $T = \sum \lambda_i e_i \otimes \bar{e}_i'$ where $\{e_i\}$ and $\{e_i'\}$ are orthonormal sequences, $\lambda_i > 0$ and $\sum \lambda_i < \infty$, then $\sqrt{T^*T} = \sum \lambda_i e_i' \otimes \bar{e}_i'$ and $\sqrt{TT^*} = \sum \lambda_i e_i \otimes \bar{e}_i$. In particular, the nonzero eigenvalues of $\sqrt{T^*T}$ and $\sqrt{TT^*}$ are the λ_i and their multiplicities are $|\{j : \lambda_j = \lambda_i\}|$.*

PROOF. Suppose $T = \sum v_i \otimes \bar{w}_i$ where $\sum ||v_i||^2 < \infty$ and $\sum ||w_i||^2 < \infty$. Then $AT = \sum(Av_i) \otimes \bar{w}_i$ and $TA = \sum v_i \otimes \overline{A^*w_i}$; and since $\sum ||Av_i||^2 \leq ||A||^2 \sum ||v_i||^2 < \infty$ and $\sum ||A^*w_i||^2 \leq ||A^*||^2 \sum ||w_i||^2 < \infty$, we see both AT and TA are trace class operators. This shows (a).

Now for (b), if T is trace class, then by part (a), T^*T is positive and trace class. Hence $T^*T = \sum v_i \otimes \bar{w}_i$ where $\sum (||v_i||^2 + ||w_i||^2) < \infty$. This implies T^*T is a compact linear operator. Indeed, let $\{v_j'\}$ be a bounded sequence of vectors in $\mathbb{H}$. Then for each i,

$j \mapsto \langle v'_j, w_i \rangle$ is a bounded sequence of complex numbers. Using the Cantor diagonalization process, there is a subsequence v'_{j_k} such that $k \mapsto \langle v'_{j_k}, w_i \rangle$ converges for each i. Thus for each n, the sequence $k \mapsto \sum_{i=1}^{n} v_i \otimes \bar{w}_i(v'_{j_k})$ converges. Now if $\epsilon > 0$, choose n such that

$$\left\| \sum_{i=n+1}^{\infty} v_i \otimes \bar{w}_i \right\| \leq \left(\sum_{i=n+1}^{\infty} \|v_i\|^2 \right)^{1/2} \left(\sum_{i=n+1}^{\infty} \|w_i\|^2 \right)^{1/2} < \frac{\epsilon}{4 \sup_j \|v'_j\|}.$$

For this particular n, pick N such that

$$\left\| \sum_{i=1}^{n} v_i \otimes \bar{w}_i(v'_{j_k}) - \sum_{i=1}^{n} v_i \otimes \bar{w}_i(v'_{j_l}) \right\| < \frac{\epsilon}{2}$$

for $k, l > N$. Then for $k, l > N$, one has

$$\left\| \sum_{i=1}^{\infty} v_i \otimes \bar{w}_i(v'_{j_k}) - \sum_{i=1}^{\infty} v_i \otimes \bar{w}_i(v'_{j_l}) \right\| \leq \left\| \sum_{i=1}^{n} v_i \otimes \bar{w}_i(v'_{j_k}) - \sum_{i=1}^{n} v_i \otimes \bar{w}_i(v'_{j_l}) \right\| +$$

$$\left\| \sum_{i=n+1}^{\infty} v_i \otimes \bar{w}_i(v'_{j_k}) - \sum_{i=n+1}^{\infty} v_i \otimes \bar{w}_i(v'_{j_l}) \right\|$$

$$< \frac{\epsilon}{2} + \left\| \sum_{i=n+1}^{\infty} v_i \otimes \bar{w}_i(v'_{j_k}) \right\| + \left\| \sum_{i=n+1}^{\infty} v_i \otimes \bar{w}_i(v'_{j_l}) \right\|$$

$$\leq \frac{\epsilon}{2} + 2 \left\| \sum_{i=n+1}^{\infty} v_i \otimes \bar{w}_i \right\| \sup_j \|v_j\|$$

$$\leq \frac{\epsilon}{2} + \frac{\epsilon}{2} = \epsilon.$$

Hence T^*T is compact and positive. Theorem II.6 implies there is a decreasing sequence $\{\mu_i\}$ of positive numbers and an orthonormal sequence $\{e'_i\}$ of vectors in $\mathbb{H}$ such that $T^*T = \sum \mu_i e'_i \otimes \bar{e}'_i$. Set $\lambda_i = \sqrt{\mu_i}$. Then $|T| = \sqrt{T^*T} = \sum \lambda_i e'_i \otimes \bar{e}'_i$. Define U on the range of $|T|$ by $U|T|v = Tv$, and extend U linearly to $\mathbb{H}$ by requiring $U = 0$ on the orthogonal complement of the range of $|T|$. Then U is a partial isometry from $\overline{\mathrm{Rang}(|T|)}$ onto $\overline{\mathrm{Rang}(T)}$, and $T = U|T|$ is the polar decomposition of T. Now $U^*T = |T|$; and thus by part (a), $|T|$ is trace class. By Proposition III.8, if $\{e'_i\}$ is extended to a complete orthonormal basis $\{e'_\alpha\}$ of $\mathbb{H}$, then $\mathrm{Tr}(|T|) = \sum_\alpha \langle (\sum \lambda_i e'_i \otimes \bar{e}'_i) e'_\alpha, e'_\alpha \rangle = \sum \lambda_i$. Thus $\sum \lambda_i < \infty$. Define

$e_i = U e_i'$. For $v \in \mathbb{H}$, one has

$$
\begin{aligned}
Tv &= U|T|v \\
&= U\left(\sum \lambda_i e_i' \otimes \bar{e}_i'\right) v \\
&= U\left(\sum \lambda_i \langle v, e_i' \rangle e_i'\right) \\
&= \sum \lambda_i \langle v, e_i' \rangle e_i \\
&= \sum_i \lambda_i e_i \otimes e_i'(v).
\end{aligned}
$$

This gives $T = \sum \lambda_i e_i \otimes \bar{e}_i'$, and thus (b) holds.

For (c), suppose $T = \sum \lambda_i e_i \otimes \bar{e}_i'$ where e_i and e_i' are orthonormal sequences, $\lambda_i > 0$, and $\sum \lambda_i < \infty$. Then $T^* = \sum \lambda_i e_i' \otimes \bar{e}_i$. Thus $T^*T = \sum \lambda_i^2 e_i' \otimes \bar{e}_i'$ and $TT^* = \sum \lambda_i^2 e_i \otimes \bar{e}_i$. Hence $\sqrt{T^*T} = \sum \lambda_i e_i' \otimes \bar{e}_i'$ and $\sqrt{TT^*} = \sum \lambda_i e_i \otimes \bar{e}_i$. Clearly the elements e_i' and e_i are eigenvectors for $\sqrt{T^*T}$ and $\sqrt{TT^*}$, respectively with λ_i being their corresponding eigenvalues. $\square$

We have seen $B_1(\mathbb{H})$ is a two sided ideal in $B(\mathbb{H})$. It is a $*$ ideal for the adjoint of $\sum v_i \otimes \bar{w}_i$ is $\sum w_i \otimes \bar{v}_i$. We define a norm on $B_1(\mathbb{H})$ by $||T||_1 = \text{Tr}(|T|)$.

THEOREM III.5. *The mapping of $B_1(\mathbb{H})$ into $B_*(\mathbb{H})$ defined by $T \mapsto f$ where $f(A) = \text{Tr}(AT)$ is an isometry of $B_1(\mathbb{H})$ onto the predual $B_*(\mathbb{H})$ of $B(\mathbb{H})$. Moreover, $||T||_1 \geq ||T||$.*

PROOF. Let $T \mapsto f$ be the natural linear isomorphism of $B_1(\mathbb{H})$ onto $B_*(\mathbb{H})$. Hence if $T = \sum v_i \otimes \bar{w}_i$ where $\sum ||v_i||^2 < \infty$ and $\sum ||w_i||^2 < \infty$, then $f(A) = \sum \langle Av_i, w_i \rangle$. But $AT = \sum Av_i \otimes \bar{w}_i$. Hence $\text{Tr}(AT) = \sum \langle Av_i, w_i \rangle$. This shows $T \mapsto f$ where $f(A) = \text{Tr}(AT)$ is the natural isomorphism of $B_1(\mathbb{H})$ onto $B_*(\mathbb{H})$.

We show $T \mapsto f$ is an isometry. By (b) and (c) of Proposition III.9, $T = \sum \lambda_i e_i \otimes \bar{e}_i'$ where the λ_i enumerate the eigenvalues of the operator $|T| = \sqrt{T^*T}$ according to their multiplicities.

Thus $\text{Tr}(|T|) = \sum \lambda_i$. Now if $A \in B(\mathbb{H})$, we have

$$
\begin{aligned}
|f(A)| &= |\text{Tr}(AT)| \\
&= |\text{Tr}(\sum \lambda_i Ae_i \otimes \bar{e}'_i)| \\
&= |\sum_i \lambda_i \langle Ae_i, e'_i \rangle| \\
&\leq \sum_i \lambda_i |\langle Ae_i, e'_i \rangle| \\
&\leq ||A|| \sum_i \lambda_i \\
&= ||A|| \text{Tr}(|T|),
\end{aligned}
$$

and thus $||f|| \leq \text{Tr}(|T|)$. Let A be the operator $\sum e'_i \otimes \bar{e}_i$. Clearly $||A|| \leq 1$, and since $AT = \sum \lambda_i e'_i \otimes \bar{e}'_i$, $f(A) = \sum \lambda_i$. This gives $||f|| \geq \sum \lambda_i = \text{Tr}(|T|)$, and thus $||f|| = ||T||_1$.

Note $||T|| \leq ||T||_1$ if $T \in B_1(\mathbb{H})$. Indeed, if $T = \sum \lambda_i e_i \otimes \bar{e}'_i$ where $\{e_i\}$ and $\{e'_i\}$ are orthonormal sequences, $\lambda_i > 0$ and $\sum \lambda_i < \infty$, then $Tv = \sum \lambda_i \langle v, e'_i \rangle e_i$ implies $||Tv||^2 = \sum \lambda_i^2 |\langle v, e'_i \rangle|^2 \leq ||v||^2 \sum \lambda_i^2$. Since all λ_i are positive,

$$
||T|| \leq \sqrt{\sum \lambda_i^2} \leq \sum \lambda_i = ||T||_1.
$$

□

THEOREM III.6. *A bounded linear operator T on $\mathbb{H}$ is trace class iff there is an orthonormal basis $\{e_\alpha\}$ of $\mathbb{H}$ such that $\sum_\alpha \langle |T| e_\alpha, e_\alpha \rangle < \infty$.*

PROOF. Suppose $\sum_\alpha \langle |T| e_\alpha, e_\alpha \rangle < \infty$. Define ϕ on $B(\mathbb{H})$ by

$$
\phi(A) = \sum \langle A|T| e_\alpha, e_\alpha \rangle.
$$

Note $||\phi|| \leq \sum \langle |T| e_\alpha, e_\alpha \rangle$. We claim $\phi \in B_*(\mathbb{H})$. Indeed, for each finite subset F consisting of α's, define $\phi_F(A) = \sum_{\alpha \in F} \langle A|T| e_\alpha, e_\alpha \rangle$. Clearly each $\phi_F \in B_*(\mathbb{H})$. Order the finite sets F by inclusion. Note

$$
\lim_F ||\phi - \phi_F|| = 0.
$$

Indeed, $||\phi - \phi_F|| \leq \sum_{\alpha \notin F} \langle |T| e_\alpha, e_\alpha \rangle \to 0$ as F increases. But by Theorem III.3, $B_*(\mathbb{H})$ is norm closed. Hence $\phi \in B_*(\mathbb{H})$. Theorem

III.5 implies there is an operator $S \in B_1(\mathbb{H})$ such that $\phi(v \otimes \bar{w}) = \langle Sv, w \rangle$ for $v, w \in \mathbb{H}$. But

$$
\begin{aligned}
\phi(v \otimes \bar{w}) &= \sum \langle (v \otimes \bar{w}) | T | e_\alpha, e_\alpha \rangle \\
&= \sum \langle |T| e_\alpha, w \rangle \langle v, e_\alpha \rangle \\
&= \sum \langle e_\alpha, |T| w \rangle \langle v, e_\alpha \rangle \\
&= \sum \langle v, e_\alpha \rangle \overline{\langle |T| w, e_\alpha \rangle} \\
&= \langle v, |T| w \rangle \\
&= \langle |T| v, w \rangle.
\end{aligned}
$$

This implies $|T| = S \in B_1(\mathbb{H})$. Now $T = U|T|$ where U is a partial isometry. By (a) of Proposition III.9, T is trace class.

Conversely, if $T \in B_1(\mathbb{H})$, then $|T|$ is trace class. Proposition III.8 implies $\sum \langle |T| e_\alpha, e_\alpha \rangle = \mathrm{Tr}(|T|) < \infty$ for any orthonormal basis $\{e_\alpha\}$. $\square$

COROLLARY III.8. *A linear operator T is trace class iff there is a complete orthonormal basis $\{e_\alpha\}$ such that for any complete orthonormal basis $\{e'_\alpha\}$, one has*

$$
\sum |\langle T e_\alpha, e'_\alpha \rangle| < \infty.
$$

Moreover,

$$
\|T\|_1 = \max_{\{e'_\alpha\}} \sum |\langle T e_\alpha, e'_\alpha \rangle|.
$$

PROOF. Suppose T is trace class. Let f be the corresponding σ-weakly continuous linear functional on $B(\mathbb{H})$. Thus $f(A) = \mathrm{Tr}(AT)$. Choose c_α with $|c_\alpha| = 1$ and $c_\alpha \langle T e_\alpha, e'_\alpha \rangle \geq 0$. Let U be the unitary operator defined by $U = \sum c_\alpha e_\alpha \otimes \bar{e}'_\alpha$. Then

$$
\begin{aligned}
f(U) &= \sum_\alpha \langle U T e_\alpha, e_\alpha \rangle \\
&= \sum_\alpha \langle T e_\alpha, U^* e_\alpha \rangle \\
&= \sum_\alpha \langle T e_\alpha, \bar{c}_\alpha e'_\alpha \rangle \\
&= \sum_\alpha c_\alpha \langle T e_\alpha, e'_\alpha \rangle \\
&= \sum_\alpha |\langle T e_\alpha, e'_\alpha \rangle|.
\end{aligned}
$$

We thus see $\sum_\alpha |\langle T e_\alpha, e'_\alpha \rangle| = |f(U)| \leq \|f\| = \|T\|_1$.

Conversely, suppose $\sum_\alpha |\langle Te_\alpha, e'_\alpha \rangle| < \infty$ for any basis $\{e'_\alpha\}$. Let $T = U|T|$ be a polar decomposition of T where U is some unitary operator. Define $e'_\alpha = Ue_\alpha$ for each α. Then

$$\sum \langle |T|e_\alpha, e_\alpha \rangle = \sum |\langle |T|e_\alpha, U^*e'_\alpha \rangle|$$
$$= \sum |\langle U|T|e_\alpha, e'_\alpha \rangle|$$
$$= \sum |\langle Te_\alpha, e'_\alpha \rangle|$$
$$< \infty.$$

Theorem III.6 implies T is trace class. Moreover,

$$\|T\|_1 = \max_{\{e'_\alpha\}} \sum |\langle Te_\alpha, e'_\alpha \rangle|.$$

$\square$

COROLLARY III.9. *Let T be trace class, and let U be unitary. Then*

$$\|UT\|_1 = \|TU\|_1.$$

5. The Kaplansky Density Theorem

THEOREM III.7. *Let A be a von Neumann algebra on a Hilbert space $\mathbb{H}$. Suppose M is a $*$ subalgebra of A which is strongly dense in A. Then $M \cap S$ is strongly dense in $A \cap S$ where S is the unit ball in $B(\mathbb{H})$.*

PROOF. Since the strong closure of M contains the norm closure and the norm closure of $M \cap S$ is the intersection of S and the norm closure of M, we may assume M is norm closed. Hence M is a C* algebra.

Consider the algebra $A_2 = A \otimes B(\mathbb{C}^2)$ which consists of all operators

$$\begin{pmatrix} A_{1,1} & A_{1,2} \\ A_{2,1} & A_{2,2} \end{pmatrix} \text{ where } A_{i,j} \in A.$$

This is a von Neumann algebra on $2\mathbb{H} = \mathbb{H} \otimes \mathbb{C}^2 = \mathbb{H} \oplus \mathbb{H}$. Similarly, we can form $M_2 = M \otimes B(\mathbb{C}^2)$. It is a $*$ subalgebra of A_2. It is strongly dense in A_2. Indeed, if $(A_{i,j}) \in A_2$, then there exist a directed set of $\alpha's$ and nets $B^\alpha_{i,j}$ which converge strongly to $A_{i,j}$ for $i, j \in \{1, 2\}$. It follows that

$$\begin{pmatrix} B^\alpha_{1,1} & B^\alpha_{1,2} \\ B^\alpha_{2,1} & B^\alpha_{2,2} \end{pmatrix} \rightarrow \begin{pmatrix} A_{1,1} & A_{1,2} \\ A_{2,1} & A_{2,2} \end{pmatrix}$$

strongly on $\mathbb{H} \otimes \mathbb{C}^2$. Moreover, since $||B_{r,s}|| \leq ||(B_{i,j})|| \leq ||B_{1,1}|| + ||B_{1,2}|| + ||B_{2,1}|| + ||B_{2,2}||$ for each r and s, the algebra $\mathcal{M}_2$ is norm closed and thus is a C* algebra.

Since $\mathcal{M}_2$ is strongly dense in $\mathcal{A}_2$, it is weakly dense in $\mathcal{A}_2$. Thus every self adjoint operator in $\mathcal{A}_2$ is a weak limit of a net in $\mathcal{M}_2$. Since $T \mapsto \frac{T+T^*}{2}$ is weakly continuous, every self adjoint operator in $\mathcal{A}_2$ is a weak limit of a net of self adjoint operators in $\mathcal{M}_2$. Since the set of self adjoint operators form a convex set, it follows by Corollary III.7 that every self adjoint element in $\mathcal{A}_2$ is a strong limit of self adjoint elements in $\mathcal{M}_2$.

Now let $A \in \mathcal{A} \cap S$. Then the operator $C' = \begin{pmatrix} 0 & A \\ A^* & 0 \end{pmatrix}$ is self adjoint. Also

$$||C' \begin{pmatrix} v \\ w \end{pmatrix}||^2 = || \begin{pmatrix} Aw \\ A^*v \end{pmatrix} ||^2 = ||Aw||^2 + ||A^*v||^2 \leq ||v||^2 + ||w||^2 = || \begin{pmatrix} v \\ w \end{pmatrix} ||^2.$$

Hence C' is in the unit ball of $B(2\mathbb{H})$.

Set $h(t) = \frac{2t}{1+t^2}$ for $t \in [-1,1]$. h is a strictly increasing mapping of [-1,1] onto [-1,1]. Its inverse is $h^{-1}(t) = \frac{t}{1+\sqrt{1-t^2}}$. In particular $C = h^{-1}(C')$ is a self adjoint operator in $\mathcal{A}_2$ and thus is a strong limit of a net $B_\alpha \in \mathcal{M}_2$ of self adjoint operators. Set $B'_\alpha = 2B_\alpha(I + B_\alpha^2)^{-1}$. Note $B'_\alpha \in \mathcal{M}_2$. Moreover, since $|\frac{2t}{1+t^2}| \leq 1$ for all real t, $||B'_\alpha|| \leq 1$ for all α. Also

$$C' - B'_\alpha = h(C) - h(B_\alpha)$$
$$= 2(I + B_\alpha^2)^{-1}((I + B_\alpha^2)C - B_\alpha(I + C^2))(I + C^2)^{-1}$$
$$= 2(I + B_\alpha^2)^{-1}(C - B_\alpha)(I + C^2)^{-1}$$
$$\quad + 2(I + B_\alpha^2)^{-1}(B_\alpha(B_\alpha - C)C)(I + C^2)^{-1}$$
$$= 2(I + B_\alpha^2)^{-1}(C - B_\alpha)(I + C^2)^{-1} + \frac{1}{2}B'_\alpha(B_\alpha - C)C'.$$

Hence for each $v \in 2\mathbb{H}$,

$$\lim_\alpha ||C'v - B'_\alpha v|| \leq 2 \lim_\alpha ||(C - B_\alpha)(I + C^2)^{-1}v|| + \frac{1}{2} \lim_\alpha ||(B_\alpha - C)C'v|| = 0.$$

Thus B'_α converges strongly to C'. But $B'_\alpha = \begin{pmatrix} B_{1,1}^\alpha & B_{1,2}^\alpha \\ B_{2,1}^\alpha & B_{2,2}^\alpha \end{pmatrix}$ for operators $B_{i,j}^\alpha \in \mathcal{M}$. It follows that $B_{1,2}^\alpha$ converges to A strongly and $||B_{1,2}^\alpha|| \leq ||B'_\alpha|| \leq 1$. $\square$

6. Boolean Algebras of Projections

If $\mathbb{H}$ is a separable Hilbert space, the family of all orthogonal projections on $\mathbb{H}$ form a lattice with operations $P \vee Q$ being the orthogonal projection onto smallest closed subspace containing both the range of P and Q, $P \wedge Q$ being the orthogonal projection onto the intersections of the ranges of P and Q, and $P' = I - P$. The reason that this is not a Boolean algebra is that the distributive law fails. Indeed, we note that a Boolean algebra of orthogonal projections on a Hilbert space must contain commuting operators. More specifically, note that if P and Q are orthogonal projections in a Boolean algebra, then $P = (P \wedge Q) \vee (P \wedge (I - Q)) = (P \wedge Q) + (P \wedge (I - Q))$. Thus $PQ = (P \wedge Q)Q = P \wedge Q = Q(P \wedge Q) = QP$.

THEOREM III.8. *If $\mathcal{B}$ is a σ-Boolean algebra of projections on a separable Hilbert space $\mathbb{H}$, then $\mathcal{B}$ is the range of a projection valued measure P on a standard Borel space.*

PROOF. Let $\{Q_a\}$ be the family of atoms in $\mathcal{B}$. Since $\mathbb{H}$ is separable, this family is countable; and thus $Q = \vee_a Q_a \in \mathcal{B}$. Thus the Boolean algebra $\mathcal{B}^Q$ on the Hilbert space $Q\mathbb{H}$ consisting of all projections of form QS where $S \in \mathcal{B}$ is the range of a projection valued measure based on a countable set. The 'complementary' algebra $\mathcal{B}^{I-Q}$ is clearly atom free. By using the notion of a direct sum of projection valued measures based on the disjoint Borel union of their base spaces, it follows that to prove the theorem we may assume $\mathcal{B}$ is atom free.

The argument in the proof of Proposition III.1 shows that the Boolean algebra $\mathcal{B}$ has a separating vector v. Thus the function defined on $\mathcal{B}$ by $\mu(Q) = \langle Qv, v \rangle$ is a measure on the σ-algebra. By Lemma III.3, $\mathcal{B}$ contains a countable set $\{Q_i\}_{i=1}^{\infty}$ which is dense in $\mathcal{B}$ in the strong operator topology. This set is also dense in the metric ρ defined by μ. Indeed,

$$
\begin{aligned}
\rho(Q,Q_i) &= \mu(Q(I - Q_i)) + \mu(Q_i(I - Q)) \\
&= \langle Qv, v \rangle - \langle QQ_iv, v \rangle + \langle Q_iv, v \rangle - \langle Q_iQv, v \rangle \\
&= \langle Q^2v, v \rangle - \langle Q_iQv, v \rangle + \langle Q_iv, v \rangle - \langle Qv, v \rangle + \langle Q^2v, v \rangle - \langle Q_iQv, v \rangle \\
&= 2\langle (Q - Q_i)Qv, v \rangle + \langle (Q_i - Q)v, v \rangle \\
&\leq 3\|(Q - Q_i)v\| \, \|v\|
\end{aligned}
$$

which is small if Q_i is close to Q in the strong operator topology. Hence $\mathcal{A}$ is a σ-Boolean algebra with a separable atom free measure

μ. Theorem I.25 implies $\mathcal{A}$ is the range of an isomorphism P defined on the measure algebra $M(m)$. $\square$

The notions of multiplicity, cyclic vectors, and cyclicly orthogonal vectors depend only on the range of the projection valued measure and not on the particular mapping P. Thus the results in Section 1 hold for any σ-algebra of orthogonal projections on a separable Hilbert space.

7. The Double Commutant Theorem

The commutant of a set W of bounded operators is the algebra W' consisting of those bounded operators which commute with every member of W. If W is $*$ closed, that is $A^* \in W$ whenever $A \in W$, then W' is a von Neumann algebra.

THEOREM III.9 (VON NEUMANN). *Let $\mathcal{A}$ be a σ-strongly closed $*$ algebra of operators on a Hilbert space $\mathbb{H}$. Then there is an orthogonal projection P in $\mathcal{A} \cap \mathcal{A}'$ with $PA = A$ for all $A \in \mathcal{A}$ and $\mathcal{A}'' = \mathcal{A} \oplus \mathbb{C}(I - P)$. In particular, if $\mathcal{A}$ is a von Neumann algebra, then $\mathcal{A}'' = \mathcal{A}$. (Here P is the orthogonal projection onto the closure of the linear span of all vectors Av where $A \in \mathcal{A}$ and $v \in \mathbb{H}$.)*

PROOF. Let P be the orthogonal projection of $\mathbb{H}$ onto the closure of the linear span of all vectors Av where $A \in \mathcal{A}$ and $v \in \mathbb{H}$. Thus $PAv = Av$ for all A and v. This gives $PA = A$ for all A in $\mathcal{A}$. Taking adjoints gives $AP = P$ for all $A \in \mathcal{A}$. Thus $P \in \mathcal{A}'$. But if $B \in \mathcal{A}'$, then $BAv = ABv$ for all $A \in \mathcal{A}$ and $v \in \mathbb{H}$. Hence $PBAv = ABv$. This implies $PBA = AB = BA = BPA$. This gives $PBPA = BPA$ for all $A \in \mathcal{A}$, and hence $PBP = BP$ for all B. Thus $P \in \mathcal{A}''$.

Let $B \in \mathcal{A}''$. Then $PB \in \mathcal{A}''$. Let $v_1, v_2, \ldots$ be a sequence in $\mathbb{H}$ satisfying $\sum \|v_k\|^2 < \infty$. Let $\mathbb{H}_\infty$ be the Hilbert space of all sequences (w_k) of vectors in $\mathbb{H}$ satisfying $\sum \|w_k\|^2 < \infty$. For $A \in \mathcal{A}$, let A_∞ be the operator on $\mathbb{H}_\infty$ defined by $A_\infty(v_k) = (Av_k)$. Then $\mathcal{A}_\infty = \{A_\infty : A \in \mathcal{A}\}$ is a $*$ algebra of operators. It is an easy matter to check that $(\mathcal{A}_\infty)'' = (\mathcal{A}'')_\infty$. Set $v_\infty = (v_k) \in \mathbb{H}_\infty$.

The closure $\mathbb{K}$ of the collection of vectors $(A_\infty + \lambda I)v_\infty$ where $A \in \mathcal{A}$ and $\lambda \in \mathbb{C}$ is invariant under $\mathcal{A}_\infty$. Hence if Q is the orthogonal projection onto $\mathbb{K}$, $Q \in (\mathcal{A}_\infty)'$. Since P commutes with B, and P and B are in $\mathcal{A}''$, we have $QB_\infty P_\infty = P_\infty B_\infty Q$, and the vector $P_\infty B_\infty v_\infty$ is in the range of Q. Hence there is a sequence $A_n \in \mathcal{A}$ with $(A_n)_\infty v_\infty \to P_\infty B_\infty v_\infty$. This implies $\sum_k \|A_n v_k - PBv_k\|^2 \to 0$

as $n \to \infty$. Thus PB is in the σ-strong closure of $\mathcal{A}$. Since $\mathcal{A}$ is σ-strongly closed, $PB \in \mathcal{A}$. Taking $B = I$ gives $P \in \mathcal{A}$.

Thus $B = PB + (I - P)B$ where $PB \in \mathcal{A}$. Consider $(I - P)B$. It belongs to $\mathcal{A}''$. But all operators of form $(I - P)C(I - P)$ commute with $\mathcal{A}$, for P is an identity in $\mathcal{A}$. Hence $(I - P)B(I - P)$ commutes with all operators of form $(I - P)C(I - P)$. This implies $(I - P)B = \lambda(I - P)$ for some scalar λ. $\square$

COROLLARY III.10. *Let* $\mathcal{A}$ *be a* σ-*strongly closed* $*$ *subalgebra of* $B(\mathbb{H})$. *Then* $\mathcal{A}$ *is strongly closed.*

PROOF. Let P be the identity orthogonal projection for $\mathcal{A}$. Then $\mathcal{A}'' = \mathcal{A} \oplus \mathbb{C}(I - P)$. Note $\mathcal{A}''$ is strongly closed. Thus if A_i is a net in $\mathcal{A}$ converging strongly to B in $B(\mathbb{H})$, then $B \in \mathcal{A}''$. But $B = A + \lambda(I - P)$ for some scalar λ and some $A \in \mathcal{A}$, and $A_i = A_i P$ converges strongly to BP. Hence $BP = B$. This implies $B = A$, and thus $\mathcal{A}$ is strongly closed. $\square$

LEMMA III.7. *If* T *commutes with all the orthogonal projections in a von Neumann algebra* $\mathcal{A}$, *then* $T \in \mathcal{A}'$. *If* T *commutes with all the unitary operators in a von Neumann algebra* $\mathcal{A}$, *then* $T \in \mathcal{A}'$.

PROOF. Suppose T commutes with all the orthogonal projections P in the von Neumann algebra $\mathcal{A}$. Let A be a self adjoint operator in $\mathcal{A}$. By the spectral theorem (Theorem II.5), A is in the norm closure of the linear span of all the projections commuting with the bounded operators commuting with A. Thus A is in the strong closure of the linear span of the orthogonal projections in $\mathcal{A}''$. $\mathcal{A}'' = \mathcal{A}$ implies A is in the strong closure of the linear span of the orthogonal projections in $\mathcal{A}$. This implies T commutes with A; i.e., T commutes with every self adjoint operator in $\mathcal{A}$. Since every bounded linear operator is a linear combination of two self adjoint operators, T is in $\mathcal{A}'$.

The last statement follows from the first; for every orthogonal projection $P = \frac{1}{2}(2P - I) + \frac{1}{2}I$, and $2P - I$ and I are unitary. $\square$

We thus see every von Neumann algebra is the strong closure of the linear span of its orthogonal projections.

THEOREM III.10. *Let* $\mathcal{F}$ *be a directed set of positive operators in* $B(\mathbb{H})$ *and suppose* $\sup_{A \in \mathcal{F}} \|A\| < \infty$. *Then* $\lim_{A \in \mathcal{F}} A = A_0$ *in the strong operator topology. Moreover,* $A_0 = \sup_{A \in \mathcal{F}} A$.

PROOF. Theorem III.4 implies the weak topologies on balls in $B(\mathbb{H})$ are compact. Thus a subnet of $\mathcal{F}$ converges in the weak operator topology to an operator A_0. Since $\mathcal{F}$ is a directed set of positive operators, this implies $\lim_{\mathcal{F}} \langle Av, v \rangle = \langle A_0 v, v \rangle$ for each vector v in $\mathbb{H}$. Thus $A_0 \geq 0$ and $\langle Av, v \rangle \leq \langle A_0 v, v \rangle$ for all $v \in \mathbb{H}$. Thus $A_0 = \sup_{A \in \mathcal{F}} A$. Also note

$$\begin{aligned}
||(A_0 - A)v||^2 &= ||(A_0 - A)^{1/2}(A_0 - A)^{1/2}v||^2 \\
&\leq ||(A_0 - A)^{1/2}||^2 \, ||(A_0 - A)^{1/2}v||^2 \\
&\leq M\langle (A_0 - A)v, v \rangle \to 0
\end{aligned}$$

as A increases. $\square$

PROPOSITION III.10. *Let $\mathcal{A}$ be a σ-strongly $*$ closed $*$ algebra of bounded linear operators on a Hilbert space $\mathbb{H}$. Then $\mathcal{A}$ is strongly closed.*

PROOF. By replacing $\mathbb{H}$ by the closure of $\mathcal{A}\mathbb{H}$, we may assume $\mathcal{A}\mathbb{H}$ is dense in $\mathbb{H}$.

Let $\widetilde{\mathbb{H}} = l_2(\mathbb{H}) = \{\tilde{v} = \{v_n\}_{n=1}^\infty : \sum ||v_n||^2 < \infty\}$. Then $\widetilde{\mathbb{H}}$ is a Hilbert space with inner product $\langle \tilde{u}, \tilde{w} \rangle = \sum \langle v_n, w_n \rangle$. For each $T \in B(\mathbb{H})$, let $\tilde{T}$ be the bounded operator on $\widetilde{\mathbb{H}}$ defined by $\tilde{T}\tilde{v} = \{Tv_n\}_{n=1}^\infty$. Note $\tilde{\mathcal{A}}$ is a $*$ algebra on $\widetilde{\mathbb{H}}$, and $\tilde{\mathcal{A}}$ is strongly $*$ closed. Indeed, suppose $T_i \in \mathcal{A}$ and $\tilde{T}_i$ converges strongly $*$ to an operator S on $\widetilde{\mathbb{H}}$. Then T_i converges pointwise to an operator T on $\mathbb{H}$ and $S = \tilde{T}$. Since $\tilde{T}_i$ converges strongly $*$ to $\tilde{T}$, one has $\tilde{T}_i\tilde{v} \to \tilde{T}\tilde{v}$ and $\tilde{T}_i^*\tilde{v} \to \tilde{T}^*\tilde{v}$ in $\widetilde{\mathbb{H}}$ for all $\tilde{v} \in \widetilde{\mathbb{H}}$. In particular,

$$\sum ||T_i v_n - T v_n||^2 \to 0 \text{ and}$$
$$\sum ||T_i^* v_n - T^* v_n||^2 \to 0$$

for all $\tilde{v} = (v_n)_{n=1}^\infty$ with $\sum ||v_n||^2 < \infty$ as i increases. Thus T is the σ-strong $*$ limit of the T_i, and we see $T \in \mathcal{A}$. By Corollary III.7, $\tilde{\mathcal{A}}$ is strongly closed. One also has $\tilde{\mathcal{A}}\widetilde{\mathbb{H}}$ is dense in $\widetilde{\mathbb{H}}$. Indeed, if $\tilde{w} \perp \tilde{\mathcal{A}}\widetilde{\mathbb{H}}$, then $\langle w_n, Av \rangle = 0$ for each n and every $A \in \mathcal{A}$ and every $v \in \mathbb{H}$. Since $\mathcal{A}\mathbb{H}$ is dense in $\mathbb{H}$, $w_n = 0$ for all n; and thus $w = 0$. Theorem III.9 then implies $\tilde{\mathcal{A}}$ is a von Neumann algebra. Hence $\left(\tilde{\mathcal{A}}\right)'' = \tilde{\mathcal{A}}$. But $\left(\tilde{\mathcal{A}}\right)'' = \widetilde{(\mathcal{A}'')}$. Indeed, $\left(\tilde{\mathcal{A}}\right)'$ consists of all bounded operators having matrix $[T_{m,n}]_{m,n}$ where $T_{m,n} \in \mathcal{A}'$. Here $[T_{m,n}](\tilde{v}) = \{\sum T_{m,n} v_n\}_{m=1}^\infty$. Thus $E_{i,j} = [\delta_{m,i}\delta_{n,j}I] \in \left(\tilde{\mathcal{A}}\right)'$. Let

$A \in \left(\tilde{\mathcal{A}}\right)''$. Then $AE_{i,j} = E_{i,j}A$ for all i, j. Hence A leaves each of the subspaces $E_{i,i}\tilde{\mathbb{H}}$ invariant; and if $v \mapsto U_i v = \{\delta_{n,i}v\}_{n=1}^{\infty}$ is the natural unitary correspondence between $\mathbb{H}$ and $E_{i,i}\tilde{\mathbb{H}}$, we have $U_i^* A U_i v = U_i^* A E_{i,j} U_j v = U_i^* E_{i,j} A U_j = U_j^* A U_j$ for all i, j. Thus A has form $\tilde{D}$ where D is a bounded operator on $\mathbb{H}$. Since $\tilde{D} \cdot [\delta_{m,n}T] = [\delta_{m,n}T] \cdot \tilde{D}$ for each m and n and $T \in \mathcal{A}'$, we see $DT = TD$ for all $T \in \mathcal{A}'$. Thus $D \in \mathcal{A}''$, and we see $\left(\tilde{\mathcal{A}}\right)'' = \widetilde{(\mathcal{A}'')}$. Since $\tilde{\mathcal{A}} = \left(\tilde{\mathcal{A}}\right)''$, $\mathcal{A} = \mathcal{A}''$. Since commutants are always strongly closed, $\mathcal{A}$ is strongly closed. $\square$

COROLLARY III.11. *Let $\mathcal{A}$ be a $*$ algebra of bounded operators containing the identity. The following are equivalent:*

   (a) $\mathcal{A} = \mathcal{A}''$
   (b) $\mathcal{A}$ *is strongly closed*
   (c) $\mathcal{A}$ *is weakly closed*
   (d) $\mathcal{A}$ *is σ-strongly $*$ closed*
   (e) $\mathcal{A}$ *is σ-strongly closed*
   (f) $\mathcal{A}$ *is σ-weakly closed.*

PROOF. These follow from the double commutant theorem, the fact that $\mathcal{A}''$ is σ-strongly $*$ closed, Corollary III.6 and Corollary III.7. $\square$

PROPOSITION III.11. *Let $\mathcal{M}$ be a von Neumann algebra, and let $\mathcal{N}$ be a left ideal. Then there is a unique orthogonal projection $E \in \mathcal{M}$ such that*
$$\bar{\mathcal{N}}^w = \mathcal{M}E = \bar{\mathcal{N}}^{\sigma w}.$$

Moreover, there is an increasing net of positive operators U_λ in $\mathcal{N}$ such that U_λ converges strongly to E. If $\mathcal{N}$ is a two sided ideal, then E is in the center of $\mathcal{M}$.

PROOF. By Corollary II.4, there is an increasing net U_λ in $\mathcal{N}$ with $0 \leq U_\lambda \leq I$ so that $X - XU_\lambda \to 0$ in norm for all $X \in \mathcal{N}$. Theorem III.10 implies U_λ converges strongly to a bounded operator E. Since the net U_λ is bounded, we also have U_λ converges σ-strongly to E. Hence E belongs to the weak closure $\bar{\mathcal{N}}^w$ and the σ-weak closure $\bar{\mathcal{N}}^{\sigma w}$ of $\mathcal{N}$. Moreover, both $\bar{\mathcal{N}}^w$ and $\bar{\mathcal{N}}^{\sigma w}$ are left ideals. Hence $\mathcal{M}E \subseteq \bar{\mathcal{N}}^w$ and $\mathcal{M}E \subseteq \bar{\mathcal{N}}^{\sigma w}$. Now if $T \in \mathcal{N}$, then $TU_\lambda \to T$. Thus $TE = T$ for all $T \in \mathcal{N}$. Hence $TE = T$ for all $T \in \bar{\mathcal{N}}^w$ and for all $T \in \bar{\mathcal{N}}^{\sigma w}$. In particular E is a projection, and $\bar{\mathcal{N}}^w \subseteq \mathcal{M}E$ and $\bar{\mathcal{N}}^{\sigma w} \subseteq \mathcal{M}E$. Thus $\bar{\mathcal{N}}^w = \mathcal{M}E = \bar{\mathcal{N}}^{\sigma w}$. To see E is unique, suppose F is another such orthogonal projection. Then $BE = B$ and $BF = F$ if $B \in \bar{\mathcal{N}}^w$. In particular, $FE = F$ and $EF = E$. Hence $E = E^* = FE = F$.

Suppose $\mathcal{N}$ is a two sided ideal. Corollary III.7 implies $\bar{\mathcal{N}}^w$ is a two sided strongly closed ideal. Since $\mathcal{M}$ is strongly closed, it is norm closed. By Corollary II.5, $\mathcal{M}E$ is a $*$ ideal. But then E is the identity in $\mathcal{M}E$. Thus $EA = EAE = AE$ for all $A \in \mathcal{M}$, and E is central. $\square$

8. Commutative von Neumann Algebras

THEOREM III.11. *Let $\mathcal{A}$ be a commutative von Neumann algebra on a Hilbert space $\mathbb{H}$ with Gelfand spectrum Δ. Then there is a regular projection valued measure P on Δ with values orthogonal projections on $\mathbb{H}$ such that $\mathcal{A}$ consists of all operators $\int f(x)\, dP(x)$ where f is a bounded Borel function on Δ.*

REMARK. We define the operator $M_f = \int f(x)\, dP(x)$ weakly as follows:
$$\langle M_f v, w \rangle = \int f(x)\, d\langle P(x)v, w \rangle$$
for f a bounded Borel function on Δ. Recall a projection valued measure is regular if $E \mapsto \mu_{v,w}(E) = \langle P(E)v, w \rangle$ is a regular Borel measure on Δ for each v and w.

PROOF. Let $\phi : C(\Delta) \mapsto \mathcal{A}$ be the inverse of the Gelfand transform. Then Φ is a representation of the commutative C* algebra $\mathcal{A}$. By Theorem II.4, there is a regular projection valued measure P on Δ such that $\phi(f) = \int f(x)\, dP(x)$ for $f \in C(\Delta)$. Thus $\mathcal{A} = \{M_f : f \in C(\Delta)\}$. Suppose f is a bounded Borel function on Δ. We claim $M_f \in C(\Delta)$. It suffices to show M_f is in the double commutant of $\mathcal{A}$. Let U be a unitary operator in $\mathcal{A}'$. Then $UM_fU^{-1} = M_f$ for $f \in C(\Delta)$. This implies
$$\langle UM_fU^{-1}v, w \rangle = \langle M_f U^{-1}v, U^{-1}w \rangle$$
$$= \int f(x)\, d\mu_{U^{-1}v, U^{-1}w}(x)$$
$$= \int f(x)\, d\mu_{v,w}(x)$$
$$= \langle M_f v, w \rangle$$
for all $f \in C(\Delta)$. By regularity, $\mu_{v,w}(E) = \mu_{U^{-1}v, U^{-1}w}(E)$ for all Borel sets E and all vectors v and w. Thus $\langle P(E)U^{-1}v, U^{-1}w \rangle = \langle P(E)v, w \rangle$ for all v and w. Hence $UP(E)U^{-1} = P(E)$. By Lemma

III.2, $P(E)$ belongs to $\mathcal{A}''$. The double commutant theorem implies $P(E) \in \mathcal{A}$ for all E. Hence if $T \in \mathcal{A}'$ and f is a bounded Borel function,

$$\begin{aligned}
\langle M_f Tv, w \rangle &= \langle Tv, M_{\bar{f}} w \rangle \\
&= \overline{\langle M_{\bar{f}} w, Tv \rangle} \\
&= \overline{\int \bar{f}\, d\mu_{w,Tv}} \\
&= \int f\, d\bar{\mu}_{T^*w,v} \\
&= \int f\, d\mu_{v,T^*w} \\
&= \langle M_f v, T^*w \rangle \\
&= \langle TM_f v, w \rangle,
\end{aligned}$$

for

$$\mu_{w,Tv}(E) = \langle P(E)w, Tv \rangle = \langle T^*P(E)w, v \rangle = \langle P(E)T^*w, v \rangle = \mu_{T^*w,v}(E).$$

Thus $M_f \in \mathcal{A}''$, and we have $M_f \in \mathcal{A}$. $\square$

9. Hilbert Bundles and Direct Integrals

DEFINITION. A **Hilbert bundle** over a standard Borel space S is Borel space $\mathcal{H}$ and an onto Borel mapping $p : \mathcal{H} \to S$ such that the following are true:

(1) $\mathcal{H}_s = p^{-1}(s)$ is a Hilbert space with an Hermitian form $\langle \cdot, \cdot \rangle_s$ for each $s \in S$.

(2) There exists a sequence $\{f_n\}_{n=1}^{\infty}$ of Borel functions from S into $\mathcal{H}$ with $p \circ f_n(s) = s$ for all s such that

(2a) the linear span of $\{f_n(s)\}$ is dense in $\mathcal{H}_s$ for each s;

(2b) for each m and n, the function $s \mapsto \langle f_m(s), f_n(s) \rangle_s$ is Borel; and

(2c) a function F from a Borel space X into $\mathcal{H}$ is Borel if and only if the functions $p \circ F$ and the functions

$$x \mapsto \langle F(x), f_n(p(F(x))) \rangle_{p(F(x))}$$

are Borel for all n.

We give some examples of Hilbert bundles. The first is a direct sum of trivial Hilbert bundles. For $n \in \{\infty\} \cup \mathbf{N}$, let S_n be a standard Borel space and let $\mathbb{C}_n$ be the standard Hilbert space of dimension n, i.e., of countable dimension in the infinite case. Set $\mathcal{H}_n$ to be the standard Borel space $S_n \times \mathbb{C}_n$ with the product Borel structure. Let p_n be the projection onto the first coordinate and define $f_j(s) = (s, e_j)$ where $e_1, \ldots, e_n$ is an orthonormal basis of $\mathbb{C}_n$. $p_n^{-1}(s) = \{s\} \times \mathbb{C}_n$ is naturally identified with the Hilbert space $\mathbb{C}_n$. (1), (2a), and (2b) are clearly satisfied. Moreover, (2c) follows from Lemma III.5. Hence $\mathcal{H}_n$ is a Hilbert bundle.

Now set $\mathcal{H}$ to be the disjoint Borel sum of the standard Borel spaces $\mathcal{H}_n$, and set $S = \cup_n S_n$. Let p be the projection onto S defined by $p = p_n$ on S_n. Then $\mathcal{H}$ is a Hilbert bundle which is the direct sum of the trivial bundles $\mathcal{H}_n$. We shall see shortly that up to isomorphism all Hilbert bundles are of this form.

A **Borel section** of a Hilbert bundle $\mathcal{H}$ is a mapping f from S into $\mathcal{H}$ satisfying $p \circ f = id_S$ and $s \mapsto \langle f(s), f_n(s) \rangle_s$ is a Borel function for each n. Two Hilbert bundles $\mathcal{H}$ and $\mathcal{K}$ over S are isomorphic if there exists a mapping $s \mapsto U(s)$, where each $U(s)$ is a unitary isomorphism from $\mathcal{H}_s$ onto $\mathcal{K}_s$, satisfying $s \mapsto U(s)f(s)$ is a Borel section of $\mathcal{K}$ if and only if f is a Borel section of $\mathcal{H}$.

THEOREM III.12. *Every Hilbert bundle $\mathcal{H}$ is isomorphic to a direct sum of trivial Hilbert bundles. In particular, every Hilbert bundle is a standard Borel space.*

PROOF. Let $\{f_n\}$ be the sequence of functions satisfying the conditions of (2). By condition (2b), the set of s where $f_n(s) = 0$ is a Borel set. Using cutting and pasting and the Gram-Schmidt orthonormalization process and the conditions of (2), one may replace the functions f_n by functions e_n that have the additional property that $e_m(s) \neq 0$ and $e_{m+1}(s) = 0$ implies $e_1(s), \ldots, e_m(s)$ is a orthonormal basis of $\mathcal{H}_s$. Thus $S_n = \{s : e_n(s) \neq 0 \text{ and } e_{n+1}(s) = 0\}$ is a Borel set, and $\mathcal{H}$ is the Borel direct sum of the Hilbert bundles $\mathcal{H}_n = p^{-1}(S_n)$ over S_n. We may therefore assume $S = S_n$. Let $u_1, \ldots, u_n$ be an orthonormal basis of $\mathbb{C}_n$. Define $U(\sum_i a_i e_i(s)) = (s, \sum_i a_i u_i)$. Conditions (2b) and (2c) along with Lemma III.5 show U is a Borel isomorphism of $\mathcal{H}_n$ onto $S \times \mathbb{C}_n$. Finally set $U(s) = U|_{p^{-1}(s)}$. □

DEFINITION. Let $\mathcal{H}$ be a Hilbert bundle over S. Suppose μ is a σ-finite measure on S. The direct integral $\int^{\oplus} \mathcal{H}_s \, d\mu(s)$ is the Hilbert space consisting of all Borel sections f with $\int \|f(s)\|_s^2 \, d\mu(s) < \infty$. It is called the **direct integral of the Hilbert bundle $\mathcal{H}$** with respect to μ.

An important class of Hilbert bundles occurs when one considers disintegrations of measures. Let X and S be standard Borel spaces and suppose $s \mapsto \mu_s$ is a Borel map into the space of finite measures on X. Thus $s \mapsto \mu_s(E)$ is a Borel function for each Borel subset E of X. Define $\mu * (X \times \mathbb{C})$ to be the set consisting of all pairs (s, f) where $f \in L^2(\mu_s)$. Set p to be the natural projection onto S. Then clearly $p^{-1}(s)$ is isomorphic to $L^2(\mu_s)$ and hence is a Hilbert space. Let $\{A_n\}$ be a countable algebra of sets on X which generate the Borel sets. Set $f_n(s) = (s, 1_{A_n})$. Then (2a) and (2b) are satisfied. Define a Borel structure on $\mu * (X \times \mathbb{C})$ by taking it to be the smallest σ-algebra such that the mappings $(s, f) \mapsto s$ and $(s, f) \mapsto \int_{A_n} f(x) \, d\mu_s(x)$, $n \geq 1$ are Borel. In particular, a mapping $y \mapsto (s(y), f_y)$ is Borel on a Borel space Y iff s is a Borel mapping from Y into S and $y \mapsto \int_{A_n} f_y(x) \, d\mu_{s(y)}(x)$ is Borel for all n. This gives (2c) and thus $\mu * (X \times \mathbb{C})$ is a Hilbert bundle.

A slight generalization of this method can be defined in terms of direct integrals of Hilbert bundles. Let $\mathcal{H}$ be a Hilbert bundle over X. Let $s \mapsto \mu_s$ be as before. Define $\mu * \mathcal{H}$ to be the set consisting of all pairs (s, f) where $f \in \int^{\oplus} \mathcal{H}_x d\mu_s(x)$, and let p be the mapping sending (s, f) to s. We note we may take the functions satisfying (1) and (2) for the Hilbert bundle $\mathcal{H}$ to be bounded; that is $\|f_n(x)\|_x \leq 1$ for all $x \in X$. Define functions $f_{m,n}$ on S by $(f_{m,n}(s))(x) = 1_{A_m}(x) f_n(x)$ and give $\mu * \mathcal{H}$ the smallest Borel structure for which the mappings $(s, f) \mapsto s$ and $(s, f) \mapsto \int \langle f(x), f_{m,n}(x) \rangle \, d\mu_s(x)$ are Borel for all m and n. Again a function $y \mapsto (s(y), f_y)$ is Borel iff the functions $y \mapsto s(y)$ and $y \mapsto \int \langle f_y(x), f_{m,n}(x) \rangle \, d\mu_{s(y)}(x)$ are all Borel, and $\mu * \mathcal{H}$ is a Hilbert bundle over S. We note that the σ-algebra of Borel sets on $\mu * \mathcal{H}$ is the smallest σ-algebra making the maps $(s, f) \mapsto s$ and $(s, f) \mapsto \int_{A_m} \langle f(x), f_n(x) \rangle \, d\mu_s(x)$ all Borel.

Let $\mathcal{H}$ and $\mathcal{K}$ be Hilbert bundles over S. A **Borel field** of bounded operators from $\mathcal{H}$ to $\mathcal{K}$ on S is a function $s \mapsto A(s)$ where each $A(s)$ is a bounded operator from $\mathcal{H}_s$ into $\mathcal{K}_s$ such that if f is a Borel section from S into $\mathcal{H}$, then $s \mapsto A(s)f(s)$ is a Borel section of S into $\mathcal{K}$. It is an easy matter to check that A is Borel whenever Af_n is Borel for the f_n satisfying condition (2).

Suppose μ is a σ-finite measure on S and A is a Borel field of bounded operators from $\mathcal{H}$ to $\mathcal{K}$ such that the essential supremum of the operator norms $\|A(s)\|_s$ with respect to μ is finite. Then the operator M_A defined by $(M_A f)(s) = A(s)f(s)$ is a bounded operator from the Hilbert space $\int^{\oplus} \mathcal{H}_s \, d\mu(s)$ into the Hilbert space $\int^{\oplus} \mathcal{K}_s \, d\mu(s)$ and is called the **direct integral** of A over S with respect to μ. It is also denoted by $\int^{\oplus} A(s) \, d\mu(s)$.

We remark that if $\mathcal{H}$ and $\mathcal{K}$ are trivial, then A is Borel if and only if $s \mapsto A(s)$ is weakly Borel or equivalently strongly Borel as operators from $\mathbb{C}_n$ into $\mathbb{C}_m$. Since all Hilbert bundles are isomorphic to a sum of bundles isomorphic to trivial bundles, one can always consider the trivial case when establishing results about direct integrals.

PROPOSITION III.12. *Direct integrals of bounded Borel fields of operators satisfy*

(1) $|| \int^{\oplus} A(s)\, d\mu(s)|| = ||M_A|| = ||A||_{\infty} = ess\ sup\, ||A(s)||$,

(2) $\left(\int^{\oplus} A(s)\, d\mu(s)\right)^{*} = \int^{\oplus} A(s)^{*}\, d\mu(s)$,

(3) $\int^{\oplus} (A(s) + B(s))\, d\mu(s) = \int^{\oplus} A(s)\, d\mu(s) + \int^{\oplus} B(s)\, d\mu(s)$,

(4) $\int^{\oplus} A(s)\, d\mu(s) \int^{\oplus} B(s)\, d\mu(s) = \int^{\oplus} A(s)B(s)\, d\mu(s)$,

(5) $\int^{\oplus} A(s)\, d\mu(s)$ *is the identity operator iff* $A(s) = I$ *for a.e. s,*

(6) $\int^{\oplus} U(s)\, d\mu(s)$ *is unitary iff* $U(s)$ *is unitary for a.e. s, and*

(7) $\int^{\oplus} Q(s)\, d\mu(s)$ *is an orthogonal projection iff* $Q(s)$ *is an orthogonal projection for a.e. s.*

PROOF. We may assume the Hilbert bundles are trivial. It is clear that $||M_A|| \leq ||A||_{\infty}$. Let $\epsilon > 0$. Then $E = \{s : ||A(s)|| > ||A||_{\infty} - \epsilon\}$ has μ positive measure. Moreover, since $s \mapsto A(s)$ weakly Borel implies $s \mapsto A(s)$ is strongly Borel, the set $\{(s, v) : ||A(s)v|| > (||A||_{\infty} - \epsilon)||v||\}$ is a Borel subset whose projection onto S is E. The von Neumann selection lemma (see Theorem I.13) implies there is a subset E_0 of E having positive measure and a strongly Borel map f on E_0 such that $||A(s)f(s)|| > (||A||_{\infty} - \epsilon)||f(s)||$ for $s \in E_0$. Define $f(s) = 0$ for $s \notin E_0$. Then $||M_A f|| > (||A||_{\infty} - \epsilon)||\,||f||$. Hence $||M_A|| > ||A||_{\infty} - \epsilon$. The other statements follow from similar arguments. $\square$

DEFINITION. Let $\mathcal{H}$ be a Hilbert bundle over S, and let μ be a σ-finite measure on S. The projection valued measure defined by $P(E) = 1_E(s)f(s)$ is the **canonical projection valued measure** for the direct integral of the Hilbert bundle $\mathcal{H}$ with respect to μ.

If the Hilbert bundle $\mathcal{H}$ is trivial, then the canonical projection valued measure is identified with the canonical projection valued measure on $L^2(S, \mu, \mathbb{C}^k)$.

THEOREM III.13. *Let P be a projection valued measure on the standard Borel space S which operates on the separable Hilbert space $\mathbb{H}$. Then there are a Hilbert Bundle $\mathcal{H}$ over S, a finite measure μ on S, and a unitary isomorphism of $\mathbb{H}$ onto $\int^{\oplus} \mathcal{H}_s\, d\mu(s)$ such that*

$E \mapsto UP_E U^{-1}$ *is the canonical projection valued on this direct integral of Hilbert spaces. Moreover, the measure class of μ is unique.*

PROOF. This follows easily from Theorem III.1 and Theorem III.2 and the fact that $\mu(E) = 0$ precisely when $P(E) = 0$. $\square$

THEOREM III.14. *Let P be the canonical projection valued measure on $\int^{\oplus} \mathcal{H}_s \, d\mu(s)$. Let $\mathcal{A}$ be the smallest von Neumann algebra containing all the operators $P(E)$, E a Borel subset of S. Then:*

  (1) $\mathcal{A} = \{\int^{\oplus} f(s) I_s \, d\mu(s) : f \in L^{\infty}(S)\}$ *and*
  (2) $\mathcal{A}' = \{\int^{\oplus} A(s) \, d\mu(s) : s \mapsto A(s)$ *is any essentially bounded Borel field of operators}.*

PROOF. Using Theorem III.12 we may assume the Hilbert bundle is trivial. Hence P is the projection valued measure defined on $L^2(S, \mu, \mathbb{H})$ defined by $P_E f = 1_E f$. We show (2) first. Since the operators $M_A = \int^{\oplus} A(s) \, d\mu(s)$ commute with the projections $P(E)$, they belong to $\mathcal{A}'$. Suppose $B \in \mathcal{A}'$. Let $w, v \in \mathbb{H}$. Define a complex measure $\mu_{w,v}$ on S by $\mu_{w,v}(E) = \langle P_E Bw, v \rangle$. Here w and v are the constant functions on S with value w and v, respectively. The measure $\mu_{w,v}$ is absolutely continuous relative to μ. Let $s \mapsto \langle Bw, v \rangle(s)$ be the Radon-Nikodym derivative. Since $|\mu_{w,v}(E)| \leq ||w|| ||v|| ||B|| \mu(E)$ for all E, it follows that

(1) $$|\langle Bw, v \rangle(s)| \leq ||B|| ||w|| ||v|| \text{ a.e. } s.$$

Let W be a countable dense subset of $\mathbb{H}$ which is a linear space over the ring consisting of the complex numbers with rational real and imaginary components. Since $\mu_{w,v}$ is conjugate linear in w and v, the a.e. uniqueness of Radon-Nikodym derivatives implies there exists a conull Borel subset S_0 of S such that $\langle Bw, v \rangle(s)$ is conjugate bilinear on W for each $s \in S_0$. By (1), we may also assume $|\langle Bw, v \rangle(s)| \leq ||B|| ||w|| ||v||$ for $w, v \in W$ and $s \in S_0$. Thus for $s \in S_0$, the function $(w, v) \mapsto \langle Bw, v \rangle(s)$ is uniformly bicontinuous on the dense subset $W \times W$ of $\mathbb{H} \times \mathbb{H}$. It follows that for s in S_0, one can extend this form to a conjugate linear form on $\mathbb{H}$ with norm no larger than $||B||$. Moreover, by the Lebesgue dominated convergence theorem, $\langle Bw, v \rangle(s)$ is the Radon-Nikodym derivative of $\mu_{w,v}$ for each w and v in $\mathbb{H}$. Define $\langle Bw, v \rangle(s) = 0$ for $s \notin S_0$. Hence for each s, there is a bounded linear operator $B(s)$ on $\mathbb{H}$ with $||B(s)|| \leq ||B||$ such that $\langle B(s)w, v \rangle = \langle Bw, v \rangle(s)$ for all w, v, and s. The mapping $s \mapsto B(s)$ is weakly Borel. Note $\langle (M_B(1_E w), 1_F v \rangle = \int_{E \cap F} \langle B(s)w, v \rangle \, d\mu(s) = (P(E \cap F)Bw, v \rangle =$

$\langle BP(E)w, P(F)w \rangle$. This implies $\langle M_B f, g \rangle = \langle Bf, g \rangle$ for all simple functions f, g in $L^2(S, \mu, \mathbb{H})$. Thus $M_B = B$.

To prove (1), note that $\mathcal{A}$ is abelian, and thus $\mathcal{A}' \supset \mathcal{A}$. Suppose $B \in \mathcal{A}$. Then $B \in \mathcal{A}'$; and hence by (2), $B = \int^{\oplus} B(s) \, d\mu(s)$. But since M_B is in $\mathcal{A}''$, (2) implies $\int^{\oplus} (AB(s) - B(s)A) \, d\mu(s) = 0$ for each bounded operator A on $\mathbb{H}$. By (1) of Proposition III.12, $AB(s) - B(s)A = 0$ for a.e. s. Let $\{A_n\}$ be a countable set which is strongly operator dense in the unit ball of bounded operators. It follows that there is a conull Borel subset S_0 such that $A_n B(s) = B(s) A_n$ for all n and all s in S_0. This implies that for each $s \in S_0$, $B(s)$ commutes with all bounded operators and thus has form $f(s)I$. Define $f(s) = 0$ for $s \notin S_0$. Then $B = \int^{\oplus} f(s) I \, d\mu(s)$. $\square$

PROPOSITION III.13. *Let $\mathcal{H}$ and $\mathcal{K}$ be Hilbert bundles over the standard Borel space S. Let μ be a σ-finite measure on S and suppose P and Q are the canonical projection valued measures on $\int^{\oplus} \mathcal{H}_s \, d\mu(s)$ and $\int^{\oplus} \mathcal{K}_s \, d\mu(s)$. Then if T is a unitary isomorphism of $\int^{\oplus} \mathcal{H}_s \, d\mu(s)$ onto $\int^{\oplus} \mathcal{K}_s \, d\mu(s)$ satisfying $TP(E)T^{-1} = Q(E)$ for all Borel sets E, then there is a bounded Borel field $s \mapsto T(s)$ of operators such that $T(s)$ is a unitary isomorphism of $\mathcal{H}_s$ onto $\mathcal{K}_s$ for μ a.e. s and such that $(Tf)(s) = T(s)f(s)$ a.e. s for each square integrable Borel section f in $\int^{\oplus} \mathcal{H}_s \, d\mu(s)$.*

PROOF. We may assume the bundles $\mathcal{H}$ and $\mathcal{K}$ are trivial. Thus $\mathcal{H} = S \times \mathbb{H}$ and $\mathcal{K} = S \times \mathbb{K}$ where $\mathbb{H}$ and $\mathbb{K}$ are Hilbert spaces. In this case T is then a unitary isomorphism of $L^2(S, \mathbb{H})$ onto $L^2(S, \mathbb{K})$ satisfying $TP(E)T^{-1} = Q(E)$ for all E. It follows by Corollary III.2 that $\dim \mathbb{H} = \dim \mathbb{K}$. We thus may assume using appropriate isomorphisms that $\mathbb{H} = \mathbb{K}$. The result now follows by using Theorem III.14 and applying Proposition III.12. $\square$

THEOREM III.15. *Let $\mathcal{H}$ be a Hilbert bundle over X, and let $\mathcal{K}$ be a Hilbert bundle over Y where X and Y are standard Borel spaces of the same cardinality. Suppose μ is a σ-finite measure on X and ν is a σ-finite measure on Y. Suppose T is a unitary isomorphism of $\int^{\oplus} \mathcal{H}_x \, d\mu(x)$ onto $\int^{\oplus} \mathcal{K}_y \, d\nu(y)$ which carries the range of the canonical projection valued measure on $\int^{\oplus} \mathcal{H}_x \, d\mu(x)$ onto the range of the canonical projection valued measure on $\int^{\oplus} \mathcal{K}_y \, d\nu(y)$. Then there exist a Borel isomorphism φ from X onto Y such that $\varphi_* \mu \sim \nu$ and a bounded Borel field of operators $T(x) : \mathcal{H}_x \to \mathcal{K}_{\varphi(x)}$ such that $T(x)$ is a unitary isomorphism μ a.e. x and such that*

$Tf(\varphi(x)) = \frac{d\mu \circ \varphi^{-1}}{d\nu}(\varphi(x))^{1/2}T(x)f(x)$ *a.e.* x *for each square inte-grable Borel section* f *for the bundle* $\mathcal{H}$.

PROOF. Let P be the canonical projection valued measure for $\int^{\oplus} \mathcal{H}_x \, d\mu(x)$ and Q to be the canonical projection valued measure for $\int^{\oplus} \mathcal{K}_y \, d\nu(y)$. Define Φ from the measure algebra defined by ν to the measure algebra defined by μ by $T^{-1}Q(E)T = P(\Phi(E))$. Then Φ is a σ-isomorphism between these two measure algebras. By Theorem I.26, there exists a Borel isomorphism φ from X onto Y with $\varphi_*\mu \sim \nu$ satisfying $\Phi(E) = \varphi^{-1}(E)$ for all E. Let $\varphi^*\mathcal{K}$ be the Hilbert bundle over X with sections $\varphi^*\mathcal{K}_x = \mathcal{K}_{\varphi(x)}$. Define a unitary isomorphism W of $\int^{\oplus} \mathcal{K}_y \, d\nu(y)$ onto $\int^{\oplus} \varphi^*\mathcal{K}_x \, d\mu(x)$ by

$$Wf(x) = \frac{d\nu \circ \varphi}{d\mu}(x)^{1/2} f(\varphi(x)).$$

This unitary transformation satisfies $WQ(E)W^{-1} = Q'(\varphi^{-1}(E))$ where Q' is the canonical projection valued measure on $\int^{\oplus} \varphi^*\mathcal{K}_x \, d\mu(x)$. Thus the unitary transformation WT satisfies

$$\begin{aligned}
WTP(E)T^{-1}W^{-1} &= WQ(\Phi^{-1}(E))W \\
&= Q'(\phi^{-1}\Phi^{-1}(E)) \\
&= Q'(\Phi \circ \Phi^{-1}E) \\
&= Q'(E)
\end{aligned}$$

for all Borel subsets E of X. Proposition III.13 implies there is a Borel field of operators $T(x)$ such that $T(x)$ is unitary from $\mathcal{H}_x$ onto $\varphi^*\mathcal{K}_x = \mathcal{K}_{\varphi(x)}$ a.e. x and $WTf(x) = T(x)f(x)$ for $f \in \int^{\oplus} \mathcal{H}_x \, d\mu(x)$. Since

$$\frac{d\mu \circ \varphi^{-1}}{d\nu}(\varphi(x))\frac{d\nu \circ \varphi}{d\mu}(x) = 1$$

for almost all x, the result follows. $\square$

10. Borel Fields of von Neumann Algebras

DEFINITION. Let $\mathcal{H}$ be a Hilbert bundle over S, and suppose $s \mapsto \mathcal{A}(s)$ is a map such that each $\mathcal{A}(s)$ is a von Neumann algebra on $\mathcal{H}_s$. This map is said to be **Borel** if there exists a countable family $\{A_n\}$ of Borel fields of bounded operators on $\mathcal{H}$ such that for each s, the smallest von Neumann algebra containing the collection $\{A_n(s)\}_{n=1}^{\infty}$ is $\mathcal{A}(s)$.

PROPOSITION III.14. *Suppose* $\mathcal{H}$ *is a trivial Hilbert bundle* $S \times \mathbb{H}$. *Then:*

(a) If $s \mapsto \mathcal{A}(s)$ is Borel, then the set $\{(s, A) : s \in S, A \in \mathcal{A}(s)\}$ is a Borel set in the product space $S \times \mathcal{B}(\mathbb{H})$ where $\mathcal{B}(\mathbb{H})$ is the space of bounded operators on $\mathbb{H}$ with the strong operator Borel structure.

(b) Conversely, if $\mathcal{A}$ is a Borel subset of $S \times \mathcal{B}(\mathbb{H})$ such that $\mathcal{A}_s = \{A : (s, A) \in \mathcal{A}\}$ is a von Neumann algebra for each s, then $s \mapsto \mathcal{A}_s$ is Borel on S.

PROOF. We prove (b). Assume $\mathcal{A}$ is a Borel set such that $\mathcal{A}_s$ is a von Neumann algebra for each s. By Theorem III.4, the unit ball of $\mathcal{B}(\mathbb{H})$ is a compact Polish space. Choose a compatible complete separable metric; and let E_k, $k = 1, 2, \ldots$ be the weak operator closures of sets forming a countable basis for the weak operator topology on the unit ball of $\mathcal{B}(\mathbb{H})$. Set $W_k = \mathcal{A} \cap (S \times E_k)$. Then W_k is a Borel set whose vertical sections over S are weakly closed and hence weakly compact subsets of the unit ball. Let p_S be the coordinate projection onto S. Then by Theorem I.22, the set $S_k = p_S(W_k)$ is Borel; and there is a Borel function $s \mapsto A_k(s)$ on S_k such that $(s, A_k(s)) \in W_k$ for all $s \in S_k$. Define $A_k(s) = 0$ for $s \notin S_k$. To see $s \mapsto \mathcal{A}_s$ is Borel on S, it suffices to show each $\mathcal{A}_s$ is generated by the operators $A_k(s)$. Let $A \in \mathcal{A}_s$ and suppose $||A|| \leq 1$. Choose a sequence $\{k_j\}_{j=1}^{\infty}$ such that the diameters of the sets E_{k_j} tend to 0 and $\cap_{k_j} E_{k_j} = \{A\}$. Then $A_{k_j}(s)$ converges to A in the weak operator topology. Corollary III.11 implies A is in the von Neumann algebra generated by the operators $\{A_k(s)\}_{k=1}^{\infty}$.

To prove (a), first note that if $s \mapsto A_k(s)$ are Borel and for each s generate a von Neumann algebra $\mathcal{A}(s)$, then $\mathcal{A}' = \{(s, A) : AA_k(s) = A_k(s)A \text{ for all } k\}$ is a Borel set whose S sections are von Neumann algebras. Thus by (b), there are Borel functions $s \mapsto A'_k(s)$, $k = 1, 2, \ldots$ such that the commutant $\mathcal{A}(s)'$ is generated by the operators $\{A'_k(s)\}_{k=1}^{\infty}$. Let $\mathcal{A} = \{(s, A) : AA'_k(s) = A'_k(s)A \text{ for all } k\}$. Then $\mathcal{A}$ is a Borel set; and by the double commutant theorem, $\mathcal{A}_s = \mathcal{A}(s)$ for s in S. $\square$

COROLLARY III.12. If $s \mapsto \mathcal{A}(s)$ is a von Neumann algebra valued Borel function over a Hilbert bundle $\mathcal{H}$, then the function $s \mapsto \mathcal{A}(s)'$ is also Borel on S.

PROOF. We may assume the Hilbert bundle is trivial. The result was established in the proof of Proposition III.14. $\square$

COROLLARY III.13. *If $s \mapsto \mathcal{A}(s)$ is a Borel von Neumann algebra valued function on the standard Borel space S, then the set $\{s : \mathcal{A}(s) = \mathbb{C} I\}$ is a Borel set.*

PROOF. We may assume the Hilbert bundle is trivial. Thus, it suffices to note that if $s \mapsto A(s)$ is a weakly Borel operator valued function into $B(\mathbb{H})$, then the set of s such that $A(s)$ is a scalar operator is a Borel set. To see this let $\{v_i\}_{i=1}^{\infty}$ be a countable dense subset of $\mathbb{H}$. Then the operator $A(s)$ is scalar if and only if $\langle A(s)v_i, v_j \rangle \langle v_j, v_i \rangle = \langle A(s)v_j, v_i \rangle \langle v_i, v_j \rangle$ for all i and j. $\square$

Let $\mathcal{A}$ be a von Neumann algebra. By Lemma III.7, the set of orthogonal projections in $\mathcal{A}$ has the same commutant as $\mathcal{A}$. In particular, by the double commutant theorem, $\mathcal{A}$ is the von Neumann algebra generated by its orthogonal projections.

Let $\mathcal{A}$ be a von Neumann algebra. The **center** of $\mathcal{A}$ is the von Neumann algebra $\mathcal{A} \cap \mathcal{A}'$. An orthogonal projection in the center is called a **central projection**. If the only central projections are 0 and I, then $\mathcal{A}$ is said to be a **factor**. Since a von Neumann algebra is generated by its orthogonal projections, this is equivalent to the center being the set consisting of scalar multiples of the identity operator.

COROLLARY III.14. *Suppose $s \mapsto \mathcal{A}(s)$ is Borel on S. Let $\mathcal{Z}(s)$ be the center of $\mathcal{A}(s)$. Then the mapping $s \mapsto \mathcal{Z}(s)$ is Borel on S.*

PROOF. The result follows from Proposition III.14 and the fact that if $\mathcal{A}$ and $\mathcal{B}$ are Borel subsets of $S \times \mathcal{B}(\mathbb{H})$ whose S sections are von Neumann algebras, then the S sections of the Borel set $\mathcal{A} \cap \mathcal{B}$ are von Neumann algebras. $\square$

THEOREM III.16. *Let $s \mapsto \mathcal{A}(s)$ be a Borel field of von Neumann algebras on the Hilbert bundle $\mathcal{H}$. If μ is a σ-finite measure on S, then the collection of the direct integral operators $M_A = \int^{\oplus} A(s)\,d\mu(s)$ where $A(s) \in \mathcal{A}(s)$ for all s is a von Neumann algebra. Its commutant is the von Neumann algebra $\int^{\oplus} \mathcal{A}(s)'\,d\mu(s)$.*

PROOF. Let $\mathcal{M}_A$ be the collection of all operators M_A where $s \mapsto A(s)$ is a bounded Borel field of operators with $A(s) \in \mathcal{A}(s)$ for a.e. s. This collection is a $*$-algebra for $M_A^* = M_{A^*}$, $M_A M_B = M_{AB}$, and $M_{cA+B} = cM_A + M_B$. Since this algebra contains the range of the canonical projection valued measure defined on $\int^{\oplus} \mathcal{H}_s\,d\mu(s)$, by

Theorem III.14 we see that if $B \in \mathcal{M}'_A$, then $B = \int^\oplus B(s) \, d\mu(s)$ for some essentially bounded Borel function $s \mapsto B(s)$.

Let A_n, $n = 1, 2, \ldots$, be operator valued Borel functions over the Hilbert bundle $\mathcal{H}$ with the property that the operators $A_n(s)$ generate $\mathcal{A}(s)$ for each s. Since $M_B \in \mathcal{M}'_A$, Proposition III.12 implies that $B(s)A_n(s) = A_n(s)B(s)$ a.e. s for each n. Therefore $B(s) \in \mathcal{A}(s)'$ a.e. s, and the commutant of M_A is $\int^\oplus \mathcal{A}(s)' \, d\mu(s)$. Applying this result to the function $s \mapsto \mathcal{A}(s)'$ shows the double commutant of M_A is $\int^\oplus \mathcal{A}(s) \, d\mu(s)$. $\square$

PROPOSITION III.15. *Let $\mathcal{A}$ be a von Neumann algebra on the direct integral $\int^\oplus \mathcal{H}_s \, d\mu(s)$ of the Hilbert bundle $\mathcal{H}$, and suppose the center of $\mathcal{A}$ contains the range of the canonical projection valued measure on $\int^\oplus \mathcal{H}_s \, d\mu(s)$. Then $\mathcal{A} = \int^\oplus \mathcal{A}(s) \, d\mu(s)$ for some Borel von Neumann algebra valued function $s \mapsto \mathcal{A}(s)$ on S.*

PROOF. Let $\{A_n\}_{n=1}^\infty$ be a sequence of operators in $\mathcal{A}$ which is strongly dense in the unit ball of $\mathcal{A}$. Then each A_n is in the commutant of the von Neumann algebra generated by the canonical projection valued measure. By Theorem III.14, each $A_n = \int^\oplus A_n(s) \, d\mu(s)$ for some Borel function $s \mapsto A_n(s)$. Let $\mathcal{A}(s)$ be the von Neumann algebra generated by the operators $A_n(s), n = 1, 2, \ldots$. Clearly $\int^\oplus \mathcal{A}(s) \, d\mu(s) \supset \mathcal{A}$.

Next, suppose $B \in \mathcal{A}'$. Then since B commutes with the range of the canonical projection valued measure, one has $B = \int^\oplus B(s) \, d\mu(s)$. Moreover, B commuting with $\mathcal{A}$ implies that $B(s)A_n(s) = A_n(s)B(s)$ a.e. s for each n. Therefore $B(s)$ commutes with every member of $\mathcal{A}(s)$ a.e. s. Thus $B \in \left(\int^\oplus \mathcal{A}(s) \, d\mu(s) \right)'$ and $\int^\oplus \mathcal{A}(s) \, d\mu(s) \subset \mathcal{A}''$. The result follows by the double commutant theorem. $\square$

Let $\mathcal{H}$ be a Hilbert bundle over the standard Borel space S. Suppose S^* is a standard Borel space and q is a Borel mapping from S^* into S. Define $q^*\mathcal{H}$ to be the Borel subset of $S^* \times \mathcal{H}$ consisting of all pairs (s^*, v) where $v \in \mathcal{H}_{q(s^*)}$. Then this forms a Hilbert bundle over the space S^*.

PROPOSITION III.16. *Let $\mathcal{H}$ be a Hilbert bundle over S. Suppose $s \mapsto \mathcal{A}(s)$ is a Borel field of von Neumann algebras for $\mathcal{H}$. Let μ be a σ-finite measure on S, and let λ be a σ-finite measure on a standard Borel space X. Suppose $x \mapsto A(x)$ is a bounded weakly Borel function from X into the von Neumann algebra $\int^\oplus \mathcal{A}(s) \, d\mu(s)$. Let q*

be the projection mapping from $X \times S$ onto S. Then there is a Borel field $(x, s) \mapsto A(x, s)$ relative to the Hilbert bundle $q^*\mathcal{H}$ such that $A(x, s) \in \mathcal{A}(s)$ for all (x, s) and such that $A(x) = \int^{\oplus} A(x, s) \, d\mu(s)$ for λ a.e. x.

PROOF. We may assume $\mathcal{H}$ is trivial. Thus $\mathcal{H}_s = \mathbb{H}$ for all s, and $\int^{\oplus} \mathcal{H}_s \, d\mu(s) = L^2(S, \mathbb{H})$. Denote this Hilbert space by $\mathbb{H}^*$. Recall by Proposition III.14, there is a Borel set $\mathcal{A}$ in $S \times \mathbb{H}$ such that $\mathcal{A}(s) = \{T : (s, T) \in \mathcal{A}\}$ for each s in S.

Consider the trivial Hilbert bundle $\mathcal{H}^*$ over X defined by $\mathcal{H}_x^* = \mathbb{H}^*$. Then $x \mapsto A(x)$ is a Borel field for this Hilbert bundle, and $A = \int^{\oplus} A(x) \, d\lambda(x)$ is a bounded operator on $\int^{\oplus} \mathcal{H}_x^* \, d\lambda(x) = L^2(X, \mathbb{H}^*)$ commuting with the natural projection valued measure on $L^2(X, \mathbb{H}^*)$. But the mapping Φ defined on the simple functions f in $L^2(X, \mathbb{H}^*)$ by $\Phi(f)(x, s) = f(x)(s)$ is an isometry of a dense subspace of $L^2(X, \mathbb{H}^*)$ onto a dense subspace of $L^2(X \times S, \lambda \times \mu, \mathbb{H})$ which extends to an onto isometry satisfying $\Phi(f)(x, s) = f(x)(s)$ for μ a.e. s for λ a.e. x. Under this isomorphism, the operator A is carried to an operator that commutes with the natural projection valued measure on $L^2(X \times S, \mathbb{H})$. Indeed, let E and F be Borel subsets of X and S. Then, since $A(x)$ is in $\int^{\oplus} \mathcal{A}(s) \, d\mu(s)$, $A(x)$ commutes with the natural projection valued measure on $L^2(S, \mathbb{H})$, and hence

$$
\begin{aligned}
1_{E \times F}(x, s)\Phi(Af)(x, s) &= 1_E(x)1_F(s)(A(x)f(x))(s) \\
&= 1_E(x)(A(x)(1_F f(x)))(s) \\
&= A(x)(1_E(x)(1_F f(x)))(s) \text{ a.e. } s \text{ for a.e. } x.
\end{aligned}
$$

Thus $1_{E \times F}\Phi(Af) = \Phi A \Phi^{-1} 1_{E \times F} \Phi(f)$ for all $f \in L^2(X, \mathbb{H}^*)$. By Theorem III.14, there exists a Borel field $(x, s) \mapsto A(x, s)$ such that $\Phi A \Phi^{-1} f(x, s) = A(x, s)f(s, x)$. In particular, for $f \in L^2(S, \mathbb{H})$, $(A(x)f)(s) = A(x, s)f(s)$ a.e. x. Moreover, since $A(x)$ belongs to $\int^{\oplus} \mathcal{A}(s) \, d\mu(s)$, one has $A(x, s) \in \mathcal{A}(s)$ a.e. s for a.e. x. Redefine $A(x, s)$ to be 0 if $(s, A(x, s))$ is not in $\mathcal{A}$. $\square$

THEOREM III.17. *Suppose $\mathcal{H}$ is a Hilbert bundle over S and μ is a σ-finite measure on S. Furthermore, assume $s \mapsto A(s)$ is a Borel field of von Neumann algebras over S. If π is a representation of a separable C^* algebra $\mathcal{A}$ such that $\pi(a) \in \int^{\oplus} A(s) \, d\mu(s)$ for each $a \in \mathcal{A}$, then for each s there exists a representation π_s of $\mathcal{A}$ on $\mathcal{H}_s$ such that for each $a \in \mathcal{A}$, $s \mapsto \pi_s(a)$ is a Borel field over S and $\pi(a) = \int^{\oplus} \pi_s(a) \, d\mu(s)$. Moreover, the representations π_s are unique μ a.e. s.*

PROOF. We may assume $\mathcal{H}$ is trivial. Let $\mathbb{Q}[i]$ be the ring of complex numbers having rational real and imaginary parts. Since $\mathcal{A}$ is separable, there exists a countable norm dense $\mathbb{Q}[i] * $ subalgebra $\mathcal{B}$ of $\mathcal{A}$. By Corollary II.6, $||\pi(a)|| \leq ||a||$ for each $a \in \mathcal{A}$. Proposition III.12 then implies for each $b \in \mathcal{B}$ there is a Borel field $s \mapsto \pi_s(b)$ of bounded operators satisfying $||\pi_s(b)|| \leq ||b||$ for all s and $\pi(b) = \int^{\oplus} \pi_s(b) \, d\mu(s)$. Since $\pi(b^*) = \int^{\oplus} \pi_s(b)^* \, d\mu(s)$, $\pi(b_1 b_2) = \int^{\oplus} \pi_s(b_1) \pi_s(b_2) \, d\mu(s)$, and $\pi(r_1 b_1 + r_2 b_2) = \int^{\oplus} r_1 \pi_s(b_1) + r_2 \pi_s(b_2) \, d\mu(s)$ for b, b_1, and b_2 in $\mathcal{B}$ and r_1, r_2 in $\mathbb{Q}[i]$, by Proposition III.12 one sees there exists a conull Borel subset S_0 of S such that if $s \in S_0$,

$$\pi_s(r_1 b_1 + r_2 b_2) = r_1 \pi_s(b_1) + r_2 \pi_s(b_2)$$
$$\pi_s(b_1 b_2) = \pi_s(b_1) \pi_s(b_2) \text{ and}$$
$$\pi_s(b_1^*) = \pi_s(b_1)^* \text{ for } r_1, r_2 \in \mathbb{Q}[i] \text{ and } b_1, b_2 \in \mathcal{B}.$$

Since $\mathcal{B}$ is norm dense in $\mathcal{A}$ and these algebra operations are all norm continuous, there is for each s in S_0 a representation π_s of $\mathcal{A}$ extending the representation π_s on $\mathcal{B}$. Define $\pi_s(a) = 0$ for $s \notin S_0$. Since $\pi_s(a) = \lim_n \pi_s(b_n)$ where b_n is a sequence converging to a, one has $s \mapsto \pi_s(a)$ is strongly Borel for each $a \in \mathcal{A}$, and thus $\pi(a) = \int^{\oplus} \pi_s(a) \, d\mu(s)$ for all $a \in \mathcal{A}$.

The a.e. uniqueness of the representations π_s follows from the a.e. uniqueness of the decomposition $\pi(b) = \int^{\oplus} \pi_s(b) \, d\mu(s)$ for $b \in \mathcal{B}$ and the fact that two representations which are equal on a countable dense subset of $\mathcal{A}$ are equal. $\square$

11. Similar Projections in a von Neumann Algebra

Let $\mathcal{A}$ be a von Neumann algebra on a Hilbert space $\mathbb{H}$. Define an equivalence relation $\sim$ on the orthogonal projections in $\mathcal{A}$ by $P \sim Q$ if and only if there is a partial isometry $U \in \mathcal{A}$ from $P(\mathbb{H})$ onto $Q(\mathbb{H})$. Thus U is an operator in $\mathcal{A}$ satisfying $U^* U = P$ and $UU^* = Q$. The mapping $PAP \mapsto UAU^*$ can be shown to be an isomorphism between the algebras PAP and QAQ. More specifically, $\mathcal{A}_P = PAP$ is a von Neumann algebra on the Hilbert space $P(\mathbb{H})$ and $U|_{P\mathbb{H}}$ is a unitary isometry from $P(\mathbb{H})$ onto $Q(\mathbb{H})$ with $U \mathcal{A}_P U^* = \mathcal{A}_Q$. Note that if P and Q are central, then $P = U^* U U^* U = U^* Q U = Q U^* U = QP = UU^* P = UPU^* = UU^* UU^* = Q$. In particular, similar central projections are equal.

The **central cover** C_P of an orthogonal projection $P \in \mathcal{A}$ is the smallest central projection $Q \in \mathcal{A}$ satisfying $QP = P$.

Let S be a subset of a Hilbert space $\mathbb{H}$. In this section, $\langle S \rangle$ will denote the closure of the linear span of S. A vector v is a **cyclic vector** for a von Neumann algebra $\mathcal{A}$ if $\langle \mathcal{A}v \rangle = \mathbb{H}$.

LEMMA III.8. *v is a cyclic vector for $\mathcal{A}$ iff v is a separating vector for $\mathcal{A}'$.*

PROOF. If v is a cyclic vector for $\mathcal{A}$ and if B is an operator in $\mathcal{A}'$ satisfying $Bv = 0$, then $B\mathcal{A}v = 0$ and thus $B = 0$. Conversely, if v is a separating vector for $\mathcal{A}'$, then the orthogonal projection P onto $\langle \mathcal{A}v \rangle$ is in $\mathcal{A}'$. Moreover, $(I - P)v = 0$. Hence $P = I$. $\square$

PROPOSITION III.17. *Let A be in a von Neumann algebra $\mathcal{A}$. Then the orthogonal projection onto $\langle A\mathbb{H} \rangle$ is similar to the orthogonal projection onto $(\ker A)^{\perp}$.*

PROOF. By the spectral theorem, there is a unique positive operator B in the von Neumann algebra generated by A^*A satisfying $B^2 = A^*A$. Define U on the range of B by $UBv = Av$. Then $\langle UBv, UBv \rangle = \langle Av, Av \rangle = \langle A^*Av, v \rangle = \langle B^2v, v \rangle = \langle Bv, Bv \rangle$. Thus U is well defined and is an isometry. Extend U to $\langle B\mathbb{H} \rangle$ and set $U = 0$ on the orthogonal complement of the range of B. Then U is a partial isometry of $\langle B\mathbb{H} \rangle$ onto $\langle A\mathbb{H} \rangle$ whose kernel is the kernel of B. Furthermore the kernel of A is the kernel of B. But $\ker B^{\perp} = \langle B\mathbb{H} \rangle$. Thus the proof is complete once we show U is in $\mathcal{A}$.

Suppose C commutes with A and A^*. Then C commutes with B. Hence $UCBv = UBCv = ACv = CAv = CUBv$. Moreover, if v is in the kernel of B, Cv is in the kernel of B, and hence $UCv = 0 = CUv$. Since C and U commute on the range and kernel of B, we have $UC = CU$. By the double commutant theorem, U is in the von Neumann algebra generated by A and A^*. Therefore $U \in \mathcal{A}$. $\square$

COROLLARY III.15. *Orthogonal projections P and Q in a von Neumann algebra $\mathcal{A}$ have equivalent subprojections if and only if $PAQ \neq 0$.*

PROOF. Suppose U is a partial isometry with $U^*U \leq Q$ and $UU^* \leq P$. Then $U = PUQ \in P\mathcal{A}Q$.

Conversely if $PAQ \neq 0$, then P and Q have equivalent subprojections by Proposition III.17. $\square$

LEMMA III.9. *Suppose Q is an orthogonal projection in $\mathcal{A}$. Then $C_Q\mathbb{H} = \langle \mathcal{A}Q(\mathbb{H}) \rangle$.*

PROOF. Consider $K = \langle \mathcal{A}\mathcal{A}'Q\mathbb{H} \rangle$. Since K is invariant under both $\mathcal{A}$ and $\mathcal{A}'$, the orthogonal projection P onto K is in both $\mathcal{A}'$ and $\mathcal{A}''$ and hence is central. Since $PQ = Q$ and $C_Q\mathcal{A}\mathcal{A}'Qy = \mathcal{A}\mathcal{A}'C_QQy = \mathcal{A}\mathcal{A}'Qy$, it follows that $P = C_Q$. But since $I \in \mathcal{A}'$, K equals $\langle \mathcal{A}Q\mathcal{A}'\mathbb{H} \rangle = \langle \mathcal{A}Q\mathbb{H} \rangle$. $\square$

COROLLARY III.16. *If P and Q are orthogonal projections in $\mathcal{A}$, then $P \sim Q$ implies $C_P = C_Q$.*

PROOF. Suppose U is a partial isometry in $\mathcal{A}$ and $U^*U = P$ and $UU^* = Q$. Then $\mathcal{A}P\mathbb{H} = \mathcal{A}U^*U\mathbb{H} = \mathcal{A}U^*Q\mathbb{H} \subset \mathcal{A}Q\mathbb{H}$. Similarly $\mathcal{A}Q\mathbb{H} \subset \mathcal{A}P\mathbb{H}$. $\square$

COROLLARY III.17. *P and Q have nonzero equivalent subprojections if and only if $C_PC_Q \neq 0$.*

PROOF. If P and Q have equivalent nonzero subprojections, then by Corollary III.16, one has $C_PC_Q \neq 0$.

Suppose P and Q have no nonzero equivalent subprojections. It follows by Corollary III.15 that $PAQ = 0$. Lemma III.9 then implies $PC_Q = 0$. Therefore $C_PC_Q = 0$. $\square$

Suppose P and Q are orthogonal projections in a von Neumann algebra $\mathcal{A}$. Define $P \prec Q$ if $P \sim Q_1 \leq Q$ for some projection Q_1 in $\mathcal{A}$.

LEMMA III.10. *Suppose $P_2 \leq P_1 \leq P_0$ and $P_2 \sim P_0$. Then $P_1 \sim P_0$.*

PROOF. Let U be a partial isometry satisfying $U^*U = P_0$ and $UU^* = P_2$. Let $V = UP_1$. For $n = 1, 2, 3, \ldots$, set $P_{2n} = U^nU^{*n}$ and $P_{2n+1} = V^nV^{*n}$. Since $P_0U = U$ and $VP_1 = V$, these are all orthogonal projections. Furthermore,

$$\begin{aligned}
P_0 \geq P_1 &\geq P_2 \\
&\geq UP_1U^* = P_3 \\
&\geq UP_2U^* = P_4 \\
&\geq UP_3U^* = P_5 \\
&\geq \cdots .
\end{aligned}$$

Moreover,

$$(UP_{2n})^*(UP_{2n}) = P_{2n},$$
$$(UP_{2n})(UP_{2n})^* = P_{2n+2},$$
$$(UP_{2n+1})^*(UP_{2n+1}) = P_{2n+1}, \text{ and}$$
$$(UP_{2n+1})(UP_{2n+1})^* = P_{2n+3}.$$

Thus for $n = 0, 1, 2, \ldots$, we see $P_{2n} \sim P_{2n+2}$ and $P_{2n+1} \sim P_{2n+3}$. Set $P_\infty = \wedge P_n$. Then $P_0 = P_\infty + \sum_{n=0}^\infty ((P_{2n} - P_{2n+1}) + (P_{2n+1} - P_{2n+2})) \sim P_\infty + \sum_{n=0}^\infty ((P_{2n+2} - P_{2n+3}) + (P_{2n+1} - P_{2n+2})) = P_1.$ $\square$

THEOREM III.18 (SCHROEDER–BERNSTEIN). *If $P \prec Q$ and $Q \prec P$, then $P \sim Q$.*

PROOF. Suppose $P \sim P' \leq Q$ and $Q \sim Q' \leq P$. Choose a partial isometry U with $U^*U = Q$ and $UU^* = Q'$. Set $V = UP'$ and $P'' = VV^*$. Then $V^*V = P'QP' = P'$ and $P'' = VV^* \leq Q'$. Hence $P'' \leq Q' \leq P$ and $P'' \sim P' \sim P$. Lemma III.10 implies $Q' \sim P$. Hence $Q \sim P$. $\square$

THEOREM III.19 (COMPARISON THEOREM). *Let P and Q be orthogonal projections in a von Neumann algebra. Then there exists a central projection C such that $CP \prec CQ$ and $(I - C)Q \prec (I - C)P$.*

PROOF. Choose a maximal family of pairs (P_i, Q_i) of projections satisfying $P_i \leq P$, $Q_i \leq Q$, $P_i \sim Q_i$, and both families $\{P_i\}$ and $\{Q_i\}$ are orthogonal. Set $P_0 = \vee P_i$ and $Q_0 = \vee Q_i$. Then $P_0 \sim Q_0$. Set $\hat{P} = P - P_0$ and $\hat{Q} = Q - Q_0$. Then $\hat{P}$ and $\hat{Q}$ have no equivalent nonzero subprojections. By Corollary III.17, $C_{\hat{P}}C_{\hat{Q}} = 0$. Take $C = C_{\hat{Q}}$. Then $CP_0 \sim CQ_0$ and $(I - C)P_0 \sim (I - C)Q_0$. Hence $CP = CP_0 + C\hat{P} = CP_0 + C_{\hat{Q}}C_{\hat{P}}\hat{P} = CP_0 \sim CQ_0 \leq CQ$, and $(I - C)Q = (I - C)Q_0 + (I - C)\hat{Q} = (I - C)Q_0 + (I - C_{\hat{Q}})\hat{Q} = (I - C)Q_0 \sim (I - C)P_0 \leq (I - C)P.$ $\square$

12. Abelian Subalgebras and Type I von Neumann Algebras

An orthogonal projection P in a von Neumann algebra $\mathcal{A}$ is said to be **abelian** if the von Neumann algebra $\mathcal{A}_P = P\mathcal{A}P$ is abelian.

DEFINITION. A von Neumann algebra is **type I** if every nonzero central projection contains a nonzero abelian projection.

PROPOSITION III.18. *A von Neumann algebra is type I if and only if there is an abelian projection whose central cover is I.*

PROOF. Suppose P is abelian and $C_P = I$. Let Q be a nonzero central projection. Since $Q \leq C_P$, $QP \neq 0$; and thus $\mathcal{A}_{QP}$ is abelian.

Conversely, suppose $\mathcal{A}$ is type I. Choose a maximal family $\{P_i\}$ of nonzero abelian projections whose central covers C_{P_i} are orthogonal. Set $P = \sum P_i$. Then by maximality, $C_P = \sum C_{P_i}$ is the identity. Moreover, the mapping $PAP \mapsto (P_i A P_i)$ is an isomorphism of $\mathcal{A}_P$ into the direct product of the abelian algebras $\mathcal{A}_{P_i}$. Therefore P is abelian. $\square$

COROLLARY III.18. *Let Q be an orthogonal projection in a type I von Neumann algebra $\mathcal{A}$. Then there is an abelian projection $P \in Q\mathcal{A}Q$ whose central cover is C_Q.*

PROOF. Apply Proposition III.18 to the type one von Neumann algebra $C_Q \mathcal{A} C_Q$. $\square$

PROPOSITION III.19. *An abelian von Neumann algebra possessing a cyclic vector is maximal abelian.*

PROOF. Let v be a cyclic vector. Note that an operator A in $\mathcal{A}$ is 0 iff $A A v = \mathcal{A} A v$ is 0. Hence $A = 0$ iff $A v = 0$, and thus v is a separating vector for $\mathcal{A}$. By Lemma III.8 and the double commutant theorem, v is also a cyclic vector for $\mathcal{A}'$.

Define a mapping J on $\mathcal{A}v$ by $JAv = A^* v$ for $A \in \mathcal{A}$. Note if A and B are in $\mathcal{A}$, then $\langle JAv, JBv \rangle = \langle A^* v, B^* v \rangle = \langle Bv, Av \rangle$, for A and B^* commute. Therefore J is well defined and extends to a conjugate linear isometry of $\mathbb{H}$ satisfying $J^2 = I$. In particular $\langle Ja, Jb \rangle = \langle b, a \rangle$ for all a and b in $\mathbb{H}$.

Suppose $B \in \mathcal{A}'$. Then if $A \in \mathcal{A}$, one has $\langle B^* v, Av \rangle = \langle v, BAv \rangle = \langle v, ABv \rangle = \langle A^* v, Bv \rangle = \langle JAv, Bv \rangle = \langle JBv, Av \rangle$. Since v is cyclic for $\mathcal{A}$, one has $JBv = B^* v$ for all B in $\mathcal{A}'$.

Now suppose $A, B, C \in \mathcal{A}'$. Then

$$JAJB(Cv) = JAC^* B^* v = BCA^* v = BJAC^* v = BJAJ(Cv).$$

Since $\mathcal{A}'v$ is dense in $\mathbb{H}$, JAJ is in the commutant of $\mathcal{A}'$ and thus is in $\mathcal{A}$ by the double commutant theorem. In particular, the algebra $J\mathcal{A}'J$ is abelian. This implies $\mathcal{A}'$ is abelian. Therefore $\mathcal{A}'' \supset \mathcal{A}' \supset \mathcal{A}$ and hence $\mathcal{A} = \mathcal{A}'$. $\square$

THEOREM III.20. *If $\mathcal{A}$ is a type I von Neumann algebra, then $\mathcal{A}'$ is type I.*

PROOF. Assume first that $\mathcal{A}$ is abelian. Let Q be a central projection in $\mathcal{A}'$ and take v to be a nonzero vector in the range of Q. Let P be the projection onto the space $\langle \mathcal{A}v \rangle$. Then $P \in \mathcal{A}'$ and $P \leq Q$. Moreover, the abelian von Neumann algebra $\mathcal{A}_P$ has a cyclic vector and is therefore maximal abelian. Thus $(\mathcal{A}_P)' = \mathcal{A}_P$. But an easy argument shows that for $P \in \mathcal{A}'$, the commutant of $\mathcal{A}_P$ is $P\mathcal{A}'P$. Hence P is abelian.

If $\mathcal{A}$ is not abelian, let P be an abelian projection in $\mathcal{A}$ whose central cover is I. Then the mapping $B \mapsto BP$ is a *-algebra isomorphism of $\mathcal{A}'$ onto $P\mathcal{A}'P$. In fact if $BP = 0$, then since $BAP = ABP = 0$ for all A in $\mathcal{A}$, it follows by Lemma III.9 that $B = BC_P = 0$. Thus $\mathcal{A}'$ is isomorphic to the commutant of the abelian von Neumann algebra $\mathcal{A}_P$. $\square$

LEMMA III.11. *Two abelian projections are equivalent if and only if their central covers are equal.*

PROOF. By Corollary III.16, we know equivalent projections have equal central covers.

For the other direction, let P and Q be abelian projections with $C_P = C_Q$. We show $P \prec Q$. By the comparison theorem, there exists a central projection C with $CP \prec CQ$ and $(I - C)Q \prec (I - C)P$. Thus to show $P \prec Q$, we may assume $Q \prec P$. Moreover, since equivalent projections have the same central covers, we may even assume the stronger condition that $Q \leq P$. Thus Q is in the abelian von Neumann algebra $\mathcal{A}_P$. Hence if $A \in \mathcal{A}$, then $QA(P - Q) = QPAP(P - Q) = PAPQ(P - Q) = 0$. By Corollary III.15 and Corollary III.17, $C_Q C_{P-Q} = 0$. Thus $C_Q(P - Q) = 0$, and hence $Q = C_Q P$. Since $C_P = C_Q$, $Q = C_P P = P$. Hence $C_P = C_Q$ implies $P \prec Q$. By symmetry, one also obtains $Q \prec P$. Thus $P \sim Q$ by the Schroeder–Bernstein Theorem. $\square$

COROLLARY III.19. *If Q is a central projection in $\mathcal{A}_P$, then $Q = C_Q P$.*

COROLLARY III.20. *Let $\mathcal{A}$ be a von Neumann algebra. Suppose Q is a nonzero abelian projection in $\mathcal{A}$. Then $C_Q \mathcal{A}$ is a type I von Neumann algebra on $C_Q \mathbb{H}$.*

PROOF. Let P be a nonzero projection in $C_Q \mathcal{A}$. We have $P \leq C_Q$. Hence $C_P C_Q = C_P$. By Corollary III.17, Q and P have equivalent nonzero subprojections. Since Q is abelian, P contains a nonzero abelian projection. Hence $C_Q \mathcal{A}$ is type I. $\square$

PROPOSITION III.20. *Let $\mathcal{A}$ be a von Neumann algebra. Then there is a largest central projection C such that $C\mathcal{A}$ is type I. Moreover, $C\mathcal{A}'$ is type I; and if Q is any nonzero central projection with $Q \leq I - C$, then $Q\mathcal{A}$ and $Q\mathcal{A}'$ are nontype I.*

PROOF. By the Hausdorff maximality principle, there is a maximal central projection C with $C\mathcal{A}$ type I. Since $C\mathcal{A}' = (C\mathcal{A})'$ on $C\mathbb{H}$, Theorem III.20 implies $C\mathcal{A}'$ is type I. Now if $Q \neq 0$ is central and $Q \leq I - C$, then $Q\mathcal{A}$ is nontype I, for otherwise $C + Q$ would be a larger central projection with $(C + Q)\mathcal{A}$ type I. Similarly, by the double commutant theorem, $Q\mathcal{A}'$ is nontype I. $\square$

13. Homogeneous Projections of Degree n

DEFINITION. A type I von Neumann algebra $\mathcal{A}$ is **homogeneous** of degree n if there exists an orthogonal collection $\{P_i\}$ of equivalent abelian projections whose cardinality is n and whose sum is the identity. A central projection P in $\mathcal{A}$ is homogeneous if the von Neumann algebra $\mathcal{A}_P$ is homogeneous.

If the Hilbert space of the von Neumann algebra is separable, then the only possible infinite degree is countable and this is denoted by $n = \infty$.

Let n be a cardinal number and $\mathcal{A}$ be an abelian von Neumann algebra on a Hilbert space $\mathbb{H}$. Let $\mathbb{H}_n$ be a Hilbert space of dimension n. The smallest von Neumann algebra on $\mathbb{H} \otimes \mathbb{H}_n$ containing the operators $A \otimes B$ for $A \in \mathcal{A}$ and $B \in \mathcal{B}(\mathbb{H}_n)$ is denoted by $\mathcal{A} \otimes \mathcal{B}(\mathbb{H}_n)$. The Hilbert space $\mathbb{H} \otimes \mathbb{H}_n$ is isomorphic to $n\mathbb{H}$, the Hilbert space sum of n copies of $\mathbb{H}$. Using this isomorphism, $\mathcal{A} \otimes \mathcal{B}(\mathbb{H}_n)$ is isomorphic to $\mathcal{M}_n(\mathcal{A})$, which is the collection of all bounded operators whose matrices relative to the natural direct sum decomposition of $n\mathbb{H}$ have entries in $\mathcal{A}$. It is easy to check that the commutant of $\mathcal{A} \otimes \mathcal{B}(\mathbb{H}_n)$ is $\mathcal{A}' \otimes \mathbb{C}I_n$ where $\mathbb{C}I_n$ is the algebra of scalar operators on $\mathbb{H}_n$. Using the identification with $n\mathbb{H}$, the commutant is $\mathcal{D}_n(\mathcal{A}')$, the collection of all diagonal operators $\{nA' : A' \in \mathcal{A}'\}$ where nA' acts on $n\mathbb{H}$ by $(nA')(v_j) = (A'v_j)$. This algebra will also be denoted by $n\mathcal{A}'$. In particular, $n\mathcal{B}(\mathbb{H})$ is the von Neumann algebra of all diagonal operators nA where A is a bounded operator on $\mathbb{H}$.

THEOREM III.21. *If $\mathcal{A}$ is a homogeneous von Neumann algebra of degree n on a Hilbert space $\mathbb{H}$, then there exist a Hilbert space $\mathbb{H}_0$ and a unitary isomorphism U of $\mathbb{H}$ onto $n\mathbb{H}_0$ such that $U\mathcal{A}U^{-1} = \mathcal{M}_n(\mathcal{C})$*

where C is an abelian von Neumann algebra acting on $\mathbb{H}_0$ isomorphic to the center of $\mathcal{A}$.

PROOF. Let $\{P_i\}$ be a collection of n pairwise orthogonal abelian projections satisfying $P_i \sim P_j$ and $\sum P_i = I$. Let P be a fixed element belonging to this collection, and set $\mathbb{H}_0 = P(\mathbb{H})$. Choose $U_i \in \mathcal{A}$ satisfying $U_i^* U_i = P$ and $U_i U_i^* = P_i$.

Define a unitary isomorphism of $\mathbb{H}$ onto $n\mathbb{H}_0$ by

$$W(v) = W(\sum P_i v) = (U_i^* P_i v).$$

Then if $A \in \mathcal{A}$,

$$W(U_i A U_j^*)W^{-1}(v_k) = (w_l)$$

where $w_l = U_l^* U_i A U_j^* (\sum_k U_k v_k) = \delta_{l,i}\delta_{j,k} P A P v_k$. From this, one sees $W(U_i A U_j^*)W^{-1}$ is the operator $M_{i,j}(PAP)$ on $n\mathbb{H}_0$ whose matrix relative to the Hilbert sum decomposition $n\mathbb{H}_0$ has nonzero entry PAP in the i,j position.

Next note that $A = \sum_{i,j} P_i A P_j$ in the strong operator topology. But $P_i A P_j = U_i U_i^* A U_j U_j^* = U_i P U_i^* A U_j P U_j^*$. This implies WAW^{-1} is in the algebra $\mathcal{M}_n(\mathcal{A}_P)$. It follows that $WAW^{-1} = \mathcal{M}_n(\mathcal{A}_P)$.

To complete the proof, note that the mapping $Z \mapsto ZP$ is a *-algebra homomorphism of the center of $\mathcal{A}$ into the center of $\mathcal{A}_P$. By Corollary III.19, its range, which is a von Neumann algebra, contains every orthogonal projection in $\mathcal{A}_P$. Therefore the mapping is onto. Moreover, if $ZP = 0$, then $ZAP = 0$ which by Lemma III.9 implies $ZC_P = 0$. By Corollary III.16, $C_{P_i} = C_P$ for every i. Since $\sum P_i = I$, $C_P = I$. Therefore $Z = 0$, and the center of $\mathcal{A}$ is *-isomorphic to $\mathcal{A}_P$. □

COROLLARY III.21. *The degree of a homogeneous central projection is unique.*

PROPOSITION III.21.

(a) *Any sum of orthogonal homogeneous central projections of degree n is homogeneous and has degree n.*

(b) *Any nonzero central subprojection of a homogeneous projection of degree n is homogeneous and has degree n.*

PROOF. (a) Let P_i be orthogonal, homogeneous projections of degree n. For each i, let $\{Q_{i,j}\}$ be an orthogonal family of n equivalent abelian projections whose sum is P_i. Set $Q_j = \sum_i Q_{i,j}$. Then $C_{Q_j} = \sum_i C_{Q_{i,j}} = \sum P_i$. Thus (a) will follow by Lemma III.11 if each $\mathcal{A}_{Q_j}$ is abelian. But by the proof of Proposition III.18, the mapping

$Q_j A Q_j \mapsto (Q_{i,j} A Q_{i,j})$ is an algebra isomorphism into a direct product of abelian algebras.

(b) Let P be a homogeneous projection of degree n and Q be a nonzero central subprojection. Choose a family of cardinality n consisting of equivalent orthogonal abelian projections P_i whose sum is P. Set $Q_i = QP_i$. Each Q_i is abelian and $\sum Q_i = Q$. Furthermore, $C_{Q_i} = QC_{P_i} = QC_{P_j} = C_{Q_j}$ for all i and j. Lemma III.11 implies all the Q_i are equivalent. $\square$

14. Structure Theorem for General Type I Algebras

THEOREM III.22. *Let A be a Type I von Neumann algebra. Then for each cardinal number n, there exists a unique central projection P_n such that $P_n = 0$ or P_n is the maximum homogeneous projection in A of degree n. Moreover, $\sum P_n = I$.*

PROOF. By a maximality argument using Proposition III.21 and the uniqueness of degrees guaranteed by Corollary III.21, it suffices to show every type I von Neumann algebra has a nonzero homogeneous projection. Using Proposition III.18, one can find a maximal orthogonal family of abelian projections P_i satisfying $C_{P_i} = I$. Let $P = \sum P_i$. Then $C_{I-P} \neq I$; for otherwise, by Corollary III.18, the projection $I - P$ contains an abelian projection whose central cover is I. Let Z be the central projection $I - C_{I-P}$. Then $Z \neq 0$, and $C_{ZP_i} = ZC_{P_i} = Z$ for all i. The projections ZP_i are clearly abelian. Lemma III.11 implies they are all equivalent. Since $ZP = (I - C_{I-P})P = P$, $\sum ZP_i = P$ and Z is homogeneous. $\square$

15. Structure Theorem for Type I Algebras — Separable Case

THEOREM III.23. *Let A be a type I von Neumann algebra on a separable Hilbert space $\mathbb{H}$. Then there are a Hilbert bundle $\mathcal{H}$ over a standard Borel S, a finite measure μ on S, a Borel function n on S with values in $\{\infty, 1, 2, \ldots\}$, and a unitary isomorphism of $\mathbb{H}$ onto $\int^{\oplus} n(s)\mathcal{H}_s \, d\mu(s)$ such that UAU^{-1} is the von Neumann algebra $\int^{\oplus} M_{n(s)}(\mathbb{C}) \, d\mu(s)$. In particular this isomorphism identifies the central projections in A with the range of the canonical projection valued measure on $\int^{\oplus} n(s)\mathcal{H}_s \, d\mu(s)$. Moreover, if $(S', \mu', n', \mathcal{H}')$ defines another such decomposition, and S and S' are both uncountable, then there exists a Borel isomorphism φ from S onto S' with $\varphi_* \mu \sim \mu'$ such that $\dim \mathcal{H}_s = \dim \mathcal{H}'_{\varphi(s)}$ and $n(s) = n'(\varphi(s))$ for μ a.e. s.*

PROOF. We may assume the algebra $\mathcal{A}$ is homogeneous of degree m. By Theorem III.21, we may also assume $\mathcal{A} = \mathcal{M}_m(\mathcal{C})$ where $\mathcal{C}$ is an abelian von Neumann algebra acting on a Hilbert space $\mathbb{H}$ which is isomorphic to the center of $\mathcal{A}$. Let $\mathcal{B}$ be the Boolean algebra of projections in the algebra $\mathcal{C}$. By Theorem III.8, $\mathcal{B}$ is the range of a projection valued measure P based on a standard Borel space S. By Theorem III.13, we may assume $\mathbb{H} = \int^{\oplus} \mathcal{H}_s \, d\mu(s)$ and P is the canonical projection valued measure on $\int^{\oplus} \mathcal{H}_s \, d\mu(s)$. Since $\mathcal{C}$ is generated by $\mathcal{B}$, Theorem III.14 implies $\mathcal{C} = \int^{\oplus} \mathbb{C}_s \, d\mu(s)$ where $\mathbb{C}_s$ is the von Neumann algebra on $\mathcal{H}_s$ consisting of all scalar multiples of the identity. Moreover, it follows that $\mathcal{M}_m(\mathcal{C}) = \mathcal{M}_m(\int^{\oplus} \mathbb{C}_s \, d\mu(s))$ can be identified with $\int^{\oplus} \mathcal{M}_m(\mathbb{C}_s) \, d\mu(s)$ acting on the direct integral $\int^{\oplus} m\mathcal{H}_s \, d\mu(s)$. By Theorem III.16 and Corollary III.14, the center of this algebra is $\int^{\oplus} \mathcal{Z}(s) \, d\mu(s)$ where $\mathcal{Z}(s)$ is the center of the von Neumann algebra $\mathcal{M}_m(\mathbb{C})$. Therefore $\mathcal{Z}(s)$ is the algebra of scalar operators, and the central projections in this algebra are in the range of the canonical projection valued measure.

To see uniqueness, note that if $\mathcal{A}$ is a homogeneous von Neumann algebra of degree m, and one has another decomposition given by $(S', \mu', n', \mathcal{H}')$ whose central projections are given by the range of the natural projection valued measure on $\int^{\oplus} \mathcal{H}'(s') \, d\mu'(s')$, then $\mathcal{C}' = \int^{\oplus} \mathbb{C}_{s'} \, d\mu'(s')$ where $\mathbb{C}_{s'}$ is the algebra of scalar operators acting on $\mathcal{H}_{s'}$ is isomorphic to the center of $\mathcal{A}$ and thus isomorphic to $\mathcal{C}$. Moreover, if $S'_k = \{s' : n'(s') = k\}$ has μ' positive measure, the canonical central projection defined by the set S'_k is homogeneous of degree k. By the uniqueness of the degree of homogeneity, it follows that $n'(s') = m$ a.e. μ'.

Since the Boolean algebra of central projections in $\mathcal{A}$ can be identified with the ranges of the canonical projection valued measures P on S and P' on S', there is a corresponding σ-isomorphism from the measure algebra $M(\mu')$ onto the measure algebra $M(\mu)$. Theorem I.23 implies there is a Borel isomorphism φ from S onto S' such that if U is the unitary transformation identifying the von Neumann algebras $\int^{\oplus} \mathcal{M}_m(\mathbb{C}_s) \, d\mu(s)$ with $\int^{\oplus} \mathcal{M}_m(\mathbb{C}_{s'}) \, d\mu'(s')$, then $UP(E)U^* = P'(\varphi(E))$ for all Borel subsets E of S. Thus if $E_k = \{s : m \dim \mathcal{H}_s = k\}$, a Borel subset that defines the largest projection of multiplicity k for P, then $\varphi(E_k)$ defines the largest projection of multiplicity k for P'. Hence $m \dim \mathcal{H}_{\varphi(s)} = k$ a.e. s on E_k. It follows that $\dim \mathcal{H}_s = \dim \mathcal{H}_{\varphi(s)}$ for μ a.e. s. $\square$

By Theorem III.16, Theorem III.20, and Theorem III.23, we obtain the following structure theorem for type I von Neumann algebras on separable Hilbert spaces.

THEOREM III.24 (STRUCTURE THEOREM). *Let $\mathcal{A}$ be a type I von Neumann algebra on a separable Hilbert space $\mathbb{H}$. Then there are a Hilbert bundle $\mathcal{H}$ over a standard Borel S, a finite measure μ on S, a Borel function n on S with values in $\{\infty, 1, 2, \ldots\}$ and a unitary isomorphism of $\mathbb{H}$ onto $\int^{\oplus} n(s)\mathcal{H}_s \, d\mu(s)$ such that $U\mathcal{A}U^{-1}$ is the von Neumann algebra $\int^{\oplus} n(s)\mathcal{B}(\mathcal{H}_s) \, d\mu(s)$. In particular, this isomorphism identifies the central projections in $\mathcal{A}$ with the range of the canonical projection valued measure on $\int^{\oplus} n(s)\mathcal{H}_s \, d\mu(s)$. Moreover, if S and S' are uncountable and $(S', \mu', n', \mathcal{H}')$ gives another such decomposition, then there exists a Borel isomorphism φ from S onto S' with $\varphi_* \mu \sim \mu'$ such that $\dim \mathcal{H}_s = \dim \mathcal{H}'_{\varphi(s)}$ and $n(s) = n'(\varphi(s))$ for a.e. s.*

COROLLARY III.22. *Suppose $\mathcal{A}$ is a type I von Neumann algebra on a separable Hilbert space. Then for each $k = \infty, 1, 2, \cdots$ there are unique central projections P_k, Q_k in $\mathcal{A}$ such that whenever U is a unitary operator from the Hilbert space of $\mathcal{A}$ onto $\int^{\oplus} n(s)\mathcal{H}_s \, d\mu(x)$ which carries the Boolean algebra of central projections of $\mathcal{A}$ onto the range of the canonical projection valued measure P for $\int^{\oplus} n(s)\mathcal{H}_s \, d\mu(x)$ and carries $\mathcal{A}$ onto $\int^{\oplus} n(s)\mathcal{B}(\mathcal{H}_s) \, d\mu(x)$, then $UP_kU^{-1} = P(\{s : n(s) = k\})$ and $UQ_kU^{-1} = P(\{s : n(s) \dim(\mathcal{H}_s) = k\})$. In particular P_k and Q_k commute with every unitary operator on the Hilbert space of $\mathcal{A}$ which normalizes $\mathcal{A}$.*

PROOF. P_k is a homogeneous projection of degree k. Proposition III.21 and Theorem III.22 imply it is the unique maximum central projection of degree k. Let $\mathcal{B}$ be the Boolean algebra of central projections in $\mathcal{A}$. Then Q_k is the largest projection in $\mathcal{B}$ having multiplicity k. It is also unique. $\square$

The next theorem is used extensively in showing von Neumann algebras are type I.

THEOREM III.25. *If a von Neumann algebra is generated by compact operators, then it is type I.*

PROOF. Suppose $\mathcal{A}$ is generated by compact operators. It follows by the spectral theorem for compact operators that $\mathcal{A}$ is generated by projections of finite rank. In particular if P is a nonzero projection in $\mathcal{A}$, then there exists a projection $Q \leq P$ with smallest possible rank.

Therefore QAQ is isomorphic to the algebra of scalar operators and thus is abelian. $\square$

16. Primary Representations of a C* Algebra

We recall that a von Neumann algebra is a factor if its center is trivial. A representation π of a C* algebra is **primary** if the von Neumann algebra $\pi(A)''$ is a factor. A C* algebra A is **type I** if all of its representations are type I.

PROPOSITION III.22. *Let π_i be type I representations of a C* algebra A. Then $\pi = \oplus\pi_i$ is a type I representation.*

PROOF. Let $\mathbb{H}_i$ be the Hilbert spaces for the representations π_i. Set C to be the largest central type I projection for π. If $C \neq I$, set $Q = I - C$. Let P_i be the orthogonal projection of $\mathbb{H} = \oplus\mathbb{H}_j$ onto $\mathbb{H}_i$. Then $P_i \in \pi(A)'$ for all i. Note $QP_i \neq 0$ for some i. Since π_i is type I, $\pi_i(A)'$ is a type I von Neumann algebra; and thus there is a nonzero abelian projection R in $\pi_i(A)'$ with $R \leq QP_i$. Define R' to be 0 on $\mathbb{H}_j$ for $j \neq i$ and to be R on $\mathbb{H}_i$. Then R' is an abelian projection in $\pi(A)'$ with $R' \leq Q$. By Corollary III.20, $C_{R'}\pi(A)'$ is type I and $C_{R'} \leq Q$. Since $C_{R'}$ is nonzero, this is a contradiction to Proposition III.20. $\square$

LEMMA III.12. *Let A be a C* algebra, and let π be a primary representation of A on a Hilbert space $\mathbb{H}$. Let P be a nonzero orthogonal projection in $\pi(A)'$. Then $\pi(a)P = 0$ iff $\pi(a) = 0$.*

PROOF. Suppose $\pi(a) \neq 0$ and there are nonzero P in $\pi(A)'$ with $P\pi(a) = 0$. By Zorn's Lemma, there is a nonempty maximal family $\mathcal{F}$ of nonzero orthogonal projections P in $\pi(A)'$ with $P\pi(a) = 0$. We claim $Q = \sum_{P \in \mathcal{F}} P = I$. If not, then $I - Q$ is a nonzero projection in $\pi(A)'$ which has central cover I in $\pi(A)'$. Since $C_P = I$, Corollary III.17 implies there is a nonzero partial isometry U in $\pi(A)'$ such that $U^*U \leq P$ and $UU^* \leq I - Q$. Thus $UU^*\pi(a) = U\pi(a)U^* = U\pi(a)PU^* = 0$. Thus $\mathcal{F} = \mathcal{F} \cup \{UU^*\}$ is a larger family. Thus $Q = I$ and $\pi(a) = 0$. This is a contradiction. $\square$

PROPOSITION III.23. *Suppose A is a separable C* algebra. Let π be a primary representation equal to a direct integral $\int^{\oplus} \pi_s(a)\, d\mu(s)$ of representations π_s on a Hilbert bundle $\mathcal{H}$ over S. Then for μ a.e. s, $\ker \pi_s = \ker \pi$.*

PROOF. Let P be the canonical projection valued measure on $\int^\oplus \mathcal{H}_s \, d\mu(s)$. Then $P(E) \in \pi(\mathcal{A})'$ for each Borel subset E of S. In particular, Lemma III.12 and Proposition II.4 imply $||\pi(a)P(E)|| = ||\pi(a)||$ for all $a \in \mathcal{A}$ if $\mu(E) > 0$. Thus by Proposition III.12, $||\pi(a)|| = \text{ess sup}_{s\in E}||\pi_s(a)||$ for each Borel subset E of positive measure. Let a_i form a countable dense subset of $\mathcal{A}$. It follows that for each i, $\{s : ||\pi_s(a_i)|| \neq ||\pi(a_i)||\}$ has measure 0. Thus $S_0 = \{s : ||\pi_s(a_i)|| = ||\pi(a_i)|| \text{ for all } i\}$ is a conull Borel subset. Since all representations of a C* algebra are norm continuous, one has $||\pi_s(a)|| = ||\pi(a)||$ for all a for all $s \in S_0$. $\square$

PROPOSITION III.24. *Let $\mathcal{A}$ be a separable C* algebra. Then $\mathcal{A}$ is type I iff every primary representation of $\mathcal{A}$ on a separable Hilbert space is type I.*

PROOF. Note every representation is equivalent to a direct sum of cyclic representations π_i. In particular, if $\mathcal{A}$ is a separable C* algebra and π_i is a cyclic representation, the Hilbert space $\mathbb{H}_i$ for π_i is separable. Thus by Proposition III.22, to show every representation is type I, it suffices to show representations on separable Hilbert spaces are type I. Suppose every separable primary representation of $\mathcal{A}$ is type I. Let π be a representation on a separable Hilbert space $\mathcal{H}$. By Theorem III.8, there is a projection valued measure P based on a standard Borel set S such that the range of P is the σ-Boolean algebra of central projections in $\pi(\mathcal{A})''$. By Theorem III.13, we may assume P is the canonical projection valued measure on the direct integral $\int^\oplus \mathcal{H}_s \, d\mu(s)$ of a Hilbert bundle $\mathcal{H}$ over S. By Theorems III.14 and III.17, $\pi = \int \pi_s \, d\mu(s)$ where each π_s is a representation of $\mathcal{A}$ on $\mathcal{H}_s$.

Now note $s \mapsto \pi_s(\mathcal{A})''$ is a Borel field of von Neumann algebras. Indeed if a_i form a countable dense subset of $\mathcal{A}$, then for each i the mapping $s \mapsto \pi_s(a_i)$ is a Borel field of bounded operators and $\pi_s(\mathcal{A})''$ is the von Neumann algebra generated by the operators $I, \pi(a_i), i = 1, 2, \cdots$. By Corollary III.14, $s \mapsto \mathcal{Z}(s) = \pi_s(\mathcal{A})'' \cap \pi_s(\mathcal{A})'$ is Borel; and thus by Corollary III.13, $S_0 = \{s : \pi_s \text{ is not primary}\}$ is a Borel set. We show it is null. Suppose not. For convenience, we assume the Hilbert bundle $\mathcal{H}$ is trivial. It follows by Proposition III.14 that the subset $\{(s, Q) : s \in S_0, \quad Q \neq I \text{ or } 0, Q \in \mathcal{Z}(s), \quad Q^2 = Q, \quad Q^* = Q\}$ of $S \times B(\mathbb{H})$ is Borel. Since the mapping $(s, Q) \mapsto s$ maps this set onto S_0, Theorem I.14 implies there exists a Borel function $s \mapsto Q(s)$ on S such that $Q(s)$ is a projection in $\mathcal{Z}(s)$ for all s and $Q(s) \neq 0$ and $Q(s) \neq I$ on a set of positive measure. Hence the operator $Q =$

$\int^{\oplus} Q(s)\,d\mu(s)$ is a central projection in $\pi(\mathcal{A})''$. But then $Q = P(E)$ for some Borel subset E of S. This gives $Q(s) = 0$ a.e. for $s \notin E$ and $Q(s) = I$ a.e. for $s \in E$, a contradiction. Thus a.e. π_s is primary. By redefining $\pi_s = 0$ if π_s is not primary, we have π_s is primary for all s.

We next claim every nonzero projection P in $\pi(\mathcal{A})'$ contains a nonzero abelian projection. Indeed, let $P = \int^{\oplus} P(s)\,d\mu(s)$ where $P(s) \in \pi_s(\mathcal{A})'$ for all s. Consider $\{(s,Q) : P(s) \neq 0, Q \neq 0, Q \leq P(s), Q \text{ is abelian}\}$. This set is Borel. (To see this one may assume the Hilbert bundle is trivial. One then need only note the set $\{(s,Q) : Q \text{ is an abelian projection in } \pi_s(\mathcal{A})'\}$ is Borel.) Then again by Theorem I.14, there is a Borel mapping $s \mapsto Q(s)$ with $Q(s) \leq P(s)$ such that $Q(s)$ is abelian in $\pi_s(\mathcal{A})'$ for all s and Q(s) is not zero on a set of positive measure. Let $Q = \int^{\oplus} Q(s)\,d\mu(s)$. Then $Q \leq P$ and $Q\pi(\mathcal{A})'Q = \int Q(s)\pi_s(\mathcal{A})'Q(s)\,d\mu(s)$ is an abelian algebra. Hence $\pi(\mathcal{A})'$ is type I. By Theorem III.20, $\pi(\mathcal{A})''$ is type I. $\square$

CHAPTER IV

GROUPS AND GROUP ACTIONS

In group representation theory a group action by linear transformations is the fundamental notion. In this chapter we develop some aspects related to this notion.

1. Polish Groups

A **standard Borel group** is a group G with a standard Borel structure with the additional property that the mapping $(x, y) \mapsto xy^{-1}$ is a Borel function on $G \times G$ where $G \times G$ has the product Borel structure. This is equivalent to both the mappings $(x, y) \mapsto xy$ and $x \mapsto x^{-1}$ being Borel. A **Polish group** is a topological group whose topology is Polish. Of course, the Borel structure induced by a Polish group makes the Polish group a standard Borel group. A topological group is standard if the Borel structure induced by the topology is standard.

An important example of a Polish group is the unitary group on a separable Hilbert space. The topology on the unitary group is the strong-$*$ topology. It is the topology having the property that a net A_η converges to A iff A_η converges strongly to A and A_η^* converges strongly to A^*.

LEMMA IV.1. *Suppose A_η and B_η are nets of bounded operators on a Hilbert space which converge strongly to the operators A and B. Then if A_η is uniformly norm bounded, then $A_\eta B_\eta$ converges strongly to AB.*

PROOF. Take M with $||A_\eta|| \leq M$ for all n. Then $||(A_\eta B_\eta - AB)x|| \leq ||(A_\eta B_\eta - A_\eta B)x|| + ||(A_\eta B - AB)x|| \leq ||A_\eta|| ||(B_\eta - B)x|| +$

135

$||(A_\eta - A)Bx||$. Hence $||(A_\eta B_\eta - AB)x|| \leq M||(B_\eta - B)x|| + ||(A_\eta - A)Bx||$ which converges to 0 as η becomes large. $\square$

PROPOSITION IV.1. *The unitary group* $\mathcal{U}(\mathbb{H})$ *of a separable Hilbert space* $\mathbb{H}$ *is a Polish group under the strong-* *operator topology.*

PROOF. The continuity of $U \mapsto U^{-1}$ is immediate since $U^* = U^{-1}$ and $U \mapsto U^*$ is continuous in the strong-* topology.

Continuity of multiplication $(U, V) \mapsto UV$ follows from Lemma IV.1. Indeed, if $U_\eta \to U$ and $V_\eta \to V$ in the strong-* topology, then $U_\eta \to U$, $V_\eta \to V$, $U_\eta^* \to U^*$ and $V_\eta^* \to V^*$ strongly. Hence $U_\eta V_\eta \to UV$ and $V_\eta^* U_\eta^* \to V^* U^*$ strongly. Thus $U_\eta V_\eta \to UV$ in the strong-* topology.

By Lemma III.3, the unit ball in $B(\mathbb{H})$ is Polish in the strong-* topology. Moreover, $\mathcal{U}(\mathbb{H})$ is strongly-* closed. Indeed, if $U_n \in U$ converges strongly-* to $U \in B(\mathbb{H})$, then U_n^* converges strongly-* to U^*, and thus by Lemma IV.1, $UU^* = \lim U_n U_n^* = I = \lim U_n^* U_n = U^* U$. $\square$

A left invariant metric on a group is a metric ρ with the property $\rho(gx, gy) = \rho(x, y)$ for all x, y, and g in the group.

THEOREM IV.1. *Every second countable locally compact group has a complete left invariant metric which generates the topology.*

PROOF. Let D be the set of positive rational dyadics. Every element in D strictly between 0 and 1 can be expressed either as a finite sum $\sum \frac{1}{2^{l_j}}$ or as $.a_1 a_2 \ldots a_n$ where a_i is either 0 or 1.

Let $\{U_n\}_{n=1}^{\infty}$ be a decreasing sequence of compact symmetric neighborhoods of the identity satisfying

$$(a) \quad U_{n+1} U_{n+1} \subset U_n$$
$$(b) \quad \cap_{n=1}^{\infty} U_n = \{e\}$$

Set $U_n^0 = \{e\}$ and $U_n^1 = U_n$. If $r = .b_1 b_2 \ldots b_n D$, set $V_r = U_1^{b_1} U_2^{b_2} \ldots U_n^{b_n}$. If $r \geq 1$, set $V_r = G$. Note $V_{2^{-k}} = U_k$, and if $r < 1 \leq s$, then $V_r \subset V_s$. If $r < s < 1$, then $r = .b_1 b_2 \ldots b_n$ and $s = c_1 c_2 \ldots c_n$ where $b_1 = c_1, \ldots, b_{k-1} = c_{k-1}$, and $b_k = 0$ and $c_k = 1$ for some $k \in \{1, 2, \ldots, n\}$. Thus $V_r = U_1^{b_1} \ldots U_{k-1}^{b_{k-1}} U_{k+1}^{b_{k+1}} \ldots U_n^{b_n}$. But by (a) $U_{k+1}^{b_{k+1}} \ldots U_n^{b_n} \subset U_k$. Hence $V_r \subset U_1^{c_1} \ldots U_{k-1}^{c_{k-1}} U_k \subset V_s$. Hence for any two positive dyadic rationals with $r < s$, one has $V_r \subset V_s$.

The definition of V_r implies $V_r U_k = V_{r+2^{-k}}$ whenever $r = .a_1 a_2 \ldots a_l$ with $k > l$. Now suppose $r = .a_1 a_2 \ldots a_l$ with $k \leq l$. Set $r_1 =$

$\frac{1}{2^{k-1}} - .00 \ldots 0 a_k a_{k+1} \ldots a_l$. Then $V_r U_k \subset V_{r+r_1} U_k \subset V_{r+r_1+2^{-k}} \subset V_{r+2^{-(k-1)}+2^{-k}}$. Hence in every case $V_r U_k \subset V_{r+2^{-(k-2)}}$.

Define a function ϕ on G with values between 0 and 1 by $\phi(x) = \inf\{r : x \in V_r\}$. Note $\phi(x) = 0$ iff $x = e$. Set $\rho(x,y) = \sup_{z \in G} |\phi(zx) - \phi(zy)|$. Then ρ is a left invariant metric on G.

Let $u \in U_k$. Then $zx \in V_r$ implies $zxu \in V_{r+\frac{1}{2^{k-2}}}$; and since $U_k = U_k^{-1}$, one has $zxu \in V_r$ implies $zx \in V_{r+\frac{1}{2^{k-2}}}$. Thus $|\phi(zxu) - \phi(zx)| \leq \frac{1}{2^{k-2}}$ for all $u \in U_k$. This implies $|\rho(xu, yv) - \rho(x,y)| \leq \frac{1}{2^{k-3}}$ for u and v in U_k. Thus ρ is continuous, and the topology defined by ρ is weaker than the locally compact topology on G.

Next suppose $\rho(x,e) < \frac{1}{2^n}$. Then $|\phi(zx) - \phi(z)| < \frac{1}{2^n}$ for all z. Hence $\phi(x) < \frac{1}{2^n}$. This implies $x \in V_r$ for some $r < \frac{1}{2^n}$. But $V_r \subset U_n$ whenever $r < \frac{1}{2^n}$. Thus the topology defined by ρ is stronger than the locally compact topology on G. Therefore ρ is a metric inducing the topology.

To show ρ is complete, note that if $\{x_n\}_{n=1}^\infty$ is Cauchy, then there exists an N with $x_n x_N^{-1} \in U_1$ for n large. If follows that x_n is in the compact set $U_1 x_N$ for n large. Thus $\{x_n\}$ converges. $\square$

COROLLARY IV.1. *Let H be a closed subgroup of a second countable locally compact group G. Then the coset space $H\backslash G$ with the quotient topology is a Polish space.*

PROOF. Define a metric on $H\backslash G$ by $d(Hx, Hy) = \inf_{h \in H} \rho(x, hy)$. d is well defined by the left invariance of ρ. Moreover, $d(Hx, Hy) = 0$ implies there is a sequence $\{h_n\}$ with $h_n y \to x$. Thus, since H is closed, $xy^{-1} \in H$; and hence $Hx = Hy$. It is an easy matter to check that a set is d open in $H\backslash G$ if and only if its preimage in G is ρ open in G. To see that d is complete, note that if Hx_n is Cauchy, you could choose a subsequence Hy_k with $d(Hy_k, Hy_{k+1}) < 2^{-(k+1)}$ for each k. Consequently, you can choose a sequence $\{h_k\} \in H$ with $\rho(h_k y_k, h_{k+1} y_{k+1}) < 2^{-(k+1)}$. This implies the sequence $\{h_k y_k\}$ is Cauchy in G and thus converges to some y. Hence $Hx_n \to Hy$ and d is complete. $\square$

PROPOSITION IV.2 (BOREL CROSS SECTION).
Let G be a Polish group, and suppose H is a closed subgroup of G. Then there is a Borel function $\sigma : H\backslash G \to G$ satisfying $H\sigma(Hg) = Hg$ for all Hg.

PROOF. G is a complete metric space. The right cosets Hg are closed and hence complete. They are the orbits under the left action of H on G defined by $h \cdot g = hg$. Moreover, if U is an open subset of G, then its saturation HU is also open in G. By Theorem I.15, there is a Borel set B in G with $B \cap Hg = \{\sigma(Hg)\}$ for each Hg.

To see σ is Borel, note σ is the inverse of the Borel mapping $x \mapsto Hx$ from B onto $H \backslash G$. Since $H \backslash G$ and B are standard Borel spaces, Kuratowski's Theorem implies σ is a Borel isomorphism of $H \backslash G$ onto B. $\square$

THEOREM IV.2. *Let ϕ be a Borel homomorphism from a Polish group G into a second countable topological group H. Then ϕ is continuous.*

PROOF. By Proposition I.7, there exists a meager subset P such that ϕ is continuous on the set $R = G - P$. Suppose $x_n \to e$ as $n \to \infty$. The set $P \cup (\cup_n x_n^{-1} P)$ is meager. By the Baire Category Theorem, its complement $R \cap (\cap_n x_n^{-1} R)$ is second category and consequentially nonempty. Let y be in this set. Then $y \in R$ and $x_n y \in R$ for each n. Moreover, $x_n y \to y$ as $n \to \infty$. By the continuity of ϕ on R, it follows that $\phi(x_n y) \to \phi(y)$ as $n \to \infty$. This implies $\phi(x_n)$ converges to e, and thus ϕ is continuous at the identity. $\square$

2. Haar Integrals on Locally Compact Groups

Let G be a locally compact topological group. The **support**, $\mathrm{supp}\,(f)$, of a function f on G is the closure of the set $\{g : f(g) \neq 0\}$. Let $C_c(G)$ be the linear space consisting of all continuous real valued functions f on G whose supports are compact. If f is a function on G and $g \in G$, then the **right translate** f_g of f by g is the function defined by $f_g(x) = f(xg)$ for $x \in G$. A **positive integral** I on G is a real valued linear functional on $C_c(G)$ with the additional property that $I(f) > 0$ whenever f is nonzero and nonnegative. If in addition $I(f_g) = I(f)$ for all $f \in C_c(G)$ and $g \in G$, then the integral I is said to be **right invariant**. We note that if I is right invariant, nonnegative and nonzero, then it is positive. Indeed if $f \geq 0$ and $I(f) > 0$, then if h is a nonzero, nonnegative continuous function with compact support, there exist $c_i > 0$ and $g_i \in G$ for which $f(x) \leq \sum_{i=1}^n c_i h(xg_i)$. By right invariance of I, we would have $I(f) \leq I(h) \sum_{i=1}^n c_i$ and thus $I(h) > 0$.

PROPOSITION IV.3. *Let $f \in C_c(G)$ and suppose $\epsilon > 0$. Then there exists a neighborhood U of the identity such that $|f(xy) - f(x)| < \epsilon$ and $|f(yx) - f(x)| < \epsilon$ whenever $x \in G$ and $y \in U$.*

PROOF. Let $K = \text{supp} f$ and W be a compact neighborhood of e. For $x \in WKW$, let N_x be an open neighborhood of e such that if $z \in xN_x \cup N_x x$, then $|f(z) - f(x)| < \frac{\epsilon}{2}$. For each N_x, take an open symmetric neighborhood $N_x^{\frac{1}{2}}$ of e with $N_x^{\frac{1}{2}} N_x^{\frac{1}{2}} \subset N_x$. The sets $xN_x^{\frac{1}{2}} \cap N_x^{\frac{1}{2}} x$ for $x \in WKW$ cover the compact set WKW, and there exist finitely many x_i so that $WKW \subset \cup(x_i N_{x_i}^{\frac{1}{2}} \cap N_{x_i}^{\frac{1}{2}} x_i)$. Let U be a symmetric compact neighborhood of e contained in $\cap N_{x_i}^{\frac{1}{2}} \cap V$ where V is a neighborhood of e with $VV \subset W$. Let $y \in U$. If $x \in UKU$, then xy and yx belong to WKW. Hence $xy \in x_i N_{x_i}^{\frac{1}{2}} \cap N_{x_i}^{\frac{1}{2}} x_i$ for some i. Moreover,

$$\begin{aligned}
|f(xy) - f(x)| &= |f(x_i x_i^{-1} xy) - f(x_i) + f(x_i) - f(x_i x_i^{-1} x)| \\
&\leq |f(x_i u) - f(x_i)| + |f(x_i) - f(x_i v)| \\
&< \epsilon
\end{aligned}$$

for $u = x_i^{-1} xy \in N_{x_i}^{\frac{1}{2}} \subset N_{x_i}$ and $v = x_i^{-1} x \in N_{x_i}^{\frac{1}{2}} y^{-1} \subset N_{x_i}^{\frac{1}{2}} U \subset N_{x_i}$. Next suppose $x \notin UKU$. Then $xy \notin UK$ for otherwise $x = xyy^{-1} \in UKy^{-1} \subset UKU^{-1} = UKU$ which is a contradiction. Hence $f(xy) = f(x) = 0$ and $|f(xy) - f(x)| < \epsilon$. Hence $|f(xy) - f(x)| < \epsilon$ for all x if $y \in U$. By symmetry, $|f(yx) - f(x)| < \epsilon$ if $y \in U$. $\square$

THEOREM IV.3 (HAAR INTEGRAL). *Every locally compact group G has a positive right invariant integral. Moreover, such an integral is unique up to a positive scalar multiple.*

PROOF. Let $C_c^+(G)$ denote the subset of $C_c(G)$ consisting of all nonzero, nonnegative functions. For f and h in $C_c^+(G)$ define

$$(f; h) = \inf\{\sum_{i=1}^n c_i : \exists x_1, \ldots, x_n \in G \text{ with } h \leq \sum_{i=1}^n c_i f_{x_i}\}$$

One can establish the following properties.

(a) $(f; h_g) = (f; h)$
(b) $(f; h_1 + h_2) \leq (f; h_1) + (f; h_2)$
(c) $(f; ch) = c(f; h)$ whenever $c > 0$
(d) $h_1 \leq h_2$ implies $(f; h_1) \leq (f; h_2)$
(e) $(f; h) \leq (f; g)(g; h)$
(f) $(f; h) \geq \frac{M_h}{M_f}$ where M_h denotes the maximum value of h.

Each of these follows almost immediately from the definition of $(f; h)$. For instance, (f) follows from the fact that if $\sum_i c_i f_{x_i} \geq h$, then $\sum_i c_i M_f \geq M_h$.

Next fix a function f_0 in $C_c^+(G)$. Define $J_f(h) = \frac{(f;h)}{(f;f_0)}$. Note (e) implies

$$\frac{(f;h)}{(f;h)(h,f_0)} \le \frac{(f;h)}{(f;f_0)} \le \frac{(f;f_0)(f_0;h)}{(f;f_0)}$$

and thus

(g) $\frac{1}{(h;f_0)} \le J_f(h) \le (f_0;h)$

The following along with (b) shows the almost linearity of J_f for f with small support.

(h) Let ϵ, h_1 and h_2 be given. Then there exists a neighborhood V of e such that $J_f(h_1 + h_2) > J_f(h_1) + J_f(h_2) - \epsilon$ whenever $\operatorname{supp}(f) \subset V$.

To show this, first pick a nonnegative function $g \in C_c(G)$ with $g(x) > 0$ for $x \in \operatorname{supp} h_1 \cup \operatorname{supp} h_2$. Set $h = h_1 + h_2 + \epsilon_1 g$. Then temporarily using $\frac{0}{0} = 0$, the functions $\frac{h_1}{h}$ and $\frac{h_2}{h}$ are continuous with compact support. By Proposition IV.3, there exists a neighborhood V of e with the property that $|\frac{h_i}{h}(yx) - \frac{h_i}{h}(x)| < \epsilon_1$ for all x when $y \in V$ and $i = 1$ or 2. Now suppose $\operatorname{supp} f \subset V$ and $h(x) \le \sum c_j f(xx_j)$ for all x. Then $h_i(x) = \frac{h_i}{h}(x)h(x) \le \sum_j c_j \frac{h_i}{h}(x)f(xx_j)$. But if $f(xx_j) \ne 0$, then $xx_j \in V$; and hence $\frac{h_i}{h}(x) = \frac{h_i}{h}(xx_jx_j^{-1}) < \frac{h_i}{h}(x_j^{-1}) + \epsilon_1$. Therefore $h_i(x) \le \sum_j c_j(\frac{h_i}{h}(x_j^{-1}) + \epsilon_1)f(xx_j)$. Hence $(f;h_1) + (f;h_2) \le \sum c_j(\frac{h_1+h_2}{h}(x_j^{-1}) + 2\epsilon_1)$. Thus $(f;h_1) + (f;h_2) \le \sum_j c_j(1 + 2\epsilon_1)$. This implies $(f;h_1) + (f;h_2) \le (f,h)(1 + 2\epsilon_1)$. Thus $J_f(h_1 + h_2) + \epsilon_1 J_f(g) \ge J_f(h) \ge \frac{J_f(h_1)+J_f(h_2)}{1+2\epsilon_1}$. Since $J_f(g) \le (f_0;g)$, (h) follows by a proper choice of ϵ_1.

Let T be $\prod_{h\in C_c^+(G)}[(h;f_0)^{-1}, (f_0;h)]$ with the product topology. T is compact, and (g) implies $J_f \in T$ whenever $f \in C_c^+(G)$. Let V be a neighborhood of e. Set T_V to be the closure of the subset $\{J_f : \operatorname{supp}(f) \subset V\}$. Since $T_{V_1} \cap T_{V_2} \cap \cdots \cap T_{V_n} \supset T_{V_1\cap V_2\cap\cdots\cap V_n} \ne \emptyset$, the intersection of all the T_V's is nonempty. Choose an I in this intersection. It follows by (b), (c), and (h) that I is linear on $C_c^+(G)$. Moreover, $I(h) \ge \frac{1}{(h;f_0)} > 0$ for all $h \in C_c^+(G)$. Extend I to all of $C_c(G)$ by defining $I(h_1 - h_2) = I(h_1) - I(h_2)$ for h_1 and h_2 in $C_c^+(G)$. Then I is a positive linear functional on $C_c(G)$ which by (a) is right invariant.

Assume J is another positive right invariant linear functional on $C_c(G)$ which is nonzero. We show $J = cI$ for some c. Let $f \in C_c^+(G)$. Fix a symmetric compact neighborhood V_0 of e, and let $F \in C_c^+(G)$ satisfy $F(x) = 1$ on $V_0 \operatorname{supp} f$. For each open neighborhood V of e contained in V_0, set $M_V(f) = \sup_{x\in G, y\in V} |f(x) - f(yx)|$. For y in

V, one has $|f(x) - f(yx)| \le F(x)M_V(f)$ for all x. By Proposition IV.3, for each $\epsilon > 0$, there is a neighborhood V of e contained in V_0 such that $M_V(f) < \epsilon$. In particular, $|f(wyx) - f(yx)| \le \epsilon F(yx)$ for $w \in V$. This implies $|J_x(f(wyx)) - J_x(f(yx))| \le \epsilon J_x(F(yx))$. Hence $y \mapsto J_x(f(yx))$ is continuous.

Let h be a symmetric $(h(x^{-1}) = h(x))$ function in $C_c^+(G)$. Then

$$
\begin{aligned}
I(h)J(f) &= I_y(h(y))J_x(f(x)) \\
&= I_y J_x(h(y)f(x)) \\
&= I_y J_x h(y)(f(x) - f(yx)) + I_y J_x(h(y)f(yx)) \\
&= I_y J_x h(y)(f(x) - f(yx)) + I_y J_x(h(yx^{-1})f(y)) \\
&= I_y J_x h(y)(f(x) - f(yx)) + I_y J_x(h(xy^{-1})f(y)) \\
&= I_y J_x h(y)(f(x) - f(yx)) + I_y J_x(h(x)f(y)) \\
&= I_y J_x h(y)(f(x) - f(yx)) + J(h)I(f).
\end{aligned}
$$

This gives

$$|I(h)J(f) - J(h)I(f)| \le I_y J_x(h(y)|f(x) - f(yx)|).$$

Dividing by $I(h)I(f)$ gives

$$\left| \frac{J(f)}{I(f)} - \frac{J(h)}{I(h)} \right| \le \frac{1}{I(h)I(f)} I_y J_x(h(y)|f(x) - f(yx)|).$$

Again, let V be a neighborhood of e contained in V_0. If $y \in V$, then $|f(x) - f(yx)| \le F(x)M_V(f)$. Now, additionally, suppose h has support in V. Then

$$\left| \frac{J(f)}{I(f)} - \frac{J(h)}{I(h)} \right| \le M_f(V)\frac{J(F)}{I(f)}.$$

Hence for any f and any $\epsilon > 0$, Proposition IV.3 implies there is a neighborhood V of e contained in V_0 satisfying that $M_f(V)\frac{J(F)}{I(f)} < \epsilon$. Hence for any f_1 and f_2 in $C_c^+(G)$, there is an h for which

$$\left| \frac{J(f_i)}{I(f_i)} - \frac{J(h)}{I(h)} \right| < \epsilon.$$

This gives

$$\left| \frac{J(f_1)}{I(f_1)} - \frac{J(f_2)}{I(f_2)} \right| < 2\epsilon.$$

Since this is true for all $\epsilon > 0$ and f_1 and f_2, one has $J(f) = \frac{J(f_1)}{I(f_1)}I(f)$. $\quad\square$

A measure μ on a standard Borel group is **right invariant** if $\mu(Eg) = \mu(E)$ for every element g in G and every Borel subset E of G. It is **left invariant** if $\mu(gE) = \mu(E)$ for all g and E. A measure which is both right and left invariant is said to be **biinvariant**. A measure μ which satisfies $\mu(Eg) = 0$ if and only if $\mu(E) = 0$ is said to be **right quasi-invariant**. If it satisfies $\mu(gE) = 0$ if and only if $\mu(E) = 0$, then it is **left quasi-invariant**.

By the Riesz representation theorem, if I is a positive linear functional on $C_c(G)$, there is a unique regular Borel measure μ on G satisfying $I(f) = \int f(g) \, d\mu(g)$. We recall that a measure μ is **regular** if $\mu(K) < \infty$ for all compact sets K, $\mu(O) = \sup_{K \subset O} \mu(K)$ for all open sets O, and $\mu(E) = \inf_{E \subset O} \mu(O)$ for all Borel sets E.

THEOREM IV.4 (HAAR MEASURE).
Let G be a locally compact group. Then there exists a regular right invariant measure μ on G. Moreover, any other regular right invariant measure is a multiple of this one.

PROOF. Let I be a right invariant positive linear functional on $C_c(G)$. By the Riesz representation theorem, there exists a unique regular measure μ such that $I(f) = \int f(g) \, d\mu(g)$. Since the measure $\mu \cdot g$ defined by $\mu \cdot g(E) = \mu(Eg^{-1})$ for Borel sets E is again regular and $I(f) = I(f_g) = \int f(xg) \, d\mu(x) = \int f(x) \, d\mu \cdot g(x)$, one has $\mu \cdot g = \mu$ and μ is right invariant.

Suppose ν is another nonzero right invariant regular Borel measure on G. Then $J(f) \equiv \int f(x) \, d\nu(x)$ defines a nonnegative nonzero right invariant linear functional on $C_c(G)$. By Theorem IV.3, $J = cI$ for some $c > 0$. Again by the Riesz representation theorem, $\nu = c\mu$; and we see μ is unique up to a positive scalar multiple. $\square$

COROLLARY IV.2 (MODULAR HOMOMORPHISM).
There exists a unique continuous homomorphism δ from G into the positive reals such that if μ is a right invariant Haar measure on G, then $\mu(gE) = \delta(g)\mu(E)$ for all Borel sets E and all $g \in G$. Moreover:
(a) *$\int f(gx) \, d\mu(x) = \int f(x) \, d\mu(g^{-1}x) = \delta(g^{-1}) \int f(x) \, d\mu(x)$ for all integrable f and $g \in G$ and*
(b) *$\int f(x^{-1})\delta(x^{-1}) \, d\mu(x) = \int f(x) \, d\mu(x)$ for all integrable f.*

PROOF. The measure $E \mapsto \mu(gE)$ is a right invariant regular Borel measure. Thus there is a constant $\delta(g) > 0$ such that $\mu(gE) = \delta(g)\mu(E)$. Since $\mu(g_1 g_2 E) = \delta(g_1)\mu(g_2 E) = \delta(g_1)\delta(g_2)\mu(E)$ and $\mu(g_1 g_2 E) = \delta(g_1 g_2)\mu(E)$, we see $\delta(g_1 g_2) = \delta(g_1)\delta(g_2)$ and δ is a homomorphism into the positive real numbers. To see δ is continuous, let f be a nonnegative nonzero function in $C_c(G)$. Fix a compact symmetric neighborhood V_0 of e, and let $F(x)$ be a function in $C_c(G)$ with $0 \le F(x) \le 1$ and $F = 1$ on $V_0 \operatorname{supp} f$. Then if $V \subset V_0$, $|f(yx) - f(x)| \le F(x) \sup_{y \in V} |f(yx) - f(x)|$ for y in V and $x \in G$. Hence

$$\delta(y^{-1})I(f) - I(f) = \int f(x)\delta(y^{-1})\,d\mu(x) - \int f(x)\,d\mu(x)$$

$$= \int f(yx)\,d\mu(x) - \int f(x)\,d\mu(x)$$

$$= \int f(yx) - f(x)\,d\mu(x).$$

Hence

$$I(f)|1 - \delta(y^{-1})| \le \int |f(yx) - f(x)|\,d\mu(x)$$

$$\le I(F)\sup_{x \in G, y \in V} |f(yx) - f(x)|$$

if $y \in V$. Using Proposition IV.3, one sees $|1 - \delta(y^{-1})| < \epsilon$ for some neighborhood V of e contained in V_0. Thus δ is continuous at e and therefore is continuous.

Define $\mu'(E) = \int 1_E(x^{-1})\delta(x^{-1})\,d\mu(x)$. Then

$$\mu'(Eg) = \int 1_E((gx)^{-1})\delta(x^{-1})\,d\mu(x)$$

$$= \int 1_E((gg^{-1}x)^{-1})\delta((g^{-1}x)^{-1})\,d\mu(g^{-1}x).$$

Hence by (a),

$$\mu'(Eg) = \delta(g^{-1})\int 1_E(x^{-1})\delta((g^{-1}x)^{-1})\,d\mu(x) = \mu'(E).$$

Thus μ' is right invariant, and there is a constant $c > 0$ with $\mu' = c\mu$. Moreover,

$$\int 1_E(x^{-1})\delta(x^{-1})\,d\mu(x) = c\int 1_E(x)\,d\mu(x)$$

for Borel sets E implies

$$\int f(x^{-1})\delta(x^{-1})\,d\mu(x) = c\int f(x)\,d\mu(x)$$

for simple functions f and hence for all integrable functions. In particular,

$$\int f(x)\,d\mu(x) = \int (f((x^{-1})^{-1})\delta(x^{-1})^{-1})\delta(x^{-1})\,d\mu(x)$$

$$= c \int f(x^{-1})\delta(x^{-1})\,d\mu(x)$$

$$= c^2 \int f(x)\,d\mu(x).$$

Therefore $c = 1$. □

COROLLARY IV.3. *G is compact if and only if $\mu(G) < \infty$.*

PROOF. If G is compact, then $\mu(G) = I(1) < \infty$. Conversely, suppose G is not compact. Choose a compact neighborhood V . Then $\mu(V) > 0$ and $U = V^{-1}V$ is compact. Moreover, since finite unions of compact sets are compact, there exists a sequence g_n such that $g_{n+1} \notin \cup_{i=1}^{n} Ug_i$. Therefore $Vg_i \cap Vg_j = \emptyset$ if $i > j$, for otherwise $g_i \in V^{-1}Vg_j = Ug_j$. Hence $\mu(G) \geq \sum_n \mu(Vg_n) = \sum_n \mu(V) = \infty$. □

We shall use the mnemonics

$$d\mu(gx) = \delta(g)\,d\mu(x) \text{ and } d\mu(x^{-1}) = \delta(x^{-1})\,d\mu(x)$$

to denote the change of variable formulas (a) and (b). Note that if ν is a left invariant σ-finite measure, then $\nu(Eg) = \delta(g^{-1})\nu(E)$ and thus $d\nu(xg) = \delta(g^{-1})\,d\nu(x)$. One can also show $d\nu(x^{-1}) = \delta(x)\,d\nu(x)$. We let $\Delta(g) = \delta(g^{-1})$. Then for left Haar measures ν, we have $\nu(Eg) = \Delta(g)\nu(E)$, $d\nu(xg) = \Delta(g)d\nu(x)$ and $d\nu(x^{-1}) = \Delta(x^{-1})d\nu(x)$. In many instances we shall use the shortened notation dg for $d\mu(g)$ where μ is understood to be some right (left) invariant Haar measure on G.

The next theorem relates the Haar measures of closed subgroups of the group G to the Haar measure on G. We present it for second countable groups only and in terms of their right Haar measures. For general locally compact groups, see Bourbaki [45].

Before giving the theorem, we present a useful lemma.

LEMMA IV.2. *Let G be a locally compact group and let H be a closed subgroup. Then $f_H(Hx) = \int f(hx)\,dh$ defines a mapping $f \mapsto f_H$ from $C_c(G)$ onto $C_c(H\backslash G)$.*

PROOF. Let p be the natural projection of G onto $H\backslash G$. If C is a compact subset of $H\backslash G$ and if V is a nonempty open set in G whose closure is compact, then since the sets HVx are open in $H\backslash G$, there exists a finite sequence $x_1, \ldots, x_n$ such that C is covered by the sets HVx_i. Thus $K = \cup_{i=1}^n \overline{V}x_i$ is compact and $p(K) \supset C$.

Hence if $f \in C_c(H\backslash G)$, then there exists an open set V in G whose closure is compact and which satisfies $p(V) \supset \text{supp}(f)$. Let $t \in C_c^+(G)$ be such that $t = 1$ on $\overline{V}$. Then $t_H \in C_c^+(H\backslash G)$ satisfies $t_H > 0$ on HV. Hence if we define $\frac{0}{0} = 0$, the function $h(x) = \frac{t(x)f(Hx)}{t_H(Hx)}$ is in $C_c(G)$ and $h_H = f$. $\square$

THEOREM IV.5. *Let G be a locally compact second countable group. Let X and Y be closed subgroups of G with $X \cap Y$ compact. Suppose XY is a conull Borel subset of G. Then right Haar measures on X, Y, and G can be normalized so that*

$$\int f(g)\, dg = \iint_{X \times Y} f(xy) \frac{\delta_G(x)}{\delta_X(x)}\, dx\, dy.$$

PROOF. Define an action of $X \times Y$ on G by $g \cdot (x,y) = x^{-1}gy$. Then the mapping $((x,y),g) \mapsto g \cdot (x,y)$ is Borel. The stabilizer of e is $Z = \{(x,x) : x \in X \cap Y\}$, the orbit of e is XY, and the mapping $(x,y) \mapsto x^{-1}y$ is a one-to-one Borel isomorphism of $Z\backslash(X \times Y)$ onto XY. Let μ_G be a right Haar measure on G restricted to XY. Let φ be the inverse of the isomorphism from $Z\backslash(X \times Y)$ onto XY. Thus $\varphi(xy) = Z(x^{-1}, y)$. Then $\varphi_*\mu_G(E(x,y)) = \mu_G(\varphi^{-1}(E(x,y))) = \mu_G(x^{-1}\varphi^{-1}(E)y) = \delta_G(x^{-1})\varphi_*(\mu_G)(E)$. Now let dz be the Haar measure on the compact subgroup $X \cap Y$ of G with total measure 1. By Lemma IV.2, the mapping $f \mapsto \tilde{f}$ defined by $\tilde{f}(Z(x,y)) = \int f((z,z)(x,y))\, dz$ is a mapping from the continuous functions on $X \times Y$ with compact support onto the continuous functions on $Z\backslash(X \times Y)$ with compact support. Define an integral I by $I(f) = \int (\delta_G f)\widetilde{\,}\, d\varphi_*\mu_G(x,y)$ for f a continuous function on $X \times Y$ with compact support, where $\delta_G f$ is the function on $X \times Y$ defined by $(\delta_G f)(x,y) = \delta_G(x)f(x,y)$. We claim I is a right Haar integral on $X \times Y$. Since $\varphi_*\mu_G$ is right invariant under $\{e\} \times Y$, we need only show it is right invariant under $X \times \{e\}$. Note if $f_a(x,y) = f(xa,y)$, we see $(\delta_G f_a\tilde{)}(Z(x,y)) = \int \delta_G(zx) f(zxa, zy)\, dz$. Thus

$$
\begin{aligned}
I(f_a) &= \iint \delta_G(zx) f(zxa, zy)\, dz\, d\varphi_* \mu_G(Z(x,y)) \\
&= \iint \delta_G(zxa^{-1}) f(zx, zy)\, dz\, d\varphi_* \mu_G(Z(xa^{-1}, y)) \\
&= \iint \delta_G(zxa^{-1}) f(zx, zy)\, dz\, \delta_G(a)\, d\varphi_* \mu_G(Z(x,y)) \\
&= \int (\delta_G f\tilde{)}(Z(x,y))\, d\varphi_* \mu_G(Z(x,y)) \\
&= I(f).
\end{aligned}
$$

Hence if dx and dy are right Haar measures on X and Y and are normalized properly,

$$
\int f(x,y)\, dx\, dy = \iint \delta_G(zx) f(zx, zy)\, dz\, d\varphi_* \mu_G(Z(x,y)).
$$

Thus $\int f(x,y)\, dx\, dy = \int (\delta_G f\tilde{)} \circ \varphi(x^{-1} y)\, d\mu_G(x^{-1} y)$. This gives

$$
\int f(x,y)\, dx\, dy = \int (\delta_G f\tilde{)} \circ \varphi(g)\, d\mu_G(g).
$$

Suppose h is a nonnegative Borel function on G. Set $f(x,y) = \delta_G(x^{-1}) h(x^{-1} y)$. Then one has $(\delta_G f\tilde{)}(Z(x,y)) = \int_Z h(x^{-1} z^{-1} zy)\, dz = h(x^{-1} y)$. This gives

$$
\iint \delta_G(x^{-1}) h(x^{-1} y)\, dy\, dx = \int (\delta_G f\tilde{)} \circ \varphi(g)\, dg.
$$

But $(\delta_G f\tilde{)} \circ \varphi(x^{-1} y) = (\delta_G f\tilde{)}(Z(x,y)) = h(x^{-1} y)$ for all x, y. Thus $(\delta_G f\tilde{)} \circ \varphi(g) = h(g)$ a.e. g and we see

$$
\int h(g)\, dg = \iint \delta_G(x^{-1}) h(x^{-1} y)\, dx\, dy = \iint \delta_G(x) \delta_X(x^{-1}) h(xy)\, dx\, dy.
$$

$\square$

3. Standard Borel Groups with Quasi-invariant Measure

LEMMA IV.3. *If μ is a nonzero right quasi-invariant σ-finite measure and ν is a nonzero left quasi-invariant σ-finite measure on a standard Borel group, then μ is equivalent to ν.*

PROOF. Since μ and ν are σ-finite, we may replace them by equivalent finite measures. Thus we may assume both are finite.

Now $\mu(E) = 0$ iff $\mu(Ey^{-1}) = 0$ iff $\int \mu(Ey^{-1})\, d\nu(y) = 0$. Thus $\mu(E) = 0$ is equivalent to $\iint 1_E(xy)\, d\mu(x)\, d\nu(y) = 0$. So by Fubini's

Theorem, $\mu(E) = 0$ iff $\iint 1_E(xy)\,d\nu(y)\,d\mu(x) = \int \nu(x^{-1}E)\,d\mu(x) = 0$ which occurs by the left quasi-invariance of ν iff $\nu(E) = 0$. $\square$

PROPOSITION IV.4. *All left and all right σ-finite quasi-invariant measures on a standard Borel group are equivalent. In particular, any left quasi-invariant σ-finite measure is right quasi-invariant.*

PROOF. By Lemma IV.3, it suffices to show a right quasi-invariant measure μ has to be left quasi-invariant. Define μ^{-1} by $\mu^{-1}(E) = \mu(E^{-1})$. Clearly μ^{-1} is left quasi-invariant. By Lemma IV.3, $\mu \sim \mu^{-1}$. Thus μ is also left quasi-invariant. $\square$

A measure which is both right and left quasi-invariant is said to be quasi-invariant.

COROLLARY IV.4. *If μ is a σ-finite quasi-invariant measure on a standard Borel group, then $\mu \sim \mu^{-1}$.*

COROLLARY IV.5. *Let μ be a σ-finite right quasi-invariant measure on a standard Borel group G. If f is a real valued Borel function on G which satisfies $f(xg) = f(x)$ a.e. x for each g, then f is constant a.e. on G.*

PROOF. Assume $f(xg) = f(x)$ for μ a.e. x for each g. By Fubini's Theorem, there exists an x_0 such that $f(x_0 g) = f(x_0)$ for μ a.e. g. Then $\{x : f(x) = f(x_0)\} \supset x_0 G_0$ where G_0 is a conull Borel subset of G. But by Proposition IV.4, $x_0 G_0$ is conull in G. Thus $f(x) = f(x_0)$ for μ a.e. x. $\square$

The existence of invariant measures on locally compact groups made possible the development of abstract representation theory for these groups. The following theorem of André Weil shows these groups are in this sense the largest class possible.

THEOREM IV.6 (WEIL). *Let G be a standard Borel group with a right invariant σ-finite measure μ. Then there is a second countable locally compact topology which generates the Borel structure under which G is a topological group.*

PROOF. Let $\mathcal{H} = L^2(G, \mu)$. Then $\mathcal{H}$ is a separable Hilbert space and the unitary group $\mathcal{U}(\mathcal{H})$ is Polish in the strong-$*$ operator topology. Define a homomorphism U from G into $\mathcal{U}(\mathcal{H})$ by $U_g f(x) = f(xg)$. This homomorphism is Borel. Indeed, by Lemma III.4, it suffices to show it is weakly Borel. But $g \mapsto \langle U_g f, h \rangle$ is Borel by Fubini's Theorem for each f and h in $\mathcal{H}$. Thus U is strongly-$*$ Borel.

Suppose $U_g = I$. Let E_i be a countable generating sequence of Borel subsets of G. Since μ is σ-finite, we may assume $\mu(E_i) < \infty$ for all i. Since $U_g 1_{E_i} = 1_{E_i}$, it follows that $F_i = \{x : x \in E_i \text{ iff } x \in E_i g^{-1}\}$ is conull for every i. Thus $\cap_i F_i$ is conull. Let x be in this intersection. Choose i_k so that $\cap_k E_{i_k} = \{x\}$. Since $x \in F_{i_k}$ and $x \in E_{i_k}$ implies $xg^{-1} \in E_{i_k}$, it follows that $xg^{-1} \in \cap_k E_{i_k}$. Therefore $xg^{-1} = x$. Thus $g = e$ and U is one-to-one. By Theorem I.8, U is a Borel isomorphism of G onto the Borel subgroup $U(G)$ of $\mathcal{U}(\mathcal{H})$.

To finish the proof, we show that $U(G)$ with the relative strong-$*$ operator topology is locally compact. To do this we use U as an identification of G with $U(G)$. In particular, μ is a measure on $\mathcal{U}(\mathcal{H})$ supported on $U(G)$. By Theorem I.11, there exists a compact subset K of $U(G)$ with $0 < \mu(K) < \infty$. Let $V = K^{-1}K$. Clearly V is a compact subset of $U(G)$. If V does not contain a neighborhood of the identity, then there is a sequence g_n in G such that U_{g_n} converges to I but $U_{g_n} \notin V$ for all n. But then $U_{g_n} 1_K$ converges to 1_K in $L^2(G, \mu)$. This implies since $\mu(K) > 0$ that $Kg_n^{-1} \cap K \neq \emptyset$ for n large. In particular, $U_{g_n} \in K^{-1}K = V$ for these n. This contradiction shows $U(G)$ is locally compact. $\square$

4. Standard Borel G Spaces

DEFINITION. Let G be a standard Borel group. A **standard Borel G space** is a standard Borel space X along with a Borel mapping $(x, g) \mapsto x \cdot g$ from $X \times G$ into X which satisfies:

(a) $x \cdot e = x$ for all x and
(b) $x \cdot g_1 g_2 = (x \cdot g_1) \cdot g_2$ for $x \in X$ and g_1 and g_2 in G.

A mapping satisfying (a) and (b) is called a right G action. Thus a (right) Borel G space is a Borel space with a right Borel G action. It is standard if the space X is standard.

A Borel measure μ on X is **quasi-invariant** if $\mu(E \cdot g) = 0$ whenever $\mu(E) = 0$. It is **invariant** if $\mu(E \cdot g) = \mu(E)$ for all E and g.

A Borel subset E of X is almost invariant if $\mu(E \triangledown Eg) = 0$ for all elements g. It is said to be invariant if $Eg = E$ as sets for all g. The following proposition shows these are essentially the same.

PROPOSITION IV.5. *Suppose E is an almost invariant Borel set. Then there exists a G-invariant Borel set E' such that $\mu(E \triangledown E') = 0$.*

PROOF. Let $E' = \{x : x \cdot g \in E \quad \text{for almost all } g\}$. Then E' is the set of x such that the function

$$g \mapsto 1_E(xg^{-1})$$

is identically one almost everywhere g. Thus E' is a Borel set. Moreover, E' is invariant, and since

$$1_E(xg) = 1_E(x)$$

a.e. x for all g, we see by Fubini that for almost all x, $x \cdot g \in E$ for almost all g iff $x \in E$ and hence $\mu(E' \triangledown E) = 0$. $\square$

Suppose μ is a quasi-invariant measure on X. Then this measure is said to be **ergodic** if each almost invariant Borel set has measure 0 or its complement has measure 0.

Suppose X is a Borel G space. Then X/G, the space consisting of all orbits $[x] = x \cdot G$, is a Borel space when given the quotient Borel structure. In particular, a subset E of X/G is Borel when $\cup_{[x] \in E}[x]$ is a Borel subset of X. This Borel space is called the **orbit space** of the action of G on X.

PROPOSITION IV.6. *Suppose X is a standard Borel G space and X/G's quotient Borel structure is countably separated. Then each ergodic quasi-invariant σ-finite measure on X is supported on an orbit in X.*

PROOF. We may assume μ is finite. Let E_i, $i = 1, 2, \cdots$ be a countable separating family of Borel sets of X/G which is closed under complementation. Then if p is the projection $x \mapsto [x]$ from X onto X/G, the sets $F_i = p^{-1}(E_i)$ are G invariant Borel subsets of X having the property that $[x] = \cap_{[x] \in E_i} F_i$ for each x in X. Since μ is ergodic, $\mu(F_i) = 0$ or $\mu(X - F_i) = 0$ for each i. Let $F = \cap_{\mu(X - F_i) = 0} F_i$. Then $\mu(X - F) = 0$. To see that F is an orbit, note that if $[x] \cup [y] \subset F$ and $[x] \neq [y]$, then there is an F_i with $[x] \subset F_i$ and $[y] \cap F_i = \emptyset$. But since $\mu(F_i) = 0$ or $\mu(X - F_i) = 0$, either $[x] \cap F_i$ or $[y] \cap F_i$ is the empty set. This is a contradiction. $\square$

A space X is a **Polish** G space if both X and G are Polish and the action of G on X is Polish.

An important example of a Polish G space is the space $H \backslash G$ of right cosets of a closed subgroup H of a second countable locally compact group G. Indeed, by Corollary IV.1, $H \backslash G$ is a Polish space when given the quotient topology, and it is an easy matter to show the mapping $(Hx, g) \mapsto Hxg$ is continuous on $H\backslash G \times G$ into $H\backslash G$.

PROPOSITION IV.7. *Let G be a second countable locally compact group, and let H be a closed subgroup. Then any two quasi-invariant σ-finite measures on $H\backslash G$ are equivalent. In fact, if μ is such a measure, then $\mu(E) = 0$ iff the preimage of E in G has Haar measure 0.*

PROOF. Let μ be a σ-finite measure which is quasi-invariant. We may replace μ by an equivalent finite measure and hence may assume it is finite. Let ν be a probability measure on H equivalent to Haar measure. Define $\mu'(E) = \iint 1_E(hx) \, d\nu(h) \, d\mu(Hx)$ for Borel subsets E of G. Then μ' is a finite right quasi-invariant measure on G. Thus μ' is equivalent to Haar measure on G. But if W is a Borel subset of $H\backslash G$, then $\mu'(p^{-1}(W)) = \mu(W)$ where p is the natural projection of G onto $H\backslash G$. $\square$

COROLLARY IV.6. *If μ and ν are σ-finite invariant measures on $H\backslash G$, then there is a scalar $c > 0$ such that $\mu = c\nu$.*

PROOF. By Proposition IV.7 and the Radon-Nikodym Theorem, there is a positive Borel function F on $H\backslash G$ such that $\mu(E) = \int_E F(Hx) \, d\nu(Hx)$ for Borel subsets E in $H\backslash G$. By the invariance of μ and ν and the a.e. uniqueness of Radon-Nikodym derivatives, one has $F(Hxg) = F(Hx)$ a.e. Hx for each g. Again by Proposition IV.7, $f(x) \equiv F(Hx)$ satisfies $f(xg) = f(x)$ a.e. x for each g. By Corollary IV.5, $f(x)$ is constant a.e. x. Thus F is essentially constant. $\square$

THEOREM IV.7. *Suppose G is a second countable locally compact group and H is a closed subgroup. Then there is a nonzero invariant measure on $H\backslash G$ which is finite on compact sets iff $\delta_G(h) = \delta_H(h)$ for all h in H.*

PROOF. Assume ν is a nonzero σ-finite invariant measure on $H\backslash G$ which is finite on compact sets. Define $I(f) = \iint f(hx) \, dh \, d\nu(Hx)$ for $f \in C_c(G)$. Then I is right invariant. Thus by the Riesz representation theorem and Theorem IV.4, I is a right invariant Haar integral. Therefore, $\delta_G(h_1)I(f) = \delta_G(h_1) \int f(g) \, dg = \int f(h_1^{-1}g) \, dg = \iint f(h_1^{-1}hg) \, dh \, d\nu(Hg) = \delta_H(h_1)I(f)$; and we see $\delta_G(h_1) = \delta_H(h_1)$.

Conversely, suppose the modular functions are equal on H. Suppose $f \in C_c(G)$ and $f_H(Hg) = \int f(hg) \, dh$. Set $J(f_H) = \int f(g) \, dg$.

To see this is well defined, it suffices to show $\int f(g)\,dg = 0$ whenever $f_H = 0$. Then by Lemma IV.2, we would have a right invariant positive linear functional J on $C_c(H\backslash G)$. The Riesz representation theorem would yield a nonzero invariant measure which is finite on compact sets. Hence suppose $f_H(Hx) = \int f(hx)\,dh = 0$ for all x. Then $\iint \phi(x)f(hx)\,dh\,dx = 0$ for all $\phi \in C_c(G)$. By Fubini's Theorem and $d(h^{-1}x) = \delta_G(h^{-1})\,dx$, we see $\iint \phi(h^{-1}x)f(x)\delta_G(h^{-1})\,dx\,dh = 0$. Again by Fubini and the fact that $dh^{-1} = \delta_H(h^{-1})\,dh = \delta_G(h^{-1})\,dh$, we have $\iint \phi(hx)f(x)\,dh\,dx = 0$. Therefore $\int \phi_H(Hx)f(x)\,dx = 0$ for all $\phi \in C_c(G)$. By Lemma IV.2, there is a ϕ with $\phi_H(Hx) = 1$ for $x \in \mathrm{supp}\,(f)$. Hence $\int f(x)\,dx = 0$. $\square$

5. A Universal Borel G Space

Let G be a second countable locally compact group. We next construct a **universal G space**. This concept was introduced by G. Mackey.

Let μ be a right invariant Haar measure on G. Set $\mathcal{U}_G$ to be the set of all Borel measurable functions $f : G \to [0,1]$ identified when they are equal almost everywhere. For each compact subset K, define a pseudometric ρ_K on $\mathcal{U}_G$ by $\rho_K(f_1, f_2) = \int_K |f_1(x) - f_2(x)|\,d\mu(g)$; and let $\mathcal{U}_G$ have the topology generated by these pseudometrics. Define an action of G on $\mathcal{U}_G$ by $(f \cdot g)(x) = f(xg^{-1})$.

PROPOSITION IV.8. $\mathcal{U}_G$ is a Polish G space. Specifically, $\mathcal{U}_G$ is a Polish space and the action $(f, g) \mapsto f \cdot g$ is continuous.

PROOF. To see $\mathcal{U}_G$ is Polish, take a sequence K_n of compact sets whose interiors cover G. Define $\rho = \sum_n \frac{1}{2^n}\rho_{K_n}$. Then ρ is a metric which defines the topology on $\mathcal{U}_G$. To see that ρ is complete, note that a sequence f_n is Cauchy if and only if f_n is Cauchy in measure on each compact set. It follows that there exists an f in $\mathcal{U}_G$ such that f_n converges in measure to f on every compact set K. We thus see f_n converges to f in $\mathcal{U}(G)$. To see that $\mathcal{U}_G$ is separable, take a countable generating algebra $\mathcal{A}$ of Borel subsets of G. Then the collection of simple functions with rational values in $[0,1]$ whose level sets are in $\mathcal{A}$ is a countable dense set.

In order to see that the action is continuous, first note that if g_n converges to g and K is a compact set, then there exists a compact set K' which contains all the sets Kg_n^{-1}. Hence if f_n converges to f and g_n converges to g, then by the right invariance of μ, $\rho_K(f_n \cdot g_n, f \cdot g) \le \rho_K(f_n \cdot g_n, f \cdot g_n) + \rho_K(f \cdot g_n, f \cdot g) \le \rho_{Kg_n^{-1}}(f_n, f) + \rho_K(f \cdot g_n, f \cdot g) \le \rho_{K'}(f_n, f) + \rho_K(f \cdot g_n, f \cdot g)$. Thus $f_n \cdot g_n$ converges to $f \cdot g$ if

$\rho_K(f \cdot g_n, f \cdot g)$ converges to 0. Hence it suffices to show $g \mapsto f \cdot g$ is continuous and this we need only do at the identity.

Suppose g_n converges to e. Choose a compact set K' as before. Pick a continuous function $\phi \in \mathcal{U}_G$ such that $\rho_{K'}(f, \phi)$ is small. Then $\rho_K(f \cdot g_n, f) \leq \rho_K(f \cdot g_n, \phi \cdot g_n) + \rho_K(\phi \cdot g_n, \phi) + \rho_K(\phi, f) \leq 2\rho_{K'}(f, \phi) + \rho_K(\phi \cdot g_n, \phi)$, which is small if n is large, for $g \mapsto \phi \cdot g$ is continuous by the dominated convergence theorem. $\square$

LEMMA IV.4. *A function ϕ from a Borel space X into $\mathcal{U}_G$ is Borel if and only if $x \mapsto \int_K \phi(x)(g)f(g)\,d\mu(g)$ is Borel for all f in $\mathcal{U}_G$ and all compact subsets K.*

PROOF. Note $h \mapsto \int_K h(g)f(g)\,d\mu(g)$ is continuous on $\mathcal{U}_G$ for each f. Thus if ϕ is Borel, the map $x \mapsto \int_K \phi(x)(g)f(g)\,d\mu(g)$ is Borel.

To prove the converse, it suffices by Proposition I.4 to note that there is a countable set of functions of the form $h \mapsto \int_K h(g)f(g)\,d\mu(g)$ which separate the points of $\mathcal{U}_G$. $\square$

THEOREM IV.8. *Let X be a standard Borel G space. Then there exists a Borel isomorphism Φ from X into $\mathcal{U}_G$ satisfying $\Phi(x \cdot g) = \Phi(x) \cdot g$ for all x and g. In particular, there is a separable metric on X which makes the action continuous.*

PROOF. Since X is standard, we may assume S is a Borel subset of $[0, 1]$ with the relative Borel structure. Define $\Phi(x)(g) = x \cdot g^{-1}$. Then $\Phi(x \cdot g)(g_1) = x \cdot g \cdot g_1^{-1} = x \cdot (g_1 g^{-1})^{-1} = \Phi(x)(g_1 g^{-1}) = (\Phi(x) \cdot g)(g_1)$. Moreover, if $\Phi(x_1) = \Phi(x_2)$, then $x_1 \cdot g^{-1} = x_2 \cdot g^{-1}$ a.e. g. Choose g so that this equality holds. Then $x_1 = x_1 \cdot g^{-1} \cdot g = x_2 \cdot g^{-1} \cdot g = x_2$. Hence Φ is one-to-one. Lemma IV.4 implies Φ is Borel. Indeed, $x \mapsto \int_K \Phi(x)(g)f(g)\,d\mu(g) = \int_K (x \cdot g^{-1})f(g)\,d\mu(g)$ is Borel by Fubini's Theorem. The conclusion now is a consequence of Theorem I.8. $\square$

COROLLARY IV.7. *Suppose the group G is compact. Then there exists a Borel set E meeting each G orbit in exactly one point, and the space of orbits X/G with the quotient Borel structure is standard.*

PROOF. We may assume the action of G is continuous. Then the saturation of every open set is open, and each orbit is compact and hence complete. By Theorem I.15, there exists a Borel set E meeting each orbit $[x] = x \cdot G$ in exactly one point. Moreover, if F is a Borel subset of E, then the saturations of F and $E - F$ are disjoint complementary sets. But the saturation of any Borel set W is the image of $W \times G$ under the mapping $(x, g) \mapsto x \cdot g$. In particular these saturations are analytic. By Corollary I.2, $[F]$ is a Borel set. Using Proposition I.5, we see the space of G orbits is standard. $\square$

6. Cocycles on Borel G Spaces

DEFINITION. Let X be a Borel G space and H be a standard Borel group. A **cocycle** on X with values in H is a Borel mapping ϕ from $X \times G$ into H satisfying $\phi(x,g)\phi(xg,g') = \phi(x,gg')$ for all x, g, and g'.

THEOREM IV.9. *Suppose X is a standard Borel G space with a quasi-invariant measure μ. Then if ϕ is a Borel mapping from $X \times G$ into a standard Borel group H satisfying $\phi(x,g)\phi(xg,g') = \phi(x,gg')$ a.e. x, g, and g', then there exists a Borel cocycle ψ equal to ϕ a.e. on $X \times G$.*

PROOF. Let $F = \{(x,g) : \phi(x,gg_1)\phi(xg,g_1)^{-1}$ is constant a.e. $g_1\}$. By Proposition I.8, F is a Borel set. By Fubini's Theorem, F is conull. Moreover, if (x,g) and (xg,g') are in F, then both

$$\phi(x,gg_1)\phi(xg,g_1)^{-1} \text{ and } \phi(xg,g'g_1)\phi(xgg',g_1)^{-1}$$

are constant a.e. g_1. Thus

$$\phi(x,gg'g_1)\phi(xg,g'g_1)^{-1}\phi(xg,g'g_1)\phi(xgg',g_1)^{-1} = \phi(x,gg'g_1)\phi(xgg',g_1)^{-1}$$

is constant a.e. g_1. Hence if (x,g) and (xg,g') are in F, then (x,gg') is in F. Because of this property, we say that F is multiplicatively closed.

Now let ν be a probability measure on G equivalent to Haar measure. Redefine ϕ on F by letting $\phi(x,g)$ be the a.e constant value of the mapping $g_1 \mapsto \phi(x,gg_1)\phi(xg,g_1)^{-1}$. Note ϕ remains Borel on F, for

$$\phi(x,g) = \int \phi(x,gg_1)\phi(xg,g_1)^{-1} \, d\nu(g_1)$$

is Borel by Fubini's Theorem where here we have identified H Borel isomorphically to a Borel subset of $[0,1]$. Moreover, the argument showing that $(x,gg') \in F$ when (x,g) and (xg,g') are in F shows that the redefined ϕ satisfies $\phi(x,gg') = \phi(x,g)\phi(xg,g')$ when (x,g) and (xg,g') are in F. Furthermore, the redefined ϕ equals the original ϕ almost everywhere.

Since the Borel mapping $(x,g) \mapsto (xg,g^{-1})$ preserves the measure class of $\mu \times \nu$, the set X_0 consisting of those x with $(x,g) \in F$ and $(xg,g^{-1}) \in F$ for a.e. g is a conull Borel subset of X. We now note that X_0 has the property that whenever $x \in X_0$ and $xg \in X_0$, then

$(x, g) \in F$. In fact in this case note we can choose an h satisfying both $(x, gh) \in F$ and $(xgh, h^{-1}) \in F$. Thus, since F is multiplicatively closed, $(x, g) \in F$.

Now by Theorem IV.8, we may assume that X is an invariant Borel subset of $\mathcal{U}_G$ and that the action of G on X is the restriction of the universal G action to X. By Theorem I.11, there is a conull σ-compact subset W of X_0. Let f be the continuous function on the σ-compact set $W \times G$ defined by $f(x, g) = xg$. The image of this map is the **saturation** $[W]$ of W; that is, the smallest G invariant set containing W. In this case $[W]$ is σ-compact and hence Borel. By Corollary I.8, the function f has a Borel right inverse. This right Borel inverse of f must have form $x \mapsto (x\theta(x)^{-1}, \theta(x))$ where θ is a Borel function from $[W]$ into G. In particular, $x\theta(x)^{-1} \in W$ for all $x \in [W]$. We redefine θ so that $\theta(x) = e$ for $x \in W$.

Now define ψ by $\psi(x, g) = \phi(x\theta(x)^{-1}, \theta(x)g\theta(xg)^{-1})$ if $x \in [W]$ and $\phi(x, g) = e$ if $x \notin [W]$. Note that $\psi(x, g) = \phi(x, g)$ if $x \in W$ and hence ψ equals ϕ a.e. To see that ψ is a cocycle, note that if $x \in [W]$, then

$$
\begin{aligned}
\psi(x, g)\psi(xg, g') &= \phi(x\theta(x)^{-1}, \theta(x)g\theta(xg)^{-1})\phi(xg\theta(xg)^{-1}, \theta(xg)g'\theta(xgg')^{-1}) \\
&= \phi(x\theta(x)^{-1}, \theta(x)xgg'\theta(xgg')^{-1}) \\
&= \psi(x, gg')
\end{aligned}
$$

since $x\theta(x)^{-1}$, $xg\theta(xg)^{-1}$, and $xgg'\theta(xgg')^{-1}$ are all in X_0. That $\psi(x, g)\psi(xg, g') = \psi(x, gg')$ holds when $x \notin [W]$ is immediate. $\square$

COROLLARY IV.8. *Let G be a second countable locally compact group and assume ϕ is a Borel mapping from G into a standard Borel group satisfying $\phi(g_1 g_2) = \phi(g_1)\phi(g_2)$ a.e. g_1 and g_2. Then there exists a Borel homomorphism ψ which is equal to ϕ a.e. on G.*

Cocycles equal almost everywhere do not need to be equal everywhere; but they are more than almost everywhere equal.

PROPOSITION IV.9. *Let φ and ψ be cocycles on the standard Borel G space X. Assume μ is a quasi-invariant measure on X. If $\varphi(x, g) = \psi(x, g)$ a.e. (x, g), then for all g, $\varphi(x, g) = \psi(x, g)$ for almost all x.*

PROOF. Set $G_0 = \{g : \varphi(x, g) = \psi(x, g) \text{ for almost all } x\}$. Then G_0 is conull. Moreover, if g_1 and g_2 are in G_0, then $\varphi(x, g_1) = \psi(x, g_1)$

for a.e. x and $\varphi(x \cdot g_1, g_2) = \psi(x \cdot g_1, g_2)$ for a.e. x. It follows from the cocycle property that

$$\varphi(x, g_1 g_2) = \varphi(x, g_1) \varphi(x \cdot g_1, g_2) = \psi(x, g_1) \psi(x \cdot g_1, g_2) = \psi(x, g_1 g_2)$$

for a.e. x. Thus G_0 is closed under multiplication. Hence if $g_3 \in G$, then since $g_3 G_0^{-1}$ and G_0 are conull, there exist g_1 and g_2 in G_0 such that $g_3 g_2^{-1} = g_1$. Thus $g_3 = g_1 g_2 \in G_0^2 \subset G_0$ and hence $G_0 = G$. □

7. Unitary Representations of Topological Groups

DEFINITION. A unitary representation of a topological group G is a continuous homomorphism from G into the unitary group $\mathcal{U}(\mathbb{H})$ of some Hilbert space $\mathbb{H}$.

A unitary representation U of a second countable locally compact group G on a separable Hilbert space defines a unitary Borel action of G on the Hilbert space of U. Indeed, since the mappings $g \mapsto U(g^{-1})v$ for $v \in \mathbb{H}$ and the mappings $v \mapsto U(g)v$ for $g \in G$ are continuous, it follows by Proposition I.19 that the mapping $(v, g) \mapsto U(g^{-1})v$ is Borel on $\mathbb{H} \times G$. Conversely a unitary Borel right action of a locally compact group G comes from a unitary representation of G. In fact define $U(g)v = v \cdot g^{-1}$. Then U is a Borel homomorphism of G into the Polish group $\mathcal{U}(\mathbb{H})$. It follows by Theorem IV.2 that U is continuous.

A linear subspace W of $\mathbb{H}$ is said to be **invariant** under the representation U if $U(g)W \subset W$ for all $g \in G$. The representation U is said to be **irreducible** if the only closed invariant subspaces of $\mathbb{H}$ are 0 and $\mathbb{H}$. Two unitary representations U_1 and U_2 of G on Hilbert spaces $\mathbb{H}_1$ and $\mathbb{H}_2$ are **unitarily equivalent** if there exists a unitary isomorphism T of $\mathbb{H}_1$ onto $\mathbb{H}_2$ satisfying $T U_1(g) = U_2(g) T$ for all $g \in G$. Notationally, we write $U_1 \cong U_2$.

PROPOSITION IV.10. *A unitary representation U is irreducible iff the von Neumann algebra generated by the set $U(G) = \{U(g) : g \in G\}$ is $\mathcal{B}(\mathbb{H})$.*

PROOF. Suppose the von Neumann algebra generated by $U(G)$ is $\mathcal{B}(\mathbb{H})$. Then if P is the orthogonal projection onto a closed invariant subspace W, then $PU(g) = U(g)P$ for all g. This, by the double commutant theorem, implies that P commutes with every bounded operator. Thus $P = 0$ or $P = I$.

Conversely, if U is irreducible, then the only orthogonal projections commuting with every operator $U(g)$ are 0 and I. Thus the commutant of the von Neumann algebra generated by $U(G)$ is the algebra of scalar operators. By the double commutant theorem, the von Neumann algebra generated by $U(G)$ is $\mathcal{B}(\mathbb{H})$. $\square$

Let U and V be unitary representations of G. Then an **intertwining operator** T between U and V is a bounded linear operator from the Hilbert space for U into the Hilbert space for V satisfying

$$TU(g) = V(g)T$$

for all g. Let $R(U,V)$ denote the collection of all intertwining operators between U and V. Since $R(U,U) = U(G)'$, Proposition IV.10 is equivalent to U being irreducible iff $R(U,U) = \mathbb{C}I$.

PROPOSITION IV.11 (SCHUR'S LEMMA). *Let U and V be irreducible unitary representations of G. Then U is unitarily equivalent to V iff $R(U,V) \neq \{0\}$. Moreover, if W is a unitary operator establishing a unitary equivalence between U and V, then $R(U,V) = \mathbb{C}W$.*

PROOF. Suppose T is a nonzero intertwining operator between U and V. Since $TU(g^{-1}) = V(g^{-1})T$, taking adjoints implies $U(g)T^* = T^*V(g)$. Since this hold for all g, $T^* \in R(V,U)$. Thus $TT^* \in R(V,V)$ and $T^*T \in R(U,U)$. Since $R(V,V) = \mathbb{C}I$ and $R(U,U) = \mathbb{C}I$ where I denotes the identity operator on the appropriate Hilbert space, we see $T^*T = cI$ and $TT^* = dI$. Since $T \neq 0$, $c > 0$ and $d > 0$. Moreover, $d^2I = TT^*TT^* = TcT^* = cdI$ implies $c = d$. Hence $W = \frac{1}{\sqrt{c}}T$ is a unitary operator intertwining U and V, and U is unitarily equivalent to V.

Suppose W is a unitary operator between the irreducible representations U and V. Let $T \in R(U,V)$. Then $TW^* \in R(V,V)$. Hence $TW^* = cI$ for some c. This implies $T = cW$, and hence $R(U,V) = \mathbb{C}W$. $\square$

A unitary representation U is **type I** if the von Neumann algebra $\mathcal{A}_U$ generated by the set $U(G)$ is type I. The representation U is said to be **primary** if the von Neumann algebra $\mathcal{A}_U$ is a factor. A representation U is said to be **locally simple** if $\mathcal{A}_U'$ is abelian.

PROPOSITION IV.12. *Every primary type I unitary representation U is unitarily equivalent to a unique multiple of an irreducible unitary representation. The equivalence class of this irreducible unitary representation is unique.*

PROOF. Indeed, by the Theorems III.21 and III.22 and the double commutant theorem, the von Neumann algebra $\mathcal{A}_U$ is unitarily isomorphic to $n\mathcal{B}(\mathbb{H})$ for a unique n and some Hilbert space $\mathbb{H}$. Using this isomorphism, we may assume $U(g) \in n\mathcal{B}(\mathbb{H})$ for all g. Thus $U(g) = n\pi(g)$ where $\pi(g)$ is a unitary operator on $\mathbb{H}$. Moreover, π is a unitary representation on $\mathbb{H}$; and since $\mathcal{A}_U = n\mathcal{B}(\mathbb{H})$, we have $\mathcal{A}_\pi = \mathcal{B}(\mathbb{H})$. Thus by Proposition IV.10, the representation π is irreducible.

To see that the unitary equivalence class of π is unique, assume $n\pi$ and $n\pi'$ are unitarily equivalent. Let P_i be the projection from $n\mathbb{H}$ onto its ith component and let Q_j be the projection of $n\mathbb{H}'$ onto its jth component. Let T be a unitary operator from $n\mathbb{H}$ onto $n\mathbb{H}'$ giving a unitary equivalence between $n\pi$ and $n\pi'$. Then there are an i and j such that $M_{i,j} = Q_j T P_i$ is a nonzero partial isometry that establishes a unitary equivalence between the restriction of $n\pi$ to a nonzero invariant subspace of the representation $n\pi|_{P_i(n\mathbb{H})}$ and a nonzero invariant subspace of the representation $n\pi'|_{Q_j(n\mathbb{H}')}$. Since $n\pi|_{P_i(n\mathbb{H})}$ is unitarily equivalent to π and $n\pi'|_{Q_j(n\mathbb{H}')}$ is unitarily equivalent to π', we see that $M_{i,j}^* M_{i,j} = P_i$ and $M_{i,j} M_{i,j}^* = Q_j$, and the representations π and π' are unitarily equivalent. $\square$

PROPOSITION IV.13. *Every locally simple unitary representation on a separable Hilbert space is cyclic.*

PROOF. By Lemma III.8, it suffices to show $\mathcal{A}_U$ has a separating vector. But this follows from Proposition III.1 and Theorem III.8 and the spectral theorem. $\square$

PROPOSITION IV.14. *U is a primary locally simple representation iff it is irreducible.*

PROOF. If U is locally simple and primary, then $\mathcal{A}'_U$ is the center of $\mathcal{A}'_U$ and hence is $\mathbf{C}I$. Thus $\mathcal{A}_U = \mathcal{A}''_U = \mathcal{B}(\mathbb{H})$. Hence U is irreducible. Conversely, if U is irreducible, $\mathcal{A}'_U = \mathbf{C}I$ and thus U is primary and locally simple. $\square$

THEOREM IV.10. *Let U be a type I unitary representation on a separable Hilbert space. Then there exist unique maximal central projections $P^\infty, P^1, P^2, \ldots$, such that whenever P^n is nonzero, the representation U restricted to $P^n\mathbb{H}$ is unitarily equivalent to nU_n for some locally simple representation U_n*

PROOF. Let $\mathcal{A} = \mathcal{A}_U$. Then $\mathcal{A}'$ is type I; and by Theorem III.22, there are unique maximum central projections P_n such that the von

Neumann algebras $\mathcal{A}'_{P_n}$ are homogeneous of degree n. Since $(\mathcal{A}_P)' = \mathcal{A}'_P$ for central projections P, the result will follow if we show that a representation U is nV for some locally simple representation V if and only if $\mathcal{A}'_V$ is homogeneous of degree n. But this is clear for if $U = nV$, then $\mathcal{A} = \mathcal{D}_n(\mathcal{A}_V)$; and thus $\mathcal{A}' = \mathcal{M}_n(\mathcal{A}'_V)$, and one sees $\mathcal{A}'$ is homogeneous of degree n when $\mathcal{A}'_V$ is abelian. Conversely, if $\mathcal{A}'$ is homogeneous of degree n, then we may assume $\mathcal{A}' = \mathcal{M}_n(\mathcal{C})$ where $\mathcal{C}$ is an abelian von Neumann algebra acting on a Hilbert space $\mathbb{H}_0$. Hence $\mathcal{A} = \mathcal{D}_n(\mathcal{C}')$ where the von Neumann algebra $\mathcal{C}'$ has abelian commutant. But then it follows that $U = nV$ where the von Neumann algebra generated by the representation V is $\mathcal{C}'$. Thus V is locally simple. $\square$

8. The Convolution Measure Algebra

Let G be a locally compact Hausdorff group. Let $M(G)$ be the space of complex Baire measures on G. (These are the complex measures on the σ-algebra generated by the compact G_δ sets.) The Riesz representation theorem says that if I is a bounded integral on $C_c(G)$, then there is a unique Baire measure μ with

$$I(f) = \int f(x)\, d\mu(x).$$

Every Baire measure has a unique extension to a regular Borel measure; and if G is second countable, the Baire sets are the same as the Borel sets.

One can define a convolution of two complex measures μ and ν in $M(G)$. Namely

$$I_{\mu*\nu}(f) = \iint f(xy)\, d\mu(x)\, d\nu(y).$$

Note this is bounded for $|I_{\mu*\nu}(f)| \leq ||f||_\infty |\mu|\, |\nu|$. Hence there is a measure $\mu * \nu$ satisfying

$$\iint f(xy)\, d\mu(x)\, d\nu(y) = \int f(z)\, d\mu * \nu(x).$$

This measure is called the convolution of μ with ν. One can show that $M(G)$ under addition and convolution is an algebra. Moreover, ϵ_e the point mass at e is the identity in this algebra. Furthermore, $M(G)$ can be turned into a $*$ algebra. Namely, define $\mu^*(E) = \overline{\mu(E^{-1})}$.

Then $d\mu^*(x) = \overline{d\mu(x^{-1})}$. In particular one has

$$\int f(x)\,d\mu^*(x) = \overline{\int \bar{f}(x^{-1})\,d\mu(x)}.$$

One notes

$$|\mu^*| = |\mu|$$
$$\mu^{**} = \mu$$
$$(c\mu)^* = \bar{c}\mu^*$$
$$(\mu + \nu)^* = \mu^* + \nu^*$$
$$(\mu * \nu)^* = \nu^* * \mu^*.$$

To see the last equality, note

$$\int f(z)\,d(\mu * \nu)^*(z) = \overline{\int \bar{f}(z^{-1})\,d\mu * \nu(z)}$$
$$= \overline{\iint \bar{f}((xy)^{-1})\,d\mu(x)\,d\nu(y)}$$
$$= \overline{\iint \bar{f}(y^{-1}x^{-1})\,d\mu(x)\,d\nu(y)}$$
$$= \int \overline{\int \bar{f}(yx^{-1})\,d\mu(x)}\,d\nu^*(y)$$
$$= \iint f(yx)\,d\mu^*(x)\,d\nu^*(y)$$
$$= \iint f(z)\,d(\nu^* * \mu^*)(z).$$

Thus $M(G)$ is a $*$ algebra under convolution. Since $|I_{\mu*\nu}(f)| \le |\mu||\nu|\|f\|_\infty$, one has $|\mu * \nu| \le |\mu|\,|\nu|$; and clearly $|\epsilon_e| = 1$, $|\mu^*| = |\mu|$. Thus $M(G)$ is a normed $*$ algebra. We also note it is complete under the norm $|\mu|$. Indeed, if μ_n is Cauchy, then for each $f \in C_c(G)$, $\lim I_{\mu_n}(f)$ exists and defines a bounded linear functional on $C_c(G)$. Moreover, by the Riesz Representation Theorem, this integral is given by a unique Baire measure μ, and one can show $\mu_n \to \mu$.

In any case, $M(G)$ is a Banach $*$ algebra with identity and is called the group measure algebra of G.

Note $M(G)$ need not be a C^* algebra. Indeed, suppose there is an element $x \in G$ with $x \ne x^{-1}$. Set $\mu = \epsilon_e + \epsilon_x - \epsilon_{x^{-1}}$. Then $|\mu| = 3$, and thus $|\mu|^2 = 9$. Now $\epsilon_x^* = \epsilon_{x^{-1}}$. Thus $\mu^*\mu = -\epsilon_e + 2\epsilon_x - 2\epsilon_{x^{-1}} + \epsilon_{x^2} + \epsilon_{x^{-2}}$ which has norm at most 7.

9. The L^1 Algebra

Let dx be a right Haar measure on G. One can imbed $L^1(G)$ as a subalgebra of $M(G)$. Indeed, for $f \in L^1(G)$ define a measure $\mu_f \in M(G)$ by $d\mu_f(x) = f(x)dx$. Thus $\mu_f(E) = \int_E f(x)\,dx$. Then μ_f is a complex measure with total variation

$$|\mu_f| = \int |f(x)|\,dx = ||f||_1.$$

PROPOSITION IV.15. $L^1(G)$ is a closed two sided $*$ ideal in $M(G)$. It is thus a Banach $*$ subalgebra of $M(G)$. In particular $f^*(x) = \delta(x^{-1})\bar{f}(x^{-1})$ and

$$f * h(x) = \int f(xy^{-1})h(y)\,dy = \int f(y^{-1})h(yx)\,dy.$$

PROOF. We calculate μ_f^*.

$$\mu_f^*(E) = \bar{\mu}_f(E^{-1})$$
$$= \overline{\int_{E^{-1}} f(x)\,dx}$$
$$= \int 1_{E^{-1}}(x)\bar{f}(x)\,dx$$
$$= \int 1_E(x^{-1})\bar{f}(x)\,dx$$
$$= \int 1_E(x)\delta(x^{-1})\bar{f}(x^{-1})\,dx$$
$$= \int_E \delta(x^{-1})\bar{f}(x^{-1})\,dx.$$

Thus $\mu_f^* = \mu_{f^*}$ where $f^*(x) = \delta(x^{-1})\overline{f(x^{-1})}$. We next calculate $\mu_f * \mu_h$. Indeed

$$\int F(z)\,d\mu_f * \mu_h(z) = \iint F(xy)\,d\mu_f(x)\,d\mu_h(y)$$
$$= \iint F(xy)f(x)h(y)\,dy\,dx$$
$$= \iint F(x)f(xy^{-1})h(y)\,dy\,dx$$
$$= \int F(x)\int f(xy^{-1})h(y)\,dy\,dx$$
$$= \int F(x)f * h(x)\,dx,$$

and thus $d(\mu_f * \mu_h)(x) = f * h(x)dx$. Since $|\mu_f * \mu_h| \leq ||f||_1||h||_1$, we see

$$||f * h||_1 \leq ||f||_1||h||_1.$$

Furthermore $|\mu^*| = |\mu|$ yields $||f^*||_1 = ||f||_1$.

We now show it is an ideal. Consider $f * \mu$. It is the measure which satisfies

$$\int h(z)\, d(f * \mu)(z) = \iint h(xy)f(x)\, dx\, d\mu(y)$$
$$= \iint h(x)f(xy^{-1})\, dx\, d\mu(y)$$
$$= \int h(x) \int f(xy^{-1})\, d\mu(y)\, dx.$$

Thus

$$d(f * \mu)(x) = \left(\int f(xy^{-1})\, d\mu(y) \right) dx.$$

Hence

$$f * \mu(x) = \int f(xy^{-1})\, d\mu(y) \text{ and } ||f * \mu||_1 \leq ||f||_1|\mu|.$$

Similarly

$$\int h(x)\, d(\mu * f)(x) = \iint h(yx)\, d\mu(y)\, f(x)\, dx$$
$$= \iint h(x)f(y^{-1}x)\, d(y^{-1}x)\, d\mu(y)$$
$$= \int h(x) \int \delta(y^{-1})f(y^{-1}x)\, d\mu(y)\, dx$$
$$= \int h(x)\, \mu * f(x)\, dx$$

where

$$\mu * f(x) = \int \delta(y^{-1})f(y^{-1}x)\, d\mu(y).$$

Again $||\mu * f||_1 \leq |\mu|\, ||f||_1$. $\square$

PROPOSITION IV.16. *The algebra $L^1(G)$ has an approximate unit.*

PROOF. Let Λ be the set of all compact neighborhoods of the identity ordered by inclusion. Thus $\lambda \leq \lambda'$ if $\lambda' \subset \lambda$. For each λ, let f_λ be a continuous positive function with support inside λ for which $\int f_\lambda(x)\, dx = 1$ and $f_\lambda(x^{-1}) = f(x)$ for all x. We show $||f_\lambda * h - h||_1 \to 0$. First let h be a continuous function of compact support K. Proposition IV.3 implies there exists a compact neighborhood U of e so that $|h(yx) - h(x)| < \epsilon$ if $y \in U$. Suppose $\lambda > U$. Then

$$
\begin{aligned}
||f_\lambda * h - h||_1 &= \int |f_\lambda * h(x) - h(x)|\, dx \\
&= \int |\int f_\lambda(y)(h(yx) - h(x))\, dy|\, dx \\
&\leq \int\!\!\int_U f_\lambda(y)|h(yx) - h(x)|\, dy\, dx \\
&= \int_U f_\lambda(y) \int |h(yx) - h(x)|\, dx\, dy \\
&\leq \int f_\lambda(y) m(U^{-1}K)\epsilon dy = m(U^{-1}K)\epsilon.
\end{aligned}
$$

Now if $h \in L^1(G)$, choose $h_1 \in C_c(G)$ with $||h - h_1||_1 < \frac{\epsilon}{3}$. Then

$$
\begin{aligned}
||f_\lambda * h - h|| &= ||f_\lambda * h - f_\lambda * h_1 + f_\lambda * h_1 - h_1 + h_1 - h|| \\
&\leq ||f_\lambda||_1 ||h - h_1|| + ||f_\lambda * h_1 - h_1||_1 + ||h_1 - h||_1 \\
&< \frac{2\epsilon}{3} + ||f_\lambda * h_1 - h_1|| \\
&< \epsilon
\end{aligned}
$$

for λ large. The argument for $||h * f_\lambda - h||$ is similar. $\square$

$L^1(G)$ is a Banach $*$-algebra with an approximate unit. The universal enveloping C^* algebra of $L^1(G)$ is denoted by $C^*(G)$ and is called the group C^* algebra of the G.

LEMMA IV.5. *The natural mapping τ of $L^1(G)$ into $C^*(G)$ is one-to-one.*

PROOF. Define a representation π of $L^1(G)$ on $L^2(G)$ formally by

$$
\pi(f)h = (\delta^{1/2}f) * h.
$$

First note $\pi(f)$ is bounded. Indeed

$$|\langle \pi(f)h, h'\rangle| = |\int (\delta^{1/2}f) * h(x)\,\bar{h}'(x)dx|$$

$$= |\iint \delta(y^{-1})^{1/2}f(y^{-1})h(yx)\,dy\,\bar{h}'(x)\,dx|$$

$$= |\iint \delta(y^{-1})\delta(y)^{1/2}f(y)h(y^{-1}x)\bar{h}'(x)\,dx\,dy|$$

$$\leq \int |f(y)|\,\delta(y)^{-1/2}\int |h(y^{-1}x)\bar{h}'(x)|dxdy.$$

Hence by the Cauchy-Schwarz inequality,

$$|\langle \pi(f)h, h'\rangle| \leq \int |f(y)|\delta(y)^{-1/2}\left(\int |h(y^{-1}x)|^2\,dx\right)^{1/2}\left(\int |h'(x)|^2\,dx\right)^{1/2}dy$$

$$= \int |f(y)|\delta(y)^{-1/2}\left(\delta(y)\int |h(x)|^2\,dx\right)^{1/2}\left(\int |h'(x)|^2\,dx\right)^{1/2}dy$$

$$= \int |f(y)|\left(\int |h(x)|^2\,dx\right)^{1/2}\left(\int |h'(x)|^2\,dx\right)^{1/2}dy$$

$$= ||f||_1||h||_2||h'||_2.$$

Thus $||\pi(f)|| \leq ||f||_1$.

Next, note π is a $*$ representation of $L^1(G)$. Indeed,

$$(\delta^{1/2}f_1) * (\delta^{1/2}f_2)(g) = \int \delta^{1/2}(y^{-1})f_1(y^{-1})\delta^{1/2}(yg)f_2(yg)\,dy$$

$$= \delta^{1/2}(g)(f_1 * f_2)(g),$$

and thus

$$\pi(f_1 * f_2)(h) = (\delta^{1/2}(f_1 * f_2)) * h$$

$$= (\delta^{1/2}f_1) * (\delta^{1/2}f_2) * h$$

$$= \pi(f_1)\pi(f_2)h,$$

and since

$$\langle \pi(f^*)h, h'\rangle = \langle (\delta^{1/2}f^*) * h, h'\rangle$$

$$= \iint \delta^{1/2}(x^{-1})f^*(x^{-1})h(xg)\bar{h}'(g)\,dx\,dg$$

$$= \iint \delta^{1/2}(x^{-1})\delta(x)\bar{f}(x)h(xg)\bar{h}'(g)\,dx\,dg$$

$$= \iint \delta^{1/2}(x)\bar{f}(x)\delta(x^{-1})h(g)\bar{h}'(x^{-1}g)\,dx\,dg$$

$$= \int h(g) \int \delta^{1/2}(x)\delta(x^{-1})\bar{f}(x^{-1})\bar{h}'(xg)\,dx\,dg$$

$$= \overline{\int h(g) \int \delta(x^{-1})^{1/2} f(x^{-1}) h'(xg)\,dx\,dg}$$

$$= \langle h, (\delta^{1/2}f) * h'\rangle$$

$$= \langle h, \pi(f)h'\rangle,$$

$\pi(f^*) = \pi(f)^*$.

Next note if $f \neq 0$, then $\pi(f) \neq 0$, for there is an $h \in C_c(G)$ with $f * h \neq 0$. Indeed, take h from an approximate unit. Thus $\|\tau(f)\| \geq \|\pi(f)\| > 0$. $\square$

PROPOSITION IV.17. *The following are equivalent:*

  (a) G *is abelian.*
  (b) $L^1(G)$ *is commutative.*
  (c) $M(G)$ *is commutative.*
  (d) $C^*(G)$ *is commutative.*

PROOF. Clearly (c) implies (b) and (a) implies (c). We show (b) implies (a). Consider x and y. Let f, h be in $L^1(G)$. Then

$$f * h * \epsilon_{xy} = f * h * \epsilon_x * \epsilon_y$$
$$= h * \epsilon_x * f * \epsilon_y$$
$$= f * \epsilon_y * h * \epsilon_x$$
$$= h * f * \epsilon_y * \epsilon_x$$
$$= f * h * \epsilon_{yx}.$$

Take $f = u_\lambda$ to be an approximate unit. By taking limits, we see

$$h * \epsilon_{xy} = h * \epsilon_{yx}$$

for all $h \in L^1(G)$. But $h * \epsilon_x(y) = \int g(yz^{-1})\,d\epsilon_x(z) = g(yx^{-1})$. Hence, for each $g \in C_c(G)$, we have

$$g(zy^{-1}x^{-1}) = g(zx^{-1}y^{-1})$$

for all z. Thus $g(y^{-1}x^{-1}) = g(x^{-1}y^{-1})$ for all g. Thus $y^{-1}x^{-1} = x^{-1}y^{-1}$, and hence $xy = yx$.

Finally (b) and (d) are equivalent, for τ is one-to-one and has dense range. $\square$

Recall a mapping f from a topological space X into $B(\mathbb{H})$ is strongly continuous if for each vector $v \in \mathbb{H}$ the mapping $x \mapsto f(x)v$ is continuous.

Let π be a unitary representation of G. For each $\mu \in M(G)$, we can define a bounded operator $\pi(\mu)$ on $\mathbb{H}$ by

$$\langle \pi(\mu)v, w \rangle = \int \langle \pi(x)v, w \rangle \, d\mu(x).$$

Note $||\pi(\mu)|| \leq |\mu|$. Moreover, if $f \in L^1(G)$, then $\pi(f)$ will be used to denote the operator $\pi(\mu_f)$ where $\mu_f(E) = \int_E f(x) \, dx$.

THEOREM IV.11. *There is a one-to-one correspondence between the unitary representations of G and the nondegenerate $*$ representations of the Banach algebra $L^1(G)$. Moreover, each nondegenerate $*$ representation of $L^1(G)$ has a unique extension to $M(G)$.*

PROOF. Let π be a unitary representation of G. Extend π to $M(G)$. An easy argument shows $\pi(\mu^*) = \pi(\mu)^*$. To see $\pi(\mu * \nu) = \pi(\mu)\pi(\nu)$, note

$$\begin{aligned}
\langle \pi(\mu * \nu)v, w \rangle &= \iint \langle \pi(xy)v, w \rangle \, d\mu(x) \, d\nu(y) \\
&= \iint \langle \pi(y)v, \pi(x^{-1})w \rangle d\nu(y) \, d\mu(x) \\
&= \int \langle \pi(\nu)v, \pi(x^{-1})w \rangle d\mu(x) \\
&= \int \langle \pi(x)\pi(\mu)v, w \rangle \, d\mu(x) \\
&= \langle \pi(\mu)\pi(\nu)v, w \rangle
\end{aligned}$$

for all v, w. Hence $\pi(\mu * \nu) = \pi(\mu)\pi(\nu)$. We show π is nondegenerate on $L^1(G)$. Let u_λ be an approximate unit. Then $\pi(u_\lambda)v \to v$. Indeed,

$$\begin{aligned}
||\pi(u_\lambda)v - v|| &= || \int u_\lambda(x)\pi(x)v \, dx - v|| \\
&\leq \int u_\lambda(x)||\pi(x)v - v|| \, dx \\
&\leq \sup_{x \in \text{supp} u_\lambda} ||\pi(x)v - v|| \to 0
\end{aligned}$$

as λ increases.

Conversely, let π be a nondegenerate representation of the Banach $*$ algebra $L^1(G)$. Let $\mu \in M(G)$. Set

$$\pi(\mu) \sum_{i=1}^m \pi(f_i)v_i = \sum_{i=1}^m \pi(\mu * f_i)v_i.$$

To see this is well defined and can be extended to a bounded linear operator on all of $\mathbb{H}$, it suffices to show

$$\| \sum_{i=1}^{m} \pi(\mu * f_i)v_i \| \leq |\mu| \, \| \sum_{i=1}^{m} \pi(f_i)v_i \|.$$

Let $u_\lambda \in L^1(G)$ be an approximate unit. Then Corollary II.6 and $\mu * u_\lambda \in L^1(G)$ imply

$$\| \sum_{i=1}^{m} \pi(\mu * f_i)v_i \| = \lim_\lambda \| \sum_{i=1}^{m} \pi(\mu * u_\lambda * f_i)v_i \|$$

$$= \lim_\lambda \| \pi(\mu * \mu_\lambda) \left(\sum_{i=1}^{m} \pi(f_i)v_i \right) \|$$

$$\leq \limsup_\lambda \| \mu * u_\lambda \|_1 \, \| \sum_{i=1}^{m} \pi(f_i)v_i \|$$

$$\leq \lim_\lambda |\mu| \, \| u_\lambda \|_1 \, \| \sum_{i=1}^{m} \pi(f_i)v_i \|$$

$$= |\mu| \, \| \sum_{i=1}^{m} \pi(\mu * f_i)v_i \|.$$

Thus $\pi(\mu)$ extends to a bounded linear operator on $\mathbb{H}$ satisfying $\|\pi(\mu)\| \leq |\mu|$. Since $\pi(\mu)\pi(\nu)\pi(f)v = \pi((\mu*\nu)*f)v$ for all $f \in L^1(G)$ and $v \in \mathbb{H}$, one has $\pi(\mu)\pi(\nu) = \pi(\mu * \nu)$ for μ and ν in $M(G)$. Moreover,

$$\langle \pi(\mu)\pi(f)v, \pi(h)w \rangle = \langle \pi(\mu * f)v, \pi(h)w \rangle$$

$$= \langle v, \pi(f^* * \mu^* * h)w \rangle$$

$$= \langle \pi(f)v, \pi(\mu^*)\pi(h)w \rangle$$

for f and h in $L^1(G)$ and v and w in $\mathbb{H}$ implies $\pi(\mu^*) = \pi(\mu)^*$. Thus any nondegenerate representation of the Banach $*$ algebra $L^1(G)$ extends to a representation of $M(G)$. In particular, $\pi(\epsilon_e) = I$. Moreover, since $x \mapsto \epsilon_x$ is a homomorphism of G into $M(G)$, we see

$$\pi(x) = \pi(\epsilon_x)$$

defines a homomorphism of G into $B(\mathbb{H})$. Since $\epsilon_x^* = \epsilon_x^{-1}$, $\pi(x^{-1}) = \pi(\epsilon_x^*) = \pi(\epsilon_x)^*$. Thus $\pi(x)^{-1} = \pi(x)^*$, and each operator $\pi(x)$ is unitary. We need to show π is strongly continuous on G. Since π is nondegenerate on $L^1(G)$ and $C_c(G)$ is L^1 dense, we see π is nondegenerate on $C_c(G)$. Let x_n converge to e and let $f \in C_c(G)$. We show $\pi(x_n)\pi(f)v \to \pi(f)v$. Indeed, $\pi(x_n)\pi(f)v = \pi(\epsilon_{x_n} * f)v$

and $\epsilon_{x_n} * f(y) = \delta(x_n^{-1})f(x_n^{-1}y)$. Thus by the Lebesgue dominated convergence theorem, $\epsilon_{x_n} * f$ converges in L^1 to f. Hence

$$||\pi(x_n)\pi(f)v - \pi(f)v|| \le ||\epsilon_{x_n} * f - f||_1 ||v|| \to 0.$$

Thus $\pi(x_n)v \to v$ for v in a dense subspace of $\mathbb{H}$. But since $\pi(x_n)$ are unitary, we see if $\epsilon > 0$ and $w \in \mathbb{H}$ and v is chosen so that $\pi(x_n)v \to v$ and $||v - w|| < \frac{\epsilon}{3}$, then

$$||\pi(x_n)w - w|| \le ||\pi(x_n)w - \pi(x_n)v|| + ||\pi(x_n)v - v|| + ||v - w||$$
$$< \frac{2\epsilon}{3} + ||\pi(x_n)v - v|| < \epsilon$$

for n large.

To finish the proof, we need to show that if $\tilde{\pi}$ is defined on $M(G)$ by $\langle \tilde{\pi}(\mu)v, w \rangle = \int \langle \pi(x)v, w \rangle \, d\mu(x)$, then $\tilde{\pi}(\mu) = \pi(\mu)$. Indeed, first consider $f, h \in C_c(G)$. Then $f * h = \int f(x)\epsilon_x * h \, dx$, and this integral in $L^1(G)$ is a limit of sums of terms of form $f(x_i)\epsilon_{x_i} * h(y) \, m(E_i)$. Thus

$$\langle \pi(f)\pi(h)v, w \rangle = \langle \pi(f * h)v, w \rangle$$
$$= \langle \int f(x)\pi(\epsilon_x * h) \, dx \, v, w \rangle$$
$$= \int f(x)\langle \pi(\epsilon_x)\pi(h)v, w \rangle \, dx$$
$$= \langle \tilde{\pi}(f)\pi(h)v, w \rangle$$

for all v, w. Thus $\pi(f)\pi(h)v = \tilde{\pi}(f)\pi(h)v$ for all f and h in $C_c(G)$. Therefore, $\pi(f) = \tilde{\pi}(f)$ for $f \in C_c(G)$. Since π and $\tilde{\pi}$ are continuous, $\pi(f) = \tilde{\pi}(f)$ for $f \in L^1(G)$. Finally, $\pi(\mu)\pi(f)v = \pi(\mu * f)v = \tilde{\pi}(\mu * f)v = \tilde{\pi}(\mu)\pi(f)v$ for all $f \in L^1(G)$ and all v. This implies $\pi(\mu) = \tilde{\pi}(\mu)$. $\square$

REMARK. There are nondegenerate $*$ representations of $M(G)$ whose restrictions to $L^1(G)$ are zero. Indeed, if π is a unitary representation of G, define $\tilde{\pi}$ on $M(G)$ by $\tilde{\pi}(\mu) = \sum \mu(\{x\})\pi(x)$.

COROLLARY IV.9. *The correspondence $\pi \mapsto \pi' \mapsto \pi''$ defined by $\pi'(f) = \int_G f(x)\pi(x) \, dx$ and $\pi''(\tau(f)) = \pi'(f)$ is a natural identification of the unitary representations π of G and the nondegenerate $*$ representations π' and π'' of $L^1(G)$ and $C^*(G)$. Moreover, the von Neumann algebras $\pi(G)''$, $\pi'(L^1(G))''$, and $\pi''(C^*(G))''$ are all equal.*

PROOF. The identification of π, π' and π'' follow from Theorem IV.11 and Theorem II.9.

To see the final statement, first note $\pi(G)' = \pi'(L^1(G))'$. Indeed, $A \in \pi(G)'$ iff $A\pi(x)\pi(y)v = \pi(x)\pi(y)Av$ for all $x, y \in G$, $v \in \mathbb{H}$ iff $A\pi(x) \int f(y)\pi(y)v \, dy = \pi(x) \int f(y)\pi(y)Av \, dy$ for all x, f and v iff $A\pi(x)\pi'(f)v = \pi(x)A\pi'(f)v$ for all x, f and v iff $A\pi'(\epsilon_x * f)v = \pi'(\epsilon_x * f)Av$ for all x and f iff $A\pi'(f) = \pi'(f)A$ for all f iff $A \in \pi'(L^1(G))'$. Hence $\pi(G)'' = \pi'(L^1(G))''$. Also note $\pi'(L^1(G))$ is norm dense in $\pi''(C^*(G))$ since $\tau : L^1(G) \to C^*(G)$ has dense range and π'' is continuous. Hence $\pi'(L^1(G))'' = \pi''(C^*(G))''$. $\quad\square$

10. Positive Definite Functions

DEFINITION. Let G be a locally compact group. A continuous function ϕ on G is said to be positive definite if for each finite sequence $x_1, x_2, \cdots, x_n$ in G, the matrix $[\phi(x_i^{-1}x_j)]$ is positive semidefinite.

PROPOSITION IV.18. *Suppose ϕ is a positive definite function on G. Then*

$$|\phi(x)| \leq \phi(e) \text{ and}$$
$$\phi(x^{-1}) = \overline{\phi(x)}$$

for all $x \in G$.

PROOF. Indeed, $\begin{bmatrix} \phi(e) & \phi(x) \\ \phi(x^{-1}) & \phi(e) \end{bmatrix}$ is a positive semi-definite Hermitian matrix. Thus $\phi(x^{-1}) = \overline{\phi(x)}$. Moreover $\phi(e) \geq 0$, and $\phi(e)^2 - \phi(x)\phi(x^{-1}) = \phi(e)^2 - |\phi(x)|^2 \geq 0$. $\quad\square$

Let $M_1(G)$ be the collection of all regular complex Borel measures μ on G with $|\mu| \leq 1$. Then $M_1(G)$ is a compact subset in the weak $*$ topology. More specifically, we consider the Banach space $C_b(G)$ of bounded continuous real valued functions on G, and let $M_1(G)$ be the subspace of the unit ball in $C_b(G)^*$ defined by the complex regular measures. Then $M_1(G)$ is a closed subset in the weak $*$ topology.

The set $M_1^+(G)$ of positive measures in $M_1(G)$ is closed in the weak $*$ topology in $C_b(G)^*$ and hence is compact. Since it is convex, by the Krein–Milman Theorem, it is the closed linear hull of its extreme points.

For $\mu \in M(G)$ and $h \in C_b(G)$, define $\langle h, \mu \rangle = \int h(x) \, d\mu(x)$. Note $|\langle h, \mu \rangle| \leq \|h\|_\infty |\mu|$.

LEMMA IV.6. *Let μ be an extreme point in $M_1^+(G)$. Then $\mu = 0$ or $\mu = \epsilon_x$ for some x. Moreover for each $\nu \in M(G)$, there is a net μ_λ in the linear span of the point measures ϵ_x with*

$$\lim_\lambda \langle h, \mu_\lambda \rangle = \langle h, \nu \rangle$$

for each continuous bounded function h on G.

PROOF. Let μ be a nonzero extreme point. We note $|\mu| = \mu(G) = 1$, for otherwise $\mu = (1 - \mu(G))0 + \mu(G)\frac{\mu}{\mu(G)}$. We claim for any Borel measurable set E, $\mu(E) = 0$ or $\mu(E) = 1$. Indeed, note if $\mu(E) \in (0,1)$, then $\mu = \mu(E)\frac{\mu|_E}{\mu(E)} + \mu(G-E)\frac{\mu|_{G-E}}{\mu(G-E)}$, a contradiction. From this it follows there are compact sets K for which $\mu(K) = 1$. We claim the intersection of all such K contains at most one point. Indeed, if $x, y \in \cap K$ and $x \neq y$, then there are compact neighborhoods U_x and U_y of x and y with $U_x \cap U_y = \emptyset$. Note $\mu(U_x) = 0$ or $\mu(U_y) = 0$, for otherwise $\mu(U_x \cup U_y) = 2$. Suppose $\mu(U_x) = 0$. Then there is a compact set $K \subset G - U_x$ with $\mu(K) = 1$ and thus $x \notin \cap_{\mu(K)=1} K$. If the intersection of all K with $\mu(K) = 1$ were empty, then there would be $K_1, \cdots, K_n$ with $\mu(K_i) = 1$ and $\cap K_i = \emptyset$. This is impossible. Thus the intersection of all K with $\mu(K) = 1$ consists of one element x. We claim $\mu(\{x\}) = 1$. If not, then $\mu(\{x\}) = 0$; and there is a compact set K in $G - \{x\}$ with $\mu(K) = 1$. Hence x is in K which again is impossible. Hence $\mu = \epsilon_x$.

To see the last statement, we may assume ν is positive and $\nu(G) = 1$. Using the Krein–Milman Theorem, there is a net $\mu_\lambda \in M_1^+(G)$ in the linear span of the point masses ϵ_x with $\langle h, \mu_\lambda \rangle \to \langle h, \nu \rangle$ for all $h \in C_b(G)$. $\square$

THEOREM IV.12. *Suppose ϕ is a continuous function on G. The following are equivalent:*

(1) *ϕ is positive definite.*
(2) *$\langle \phi, \mu^* * \mu \rangle \equiv \int \phi(x) \, d\mu^* * \mu(x) = \iint \phi(x^{-1}y) \, d\bar{\mu}(x) \, d\mu(y) \geq 0$ for each $\mu \in M(G)$.*
(3) *$\langle \phi, f^* * f \rangle = \iint \phi(x^{-1}y)\bar{f}(x)f(y) \, dx \, dy \geq 0$ for each $f \in L^1(G)$.*

PROOF. We show (3) follows from (1). First assume $f \in C_c(G)$ and $K = \text{supp} f$. Let μ be Haar measure restricted to K. Then μ is finite. Hence μ is a weak $*$ limit of μ_i where each μ_i is a finite real linear combination $\sum \lambda_{i,j} \epsilon_{x_{i,j}}$ of point masses. Then $\mu_i \times \mu_i$

converges weakly to $\mu \times \mu$. (First check on products and then use the Stone–Weierstrass Theorem.) Thus

$$\langle \phi, f^* * f \rangle = \lim \iint \phi(x^{-1}y)\bar{f}(x)f(y)\, d(\mu_i \times \mu_i)(x,y)$$

$$= \lim \sum_{j,k} \phi(x_{i,j}^{-1}x_{i,k})\bar{f}(x_{i,j})f(x_{i,k})\lambda_{i,j}\lambda_{i,k} \geq 0.$$

Thus $\langle \phi, f^* * f \rangle \geq 0$ for $f \in C_c(G)$. In general, take $f_n \in C_c(G)$ converging in L^1 to f to see it holds for all $f \in L^1(G)$.

Suppose (3) holds. Let μ be a measure with compact support K. Let $f \geq 0$ be in $C_c(G)$. Note

$$\langle \phi, f^* * \mu^* * \mu * f \rangle = \iint d\mu^*(y)\, d\mu(z) \iint \phi(xyzt)f^*(x)f(t)dx\, dt \geq 0.$$

Let $\epsilon > 0$. Then using the compactness of $K^{-1} \times K$, there is a compact symmetric neighborhood λ of e such that if x and t belong to λ, then $|\phi(xyzt) - \phi(yz)| < \epsilon$ for $(y,z) \in K^{-1} \times K$. Hence if $\int f_\lambda(x)\, dx = 1$, $\mathrm{supp} f_\lambda \subset \lambda$ and $0 \leq f_\lambda$, then

$$|\langle \phi, \mu^* * \mu \rangle - \langle \phi, f_\lambda^* * \mu^* * \mu * f_\lambda \rangle|$$

$$= | \iint d\mu^*(y)\, d\mu(z) \left(\iint (\phi(yz) - \phi(xyzt))f_\lambda^*(x)f_\lambda(t)\, dx\, dt \right)$$

$$\leq \iint d|\mu^*|(y)\, d|\mu|(z) \left(\iint |\phi(yz) - \phi(xyzt)||f_\lambda^*(x)f_\lambda(t)\, dx\, dt \right)$$

$$\leq \epsilon|\mu^*|(K^{-1})\,|\mu|(K).$$

Taking $\epsilon = \frac{1}{n}$ and $\lambda = \lambda_n$ and taking a limit as $n \to \infty$ shows $\langle \phi, \mu^* * \mu \rangle \geq 0$ if μ is a measure with compact support. Because the measures in $M(G)$ are regular, each $\mu \in M(G)$ is a norm limit of a sequence μ_n of measures having compact support. In particular,

$$|\langle \phi, \mu^* * \mu \rangle - \langle \phi, \mu_n^* * \mu_n \rangle| \leq |\langle \phi, \mu^* * \mu \rangle - \langle \phi, \mu^* * \mu_n \rangle|$$

$$+ |\langle \phi, \mu^* * \mu_n \rangle - \langle \phi, \mu_n^* * \mu_n \rangle|$$

$$\leq \phi(e)|\mu|\,|\mu - \mu_n| + \phi(e)|\mu - \mu_n|\,|\mu_n| \to 0$$

as $n \to \infty$. Thus $\langle \phi, \mu^* * \mu \rangle \geq 0$ for any $\mu \in M(G)$, and (2) holds.

To see (2) implies (1), let $x_1, x_2, \cdots, x_n$ be a finite sequence in G and let $c_i \in \mathbb{C}$. Then $\mu = \sum c_i \epsilon_{x_i}$ is in $M(G)$. Since $\mu^* * \mu = \sum_{i,j} \bar{c}_i c_j \epsilon_{x_i^{-1}x_j}$, one has

$$\sum_{i,j} \bar{c}_i c_j \phi(x_i^{-1}x_j) = \langle \phi, \mu^* * \mu \rangle \geq 0,$$

and consequently ϕ is a positive definite function. $\square$

Note the dual of $L^1(G)$ is $L^\infty(G)$. Let $\phi \in L^\infty(G)$. It defines a linear functional ω_ϕ on $L^1(G)$ by $\omega_\phi(f) = \int f(x)\phi(x)\,dx$ satisfying $||I_\phi|| = ||\phi||_\infty$.

THEOREM IV.13. *Suppose* $\phi \in L^\infty(G)$. *Then the following are equivalent:*

(a) *The linear functional* ω_ϕ *on* $L^1(G)$ *is positive.*
(b) *There is a positive definite function* ψ *on* G *with* $\phi(x) = \psi(x)$ *for a.e.* x.
(c) *There is a unitary representation* π *of* G *with a cyclic vector* ξ *with* $\langle \pi(g)\xi, \xi \rangle = \phi(x)$ *for a.e.* x *in* G.

PROOF. We show (a) implies (c). By the GNS construction, there is a nondegenerate $*$-representation π of $L^1(G)$ with a cyclic vector ξ satisfying $\omega_\phi(f) = \langle \pi(f)\xi, \xi \rangle$ for each $f \in L^1(G)$. Theorem IV.11 implies there is a unitary representation π of G such that $\pi(f) = \int f(x)\pi(g)\,dg$. Thus $\omega_\phi(f) = \langle \pi(f)\xi, \xi \rangle = \int f(g)\langle \pi(g)\xi, \xi \rangle \, dg = \int f(g)\phi(g)\,dg$ for each $f \in L^1$. Hence $\phi(g) = \langle \pi(g)\xi, \xi \rangle$ for a.e. g. We note ξ is a cyclic vector for the unitary representation π. Indeed, if $v \perp \pi(G)\xi$, then $\langle \pi(f)\xi, v \rangle = \int f(g)\langle \pi(g)\xi, v \rangle \, dg = 0$ for all f. Thus v is perpendicular to the dense subspace $\pi(L^1(G))\xi$, and v must be 0.

(b) follows easily from (c), and (b) implies (a) is part of Theorem IV.12. □

COROLLARY IV.10. *Let* ϕ *be a positive definite function on the locally compact group* G. *Then* $|\phi(x) - \phi(y)|^2 \le 2\phi(e)[\phi(e) - \operatorname{Re}\phi(x^{-1}y)]$ *and* $|\phi(x) - \phi(y)|^2 \le 2\phi(e)[\phi(e) - \operatorname{Re}\phi(yx^{-1})]$.

PROOF. We may assume $\phi(x) = \langle \pi(x)\xi, \xi \rangle$ where π is a unitary representation of G. Thus $|\phi(x) - \phi(y)|^2 = |\langle \pi(y)\xi - \pi(x)\xi, \xi \rangle|^2 \le ||\xi||^2 ||\pi(y)\xi - \pi(x)\xi||^2$. Thus

$$\begin{aligned}
|\phi(x) - \phi(y)|^2 &\le \phi(e)\langle \pi(y)\xi - \pi(x)\xi, \pi(y)\xi - \pi(x)\xi \rangle \\
&= \phi(e)\left[||\pi(y)\xi||^2 + ||\pi(x)\xi||^2 - 2\operatorname{Re}\langle \pi(y)\xi, \pi(x)\xi \rangle\right] \\
&= \phi(e)\left[2\phi(e) - 2\operatorname{Re}\phi(x^{-1}y)\right],
\end{aligned}$$

and the first result holds. Since $\phi(x^{-1}) = \overline{\phi(x)}$, $|\phi(x) - \phi(y)| = |\phi(x^{-1}) - \phi(y^{-1})| \le \phi(e)\left[2\phi(e) - \phi(yx^{-1})\right]$. □

THEOREM IV.14 (GELFAND–RAIKOV). *Let G be a locally compact group. Suppose $e \neq g \in G$. Then there is an irreducible unitary representation π of G with $\pi(g) \neq I$.*

PROOF. Choose a compact neighborhood λ of the identity satisfying $\lambda g \cap \lambda = \emptyset$. Let $f_\lambda \geq 0$ have support in λ, be continuous, and have integral one. Then $x \mapsto f_\lambda(xg^{-1}) - f_\lambda(x)$ is a nonzero function $L^1(G)$. Hence by Corollaries II.13 and IV.9 and Lemma IV.5, there is an irreducible unitary representation π of G with

$$\int (f_\lambda(xg^{-1}) - f_\lambda(x))\langle \pi(x)\xi, \xi'\rangle \, dx = \int f_\lambda(x)\langle(\pi(xg) - \pi(x))\xi, \xi'\rangle \, dx \neq 0$$

for some ξ and ξ'. Hence $\pi(x)\pi(g) - \pi(x) \neq 0$ for some x. This gives $\pi(g) \neq I$. $\square$

Let $P(G)$ be the space of positive definite functions ϕ on G with $\phi(e) = 1$. Since $||\omega_\phi|| = ||\phi||_\infty = \phi(e)$, Theorem IV.13 implies the mapping $\phi \mapsto \omega_\phi$ is a one-to-one correspondence between $P(G)$ and the set of states for the Banach $*$ algebra $L^1(G)$. We give $P(G)$ the relative weak $*$ topology from the unit ball of $L^\infty(G)$. The following theorem shows this topology is the topology of uniform convergence on compact sets.

THEOREM IV.15 (RAIKOV). *The weak $*$ topology on $P(G)$ is the topology of uniform convergence on compact sets.*

PROOF. Suppose $\phi_\alpha \to \phi$ in the topology of uniform convergence on compact sets. Let $f \in L^1(G)$ and let $\epsilon > 0$. We assume $\int |f(x)| \, dx \leq 1$. Choose a compact set K such that $\int_{G-K} |f(x)| \, dx < \frac{\epsilon}{4}$. Choose α_0 such that $\alpha > \alpha_0$ implies $|\phi_\alpha(x) - \phi(x)| < \frac{\epsilon}{2}$ for $x \in K$. Then

$$|\int f(x)\phi_\alpha(x) \, dx - \int f(x)\phi(x) \, dx| \leq \int |f(x)||\phi_\alpha(x) - \phi(x)| \, dx$$

$$\leq \int_K |f(x)||\phi_\alpha(x) - \phi(x)| \, dx + 2\int_{G-K} |f(x)| \, dx$$

$$< \frac{\epsilon}{2}\int_K |f(x)| \, dx + \frac{\epsilon}{2}$$

$$\leq \epsilon$$

for $\alpha > \alpha_0$.

Suppose $\omega_{\phi_\alpha} \to \omega_\phi$ in the weak $*$ topology. We simplify notation by letting $\phi(f)$ denote $\omega_\phi(f)$. The functionals ϕ for $\phi \in P(G)$ are uniformly bounded on norm compact subsets of $L^1(G)$. Moreover, they are equicontinuous. Indeed, if $||f_1 - f_2||_1 < \epsilon$, then

$|\phi_\alpha(f_1) - \phi_\alpha(f_2)| < \epsilon$. By equicontinuity, if the net ϕ_α converges pointwise on a compact subset of $L^1(G)$, then it converges uniformly on this set. Now let K be a compact subset of G. For $f \in L^1(G)$ and $y \in G$, set $f_y(x) = f(xy^{-1})$. Each mapping $y \mapsto f_y$ is continuous. (Check first for $f \in C_c(G)$.) In particular, the set of functions $\{f_y : y \in K\}$ is a compact subset of $L^1(G)$. Thus $\phi_\alpha(f_y)$ converges uniformly to $\phi(f_y)$ on K. Then by right invariance of Haar measure, $\int f(x)\phi_\alpha(xy)\,dx \to \int f(x)\phi(xy)\,dx$ uniformly for $y \in K$. Let V be a compact neighborhood of e so that $|\phi(y) - \phi(e)| \le \epsilon$ if $y \in V$. Set $f = \frac{1_V}{m(V)}$, and choose α_0 so that $\alpha > \alpha_0$ implies $|\phi_\alpha(f) - \phi(f)| < \epsilon$. Note

$$|\phi_\alpha(f_y) - \phi_\alpha(y)| = \frac{1}{m(V)}\left| \int (\phi_\alpha(xy) - \phi_\alpha(y))1_V(x)\,dx \right|$$
$$\le \frac{1}{m(V)} \int_V |\phi_\alpha(xy) - \phi_\alpha(y)|\,dx.$$

By Corollary IV.10, $|\phi_\alpha(xy) - \phi_\alpha(y)| \le \sqrt{2}|\phi_\alpha(e) - \operatorname{Re}\phi_\alpha(x)|^{\frac{1}{2}}$. Thus

$$|\phi_\alpha(f_y) - \phi_\alpha(y)| \le \frac{\sqrt{2}}{m(V)} \int_V |\phi_\alpha(e) - \operatorname{Re}\phi_\alpha(x)|^{\frac{1}{2}}\,dx$$
$$\le \frac{\sqrt{2}}{m(V)} \left(\int_V (\phi_\alpha(e) - \operatorname{Re}\phi_\alpha(x))\,dx \right)^{\frac{1}{2}} \left(\int 1_V(x)\,dx \right)^{\frac{1}{2}}.$$

Now

$$\int_V (\phi_\alpha(e) - \operatorname{Re}\phi_\alpha(x))\,dx = m(V) \int (f(x) - \operatorname{Re}\phi_\alpha(x)f(x))\,dx$$
$$= m(V) - m(V)\operatorname{Re}\phi_\alpha(f) + m(V)\operatorname{Re}\phi(f) - m(V)\operatorname{Re}\phi(f)$$
$$= m(V)\left((1 - \operatorname{Re}\phi(f)) + \operatorname{Re}(\phi(f) - \phi_\alpha(f))\right)$$
$$\le m(V)\left(|1 - \phi(f)| + |\phi(f) - \phi_\alpha(f)|\right)$$
$$\le \int_V |\phi(e) - \phi(y)|\,dy + m(V)|\phi(f) - \phi_\alpha(f)|$$
$$\le m(V)\epsilon + m(V)\epsilon = 2m(V)\epsilon$$

for $\alpha > \alpha_0$. Hence

$$|\phi_\alpha(f_y) - \phi_\alpha(y)| \le 2\sqrt{\epsilon} \text{ for } y \in K$$

if $\alpha > \alpha_0$. Furthermore,

$$
\begin{aligned}
|\phi(f_y) - \phi(y)| &= |\int (\phi(xy) - \phi(y)) f(x) \, dx| \\
&\leq \int |\phi(xy) - \phi(y)| f(x) \, dx \\
&\leq \int_V \sqrt{2}(\phi(e) - \operatorname{Re} \phi(x))^{\frac{1}{2}} f(x) \, dx \\
&\leq \sqrt{2}\epsilon.
\end{aligned}
$$

Since $\phi_\alpha(f_y)$ converges uniformly to $\phi(f_y)$ on K, we can choose $\alpha_1 > \alpha_0$ such that $\alpha > \alpha_1$ implies $|\phi_\alpha(f_y) - \phi(f_y)| < \epsilon$ for all $y \in K$. Thus for $\alpha > \alpha_1$ and for $y \in K$, we see

$$
\begin{aligned}
|\phi_\alpha(y) - \phi(y)| &\leq |\phi_\alpha(y) - \phi_\alpha(f_y)| + |\phi_\alpha(f_y) - \phi(f_y)| + |\phi(f_y) - \phi(y)| \\
&< 2\sqrt{\epsilon} + \epsilon + \sqrt{2}\epsilon.
\end{aligned}
$$

Thus ϕ_α converges uniformly on compact sets to ϕ. $\square$

11. Borel Structures on Spaces of Representations

Our next topic will be to show that there are natural Borel structures on collections of representations. Here and in the remaining sections of this chapter, we take G to be a second countable locally compact group and all Hilbert spaces to be separable. Fix μ to be a right invariant Haar measure on G. Since G is σ-compact, μ is σ-finite. For each n, let $\mathbb{H}_n$ be a fixed Hilbert space of dimension n. Let $\mathcal{U}_n$ be the unitary group on $\mathbb{H}_n$ with the strong $*$ topology. By Theorem I.28, $\mathcal{M}(G, \mu, \mathcal{U}_n)$ is a Polish space when equipped with the topology of convergence in measure on sets of finite measure. Let $\operatorname{Hom}(G, \mathcal{U}_n)$ be the set of all continuous homomorphisms from G into $\mathcal{U}_n$. This set can be imbedded into $\mathcal{M}(G, \mu, \mathcal{U}_n)$, for two continuous homomorphisms which are equal a.e. are equal everywhere. This follows, for the set where they are equal is a conull Borel subgroup of G and hence is G. Give $\operatorname{Hom}(G, \mathcal{U}_n)$ the relative topology from $\mathcal{M}(G, \mu, \mathcal{U}_n)$.

PROPOSITION IV.19. $\operatorname{Hom}(G, \mathcal{U}_n)$ is a closed subset of $\mathcal{M}(G, \mu, \mathcal{U}_n)$ and hence is a Polish space.

PROOF. Let U_k converge in measure to a function U on each set of finite measure. By using a countable cover of G by sets of finite measure and using the Cantor diagonalization process, we may by taking the proper subsequence assume that $U_k(g)$ converges point-wise to $U(g)$ a.e. g. Then $U(g_1 g_2) = U(g_1)U(g_2)$ a.e. g_1, g_2. By Corollary IV.8, U may be redefined on a set of measure 0 in such a way that U becomes a Borel homomorphism. By Theorem IV.2, U is continuous and thus $U \in \mathrm{Hom}(G, \mathcal{U}_n)$. $\square$

PROPOSITION IV.20. U_k converges to U in $\mathrm{Hom}(G, \mathcal{U}_n)$ if and only if $U_k(f)$ converges strongly to $U(f)$ for each f in $L^1(G)$ iff $U_k(g)$ converges strongly to $U(g)$ for each g in G.

PROOF. Suppose U_k converges to U_∞ in $\mathrm{Hom}(G, \mathcal{U}_n)$. Then for f in $L^1(G)$,

$$\lim \|U_k(f)v - U_\infty(f)v\| \le \lim \int |f(x)|\, \|U_k(x)v - U_\infty(x)v\|\, dx = 0$$

by the Lebesgue dominated convergence theorem.

Suppose $U_k(f)v$ converges to $U_\infty(f)v$ for each $v \in \mathbb{H}_n$ and each f in $L^1(G)$. Note for any unitary representation V of G, one has $V(g)V(f) = \delta(g^{-1})V(_g f)$ where $_g f$ is the L^1 function on G defined by $_g f(g') = f(g^{-1}g')$. Indeed, note

$$\int f(g')V(gg')\, d\mu(g') = \delta(g^{-1})\int f(g^{-1}g')V(g')\, d\mu(g') = \delta(g^{-1})V(_g f).$$

Hence

$$\lim U_k(g)U_k(f)v = \lim U_\infty(g)U_\infty(f)v$$

for all v. Thus

$$\|U_k(g)U_\infty(f)v - U_\infty(g)U_\infty(f)v\| \le$$
$$\|U_k(g)U_\infty(f)v - U_k(g)U_k(f)v\| + \|U_k(g)U_k(f)v - U_\infty(g)U_\infty(f)v\| \le$$
$$\|U_\infty(f)v - U_k(f)v\| + \|U_k(g)U_k(f)v - U_\infty(g)U_\infty(f)v\| \to 0.$$

Hence $U_k(g)v$ converges to $U_\infty(g)v$ for all v in a dense subspace of $\mathbb{H}_0$. Since all $U_k(g)$'s and $U_\infty(g)$ are unitary, this convergence holds for all v.

Finally suppose $U_k(g)$ converges strongly to $U_\infty(g)$ for each $g \in G$. Since $U_k(g)^* = U_k(g^{-1})$, $U_k(g)$ converges in the strong $*$ topology to $U(g)$ for each g. It follows that U_k converges in measure to U_∞ on each set of finite measure. But this is the topology on $\mathrm{Hom}(G, \mathcal{U}_n)$. $\square$

PROPOSITION IV.21. Suppose X is a Borel space. A function $x \mapsto U_x$ from X into $\mathrm{Hom}(G, \mathcal{U}_n)$ is Borel if and only if $x \mapsto (U_x(g)v, w)$ is Borel for each v and w.

PROOF. Assume each function $x \mapsto (U_x(g)v, w)$ is Borel. Let v_i be a countable dense subset of $\mathbb{H}_n$. Define functions $h_{i,j}$ on $\mathcal{U}_n$ by $h_{i,j}(V) = (Vv_i, v_j)$. These functions are Borel and separate the points of $\mathcal{U}_n$. Furthermore, the functions $(x, g) \mapsto (U_x(g)v, w)$ are Borel by Proposition I.19. Thus by Fubini's Theorem, the functions $x \mapsto \int f(g)h_{i,j}(U_x(g)) \, dg$ are Borel. It follows by Proposition I.17 that the mapping $x \mapsto U_x$ is Borel from X into $\mathrm{Hom}(G, \mathcal{U}_n)$ is Borel.

Conversely, suppose the mapping $x \mapsto U_x$ is Borel. By Proposition IV.20, the map sending U to $U(g)$ is continuous from $\mathrm{Hom}(G, \mathcal{U}_n)$ into $\mathcal{U}_n$. This map is thus weakly Borel. It follows that $x \mapsto (U_x(g)v, w)$ is Borel, for it is a composition of Borel functions. $\square$

PROPOSITION IV.22. *Let X be a Borel space. Then a mapping $U : X \to \mathrm{Hom}(G, \mathcal{U}_n)$ is Borel if and only if $x \mapsto U_x(g_k)$ is Borel for a countable dense subset $\{g_k\}$ of G.*

PROOF. Note by the strong continuity of the representations in $\mathrm{Hom}(G, \mathcal{U}_n)$ and by Proposition IV.20 that the functions $U \mapsto U(g_k)$ are continuous and separate the points of $\mathrm{Hom}(G, \mathcal{U}_n)$. It follows by Proposition I.4 that a function U from X into $\mathrm{Hom}(G, \mathcal{U}_n)$ is Borel iff $x \mapsto U_x(g_k)$ is Borel for each k. $\square$

PROPOSITION IV.23. *A mapping U from X into $\mathrm{Hom}(G, \mathcal{U}_n)$ is Borel iff $x \mapsto U_x(f_k)$ is Borel for f_k in a countable dense subset of $L^1(G)$.*

PROOF. The same argument in the proof of Proposition IV.22 may be applied. $\square$

Let G_n^c be the subset of irreducible representations in $\mathrm{Hom}(G, \mathcal{U}_n)$.

THEOREM IV.16. *G_n^c is a Borel subset of $\mathrm{Hom}(G, \mathcal{U}_n)$ and hence is a standard Borel space.*

PROOF. Consider the von Neumann algebra valued mapping on $\mathrm{Hom}(G, \mathcal{U}_n)$ defined by $\mathcal{A}_U = U(G)''$. This mapping is Borel for $\mathcal{A}_U = \{U(g_k) : k \in \mathbf{N}\}''$ where g_k form a countable dense subset of G. Corollaries III.12 and III.13 imply the set of $U \in \mathrm{Hom}(G, \mathcal{U}_n)$ with $\mathcal{A}_U = \mathcal{B}(\mathbb{H})$ is Borel. $\square$

The unitary group $\mathcal{U}_n$ acts continuously on the Borel space G_n^c by $U \cdot V(g) = V^{-1}U(g)V$. The space of orbits is denoted by $\hat{G}_n$ and with the quotient Borel structure is called the n-dimensional unitary dual of G. In the case when n is finite, then since the group $\mathcal{U}_n$ is compact, it follows by Corollary IV.7 that $\hat{G}_n$ is a standard Borel space. In the case where $n = \infty$, $\hat{G}_n$ may be countably separated in which case it is standard or nonanalytic in which case Mackey uses the terminology **nonsmooth**.

We shall need the following lemma.

LEMMA IV.7. *The natural mapping $U \mapsto [U]$ from $\mathbf{G}^c$ into $\hat{\mathbf{G}}$ has the property that one-to-one images of Borel sets are Borel.*

PROOF. Since G_n^c is standard when n is finite, we see in using Kuratowski's Theorem that we may assume that we have a Borel subset E in $\mathbf{G}_\infty^c$ which is mapped injectively into the dual by the natural projection. Thus $E \subset \mathrm{Hom}(G, \mathcal{U}(\mathbb{H}))$ where $\dim \mathbb{H} = \infty$ and no two distinct representations in E are unitarily equivalent. Moreover, by the definition of the quotient Borel structure, the image of E in $\hat{\mathbf{G}}$ is Borel iff the saturation $[E] = E \cdot \mathcal{U}(\mathbb{H})$ is a Borel set in $\mathrm{Hom}(G, \mathcal{U}(\mathbb{H}))$. Let T be the subgroup of $\mathcal{U}(\mathbb{H})$ consisting of the operators cI where $|c| = 1$. Then T is compact and by Corollary IV.7, there is a Borel set $\mathcal{U}_0$ of $\mathcal{U}(\mathbb{H})$ which meets each coset TL in exactly one point. This implies the Borel mapping $(U, L) \mapsto U \cdot L$ from $E \times \mathcal{U}_0$ onto $[E]$ is one-to-one. Indeed, if $U_1 \cdot L_1 = U_2 \cdot L_2$, then $U_1 = U_2$ and $L_1 L_2^{-1} \in \mathcal{A}'_{U_1}$. Since U_1 is irreducible, $\mathcal{A}'_{U_1} = \mathbf{C}\,I$, and thus $L_1 L_2^{-1}$ is in T. Since $\mathcal{U}_0$ meets each T coset in $\mathcal{U}(\mathbb{H})$ exactly once, $L_1 = L_2$. Thus the mapping is one-to-one. By Kuratowski's Theorem, its image $[E]$ is Borel. $\square$

12. Direct Integrals for Representations

We next consider direct integral decompositions for unitary representations of a second countable locally compact group G on a separable Hilbert space.

THEOREM IV.17. *Suppose $\mathcal{H}$ is a Hilbert bundle over S and μ is a σ-finite measure on S. Furthermore, assume $s \mapsto \mathcal{A}(s)$ is a Borel field of von Neumann algebras over S. Then if U is a unitary representation of G such that $\mathcal{A}_U \subset \int^{\oplus} \mathcal{A}(s)\,d\mu(s)$, then there exists for each s a unitary representation U_s of G on $\mathcal{H}_s$ such that for each $g \in G$, $s \mapsto U_s(g)$ is a Borel field over S and $U(g) = \int^{\oplus} U_s(g)\,d\mu(s)$. Moreover, the representations U_s are unique for μ a.e. s.*

PROOF. We may assume $\mathcal{H}$ is trivial. Since $U(g)$ is Borel in g, it follows by Proposition III.16 that there exists a Borel field $(s, g) \mapsto U(s, g)$ satisfying $U(g) = \int^{\oplus} U(s, g)\,d\mu(s)$ for a.e. g in G. By Proposition III.12, we may assume $U(s, g)$ is unitary for all s and

g. Moreover, by a.e. uniqueness, we see $U(s,g_1g_2) = U(s,g_1)U(s,g_2)$ a.e. (s,g_1,g_2). Thus take S_0 to be a conull Borel subset of S such that $U(s,g_1)U(s,g_2) = U(s,g_1g_2)$ for a.e. (g_1,g_2) for each $s \in S_0$. By Corollary IV.8, there is for each $s \in S_0$ a unitary representation U_s of G with $U_s(g) = U(s,g)$ a.e. g. Define $U_s(g) = I$ for $s \notin S_0$. Proposition IV.23 implies $s \mapsto U_s \in \mathrm{Hom}(G,\mathcal{U}(\mathbb{H}))$ is Borel, for $s \mapsto U_s(f)$ is Borel for each $f \in L^1(G)$. It follows by Proposition IV.22 that $s \mapsto U_s(g)$ is Borel for each g. The a.e. uniqueness of the representations U_s follows from the fact that two representations which are equal a.e. on G are equal. □

COROLLARY IV.11. *Suppose $\mathcal{H} = S \times \mathbb{H}$ is a trivial Hilbert bundle over S. Furthermore, suppose that U_s is a unitary representation of G for each s. Then $s \mapsto U_s(g)$ is a Borel field for each g iff $s \mapsto U_s$ is Borel from S into $\mathrm{Hom}(G,\mathcal{U}(\mathbb{H}))$.*

THEOREM IV.18. *If U is a representation with $\mathcal{A}_U \subset \int^{\oplus} \mathcal{A}(s)\, d\mu(s)$, then in the direct integral decomposition $U = \int^{\oplus} U_s\, d\mu(s)$, U_s is irreducible for μ a.e. s iff the range of the canonical projection valued measure for $\int^{\oplus} \mathcal{H}_s\, d\mu(s)$ is a maximal Boolean algebra of projections in $\mathcal{A}'_U$.*

PROOF. We may assume the bundle is trivial. Suppose the range of the canonical projection valued measure is maximal. Set $\mathcal{A}(s) = \mathcal{A}_{U_s}$. Then $s \mapsto \mathcal{A}(s)$ is a Borel field of Von Neumann algebras. Let S_0 be the set consisting of those s for which $\mathcal{A}(s) = \mathcal{B}(\mathbb{H})$. By Corollaries III.12 and III.13 and the double commutant theorem, S_0 is a Borel subset of S. Proposition III.14 and Corollary III.12 imply the set E consisting of the pairs (s,Q) where s is in S_0 and Q is a nontrivial orthogonal projection in $\mathcal{A}(s)'$ is a Borel subset of $S_0 \times \mathcal{B}(\mathbb{H})$. If not almost all of the representations U_s are irreducible, then S_0 has μ positive measure. By Theorem I.13, there exists a Borel function Q on S_0 such that $(s,Q(s)) \in E$ for μ a.e. s. Define $Q(s) = I$ for $s \notin S_0$. Then $P = \int^{\oplus} Q(s)\, d\mu(s)$ is in $\mathcal{A}'_U$ and commutes with the canonical projection valued measure. Since the range of this projection valued measure is maximal in $\mathcal{A}'_U$, it follows that P belongs to the range. But this is a contradiction for in this case $Q(s)$ must be either 0 or I for a.e. s.

Conversely suppose U_s is irreducible for μ a.e. s. Let P be a projection in $\mathcal{A}'_U$ which commutes with the range of the canonical projection valued measure for $\int^{\oplus} \mathcal{H}_s\, d\mu(s)$. By Theorem III.14 and Proposition III.12, $P = \int^{\oplus} Q(s)\, d\mu(s)$ for some projection valued

Borel function Q on S. Since $P \in \mathcal{A}_U'$, it follows that for almost all s one has $Q(s)U_s(g_k) = U_s(g_k)Q(s)$ for g_k in a countable dense subset of G. Therefore, by the continuity of the representations U_s, we see that $Q(s)U_s(g) = U_s(g)Q(s)$ for all g for almost all s. But since a.e. U_s is irreducible, $Q(s) = 0$ or $Q(s) = I$ for almost all s. Thus P is in the range of the canonical projection valued measure for the direct integral $\int^\oplus \mathcal{H}_s \, d\mu(s)$. $\square$

COROLLARY IV.12. *Suppose π is a $*$ representation of a separable C^* algebra $\mathcal{A}$ such that $\pi(a) = \int^\oplus \pi_s(a) \, d\mu(s)$ relative to a direct integral decomposition of the Hilbert space of π where each π_s is a $*$ representation of the C^* algebra $\mathcal{A}$ on $\mathcal{H}_s$. Then μ a.e. π_s is irreducible iff the range of the canonical projection valued measure for $\int^\oplus \mathcal{H}_s \, d\mu(s)$ is a maximal abelian algebra of projections in $\pi(\mathcal{A})'$.*

PROOF. The proof follows the same argument as above. The only change is in the converse direction where one replaces $Q(s)U_s(g_k) = U_s(g_k)Q(s)$ a.e. s for a dense set of g_k's with $Q(s)\pi_s(a_i) = \pi_s(a_i)Q(s)$ a.e. s for a sequence a_i which is dense in $\mathcal{A}$. Then the continuity of π_s implies $Q(s) \in \pi_s(\mathcal{A})'$ for almost all s, and thus $Q(s) = 0$ or $Q(s) = I$ a.e. s. $\square$

PROPOSITION IV.24. *Suppose P is the canonical projection valued measure on $\int^\oplus \mathcal{H}_s \, d\mu(s)$. If U is a representation on $\int^\oplus H_s \, d\mu(s)$ such that each $U(g)$ commutes with every projection in the range of P, then $U(g) = \int^\oplus U_s(g) \, d\mu(s)$ where each U_s is a unitary representation of G.*

PROOF. By Theorem IV.17, it is sufficient to note that $\mathcal{A}_U \subset \int^\oplus \mathcal{B}(\mathcal{H}_s) \, d\mu(s)$. But this follows from Theorem III.14. $\square$

The next result shows the essential uniqueness of a decomposition of a unitary representation of a second countable locally compact group G relative to a given projection valued measure P.

THEOREM IV.19. *Let $\mathcal{H}$ be a Hilbert bundle over X and $\mathcal{K}$ be a Hilbert bundle over Y where X and Y are standard Borel spaces of the same cardinality. Suppose μ is a σ-finite measure on X and ν is a σ-finite measure on Y. Let P be the canonical projection valued measure on $\int^\oplus \mathcal{H}_x \, d\mu(s)$ and Q be the canonical projection valued measure on $\int^\oplus \mathcal{K}_y \, d\nu(y)$. If U is a unitary representation of G on $\int^\oplus \mathcal{H}_x \, d\mu(x)$ of form $U = \int^\oplus U_x \, d\mu(x)$ and V is a unitary representation of G on $\int^\oplus \mathcal{K}_y \, d\nu(y)$ of form $V = \int^\oplus V_y \, d\nu(y)$ and T*

*is a unitary intertwining operator between U and V which carries
the range of the canonical projection valued measure on $\int^{\oplus} \mathcal{H}_x \, d\mu(x)$
onto the range of the projection valued measure for $\int^{\oplus} \mathcal{K}_y \, d\nu(y)$, then
there exist a Borel isomorphism φ of X onto Y such that $\varphi_* \mu \sim
\nu$ and a bounded Borel field of operators $T(x)$ such that $T(x)$ is a
unitary isomorphism of $\mathcal{H}_x$ onto $\mathcal{K}_{\varphi(x)}$ for almost all x satisfying
$T(x)U_x(g)T(x)^{-1} = V_{\varphi(x)}(g)$ for all g a.e. x.*

PROOF. This is a direct application of Theorem III.15. Indeed,
the existence of $T(x)$ follows immediately from this theorem, and an
easy calculation shows $T(x)U_x(g_i)T(x)^{-1} = V_{\varphi(x)}(g_i)$ on a conull set
of X for all g_i in a countable dense subset of G. That the equality
holds for all g is a consequence of the continuity of the representations
U_x and $V_{\varphi(x)}$. □

COROLLARY IV.13. *Let $\mathcal{H}$ and $\mathcal{K}$ be Hilbert bundles over X. Suppose μ is a σ-finite measure on X. Let P be the canonical projection
valued measure on $\int^{\oplus} \mathcal{H}_x \, d\mu(s)$ and Q be the canonical projection valued measure on $\int^{\oplus} \mathcal{K}_x \, d\mu(x)$. If U is a unitary representation of G on
$\int^{\oplus} \mathcal{H}_x \, d\mu(x)$ of form $U = \int^{\oplus} U_x \, d\mu(x)$ and V is a unitary representation of G on $\int^{\oplus} \mathcal{K}_x \, d\mu(x)$ of form $V = \int^{\oplus} V_x \, d\mu(x)$ and T is a unitary
intertwining operator between U and V satisfying $TP(E)T^{-1} = Q(E)$
for every Borel subset E of X, then there exists a bounded Borel field
$x \mapsto T(x)$ of unitary operators $T(x)$ from the Hilbert bundle $\mathcal{H}$ to
the Hilbert bundle $\mathcal{K}$ such that $T(x)$ is a unitary isomorphism of $\mathcal{H}_x$
onto $\mathcal{K}_x$ for almost all x satisfying $T(x)U_x(g)T(x)^{-1} = V_{\varphi(x)}(g)$ for
all g a.e. x.*

PROOF. Note in this case Φ is the identity, and ϕ is the identity
on X. □

Suppose $L = \int^{\oplus} L^x \, d\nu(x)$ is a direct integral of unitary representations on the Hilbert space $\int^{\oplus} \mathcal{H}_x \, d\mu(x)$. Let P be the canonical
projection valued measure on this Hilbert space, and let W be a
Borel subset of X. Then L^W is the unitary representation on the
range of $P(W)$ defined by $L^W(g)P(W)v = P(W)L(g)v$. Clearly this
representation is $\int^{\oplus}_W L^x \, d\mu(x)$.

PROPOSITION IV.25. *Suppose L is a locally simple representation
on $\int^{\oplus} \mathcal{H}_x \, d\mu(x)$ having form $\int^{\oplus} L^x \, d\mu(x)$ where almost all L^x are
irreducible. Then L^W and L^V have equivalent nontrivial subrepre-
sentations iff $\mu(W \cap V) > 0$. Moreover, L^W and L^V are unitarily*

equivalent iff W and V define the same element in the measure algebra of μ.

PROOF. If $\mu(W \cap V) > 0$, then $L^{W \cap V}$ is a nontrivial subrepresentation of both L^W and L^V. Conversely, suppose L^W and L^V have nontrivial equivalent subrepresentations. Since the von Neumann algebra of operators commuting with the representation L is abelian, Theorem IV.18 implies the range of the canonical projection valued measure P for this direct integral consists of all projections commuting with L. Consequentially, every subrepresentation of L has form L^U for some Borel subset U. Thus we may assume L^W is unitarily equivalent to L^V where both W and V have positive measure. The proof will then be complete if we show $\mu(W \bigtriangleup V) = 0$. Suppose not. Then we may assume by symmetry that $\mu(V - W) > 0$. It then follows that L^{V-W} is unitarily equivalent to a subrepresentation of L^W. Thus by making W and V smaller, we may assume $W \cap V = \emptyset$. Let U be a unitary intertwining operator between the representation L^W and L^V. Define a unitary operator T on $\int^{\oplus} \mathcal{H}_x \, d\mu(x)$ by $Tv = v$ if v is not in the range of $P(W \cup V)$, $Tv = Uv$ if v is in the range of $P(W)$ and $Tv = U^{-1}v$ if v is in the range of $P(V)$. Then T belongs to A'_L; and since this is a commutative von Neumann algebra containing the range of P, it follows that T commutes with the range of P. In particular $P(W)T = TP(W)$. But $TP(W) = P(V)T$. This is a contradiction. $\square$

THEOREM IV.20. *Let L be a locally simple representation of G. Suppose L is unitarily equivalent to $\int_X^{\oplus} L_x \, d\mu(x)$ and unitarily equivalent to $\int_Y^{\oplus} M_y \, d\nu(y)$ where almost all the representations L_x are irreducible and almost all of the representations M_y are irreducible. Suppose X and Y have the same cardinality. Then there exists an almost everywhere μ uniquely defined Borel isomorphism φ from X onto Y with $\varphi_* \mu \sim \nu$ such that L_x is unitarily equivalent to $M_{\varphi(x)}$ for almost all x.*

PROOF. By Theorem IV.18, the ranges of the canonical projection valued measures on $\int^{\oplus} \mathcal{H}_x \, d\mu(x)$ and on $\int^{\oplus} \mathcal{K}_y \, d\nu(y)$ are maximal abelian algebras of projections commuting with the representations $\int^{\oplus} L_x \, d\mu(x)$ and $\int^{\oplus} M_y \, d\nu(y)$. But since these representations are locally simple, the von Neumann algebras of operators commuting with these representations are abelian. Thus these maximal algebras of projection consist of all the projections commuting with these two representations, respectively. Since $\int^{\oplus} L_x \, d\mu(x)$ and $\int^{\oplus} M_y \, d\nu(y)$ are unitarily equivalent, any unitary operator T establishing the equivalence carries the range of the canonical projection valued measure for $\int^{\oplus} L_x \, d\mu(x)$ isomorphically onto the range of the canonical projection valued measure for $\int^{\oplus} M_y \, d\nu(y)$. The result now follows by Theorem IV.19.

To see the almost everywhere uniqueness of φ, we first note that any such φ gives an intertwining operator T between the representations $\int^{\oplus} L_x \, d\mu(x)$ and $\int^{\oplus} M_y \, d\nu(y)$ satisfying $T^{-1}Q(E)T = P(\varphi^{-1}(E))$ where P and Q are the canonical projection valued measures associated with these direct integrals. Indeed, by removing a null set, we may assume L_x is unitarily equivalent to $M_{\phi(x)}$ for all x. Then using the von Neumann selection Theorem I.14, one can obtain a Borel field of unitary operators $x \mapsto T(x)$ satisfying $TL_xT^{-1} = M_{\phi(x)}$ for μ a.e. x. Define T by

$$Tf(\varphi(x)) = \frac{d\mu \circ \varphi^{-1}}{d\nu}(\varphi(x))^{1/2}T(x)f(x).$$

Thus, to see the uniqueness of ϕ, it suffices by Theorem I.26 to show that the isomorphism established between the Boolean algebra of projections commuting with the representation $\int^{\oplus} M_y \, d\nu(y)$ and the Boolean algebra of projections commuting with $\int^{\oplus} L_x \, d\mu(x)$ is independent of the unitary transformation T giving the equivalence. Thus, suppose T and U are unitary operators intertwining $\int^{\oplus} L_x \, d\mu(x)$ and $\int^{\oplus} M_y \, d\mu(y)$. Then TU^{-1} commutes with the locally simple representation $\int^{\oplus} M_y \, d\nu(y)$. Since the von Neumann algebra of all bounded operators commuting with this representation is abelian, one has $TU^{-1}Q(E)UT^{-1} = Q(E)$ for all Borel subsets E of Y. Therefore $U^{-1}Q(E)U = T^{-1}Q(E)T$ for all E, and the result is established. $\square$

Let $M \mapsto [M]$ denote the natural projection of $\hat{\mathbf{G}}^c$ into $\hat{G}$.

THEOREM IV.21. *Let L be a locally simple representation and suppose $L = \int_X^{\oplus} L_x \, d\mu(x)$ is a direct integral decomposition of L into irreducible representations where μ is a finite measure on a standard Borel space X. Define $\hat{\mu}_L$ to be the measure on the Borel sets in $\hat{G}$ given by $\hat{\mu}_L(E) = \mu(\{x : [L_x] \in E\})$. Then the measure class of $\hat{\mu}_L$ depends only on L and not the particular direct integral decomposition. Moreover, if $\hat{\mu}_L$ is a standard measure on $\hat{G}$, then $x \mapsto [L_x] \in \hat{G}$ is one-to-one on a conull Borel subset of X.*

PROOF. Suppose $L = \int_Y^{\oplus} M_y \, d\nu(y)$ where Y is a standard Borel space, ν is a finite measure, and the M_y are irreducible. By Theorem IV.20, there are conull subsets X_0 and Y_0 of X and Y and a Borel isomorphism $\phi : X_0 \to Y_0$ satisfying $\phi_* \mu \sim \nu$ and L_x is unitarily equivalent to $M_{\phi(x)}$ for all $x \in X_0$. Thus $\hat{\mu}_L(E) = 0$ iff $\mu(X_0 \cap \{x : [L_x] \in E\}) = 0$ iff $\nu(Y_0 \cap \{y : [M_y] \in E\}) = 0$ iff $\hat{\nu}_L(E) = 0$. Thus $\hat{\mu}_L$ and $\hat{\nu}_L$ are equivalent measures.

Suppose now the measure $\hat{\mu} = \hat{\mu}_L$ is standard. Take a Borel subset $\hat{G}_0$ of $\hat{G}$ which is conull and standard. Then without loss, we may replace X by the conull Borel subset $\{x : [L_x] \in \hat{G}_0\}$. Since we wish to show $x \mapsto [L_x]$ is essentially one-to-one, we may suppose $\hat{G}_0 \subseteq \hat{G}_n$ for some n; and accordingly we may replace the Hilbert bundle $x \mapsto \mathcal{H}_x$ by the trivial bundle $x \mapsto \mathbb{H}_n$ where $\dim \mathbb{H}_n = n$. By Theorem I.27, one can disintegrate μ on X over $\hat{\mu}$. Thus $\mu = \int \mu_\gamma \, d\hat{\mu}(\gamma)$, and each μ_γ is measure on $q^{-1}(\gamma)$ where $q : X \to \hat{G}_0$ is the Borel mapping $q(x) = [L_x]$. For each $m \in \mathbb{Z}$, let $Y_m = \{\gamma \in \hat{G}_0 : \mu_\gamma \text{ is type } m\}$. By Proposition I.14, each Y_m is a Borel subset of $\hat{G}_0$. Moreover, to show $x \mapsto q(x)$ is essentially one-to-one, it suffices to show $x \mapsto q(x)$ is essentially one-to-one on $X_m = q^{-1}(Y_m)$. By replacing X by X_m, we may assume μ_γ is type m for all $\gamma \in \hat{G}_0$. Let T be a standard Borel space, and let τ be a measure of type m on T. By Corollary I.22, there is an essential Borel isomorphism $\phi : X \to Y_m \times T$ such that $\phi_* \mu$ is equivalent to $\hat{\mu} \times \tau$ and such that for $\hat{\mu}$ almost all γ, $\phi : q^{-1}(\gamma) \to \{\gamma\} \times T$ is an essential Borel isomorphism relative to the measures μ_γ and $\epsilon_\gamma \times \tau$. Thus, again using Theorem IV.20, we may replace X by $Y_m \times T$, μ_x by $\epsilon_\gamma \times \tau$ and assume $L = \iint^{\oplus} L_{\gamma,t} \, d\tau(t) \, d\hat{\mu}(\gamma)$ where $[L_{\gamma,t}] \in \gamma$ for $\gamma \in Y_m \subseteq \hat{G}$. The result will follow if we show $m = 1$.

Suppose $m \neq 1$. Then we can find two disjoint Borel sets A and B in T having positive τ measure for which there is a Borel isomorphism $\psi : A \to B$ with $\psi_* \tau|_A$ equivalent to $\tau|_B$. Let E be the set of all triples (γ, t, U), where $t \in A$, and U is a nonzero operator in $B(\mathbb{H}_n)$ satisfying $\|U\| \leq 1$ and $UL(\gamma, t)(g) = L(\gamma, \psi(t))(g)U$ for all $g \in G$, $t \in A$. E is a Borel subset of $(Y_m \times A) \times B(\mathbb{H}_n)$ and its projection into $Y_m \times A$ is onto. By Theorem I.14 (the von Neumann selection theorem), there is a Borel function $U(\gamma, t)$ on $Y_m \times A$ such that $U(\gamma, t) \neq 0$ a.e. (γ, t) and for which $U(\gamma, t)L(\gamma, t)(g) = L(\gamma, \psi(t))(g)U(\gamma, t)$ for all g and $\|U(\gamma, t)\| \leq 1$. Define a bounded nonzero operator A on $L^2(Y_m \times T, \hat{\mu} \times \tau, \mathbb{H}_n)$ by

$$Af(\gamma, t) = 1_B(t) \frac{d\tau \circ \psi^{-1}}{d\tau}(t)^{1/2} U(\gamma, \psi^{-1}(t)) f(\gamma, \psi^{-1}(t)).$$

Then

$$A(L(g)f)(\gamma, t) = 1_B(t)\frac{d\tau \circ \psi^{-1}}{d\tau}(t)^{1/2}U(\gamma, \psi^{-1}(t))(L(g)f)(\gamma, \psi^{-1}(t))$$

$$= 1_B(t)\frac{d\tau \circ \psi^{-1}}{d\tau}(t)^{1/2}U(\gamma, \psi^{-1}(t))L(\gamma, \psi^{-1}(t))(g)(f(\gamma, \psi^{-1}(t)))$$

$$= 1_B(t)\frac{d\tau \circ \psi^{-1}}{d\tau}(t)^{1/2}L(\gamma, t)(g)U(\gamma, \psi^{-1}(t))(f(\gamma, \psi^{-1}(t)))$$

$$= L(\gamma, t)(g)(1_B(t)\frac{d\tau \circ \psi^{-1}}{d\tau}(t)^{1/2}U(\gamma, \psi^{-1}(t))(f(\gamma, \psi^{-1}(t))))$$

$$= L(g)(Af)(\gamma, t).$$

Thus A belongs to $\mathcal{A}'_L$, the commutant of the von Neumann algebra generated by L. But since $L = \iint^{\oplus} L(\gamma, t)\, d\hat\mu(\gamma)\, d\tau(t)$ and the range of the canonical projection valued measure for this decomposition is contained in $\mathcal{A}'_L$ and $\mathcal{A}'_L$ is commutative, we see A commutes with the canonical projection valued measure. By Theorem III.14, $A = \iint^{\oplus} A(\gamma, t)\, d\hat\mu(\gamma)\, d\tau(t)$ for some bounded Borel function $(\gamma, t) \mapsto A(\gamma, t) \in B(\mathbb{H}_n)$. In particular, $\psi(t) = t$ a.e. on A. This gives a contradiction. Thus $m = 1$ and $x \mapsto [L_x]$ is essentially one-to-one. $\square$

PROPOSITION IV.26. *Let L be a direct integral $\int L_y \, d\mu(y)$ of irreducible representations over a standard Borel space Y. If L is unitarily equivalent to nM where M is an irreducible representation, then L_y is unitarily equivalent to M for μ a.e. y.*

PROOF. We may suppose μ is atom free. Indeed, if $\mu(\{y\}) > 0$ and if U is a unitary operator with $UL(g)U^{-1} = nM(g)$ for all g and if P_k is the orthogonal projection of the Hilbert space of nM onto the kth summand, then $P_k U$ is nonzero for at least one k and is an intertwining operator between the irreducible unitary representation L_y and the irreducible unitary representation M. By Schur's Lemma, $L_y \cong M$ whenever $\mu(\{y\}) > 0$.

We may now suppose μ is atom free. Let $\mathbb{K}$ be a Hilbert space of dimension n. Then $nM = M \otimes I$ on the Hilbert space $\mathbb{H} \otimes \mathbb{K}$, and $\mathcal{A}'_{nM} = I \otimes B(\mathbb{K})$. In particular if $\mathcal{B}$ is the range of the canonical projection valued measure on $\int^{\oplus} \mathbb{H}_y \, d\mu(y)$, then by Theorem IV.18, $\mathcal{B}$ is a maximal Boolean atom free algebra of projections in $\mathcal{A}'_{nM}$; and thus $UBU^{-1} = I \otimes \tilde{\mathcal{B}}$ where $\tilde{\mathcal{B}}$ is a maximal atom free Boolean algebra of projections in $B(\mathbb{K})$. Theorems III.8 and Corollary III.4 imply $\mathbb{K}$ may be taken to be $\int^{\oplus}_{[0,1]} \mathbb{C}\, dm(x)$ where m is Lebesgue measure on $[0, 1]$ and $\tilde{\mathcal{B}}$ is the range of the canonical projection for this direct integral. In particular, U is a unitary isomorphism from $\int^{\oplus} \mathbb{H}_y \, d\mu(y)$

onto $\int^{\oplus} (\mathbb{H} \otimes \mathbb{C}) \, dm(y)$ which carries the range of the projection valued measure for $\int^{\oplus} \mathbb{H}_y \, d\mu(y)$ onto the range of the projection valued measure for $\int^{\oplus} (\mathbb{H} \otimes \mathbb{C}) \, dm(y)$ and satisfies $U L(g) U^{-1} = \int^{\oplus} M(g) dm(y)$ for each g. By Theorem IV.19, L_y is unitarily equivalent to M for μ a.e. y. $\square$

DEFINITION. A locally simple representation L is said to have **standard support** in $\hat{G}$ if there is a direct integral decomposition $\int_X^{\oplus} L_x \, d\mu(x)$ of L into irreducible unitary representations such that the measure $\hat{\mu}_L$ on $\hat{G}$ is standard.

We note by Theorem IV.21, that if L has standard support in $\hat{G}$, then every decomposition of L into a direct integral $\int_X^{\oplus} L_x \, d\mu(x)$ of irreducible representations is such that $\hat{\mu}_L$ is standard.

PROPOSITION IV.27. *Let L be a type I representation and suppose $L = \int^{\oplus} L_x \, d\mu(x)$ relative to the direct integral $\int^{\oplus} \mathcal{H}_x \, d\mu(x)$ over the standard Borel space X. Then if each L_x is irreducible and the mapping $x \mapsto [L_x]$ is one-to-one into $\hat{G}$, then L is a locally simple representation with standard support in $\hat{G}$, and the range of the canonical projection valued measure for $\int^{\oplus} \mathcal{H}_x \, d\mu(x)$ is the Boolean algebra of projections commuting with L.*

PROOF. Let P be the canonical projection valued measure for the direct integral $\int^{\oplus} \mathcal{H}_x \, d\mu(x)$. By Theorem IV.18, the range of P is a maximal Boolean algebra of orthogonal projections in the von Neumann algebra $\mathcal{A}'_L$. By Theorem IV.10, there is for each n a maximal central projection P^n in $\mathcal{A}_L$ with the property that L restricted to $P^n \mathbb{H}$ is unitarily equivalent to $n U_n$ where U_n is a locally simple representation. Moreover $\sum P^n = I$. We show $P^1 = I$.

Indeed, suppose $P^n \neq 0$ for some $n > 1$. Then since P^n is central and the range of P is maximal, it follows that there is a Borel set E_n of positive Borel measure with $P^n = P(E_n)$. By replacing X by E_n, we may suppose $P(X) = P^n = I$, and hence L has uniform multiplicity n for some $n > 1$. The von Neumann algebra $\mathcal{A}_{nU_n}$ is type I and has form $\int n B(\mathbb{H}_y) \, d\nu(y)$ where the canonical projection valued measure Q for the corresponding direct integral decomposition $\int_Y n\mathbb{H}_y \, d\nu(y)$ has range the Boolean algebra of central projections in $\mathcal{A}_{nU_n}$. By Theorem IV.18, we may write $U_n = \int_Y^{\oplus} U_y \, d\nu(y)$ where each U_y is an irreducible unitary representation.

Let T be a unitary equivalence between L and nU_n. If E is a Borel subset of Y, then $T^{-1} Q(E) T$ is in the range of P for it is central. Hence there is a σ-homomorphism Φ from the measure algebra given by (Y, ν) into the measure algebra given by (X, μ) satisfying $T^{-1}Q(E)T = P(\Phi(E))$. Theorem I.26 implies there is a Borel mapping $\phi : X \to Y$ with $\phi_*\mu \prec \nu$ satisfying $\phi^{-1}(E) = \Phi(E)$ for all Borel subsets E. By Theorem I.27, $\mu = \int \mu_y \, d\nu(y)$ and as seen in Section 9, there is a corresponding decomposition of the direct integral $\mathbb{H} = \int_X^\oplus \mathcal{H}_x \, d\mu(x)$ into $\int_Y^\oplus \int_{\phi^{-1}(y)}^\oplus \mathcal{H}_x \, d\mu_y(x) \, d\nu(y)$ where under the corresponding isomorphism, the representation L becomes $\int_Y^\oplus \int_{\phi^{-1}(y)}^\oplus L_x \, d\mu_y(x) \, d\nu(y)$. Moreover, if $\mathcal{K}_y = \int_{\phi^{-1}(y)}^\oplus \mathcal{H}_x \, d\mu_y(x)$, then the projection valued measure Q' for the decomposition $\int_Y^\oplus \mathcal{K}_y \, d\nu(y)$ has range the central projections in the von Neumann algebra generated by L, and the unitary mapping T between $\int_Y^\oplus \mathcal{K}_y \, d\nu(y)$ and $\int_Y^\oplus n\mathbb{H}_y \, d\nu(y)$ satisfies $TQ'(E)T^{-1} = Q(E)$ for all Borel sets E. By Corollary IV.13, we see nU_y is unitarily equivalent to $\int_{\phi^{-1}(y)}^\oplus L_x \, d\mu_y(x)$ for ν a.e. y. Proposition IV.26 implies L_x is unitarily equivalent to U_y for μ_y a.e. x for ν a.e. y. But since $x \mapsto [L_x]$ is one-to-one off a null set, this implies for ν a.e. y, μ_y is concentrated at a point. In particular, n must be 1.

By Lemma IV.7, the Borel mapping $x \mapsto [L_x]$ from X into $\hat{G}$ carries Borel subset of X to Borel subsets of $\hat{G}$. Set $Y = \{[L_x] : x \in X\}$. Then Y is a Borel subset of $\hat{G}$ and $x \mapsto [L_x]$ is a Borel isomorphism of X onto Y. Hence Y is a standard Borel space when given the relative Borel structure from $\hat{G}$; and since $\hat{\mu}_L(\hat{G} - Y) = 0$, $\hat{\mu}_L$ is a standard measure.

Since L is locally simple, the only maximal Boolean algebra of projections in the commuting ring $\mathcal{A}'_L$ is the algebra of all projections in this algebra. $\square$

LEMMA IV.8. *Let* $L = \int^\oplus L_x \, d\mu(x)$ *on the Hilbert space* $\int^\oplus \mathcal{H}_x \, d\mu(x)$ *and* $M = \int^\oplus M_y \, d\nu(y)$ *on the space* $\int^\oplus \mathcal{K}_y \, d\nu(y)$. *Then if there is a Borel isomorphism* φ *from* X *onto* Y *with the property that* L_x *is unitarily equivalent to* $M_{\varphi(x)}$ *for all* x *and* $\varphi_*\mu \sim \nu$, *then* L *is unitarily equivalent to* M.

PROOF. We may assume both Hilbert bundles are trivial having the same Hilbert space as fibers. Thus L acts on $L^2(X, \mu, \mathbb{H})$ and M acts on $L^2(Y, \nu, \mathbb{H})$. Let g_i, $i = 1, 2, \ldots$ be a dense subset of G. Let A be the set consisting of all pairs (x, T) where T is a unitary operator satisfying $TL_x(g_i)T^{-1} = M_{\varphi(x)}(g_i)$ for all i. Then A is a Borel set, and its projection into the first coordinate has range X. By Theorem I.14, there is a Borel map $T : X \to \mathcal{U}(\mathbb{H})$ with $T(x)L_x(g_i)T(x)^{-1} =$

$M_{\varphi(x)}(g_i)$ for all i for a.e. x. But by continuity of representations and the density of the set $\{g_i\}$, we see $T(x)L_x(g)T(x)^{-1} = M_{\varphi(x)}(g)$ for all g for a.e. x.

Define a unitary transformation T from $L^2(X, \mathbb{H})$ onto $L^2(Y, \mathbb{H})$ by

$$(Tf)(\varphi(x)) = \frac{d\mu \circ \varphi^{-1}}{d\nu}(\varphi(x))T(x)f(x).$$

Then T is a unitary equivalence between L and M. $\square$

Suppose μ is a standard Borel σ-finite measure on $\hat{G}$. Choose a conull Borel subset $\hat{G}_0$ which with the relative Borel structure is a standard Borel space. Let $\hat{G}_0^c$ be the preimage of this set in $\hat{G}^c$. Theorem I.13 implies there is a Borel function Φ on a conull subset of $\hat{G}_0$ on which $[\Phi(\gamma)] = \gamma$. Set $\mu_* = \Phi_*\mu$. Then $\int^\oplus \Phi(\gamma)\, d\mu_*(\gamma)$ is a unitary representation of G depending on μ and the selection Φ. It will be denoted by $L[\mu, \Phi]$.

PROPOSITION IV.28. *Suppose μ and ν are equivalent standard Borel measures on $\hat{G}$. Then for any selections Φ and Φ' for these measures, the unitary representations $L[\mu, \Phi]$ and $L[\nu, \Phi']$ are unitarily equivalent.*

PROOF. This is an application of Lemma IV.8 after the removal of appropriate sets of measure 0. $\square$

13. Type I Measures on $\hat{G}$

DEFINITION. A standard Borel measure μ on $\hat{G}$ is **type I** if each representation $L[\mu, \Phi]$ is type I.

LEMMA IV.9. *Let μ be a standard measure on $\hat{G}$, and suppose there is a Borel function Φ on $\hat{G}$ such that $[\phi(\gamma)] = \gamma$ for μ a.e γ and the representation $L[\mu, \Phi]$ is type I. Then μ is a type I measure.*

PROOF. Apply Lemma IV.8 using the function $\phi(\gamma) = \gamma$ on an appropriate standard conull Borel subset $\hat{G}_0$ of $\hat{G}$. $\square$

PROPOSITION IV.29. *Suppose μ_1 and μ_2 are type I measures on $\hat{G}$. Then $L[\mu_1, \Phi_1]$ is unitarily equivalent to $L[\mu_2, \Phi_2]$ iff μ_1 and μ_2 are equivalent measures.*

PROOF. Suppose the measures are equivalent. Then the identity mapping restricted to an appropriate conull standard Borel subset of $\hat{G}$ is a Borel isomorphism with the property $\Phi_1(\gamma)$ is unitarily equivalent to $\Phi_2(\gamma)$ for all γ in this conull subset. Since the measures are equivalent, it follows by Lemma IV.8 that the representations $L[\mu_1, \Phi_1]$ and $L[\mu_2, \Phi_2]$ are unitarily equivalent.

Conversely, if the type I representations $L[\mu_1, \Phi_1]$ and $L[\mu_2, \Phi_2]$ are unitarily equivalent, then by Proposition IV.27, both of these representations are locally simple. Hence by Theorem IV.20, there is on appropriate conull standard Borel subsets $\hat{G}_1$ and $\hat{G}_2$ a Borel isomorphism ψ satisfying $\Phi_2(\psi(\gamma))$ is unitarily equivalent to $\Phi_1(\gamma)$ and $\psi_* \mu_1 \sim \mu_2$. But $[\Phi_1(\gamma)] = \gamma$ and $[\Phi_2(\psi(\gamma))] = \psi(\gamma)$ for μ_1 a.e. γ. Thus $\psi(\gamma) = \gamma$ a.e. γ, and μ_1 is equivalent to μ_2. $\square$

THEOREM IV.22. *Let μ_1 and μ_2 be two type I measures on $\hat{G}$. Then $n_1 L[\mu_1, \Phi_1]$ is unitarily equivalent to $n_2 L[\mu_2, \Phi_2]$ iff $n_1 = n_2$ and $\mu_1 \sim \mu_2$.*

PROOF. Let $U_1 = n_1 L[\mu_1, \Phi_1]$ and $U_2 = n_2 L[\mu_2, \Phi_2]$. Then the von Neumann algebra generated by U_i is $\mathcal{A}_i = \int^{\oplus} n_i \mathcal{B}(\mathcal{H}_i(\gamma)) \, d\mu_i(\gamma)$ where $\gamma \mapsto \mathcal{H}_i(\gamma)$ is the Hilbert bundle defined by $\mathcal{H}_i(\gamma) = \mathcal{H}(\Phi_i(\gamma))$.

Now let T be a unitary operator giving a unitary equivalence between the representations U_1 and U_2. Let X be a standard Borel subset of $\hat{G}$ such that $\mu_1(\hat{G} - X) = \mu_2(\hat{G} - X) = 0$. Then $\mathcal{A}_i = \int_X^{\oplus} n_i \mathcal{B}(\mathcal{H}_i(\gamma)) \, d\mu_i(\gamma)$ for $i = 1, 2$. Since T provides a unitary spatial isomorphism between $\mathcal{A}_1$ and $\mathcal{A}_2$, we know by Theorem III.15 that there are a Borel isomorphism φ of X such that $\varphi_* \mu_1 \sim \mu_2$ and a Borel field $T(\gamma) : n_1 \mathcal{H}_1(\gamma) \to n_2 \mathcal{H}_2(\varphi(\gamma))$ of unitary isomorphisms such that

$$Tf(\varphi(\gamma)) = \frac{d\mu_1 \circ \varphi^{-1}}{d\mu_2}(\varphi(\gamma))^{1/2} T(\gamma) f(\gamma)$$

for each f in $\int_X^{\oplus} n_1 \mathcal{H}_1(\gamma) \, d\mu_1(\gamma)$. In particular, since $TU_1(g)T^{-1} = U_2(g)$ and $U_i(g) = \int^{\oplus} n_i \Phi_i(\gamma)(g) \, d\mu_i(\gamma)$ for $i = 1, 2$, one has

$$T(\gamma) n_1 \Phi_1(\gamma)(g) T(\gamma)^{-1} = n_2 \Phi_2(\varphi(\gamma))(g)$$

for each g for μ_1 a.e. γ. Since $n_i \Phi_i(\gamma)$ and $n_2 \Phi_2(\varphi(\gamma))$ are unitary representations and hence strongly continuous, we can conclude that $n_1 \Phi_1(\gamma)$ and $n_2 \Phi_2(\varphi(\gamma))$ are unitarily equivalent μ_1 a.e. γ. By Proposition IV.12, one has $n_1 = n_2$ and $\Phi_1(\gamma)$ and $\Phi_2(\varphi(\gamma))$ are unitarily equivalent for μ_1 a.e. γ. Thus $\gamma = [\Phi_1(\gamma)] = [\Phi_2(\varphi(\gamma))] = \varphi(\gamma)$ for μ_1 a.e. γ. Thus $\varphi_* \mu_1 = \mu_1$ and $\mu_1 \sim \mu_2$. By Proposition IV.29, $L[\mu_1, \Phi_1]$ is unitarily equivalent to $L[\mu_2, \Phi_2]$.

The converse follows immediately by Proposition IV.29. $\square$

COROLLARY IV.14. *Suppose T is a unitary equivalence between the representations $nL[\mu_1, \Phi_1]$ and $nL[\mu_2, \Phi_2]$ where μ_1 and μ_2 are type I measures on $\hat{G}$. Then if P_1 and P_2 are the canonical projection valued measures for $nL[\mu_1, \Phi_1]$ and $nL[\mu_2, \Phi_2]$, then $TP_1(E)T^{-1} = P_2(E)$ for all Borel subsets E of $\hat{G}$.*

PROOF. In the proof of Theorem IV.22, we saw T has form

$$Tf(\gamma) = \frac{d\mu_1}{d\mu_2}(\gamma)^{1/2} T(\gamma) f(\gamma) \text{ for } \gamma \in X,$$

where X is a conull standard Borel subset of $\hat{G}$ and $T(\gamma)$ is a Borel field of unitary operators between the Hilbert bundles $\gamma \mapsto n\mathcal{H}_1(\gamma)$ and $\gamma \mapsto n\mathcal{H}_2(\gamma)$ over X. ($\mathcal{H}_i(\gamma)$ is the Hilbert space for the representation $\Phi_i(\gamma)$.) Since $P_i(E) = \int_E^\oplus I_i(\gamma) \, d\mu_i(\gamma)$ where $I_i(\gamma)$ is the identity operator on $n\mathcal{H}_i(\gamma)$, a direct calculation shows $TP_1(E)T^{-1} = P_2(E)$. $\square$

DEFINITION. Two unitary representations L and M are **disjoint** if there are no nonzero bounded intertwining operators T between these two representations.

PROPOSITION IV.30. *Let μ_1 and μ_2 be two type I measures on $\hat{G}$. Then the representations $L[\mu_1, \Phi_1]$ and $L[\mu_2, \Phi_2]$ are disjoint iff μ_1 and μ_2 are mutually singular measures.*

PROOF. Suppose μ_1 and μ_2 are not mutually singular. Then there is a Borel set W with positive measure relative to both μ_1 and μ_2 with the property that μ_1 and μ_2 are equivalent when restricted to the subset W. Proposition IV.28 implies the representations $L[\mu_1, \Phi_1]^W$ and $L[\mu_2, \Phi_2]^W$ are unitarily equivalent. But any unitary intertwining operator between these two representations extends to a nonzero bounded intertwining operator between $L[\mu_1, \Phi_1]$ and $L[\mu_2, \Phi_2]$ by defining the extension to be zero on the orthogonal complement of the closed subspace which defines the subrepresentation $L[\mu_1, \Phi_1]^W$. Thus $L[\mu_1, \Phi_1]$ and $L[\mu_2, \Phi_2]$ are not disjoint.

Suppose the measures μ_1 and μ_2 are mutually singular and T is a nonzero bounded operator intertwining the representations $L[\mu_1, \Phi_1]$ and $L[\mu_2, \Phi_2]$. Since these representations are locally simple, the ranges of the canonical projection valued measures for each of these representations consist of all the projections commuting with the representations. It follows that every subrepresentation for these representations has form $L[\mu_i, \Phi_i]^W$ for some standard Borel subset W of $\hat{G}$. Thus by restricting T to the orthogonal complement of the null space of T and the representation $L[\mu_2, \Phi_2]$ to the closure of the range of T, we may assume, since both of these closed subspaces define subrepresentations, that the null space of T is trivial and the closure of the range of T is the Hilbert space $\int^\oplus \mathcal{H}_{\Phi_2(\gamma)} \, d\mu_2(\gamma)$. The operator T^*T is a positive operator with positive square root S. The mapping $Sv \mapsto Tv$ is unitary for

$$(Tv, Tv) = (T^*Tv, v) = (S^2v, v) = (Sv, Sv),$$

and thus is well defined on the range of S onto the range of T. It also intertwines. Indeed, note the operator T^*T commutes with $L[\mu_1, \Phi_1]$; and since S is in the smallest C^* algebra containing T^*T, S commutes with every $L[\mu_1, \Phi_1](g)$. Thus $L[\mu_1, \Phi_1](g)Sv = SL[\mu_1, \Phi_1](g)v \mapsto TL[\mu_1, \Phi_1](g)v = L[\mu_2, \Phi_2](g)Tv$. But the range of S is dense since its closure is the orthogonal complement of the kernel of S^*, the kernel of S^* being 0 for it is contained in the kernel of T. Thus since the range of T is dense, this operator extends to a unitary intertwining operator between the representations $L[\mu_1, \Phi_1]$ and $L[\mu_2, \Phi_2]$. By Proposition IV.29, the measures μ_1 and μ_2 are equivalent. This contradicts our assumptions and completes the proof. $\square$

THEOREM IV.23. *There is a one-to-one onto correspondence between equivalence classes of unitarily equivalent locally simple representations of G with standard support in $\hat{G}$ and the type I measure classes on $\hat{G}$. The inverse of this correspondence is the mapping $[\mu] \mapsto [L[\mu, \Phi]]$.*

PROOF. Using Propositions IV.27, IV.28, and IV.29, one sees it suffices to show that if L is locally simple with standard support in $\hat{G}$, then there exists a type I measure μ such that L is unitarily equivalent to $L[\mu, \Phi]$.

Let $L = \int^\oplus L_x \, d\lambda(x)$ be a direct integral decomposition of L into irreducible unitary representations over the standard Borel space X. Using Theorem IV.21 and removing a null set if necessary, we may assume that the mapping $x \mapsto [L_x]$ is one-to-none into $\hat{G}$. It follows by Lemma IV.7 that the images of Borel subsets of X under this mapping are Borel subsets of $\hat{G}$. Since $\hat{G}$ has the quotient Borel structure, the mapping q defined by $x \mapsto [L_x]$ is a Borel isomorphism of X onto $[L_X]$. We denote the range by $\hat{G}_0$; and note that if we let μ denote the measure $q_*\lambda$, then μ is a standard measure on $\hat{G}$. Moreover, define Φ on $\hat{G}$ by $\Phi([L_x]) = L_x$. Then $L[\mu, \Phi] = \int^\oplus L_x \, d\lambda(x)$, and thus is locally simple. It follows that μ is type I and $L = L[\mu, \Phi]$. $\square$

Whether every locally simple unitary representation of G has standard support in $\hat{G}$ is a highly delicate and measure theoretically difficult question; it remains unsettled.

CHAPTER V

INDUCED ACTIONS AND REPRESENTATIONS

In this chapter we will discuss a natural method of inducing an action of a subgroup to an action of the whole group. When the action is a unitary action, the induced action is a unitary action on a homogeneous Hilbert bundle. When one forms the corresponding unitary action on the direct integral of this Hilbert bundle, one obtains the usual notion of induced unitary representation.

PROPOSITION V.1. *Let G be a second countable locally compact group and suppose X and Y are standard right Borel G spaces. Let ν be a quasi-invariant finite measure on Y. Suppose $y \mapsto \lambda_y$ is a Borel map from Y into $M(X)$ satisfying $\lambda_y \cdot g \sim \lambda_{y \cdot g}$ for almost all (y, g) where $\lambda_y \cdot g(E) = \lambda_y(Eg^{-1})$. Then there exists a Borel mapping $y \mapsto \lambda'_y \in M(X)$ with $\lambda'_y = \lambda_y$ a.e. y satisfying $\lambda'_y \cdot g = \lambda'_{y \cdot g}$ for all (y, g).*

PROOF. Let $W = \{(y, g) : \lambda_{y \cdot g} g^{-1} \sim \lambda_y\}$. Using Theorem IV.8, we may assume the action of G on X is continuous. If f is a bounded continuous function on X, the mapping $(y, g) \mapsto \int f(xg) \, d\lambda_y(x)$ is Borel in y for fixed g and continuous in g for fixed y. Proposition I.19 implies these functions are Borel on $Y \times G$. Hence the collection $\mathcal{M}$ of sets E with $(y, g) \mapsto \lambda_y \cdot g(E)$ Borel contains the closed sets. Indeed, if d is a compatible metric and E is a closed set, then the function $f(x) = 1 - \frac{d(x, E)}{1 + d(x, E)}$ is continuous and satisfies $\lim_n f^n(x) = 1_E(x)$. Thus $(y, g) \mapsto \int 1_E(xg) \, d\lambda_y(x) = \lambda_y(Eg^{-1})$ is Borel. By the monotone convergence theorem, $\mathcal{M}$ is a monotone class. Moreover, since open sets are F_σ's, every intersection of a closed set with an open set is in $\mathcal{M}$. Thus $\mathcal{M}$ contains the algebra of finite disjoint unions of intersections of open and closed sets. Since $\mathcal{M}$ is monotone, this implies $\mathcal{M}$ contains the σ-algebra of Borel sets.

Moreover, if $(y, g) \in W$ and $(y \cdot g, h) \in W$, then $\lambda_{ygh} \cdot (gh)^{-1} = (\lambda_{ygh} h^{-1}) g^{-1} \sim \lambda_{yg} g^{-1} \sim \lambda_y$ and thus $(y, gh) \in W$. Next, let Y_0 be the set of y such that $(y, g) \in W$ and $(yg, g^{-1}) \in W$ for a.e. g. Since ν is quasi-invariant, Y_0 is conull. Furthermore, note if y and $y \cdot g$ are both in Y_0, one can choose an h such that both (y, gh) and (ygh, h^{-1}) are in W and thus $(y, g) \in W$. Again applying Theorem IV.8, we may assume Y is an invariant Borel subset of the universal G space $\mathcal{U}_G$. By Theorem I.11, there is a conull σ-compact subset F of Y_0. It follows that the saturation $[F] = F \cdot G$ is a Borel set for it is a countable union of compact sets; and using Corollary I.8, one sees the mapping $(y, g) \mapsto y \cdot g$ from F onto $[F]$ has a Borel inverse $z \mapsto (z\theta(z)^{-1}, \theta(z))$ where $\theta : [F] \to G$. Note we may take $\theta(z) = e$ if $z \in Y_0$. Define $\lambda'_y = 0$ if $y \notin [F]$, and set $\lambda'_z = \lambda_{z\theta(z)^{-1}} \cdot \theta(z)$ for $z \in [F]$. Thus $\lambda'_z = \lambda_z$ for $z \in F$ and thus $\lambda'_y = \lambda_y$ for a.e. y. Furthermore, if $z \in [F]$, then $z\theta(z)^{-1}$ and $zg\theta(zg)^{-1}$ both belong to Y_0. Thus $(z\theta(z)^{-1}, \theta(z)g\theta(zg)^{-1}) \in W$. Hence $\lambda'_z \cdot g = \lambda_{z\theta(z)^{-1}} \theta(z) \cdot g = \lambda_{z\theta(z)^{-1}} \theta(z) g\theta(zg)^{-1} \theta(zg)$ is equivalent to $\lambda'_{z \cdot g} = \lambda_{zg\theta(zg)^{-1}} \theta(zg)$. Since $\lambda'_y \cdot g = \lambda'_{y \cdot g}$ for $y \notin [F]$, we have $\lambda'_y \cdot g \sim \lambda'_{y \cdot g}$ for all y in Y. $\square$

1. The Point Transformation Group

Let X be a Polish space with a σ-finite measure μ. Recall $\mathcal{I}(X, \mu)$ is defined to be the subset of $\mathcal{M}(X, \mu, X)$ consisting of those elements given by transformations φ which are Borel isomorphisms of X and satisfy $\varphi_* \mu \sim \mu$. By Proposition I.24, this subset with the relative Borel structure is a standard Borel group. This group we call the **point transformation group** of the measure space (X, μ). It will play a role in the definition of induced representations.

PROPOSITION V.2. *Let X and Y be Polish spaces. Suppose μ is a σ-finite measure on X. Then the mapping $(f, \phi) \mapsto f \circ \phi$ is a Borel mapping from $\mathcal{M}(X, \mu, Y) \times \mathcal{I}(X, \mu)$ into $\mathcal{M}(X, \mu, Y)$.*

PROOF. This is an immediate consequence of Proposition I.23. $\square$

Next let $\mathbb{H}$ be a separable Hilbert space. Set $\mathcal{H} = L^2(X, \mu, \mathbb{H})$ and let P be the canonical projection valued measure on this space. $L^\infty(X, \mu)$ will denote the abelian von Neumann algebra of operators generated by all the operators $P(E)$. Since $\mathcal{H} = L^2(X, \mu, \mathbb{H})$ is the direct integral of the trivial Hilbert bundle $x \mapsto \mathbb{H}$ over X, Theorem

III.14 implies $L^\infty(X,\mu)$ consists of all operators $M_f h = fh$ for $f \in L^\infty(X,\mu)$. For each φ in $\mathcal{I}(X,\mu)$, define a unitary operator L_φ acting on $\mathcal{H}$ by

$$L_\varphi f(x) = \frac{d\mu \circ \varphi^{-1}}{d\mu}(x)^{1/2} f(\varphi^{-1}x).$$

Note that $L_{\varphi_1} \circ L_{\varphi_2} = L_{\varphi_1 \circ \varphi_2}$ for φ_1 and φ_2 in $\mathcal{I}(X,\mu)$.

THEOREM V.1. *Suppose U is a unitary operator on $L^2(X,\mu,\mathbb{H})$ satisfying*

$$U L^\infty(X,\mu) U^{-1} = L^\infty(X,\mu).$$

Then there exist a unique $A \in L^\infty(X,\mu,\mathcal{U}(\mathbb{H}))$ and a unique φ in $\mathcal{I}(X,\mu)$ satisfying $U = M_A L_\varphi$. Thus

$$U f(x) = \frac{d\mu \circ \varphi^{-1}}{d\mu}(x)^{1/2} A(x) f(\varphi^{-1}(x)).$$

PROOF. Suppose U normalizes $L^\infty(X,\mu)$. Conjugation by U carries the range of the canonical projection valued measure P onto itself. Theorem III.15 implies the existence of A and φ. One shows uniqueness of φ by using the fact that φ is determined almost everywhere by the measure algebra homomorphism φ^{-1} and $UP(E)U^{-1} = P(\varphi^{-1}(E))$ for all Borel sets E. Uniqueness of A then follows from (5) of Proposition III.12. $\square$

2. Unitary Extensions of G Spaces and Induced Representations

Let X be a standard Borel G space. A Borel cocycle L of the G space X into the unitary group $\mathcal{U}(\mathbb{H})$ of a separable Hilbert $\mathbb{H}$ is called a **unitary representation** of the Borel G space X. If μ is a σ-finite measure on X, then the unitary representation ind L of G **induced** by L is the representation on $L^2(X,\mu,\mathbb{H})$ defined by

$$\operatorname{ind} L(g) f(x) = \frac{d\mu \cdot g^{-1}}{d\mu}(x)^{1/2} L(x,g) f(x \cdot g)$$

where $\mu \cdot g(E) = \mu(Eg^{-1})$ for each Borel subset E of X. To see this defines a unitary representation of G, note if we set $\varphi_g(x) = xg^{-1}$, then $g \mapsto \varphi_g$ is a Borel homomorphism of G into $\mathcal{I}(X,\mu)$ and $L_{\varphi_g} f(x) = \frac{d\mu \cdot g^{-1}}{d\mu}(x)^{1/2} f(xg)$. Moreover, the mapping $g \mapsto M_{A_g}$ where $A_g(x) = L(x,g)$ is a Borel mapping into $\mathcal{U}(\mathcal{H})$. By Theorem V.1 and Proposition IV.1 and since ind $L(g) = M_{A_g} L_{\varphi_g}$, we see that ind L is a Borel function from G into $\mathcal{U}(\mathcal{H})$. To see that ind L is a representation, first note by the cocycle property of L that $M_{A_{g_1}} M_{A_{g_2} \circ \varphi_{g_1}^{-1}} = M_{A_{g_1 g_2}}$. Thus

$$
\begin{aligned}
\operatorname{ind} L(g_1) \operatorname{ind} L(g_2) &= M_{A_{g_1}} L_{\varphi_{g_1}} M_{A_{g_2}} L_{\varphi_{g_2}} \\
&= M_{A_{g_1}} L_{\varphi_{g_1}} M_{A_{g_2}} L_{\varphi_{g_1}}^{-1} L_{\varphi_{g_1 g_2}} \\
&= M_{A_{g_1}} M_{A_{g_2} \circ \varphi_{g_1}^{-1}} L_{\varphi_{g_1 g_2}} \\
&= M_{A_{g_1 g_2}} L_{\varphi_{g_1 g_2}} \\
&= \operatorname{ind} L(g_1 g_2).
\end{aligned}
$$

Hence $\operatorname{ind} L$ is a Borel homomorphism of G into $\mathcal{U}(\mathcal{H})$; it is continuous by Theorem IV.2.

Let X be a standard Borel G space. An extension of G space X is a standard Borel G space W and an onto Borel mapping $p : W \to X$ satisfying $p(x \cdot g) = p(x) \cdot g$ for all x and g. This is said to be a **unitary extension** if W is a Hilbert bundle over X and each mapping $g : W_x \to W_{x \cdot g}$ is unitary. If $X = H \backslash G$ for some closed subgroup H of G, then W is said to be a homogeneous extension.

In the case when G acts unitarily on a trivial Hilbert bundle $X \times \mathbb{H}$ over X, one sees there is a Borel mapping L from $X \times G$ into $\mathcal{U}(\mathbb{H})$ satisfying $(x, v) \cdot g = (x \cdot g, L(x, g)^{-1} v)$. An easy calculation shows that L is a unitary representation of the G space X. Furthermore, in general, if $\mathcal{H}$ is a unitary extension of the standard Borel G space X, then using Theorem III.12 and the argument given in its proof, there exist G invariant Borel sets X_∞, X_1, X_2, $\cdots$ such that $\mathcal{H}|_{X_n}$ is isomorphic to the trivial Hilbert bundle $X_n \times \mathbb{H}_n$ where $\mathbb{H}_n$ is an n-dimensional Hilbert space. In particular, the action of G on $\mathcal{H}$ can be realized on $X_n \times \mathbb{H}_n$ by $(x, v) \cdot g = (x \cdot g, L_n(x, g)^{-1} v)$ where L_n is a unitary representation of the G space X_n.

The correspondence between unitary representations L of Borel G spaces and unitary extensions suggests the following more general definition for an induced representation. Indeed, one can induce a unitary G bundle extension over X to a unitary G bundle extension over G space Y provided there is a Borel family λ_y of finite measures on X satisfying $\lambda_y \cdot g \sim \lambda_{y \cdot g}$ for all y and g.

Specifically, suppose $\mathcal{H}$ is a unitary extension of a standard Borel G space X and Y is a standard Borel G space and $y \mapsto \lambda_y$ is a Borel function from Y into $M(X)$ satisfying $\lambda_y \cdot g \sim \lambda_{y \cdot g}$ for all y and g. As seen in Section 9 of Chapter III, one may define a Hilbert bundle $\mathcal{H} * \lambda$ over Y having sections $(\mathcal{H} * \lambda)_y = \int^\oplus \mathcal{H}_x \, d\lambda_y(x)$. We define a

G action on $\mathcal{H} * \lambda$ by

$$(y, f) \cdot g = (y \cdot g, f \cdot g)$$

where

$$f \cdot g(x) = \frac{d\lambda_y \cdot g}{d\lambda_{y \cdot g}}(x)^{1/2} f(x \cdot g^{-1}) \cdot g$$

for $f \in (\mathcal{H} * \lambda)_y$. This map is unitary from $(\mathcal{H} * \lambda)_y$ to $(\mathcal{H} * \lambda)_{yg}$, and it defines an action on the Hilbert bundle $\mathcal{H} * \lambda$. We check that this action is Borel. To simplify matters, we may assume $\mathcal{H}$ is a trivial Hilbert bundle $X \times \mathbb{H}$ and the action of G on $\mathcal{H}$ is defined by $(xg, v) \cdot g^{-1} = (x, L(x, g)v)$ where L is a unitary representation of the G space X on the Hilbert space $\mathbb{H}$. In this case, the Hilbert bundle $\mathcal{H} * \lambda$ consists of all pairs (y, f) where $f \in L^2(X, \lambda_y, \mathbb{H})$. Moreover, if $\{A_m\}$ is a countable generating algebra for the Borel subsets of X and the vectors e_n are an orthonormal basis of $\mathbb{H}$, then the standard Borel structure on $\mathcal{H} * \lambda$ is the smallest Borel structure for which the mappings $(y, f) \mapsto y$ and $(y, f) \mapsto \int_{A_m} \langle f, e_n \rangle \, d\lambda_y$ are Borel. Let $\mathcal{U}$ be the unitary group on $\mathbb{H}$. Let $\mathcal{C}$ be a countable subgroup of $\mathcal{U}$ which is dense in the strong $*$ operator topology. For each U in $\mathcal{C}$, let $\bar{U}$ be the conjugate linear mapping of $\mathbb{H}$ defined by $\bar{U}(\sum a_n e_n) = \sum \bar{a}_n U e_n$. Then the functions $(y, f) \mapsto \int_A \langle Uf, f \rangle \, d\lambda_y$ for $U \in \mathcal{C} \cup \bar{\mathcal{C}}$ are Borel. Indeed, it suffices to note $(y, f) \mapsto (y, Uf)$ and $(y, f) \mapsto (y, \bar{U}f)$ are Borel mappings from $\mathcal{H} * \lambda$ into $\mathcal{H} * \lambda$. Moreover, for each e_i, define a function G_i on $\mathcal{H} * \lambda$ by $G_{i,m}(y, f) = \int_{A_m} 1_{\{x: \mathrm{Re}\langle f(x), e_i \rangle > 0\}}(x) \langle f(x), e_i \rangle^2 \, d\lambda_y$. We show these functions are Borel. The rational valued functions of form $\sum_{i=1}^k r_i 1_{A_i}$ are countable in number and form a dense subspace of $L^2(\lambda)$ for any finite measure λ on X. Let $f_1, f_2, \cdots$ enumerate this subspace. For each $(y, f) \in \mathcal{H} * \lambda$, define $F_n(y, f) = (y, f_l)$ where f_l is the first term in the sequence $f_1, f_2, \cdots$ satisfying $\|f - f_l\| < \frac{1}{n}$. We note there are Borel subsets $W_1, W_2, \cdots$ of $\mathcal{H} * Y$ such that if $(y, f) \in W_l$, then $F_n(y, f) = (y, f_l)$. In particular the F_n are Borel functions on $\mathcal{H} * \lambda$ converging pointwise to the identity function. Consequently $G_{i,m} \circ F_n$ converges pointwise to $G_{i,m}$. Hence $G_{i,m}$ is Borel if each $G_{i,m} \circ F_n$ is Borel. But the restriction of this function to W_l is the function $(y, f) \mapsto \int_{A_m} 1_{\{x: \mathrm{Re}\langle f_l(x), e_i \rangle > 0\}}(x) \langle f_l(x), e_i \rangle^2 \, d\lambda_y(x)$ which is Borel. Finally, we note the countable collection of Borel functions $(y, f) \mapsto y$, $(y, f) \mapsto \int_{A_m} \langle Uf, f \rangle \, d\lambda_y$, $U \in \mathcal{C} \cup \bar{\mathcal{C}}$, $(y, f) \mapsto G_{i,m}(y, f)$ separate the points of $\mathcal{H} * \lambda$. In fact suppose $\int_{A_m} \langle Uf, f \rangle d\lambda_y = \int_{A_m} \langle Uh, h \rangle d\lambda_y$ for all m and $U \in \mathcal{C}$. Then $\langle Uf(x), f(x) \rangle = \langle Uh(x), h(x) \rangle$ for λ_y a.e. x for each $U \in \mathcal{C}$. This implies $\langle Uf(x), f(x) \rangle = \langle Uh(x), h(x) \rangle$ for all unitary operators U on $\mathbb{H}$ for λ_y a.e. x . An easy argument then shows $f(x) = a(x)h(x)$ a.e. x for some measurable function $a(x)$

satisfying $|a(x)| = 1$ for λ_y a.e. x. Using $\bar{U}$, one then sees that $\int_{A_m} \bar{a}(x)^2 \langle \bar{U}h(x), h(x) \rangle \, d\lambda_y(x) = \int_{A_m} \langle \bar{U}h(x), h(x) \rangle \, d\lambda_y(x)$ for all m. This implies $\bar{a}(x)^2 = 1$ a.e. where $h(x) \neq 0$. Thus we may assume $a(x) = 1$ or $a(x) = -1$. Finally,

$$\int_{A_m} 1_{\{\mathrm{Re}\langle a(x)f(x), e_i \rangle > 0\}}(x) a(x)^2 \langle f(x), e_i \rangle^2 \, d\lambda_y(x) =$$

$$\int_{A_m} 1_{\{\mathrm{Re}\langle f(x), e_i \rangle > 0\}}(x) \langle f(x), e_i \rangle^2 d\lambda_y(x)$$

for each i implies $a(x) > 0$ for λ_y a.e. x where $f(x) \neq 0$. This gives $f = h$.

Since $(y, f, g) \mapsto y \cdot g$ is Borel, Proposition I.4 implies the action is Borel if the mappings $(f, y, g) \mapsto \int_{A_m} \langle U(f \cdot g), f \cdot g \rangle \, d\lambda_{y \cdot g}$, $U \in \mathcal{C} \cup \bar{\mathcal{C}}$ and the functions $(f, y, g) \mapsto G_{i,m}(y \cdot g, f \cdot g)$ are Borel. First note since $L(xg^{-1}, g)^{-1} = L(x, g^{-1})$ that for $U \in \mathcal{C} \cup \bar{\mathcal{C}}$ one has

$$\int_{A_m} \langle U(f \cdot g), f \cdot g \rangle \, d\lambda_{y \cdot g}$$

$$= \int_{A_m} \frac{d\lambda_y \cdot g}{d\lambda_{y \cdot g}}(x) \langle UL(x, g^{-1}) f(xg^{-1}), L(x, g^{-1}) f(xg^{-1}) \rangle \, d\lambda_{y \cdot g}(x)$$

$$= \int 1_{A_m}(x) \langle UL(x, g^{-1}) f(xg^{-1}), L(x, g^{-1}) f(xg^{-1}) \rangle \, d\lambda_y \cdot g(x)$$

$$= \int 1_{A_m}(xg) \langle UL(xg, g^{-1}) f(x), L(xg, g^{-1}) f(x) \rangle \, d\lambda_y(x).$$

To see this is Borel on $\mathcal{H} * \lambda \times G$, we will show it is a pointwise limit of Borel functions. Indeed, we use the sequence of functions F_n. Using the composition of the function $(y, f) \mapsto \int_{A_m} \langle U(f \cdot g), f \cdot g \rangle d\lambda_{y \cdot g}$ with F_n, we see the above functions are pointwise limits of functions

$$H_n(y, f, g) =$$

$$\sum_l 1_{W_l}(y, f) \int 1_{A_m}(xg) \langle UL(xg, g^{-1}) f_l(x), L(xg, g^{-1}) f_l(x) \rangle \, d\lambda_y(x).$$

But each H_n is Borel if $(y, f, g) \mapsto \int 1_{A_m}(xg) \langle L(x, g) UL(x, g)^{-1} f_l, f_l \rangle \, d\lambda_y$ are Borel. Since these functions are constant in the second coordinate of the triple (y, f, g), it suffices to note by Proposition I.15 that $\int h(x, g) \, d\lambda_y(x)$ is Borel on $Y \times G$ for any bounded Borel function h on $X \times G$.

Finally the functions $G_{i,m}(y \cdot g, f \cdot g)$ are given by

$$G_{i,m}(y \cdot g, f \cdot g) =$$
$$\int_{A_m} 1_{\{\text{Re}\langle f \cdot g, e_i \rangle > 0\}}(x) \langle f \cdot g(x), e_i \rangle^2 \, d\lambda_{y \cdot g}$$

$$= \int_{A_m} 1_{\{\text{Re}\langle f \cdot g, e_i \rangle > 0\}}(x) \frac{d\lambda_y \cdot g}{d\lambda_{y \cdot g}}(x) \langle L(x, g^{-1}) f(xg^{-1}), e_i \rangle^2 \, d\lambda_{y \cdot g}(x)$$

$$= \int 1_{A_m}(x) 1_{\{\text{Re}\langle f \cdot g, e_i \rangle > 0\}}(x) \langle L(x, g^{-1}) f(xg^{-1}), e_i \rangle^2 d\lambda_y \cdot g(x)$$

$$= \int 1_{A_m(xg)} 1_{\{\text{Re}\langle f \cdot g, e_i \rangle > 0\}}(xg) \langle L(xg, g^{-1}) f(x), e_i \rangle^2 \, d\lambda_y(x)$$

which again is a limit of a sum of functions of form

$$1_{W_l}(y, f) \int 1_{A_m(xg)} 1_{\{\text{Re}\langle f_l \cdot g(x), e_i \rangle > 0\}} \langle L(xg, g^{-1}) f_l(x), e_i \rangle^2 \, d\lambda_y(x).$$

These functions are Borel by Proposition I.15.

The unitary extension $\mathcal{H} * \lambda$ for G space Y is denoted by $\text{ind}_X^Y \mathcal{H}$ and is called the unitary extension induced by $\mathcal{H}$ from X to Y. We note that if Y consists of one point, then $\lambda_y = \mu$ where $Y = \{y\}$ and μ is a quasi-invariant measure on X. Moreover, we denote the induced representation on the corresponding Hilbert space $\int^{\oplus} \mathcal{H}_x d\mu(x)$ by $\text{ind} \, \mathcal{H}$ and note one has

$$(\text{ind} \, \mathcal{H})_g f(x) = \frac{d\mu g^{-1}}{d\mu}(x)^{1/2} f(x \cdot g) \cdot g^{-1}.$$

In particular, if $\mathcal{H} = X \times \mathbb{H}$ and the action is given by

$$(x, v) \cdot g = (xg, L(x, g)^{-1} v),$$

then

$$(\text{ind} \, \mathcal{H})_g f(x) = \frac{d\mu g^{-1}}{d\mu}(x)^{1/2} L(x, g) f(x \cdot g)$$

which is just the representation $\text{ind} \, L$ described earlier.

PROPOSITION V.3. *Let $\mathcal{H}$ be a Hilbert bundle over X and suppose $y \mapsto \lambda_y$ and $y \mapsto \lambda'_y$ are Borel finite measure valued maps on standard Borel space Y. Suppose $\lambda_y \sim \lambda'_y$ for all y. Then $\mathcal{H} * \lambda$ and $\mathcal{H} * \lambda'$ are isomorphic Hilbert bundles over Y. In particular, if these are unitary extensions, and $\lambda_y \cdot g \sim \lambda_{y \cdot g}$ and $\lambda'_y \cdot g \sim \lambda'_{y \cdot g}$ for all y and g, then $\mathcal{H} * \lambda$ and $\mathcal{H} * \lambda'$ are isomorphic unitary extensions of Y.*

PROOF. We may assume $\mathcal{H}$ is trivial. Define a mapping Φ from $\mathcal{H} * \lambda$ into $\mathcal{H} * \lambda'$ by $\Phi(y, f) = (y, \left(\frac{d\lambda_y}{d\lambda'_y}\right)^{1/2} f)$. We show Φ is Borel. As seen in the prior discussion, it is sufficient to show the functions $(y, f) \mapsto \int_A \langle U \left(\frac{d\lambda_y}{d\lambda'_y}\right)^{1/2} f, \left(\frac{d\lambda_y}{d\lambda'_y}\right)^{1/2} f \rangle \, d\lambda'_y$ for U a unitary or antiunitary operator on $\mathbb{H}$ and A a Borel subset of X and the functions

$$(y, f) \mapsto \int_A 1_{\{\mathrm{Re}\langle f(x), e_i \rangle > 0\}}(x) \langle \left(\frac{d\lambda_y}{d\lambda'_y}\right)^{\frac{1}{2}} f(x), e_i \rangle^2 \, d\lambda'_y(x)$$

are Borel. But these are the functions

$$(y, f) \mapsto \int_A \langle Uf, f \rangle \, d\lambda_y$$

and

$$(y, f) \mapsto \int_A 1_{\{\mathrm{Re}\langle f(x), e_i \rangle > 0\}}(x) \langle f(x), e_i \rangle^2 \, d\lambda_y(x)$$

which we showed to be Borel.

The inverse of the map Φ by symmetry is also seen to be Borel. If these are unitary extensions, one notes these mappings preserve the G actions. $\square$

3. Inducing in Stages

Suppose X, Y and Z are standard Borel spaces and $p : X \to Y$ and $q : Y \to Z$ are Borel mappings. Suppose $y \mapsto \mu_y$ is a Borel map into the space of probability measures on X satisfying $\mu_y(X - p^{-1}(y)) = 0$ for all y and ν is a finite measure on Y. Let $z \mapsto \nu_z$ be a Borel function into the finite measures on Y with the property that $\nu(E) = \int \nu_z(E) \, dq_* \nu(z)$ for all Borel subsets E of Y and $\nu_z(Y - q^{-1}(z)) = 0$ for all z. Define

$$\lambda_z(E) = \int \mu_y(E) \, d\nu_z(y)$$

for Borel subsets E of X. By Proposition I.15, $z \mapsto \lambda_z$ is a Borel function into the space $M(X)$ of finite measures on X.

PROPOSITION V.4. Let $\mathcal{H}$ be a Hilbert bundle over X. Then $\mathcal{H} * \lambda$ is isomorphic to $(\mathcal{H} * \mu) * \nu$.

PROOF. We may assume $\mathcal{H}$ is trivial. Thus $\mathcal{H} = X \times \mathbb{H}$. Hence $(\mathcal{H} * \mu)_y = L^2(X, \mu_y, \mathbb{H})$ and $(\mathcal{H} * \lambda)_z = L^2(X, \lambda_z, \mathbb{H})$. Define a mapping Φ from $\mathcal{H} * \lambda$ into $(\mathcal{H} * \mu) * \nu$ by

$$\Phi(z, f) = (z, y \mapsto (y, f|_{q^{-1}(y)})).$$

It is a direct argument to show this isomorphism is Borel. $\square$

THEOREM V.2 (INDUCING IN STAGES). *Assume X, Y, and Z are standard Borel G spaces and $p : X \to Y$ and $q : Y \to Z$ are onto Borel equivariant functions. Furthermore, assume $y \mapsto \mu_y$ is a bounded Borel function into $M(X)$ and $z \mapsto \nu_z$ is a Borel function into $M(Y)$ satisfying $\mu_y(X - p^{-1}(y)) = 0$ for all y, $\nu_z(Y - q^{-1}(z)) = 0$ for all z, and $\mu_y \cdot g \sim \mu_{y \cdot g}$, $\nu_z \cdot g \sim \nu_{z \cdot g}$ for all y, z, and g. Let $\mathcal{H}$ be a G unitary Hilbert bundle over X. Set $\lambda_z = \int \mu_y \, d\nu_z(y)$. Then the unitary G Hilbert bundle $\operatorname{ind}_Y^Z \operatorname{ind}_X^Y \mathcal{H}$ is G isomorphic to the G unitary Hilbert bundle $\operatorname{ind}_X^Z \mathcal{H}$.*

PROOF. By Proposition V.4, these are isomorphic as Hilbert bundles. A calculation using the isomorphism given in the proof of Proposition V.4 shows this isomorphism preserves the G actions. $\square$

4. Induced Group Actions

We now present the notion of an induced group action. This process constructs actions of a larger group in terms of the actions of a subgroup. More specifically an action of a subgroup H produces an extension of the natural action of G on $H \backslash G$.

Suppose H is a closed subgroup of G and Y is a right standard Borel H space. Consider the set $Y \times G$ with equivalence relation defined by $(y \cdot h, g) \sim (y, hg)$. Let $Y \times_H G$ be the space of equivalence classes under this equivalence relation with the quotient Borel structure. Define an action of G on $Y \times G$ by $(y, g_1) \cdot g_2 = (y, g_1 g_2)$. This action preserves the equivalence classes and thus defines an action of G on $Y \times_H G$. Moreover, the map $p : Y \times_H G \to H \backslash G$ defined by $(y, g) \mapsto Hg$ is G equivariant.

PROPOSITION V.5. *$Y \times_H G$ is a standard Borel space.*

PROOF. By Proposition IV.2, there is a Borel cross section σ of the right coset space $H \backslash G$. Thus $\sigma : H \backslash G \to G$ is a Borel function and $H\sigma(x) = x$ for all $x \in H \backslash G$. Define a mapping F from $Y \times H \backslash G \to Y \times_H G$ by $F(y, x) = [(y, \sigma(x))]$ where the square bracket indicates the equivalence class. F is one-to-one and onto. Moreover, $F^{-1}([y, g]) = F^{-1}([y, g\sigma(Hg)^{-1}\sigma(Hg)]) = F^{-1}([yg\sigma(Hg)^{-1}, \sigma(Hg)])$. Thus $F^{-1}([y, g]) = (yg\sigma(Hg)^{-1}, Hg)$. But this as a function of y and g is Borel. Since $Y \times_H G$ has the quotient Borel structure, it follows that F^{-1} is a Borel function. Hence $F(E)$ is a Borel subset of $Y \times_H G$ for every Borel subset E of $Y \times H \backslash G$. In particular, since $Y \times H \backslash G$ is a countably separated Borel space, $Y \times_H G$ is countably separated.

By Theorem I.10, the Borel space $Y \times_H G$ is analytic. From Theorem I.9, we obtain the conclusion. $\square$

Note: A simple consequence of the above argument is that a function f from a Borel space X into $Y \times_H G$ is Borel iff it has form $f(x) = [(y(x), g(x))]$ for some Borel functions $y : X \to Y$ and $g : X \to G$.

The isomorphism F can be used to move the action of G on $Y \times_H G$ to an isomorphic action on $Y \times H\backslash G$. Indeed, $F^{-1}([y, \sigma(x)g]) = (y \cdot \sigma(x)g\sigma(xg)^{-1}, xg)$. In particular, the action of G on $Y \times_H G$ and the natural projection map p from $Y \times_H G$ onto $H\backslash G$ are Borel. Hence we have the following theorem.

THEOREM V.3. *$Y \times_H G$ is a homogeneous Borel extension of $H\backslash G$. Moreover, every Borel G space extension of $H\backslash G$ is isomorphic to $Y \times_H G$ for some standard Borel H space Y.*

PROOF. Let $p : X \to H\backslash G$ be an extension. Then $Y = p^{-1}(He)$ is a standard Borel H space. Moreover, the mapping $(y, g) \mapsto y \cdot g$ is a Borel mapping which is constant on the equivalence classes defined by the action $(y, g) \cdot h = (y \cdot h, h^{-1}g)$. This mapping factors to a Borel mapping G space isomorphism from $Y \times_H G$ onto X. $\square$

Suppose now that $\mathbb{K}$ is a Hilbert space and H acts on $\mathbb{K}$ as a unitary representation. Then $\mathbb{K} \times_H G$ is a Hilbert bundle over $H\backslash G$ and the induced G action on this space is unitary. To see this note if p is the natural projection onto $H\backslash G$ and $\{e_n\}$ form an orthonormal basis of $\mathbb{K}$ and σ is a Borel cross section from $H\backslash G$ into G, then the functions $f_n(Hg) = [(e_n, \sigma(Hg))]$ satisfy the conditions for a Hilbert bundle. Indeed, note that a mapping f from a standard Borel space X into $\mathbb{K} \times_H G$ is Borel iff it has form $f(x) = [(v(x), g(x))]$ for some Borel functions $v(\cdot)$ and $g(\cdot)$ on X. Hence f is Borel iff $\langle f(x), f_n(p(f(x))) \rangle$ are Borel for all n and $p(f(x)) = Hg(x)$ is Borel.

Moreover, if one uses the isomorphism F from $\mathbb{K} \times H\backslash G$ onto $\mathbb{K} \times_H G$ defined by $F(v, x) = [(v, \sigma(x))]$, then F^{-1} is an isomorphism of $\mathbb{K} \times_H G$ onto the trivial bundle over $H\backslash G$ with fibers isomorphic to $\mathbb{K}$. Moreover, if $v \cdot h = L(h^{-1})v$ where L is the continuous homomorphism of H into $\mathcal{U}(\mathbb{K})$ defining the unitary action, then $(v, x) \cdot g = (L(\sigma(x)g\sigma(xg)^{-1})^{-1}v, xg)$.

The same argument as used in the proof of Theorem V.3 can be used to establish the following corollary.

COROLLARY V.1. *Let $\mathcal{H}$ be a Hilbert bundle over the homogeneous space $H\backslash G$ and suppose $\mathcal{H}$ is a unitary Borel G space over $H\backslash G$. Then there exists a unitary Borel H action on a Hilbert space $\mathbb{K}$ such that $\mathcal{H}$ is isomorphic to the homogeneous Hilbert bundle $\mathbb{K} \times_H G$ with its natural G action.*

5. Inducing Group Actions in Stages

We next consider inducing in stages.

PROPOSITION V.6. *Assume H and K are closed subgroups of G and $H \subset K \subset G$. Furthermore, assume that Y is a standard Borel H space. Then there is a natural Borel G space isomorphism of $Y \times_H G$ onto $(Y \times_H K) \times_K G$.*

PROOF. Consider the Borel space $Y \times K \times G$ equipped with an equivalence relation defined by:

$$(y, k, g) \sim (yh, h^{-1}kk_1, k_1^{-1}g)$$

for all (h, k_1) in $H \times K$. Then the equivalence classes under this action are isomorphic to the equivalence classes in $(Y \times_H K) \times_K G$ under the identification $[([(y, k)], g)] \mapsto [(y, k, g)]$. Moreover, they are isomorphic to the equivalence classes of $Y \times_H G$ under the identification $[(y, g)] \mapsto [(y, e, g)]$. It is easy to check that the resulting isomorphism from $Y \times_H G$ onto $(Y \times_H K) \times_K G$ preserves the G actions. $\square$

REMARK. Suppose $\mathcal{H}$ is a unitary Hilbert bundle over H space Y. Let $p : \mathcal{H} \to Y$ be the corresponding H equivariant projection. Then one may consider the collection of all points (v, g) where $v \in \mathcal{H}_y$ with equivalence relation $(v, g) \sim (v \cdot h, h^{-1}g)$. The space $\mathcal{H} \times_H G$ of equivalence classes with the quotient Borel structure is a standard Borel G space and the mapping $(v, g) \mapsto [(p(v), g)]$ factors to a G equivariant projection mapping $P : \mathcal{H} \times_H G \to Y \times_H G$. Moreover, the space becomes a Hilbert bundle over $Y \times_H G$ if one defines addition and a Hermitian structure on the equivalence classes by

$$c_1[(v, g)] + c_2[(w, g)] = [(c_1 v + c_2 w, g)] \text{ for } v, w \in p^{-1}(y) \text{ and}$$

$$([v, g], [w, g]) = (v, w)_{p(y)}.$$

In the case when $\mathcal{H}$ is a trivial Hilbert bundle, one can show $\mathcal{H} \times_H G$ is isomorphic to a trivial Hilbert bundle over $Y \times_H G$.

We also note that if $H \subset K \subset G$ are closed subgroups of G and Y is a Borel H space, then by Proposition V.6, the G space $Y \times_H G$ is naturally isomorphic to $(Y \times_H K) \times_K G$. Again using Proposition V.6, if $\mathcal{K}$ is a unitary Hilbert bundle over an H space Y, then the G unitary Hilbert bundle $\mathcal{K} \times_H G$ is isomorphic as a G space to $(\mathcal{K} \times_H K) \times_K G$; and using the natural identification of $Y \times_H G$ with $(\mathcal{K} \times_H K) \times_K G$, this can easily be checked to be a Hilbert bundle isomorphism over G spaces. In particular if Y is trivial, then $\mathcal{K} = \mathbb{K}$ is a Hilbert space and $\mathbb{K} \times_H K$ is a K unitary Hilbert bundle over $H\backslash K$. Thus $(\mathbb{K} \times_H K) \times_K G$ is a unitary G Hilbert bundle over $H\backslash K \times_K G$. We hence have the following two Corollaries.

COROLLARY V.2. *Suppose $\mathcal{K}$ is a unitary Hilbert bundle over H space Y. Then the G unitary Hilbert bundle $\mathcal{K} \times_H G$ over $Y \times_H G$ is G unitary Borel isomorphic to the Hilbert bundle $(\mathcal{K} \times_H K) \times_K G$ over $(Y \times_H K) \times_K G$.*

COROLLARY V.3. *Suppose $\mathbb{K}$ is a unitary H space and $H \subset K \subset G$ are closed subgroups of G. Then the G unitary Hilbert bundle $(\mathbb{K} \times_H K) \times_K G$ over $H\backslash K \times_K G$ is G unitary Borel isomorphic to the Hilbert bundle $\mathbb{K} \times_H G$ over $H\backslash G$.*

PROPOSITION V.7. *Suppose $\mathcal{H}$ and $\mathcal{K}$ are H unitary Hilbert bundles over a standard Borel H space X. Then the induced unitary G bundles $\mathcal{H} \times_H G$ and $\mathcal{K} \times_H G$ are isomorphic over $X \times_H G$ if and only if the unitary H bundles $\mathcal{H}$ and $\mathcal{K}$ are isomorphic over X.*

PROOF. Assume the induced G unitary Hilbert bundles are isomorphic over $X \times_H G$. Note each vector over the element $[(x,e)]$ in $X \times_H G$ has a unique representative of form (v,e), $v \in p^{-1}(x)$. Hence if Ψ is a G space isomorphism of the induced unitary Hilbert bundles, then $\Psi([(v,e)]) = [(v',e)]$ for a unique vector v' over x in $\mathcal{K}$ for each vector v over x in $\mathcal{H}$. The mapping Φ defined by $\Psi([(v,e)]) = [(\Phi(v),e)]$ is Borel and since $\Psi([(v\cdot h,e)]) = \Psi([(v,h)] = \Psi([(v,e)] \cdot h) = [(\Phi(v),e)] \cdot h = [(\Phi(v),h)] = [(\Phi(v)h,e)]$, we see that $\Phi(v \cdot h) = \Phi(v) \cdot h$. Thus Φ preserves the unitary H action. Φ^{-1} is Borel by the same argument. $\square$

COROLLARY V.4. *Let $\mathbb{H}$ and $\mathbb{K}$ be unitary H spaces. Then the unitary G bundles $\mathbb{H} \times_H G$ and $\mathbb{K} \times_H G$ are isomorphic over $H\backslash G$ iff the corresponding unitary representations of H on $\mathbb{H}$ and $\mathbb{K}$ are equivalent.*

6. Mackey's Imprimitivity Theorem

In our discussion thus far we have mainly considered Borel actions with no quasi-invariant measure. To consider group actions with quasi-invariant measure, we need a different notion of equivalence. Two G spaces X and Y with respective quasi-invariant measures μ and ν are said to be essentially isomorphic if there exists an essentially one-to-one Borel function $\varphi : X \to Y$ such that $\mu \circ \varphi^{-1} \sim \nu$ and

$$\varphi(x \cdot g) = \varphi(x) \cdot g$$

a.e. x for each g.

If U denotes the representation $\operatorname{ind} \mathcal{H}$ on $\int^{\oplus} \mathcal{H}_x \, d\mu(x)$ and P is the associated canonical projection valued measure on $\int^{\oplus} \mathcal{H}_x \, d\mu(x)$, then

$$U_{g^{-1}} P(E) U_g = P(Eg)$$

for all Borel subsets E of X and each element g in G. Mackey calls a unitary representation U and a projection valued measure P based on a Borel G space X which satisfy the above condition a **system of imprimitivity**. Mackey's celebrated imprimitivity theorem states the above construction is the way all such systems arise.

THEOREM V.4. *Let (U, P) be a system of imprimitivity based on the standard Borel G space X. Then there is a unitary extension $\mathcal{H}$ of the G space X and a quasi-invariant measure μ on X such that the pair (U, P) is unitarily equivalent to the pair $(\operatorname{ind} \mathcal{H}, Q)$ where Q is the canonical projection valued measure on $\int^{\oplus} \mathcal{H}_x \, d\mu(x)$.*

PROOF. For $k = \infty, 1, 2, \cdots$, let $P(E_k)$ be the largest projection of multiplicity k. By Theorem III.2, these are orthogonal and sum to be the identity. Using Proposition III.2, we may assume the E_k's are disjoint and have union X. Theorem III.1 implies there is for each k a probability measure μ_k on E_k such that $P(E) = 0$ for a Borel subset E of E_k iff $\mu_k(E) = 0$. Letting μ be the disjoint sum of these measures on X, we obtain a σ-finite measure μ on X with the property that $P(E) = 0$ iff $\mu(E) = 0$. The measure μ is quasi-invariant. By Proposition III.3, $U_g^{-1} P(E_k) U_g = P(E_k)$ for all g for each k. Thus $P(E_k) = P(E_k g)$ for all g. Proposition IV.5 implies there is for each k a G-invariant Borel set E_k' such that $\mu(E_k \triangledown E_k') = 0$. The sets E_k' may be taken to be disjoint with union X. Thus we may assume that the projection valued measure P on X has uniform multiplicity. Again referring to Theorem III.1, we may also assume the Hilbert space for the representation U is $L^2(X, \mu, \mathbb{H})$ and the projection valued measure P is given by

$$P(E)f = 1_E f.$$

Set $\varphi_g(x) = xg^{-1}$. Then $\varphi_{g_1} \circ \varphi_{g_2} = \varphi_{g_1 g_2}$ and the mapping $g \mapsto \varphi_g$ is Borel from G into $\mathcal{I}(X, \mu)$. Thus $g \mapsto L_{\varphi_g}$ is a strongly Borel mapping into the unitary group of $L^2(X, \mu, \mathbb{H})$ which satisfies

$$L_{\varphi_g} P(E) L_{\varphi_g}^{-1} = P(\varphi_g(E)) = P(Eg^{-1}).$$

Thus $g \mapsto A(g) \equiv U_g L_{\varphi_g}^{-1}$ is a strongly Borel map into the unitary group of $L^2(X, \mu, \mathbb{H})$ satisfying

$$A(g)P(E)A(g)^{-1} = P(E)$$

for all E and g.

By Proposition III.13, for each g there is a unitary Borel field $x \mapsto A(g, x)$ acting on $\mathbb{H}$ such that

$$A(g)f(x) = A(g, x)f(x)$$

for $f \in L^2(X, \mu, \mathbb{H})$. In particular, $g \mapsto \int (A(g, x)f(x), h(x)) \, d\mu(x)$ is Borel for all $f, h \in L^2(X, \mu, \mathbb{H})$. Hence, since $A(g) = \int^{\oplus} A(g, x) \, d\mu(x)$, Proposition III.16 implies there is a Borel field $(x, g) \mapsto R(x, g)$ from $X \times G$ into $B(\mathbb{H})$ such that $R(x, g) = A(g, x)$ a.e. x a.e. g. By redefining R on a set of measure 0, we may assume $R(x, g) \in \mathcal{U}(\mathbb{H})$ for all x and g.

Next note since $U_{g_1} U_{g_2} = U_{g_1 g_2}$ that

$$\begin{aligned}
A(g_1 g_2) L_{\varphi_{g_1 g_2}} &= A(g_1) L_{\varphi_{g_1}} A(g_2) L_{\varphi_{g_2}} \\
&= A(g_1) L_{\varphi_{g_1}} A(g_2) L_{\varphi_{g_1}}^{-1} L_{\varphi_{g_1 g_2}}.
\end{aligned}$$

Thus since

$$L_{\varphi_{g_1}} A(g_2) L_{\varphi_{g_1}}^{-1} f(x) = A(g_2, x \cdot g_1)f(x),$$

we see

$$A(g_1 g_2, x) = A(g_1, x)A(g_2, x \cdot g_1)$$

a.e. x for each g_1 and g_2. This implies

$$R(x, g_1 g_2) = R(x, g_1)R(x \cdot g_1, g_2)$$

a.e. (x, g_1, g_2). By Theorem IV.9, there exists a Borel cocycle representation $L(x, g)$ on the Borel G space X such that

$$A(g, x) = L(x, g) \text{ a.e. } x \text{ a.e. } g.$$

Since $U(g) = A(g)L_{\varphi_g}$, one sees $U(g) = \text{ind}\, L_g$ for almost all g. Proposition IV.9 then yields $U = \text{ind}\, L$; and since trivial Hilbert bundle unitary extensions of X are defined by unitary cocycle representations, the result is established. $\square$

COROLLARY V.5. *Suppose X is a standard Borel G space with a σ-finite quasi-invariant measure μ. Let $\mathbb{H}$ be a separable Hilbert space and suppose P is the canonical projection valued measure on $L^2(X, \mu, \mathbb{H})$. Then if U is a unitary representation of G on $L^2(X, \mu, \mathbb{H})$ such that the pair (U, P) form a system of imprimitivity, then there is a unitary cocycle representation L of the Borel G space X such that $U = \text{ind}_{X \times G}^G L$. More specifically*

$$U(g)f(x) = \frac{d\mu \cdot g^{-1}}{d\mu}(x)^{1/2} L(x, g) f(x \cdot g)$$

for all $f \in L^2(X, \mu, \mathbb{H})$ and all $g \in G$.

7. The Subgroup Theorem

The following theorem is a generalization of Mackey's subgroup Theorem. We suppose X is a standard Borel G space with a σ-finite quasi-invariant measure μ. Let L be a unitary representation of the G space X on the separable Hilbert space $\mathbb{H}$ and suppose H is a closed subgroup of G. We are interested in the restriction of $\text{ind}_{X \times G}^G L$ to H. We first note the following obvious fact:

$$(\text{ind}_{X \times G}^G L)\,|_H = \text{ind}_{X \times H}^H (L|_H)$$

where $L|_H$ is the unitary representation of the H space X defined by $L|_H(x, h) = L(x, h)$.

We next assume Y is a standard Borel space and p is a Borel function from X into Y which is H invariant. Hence each $p^{-1}(y)$ is a standard Borel H space. Moreover, if ν is a σ-finite measure on Y equivalent to the measure $p_*\mu$, Theorem I.27 implies there is a disintegration $\int \mu_y d\nu(y)$ of the measure of μ over the fibers of p satisfying $\mu_y(X - p^{-1}(y)) = 0$ for all y. Moreover, $\mu_y \cdot h \sim \mu_y$ a.e. y for each h. Indeed, since $\mu \cdot h \sim \mu$, there is a positive Borel function f on X such that $\mu \cdot h(E) = \int_E f(x)\, d\mu(x) = \int \int f(x)\, 1_E(x)\, d\mu_y(x)\, d\nu(y)$ for all Borel sets E. Since $\mu \cdot h(E) = \int \mu_y(Eh^{-1})\, d\nu(y)$, Theorem I.27 (3) implies

$$\mu_y \cdot h(E) = \int_E f(x)\, d\mu_y(x)$$

for all E for a.e. y. Since $f(x) > 0$ for μ a.e. x, one has $f(x) > 0$ for μ_y a.e. x for ν a.e. y. This implies $\mu_y \cdot h \sim \mu_y$ for ν a.e. y for all h. By Fubini's theorem, there is a conull Borel subset Y_0 of Y such that $\mu_y \cdot h \sim \mu_y$ a.e. h for each $y \in Y_0$. Since the set $\{h : \mu_y \cdot h \sim \mu_y\}$ is a group, we see μ_y is quasi-invariant under H for $y \in Y_0$. By redefining $\mu_y = 0$ for $y \notin Y_0$, we see there is a disintegration of μ into H quasi-invariant measures μ_y on $p^{-1}(y)$ and these measure are unique a.e. y.

THEOREM V.5 (SUBGROUP THEOREM). *Let L be a unitary representation of the standard Borel G space X on the Hilbert space $\mathbb{H}$. Suppose μ is a G quasi-invariant measure on X and H is a closed subgroup of G. Then if $p : X \to Y$ is a Borel H invariant mapping of X into a standard Borel space Y and $\int \mu_y \, d\nu(y)$ is a disintegration of μ into H quasi-invariant measures where ν is a σ-finite measure on Y equivalent to $p_* \nu$, then relative to the direct integral decomposition $L^2(X, \mathbb{H}) = \int^{\oplus} L^2(p^{-1}(y), \mu_y, \mathbb{H}) \, d\nu(y)$, one has*

$$ ind^G_{X \times G} L|_H = \int^{\oplus} ind^H_{p^{-1}(y) \times H} L^y \, d\nu(y) $$

where L^y is the unitary representation of the H space $p^{-1}(y)$ defined by $L^y(x, h) = L(x, h)$.

Recall if L is a unitary representation of H on the Hilbert space $\mathbb{H}$ and σ is a Borel cross section from $H \backslash G$ into G, then the mapping $R_L(x, g) = L(\sigma(x) g \sigma(xg)^{-1})$ is a unitary representation of the G space $H \backslash G$. Let μ be a σ-finite measure on $H \backslash G$ which is quasi-invariant under right translation by G. Then ind R_L is called a unitary representation of G induced by the representation L on the subgroup H. We denote it by $ind^G_H L$ or more precisely by $ind^G_H(L, \sigma, \mu)$. One has the following uniqueness property for induction.

THEOREM V.6. *Suppose R and R' are unitary representations of the standard Borel G space X. Suppose μ and μ' are σ-finite quasi-invariant measures on X. Let P and P' be the canonical projection valued measures on $L^2(X, \mu, \mathbb{H})$ and $L^2(X, \mu', \mathbb{H}')$. Then there is a unitary operator U from $L^2(X, \mu, \mathbb{H})$ onto $L^2(X, \mu', \mathbb{H}')$ such that $U ind(R, \mu)(g) U^{-1} = ind(R', \mu')(g)$ for all g and $U P(E) U^{-1} = P'(E)$ for all E iff $\mu \sim \mu'$ and there is a Borel field $A(x)$ of unitary operators from $\mathbb{H}$ to $\mathbb{H}'$ with $A(x) R(x, g) = R'(x, g) A(x \cdot g)$ for μ a.e. x for all g.*

PROOF. Suppose such a U exists. Since $\mu(E) = 0$ iff $P(E) = 0$ iff $P'(E) = 0$ iff $\mu'(E) = 0$, the measure μ is equivalent to the measure

μ'. Thus the operator W from $L^2(X, \mu', \mathbb{H}')$ to $L^2(X, \mu, \mathbb{H}')$ defined by $W(f)(x) = \frac{d\mu'}{d\mu}(x)^{1/2}f(x)$ is a unitary isomorphism satisfying

$$WP'(E)W^{-1} = P''(E),$$

where P'' is the canonical projection valued measure on $L^2(X, \mu, \mathbb{H}')$. Thus the operator $T = WU$ is a unitary isomorphism of $L^2(X, \mu, \mathbb{H})$ onto $L^2(X, \mu, \mathbb{H}')$ which satisfies $TP(E)T^{-1} = P''(E)$. By Proposition III.13, there is a strongly Borel field $A(x)$ of unitary operators from $\mathbb{H}$ onto $\mathbb{H}'$ with $Tf(x) = A(x)f(x)$ a.e. x for each f. In particular, $Uf(x) = \frac{d\mu}{d\mu'}(x)^{1/2}A(x)f(x)$ for all f in $L^2(X, \mu, \mathbb{H})$. But since $U\mathrm{ind}(R, \mu)(g)U^{-1} = \mathrm{ind}(R', \mu')(g)$ for each g, one has

$$U\mathrm{ind}(R, \mu)(g)U^{-1}f(x) = \frac{d\mu' \cdot g^{-1}}{d\mu'}(x)^{1/2}A(x)R(x,g)A(x \cdot g)^{-1}f(x \cdot g)$$

$$= \frac{d\mu' \cdot g^{-1}}{d\mu'}(x)^{1/2}R'(x,g)f(x \cdot g) \quad \text{a.e. } x$$

for each f. From this one can conclude that $A(x)R(x,g)A(x \cdot g)^{-1} = R'(x,g)$ a.e x for each g. To see the converse, define U by $Uf(x) = \frac{d\mu}{d\mu'}(x)^{1/2}A(x)f(x)$. $\square$

LEMMA V.1. *Suppose R and R' are unitary representations of $H\backslash G \times G$ and $R(x,g) = R'(x,g)$ a.e. x for each g. Then there is a g_0 such that*

$$R(Hg_0, g_0^{-1}hg_0) = R'(Hg_0, g_0^{-1}hg_0) \text{ for all } h.$$

PROOF. Let $M = \{(x,g) : R(x,g) = R'(x,g)\}$. Then if (x,g) and $(x \cdot g, g')$ are in M, then (x, gg') is in M. Let X be the set of all x in $H\backslash G$ such that (x,g) and (xg, g^{-1}) are in M a.e. g. X is a conull Borel subset of $H\backslash G$. Moreover, suppose x and $x \cdot g$ are in X. Then there is a g_1 such that both (x, gg_1) and (xgg_1, g_1^{-1}) are in M. Thus (x,g) is in M. Let $Hg_0 \in X$. Then $Hg_0(g_0^{-1}hg_0)$ is also in X. Hence $(Hg_0, g_0^{-1}hg_0) \in M$ for all h and the result follows. $\square$

COROLLARY V.6. *Suppose L and L' are unitary representations of a closed subgroup H of G on Hilbert spaces $\mathbb{H}$ and $\mathbb{H}'$. Let μ and μ' be σ-finite quasi-invariant measures on $H\backslash G$ and let σ and σ' be Borel cross sections. Then there is a unitary isomorphism U from $L^2(H\backslash G, \mu, \mathbb{H})$ onto $L^2(H\backslash G, \mu', \mathbb{H}')$ such that $UP(E)U^{-1} = P'(E)$ and $U\mathrm{ind}(L, \sigma, \mu)(g)U^{-1} = \mathrm{ind}_H^G(L', \sigma', \mu')(g)$ for all g iff the representation L is unitarily equivalent to L'.*

PROOF. Assume such a U exists. By Theorem V.6, there is a Borel field $A(x)$ of unitary operators from $\mathbb{H}$ onto $\mathbb{H}'$ satisfying

$$A(x)L(\sigma(x)g\sigma(xg)^{-1})A(x \cdot g)^{-1} = L'(\sigma'(x)g\sigma'(xg)^{-1})$$

a.e. x for each g. Set $R(x,g) = A(x)L(\sigma(x)g\sigma(xg)^{-1})A(x \cdot g)^{-1}$ and $R'(x,g) = L'(\sigma'(x)g\sigma'(xg)^{-1})$. Then these are unitary representations of the G space $H\backslash G$. By Lemma V.1, there is a g_0 such that $R(Hg_0, g_0^{-1}hg_0) = R'(Hg_0, g_0^{-1}hg_0)$ for all h. Furthermore, $\sigma(Hg_0) = h_0\sigma'(Hg_0)$ for some h_0. Combining these one sees

$$A(Hg_0)L(h_0)L(\sigma'(Hg_0)g_0^{-1}hg_0\sigma'(Hg_0)^{-1})L(h_0)^{-1}A(Hg_0)^{-1} =$$
$$L'(\sigma'(Hg_0)g_0^{-1}hg_0\sigma'(Hg_0)^{-1})$$

for all h. But since $\sigma'(Hg_0)g_0^{-1} \in H$, every element in H has form

$$\sigma'(Hg_0)g_0^{-1}hg_0\sigma'(Hg_0)^{-1}$$

and thus

$$A(Hg_0)L(h_0)L(h)L(h_0)^{-1}A(Hg_0)^{-1} = L'(h) \text{ for all } h.$$

Conversely, suppose L and L' are equivalent unitary representations of H. Choose a unitary operator A from $\mathbb{H}$ to $\mathbb{H}'$ such that $AL(h) = L'(h)A$. Note $\sigma'(x) = h(x)\sigma(x)$ where h is a Borel function from $H\backslash G$ into H. Define $A(x) = AL(h(x))$. Then

$$\begin{aligned}
A(x)R_L(x,g) &= AL(h(x))L(\sigma(x)g\sigma(xg)^{-1}) \\
&= AL(h(x)\sigma(x)g\sigma(xg)^{-1}h(xg)^{-1})L(h(xg)) \\
&= L'(h(x)\sigma(x)g\sigma(xg)^{-1}h(xg)^{-1})AL(h(xg)) \\
&= L'(\sigma'(x)g\sigma'(xg)^{-1})A(xg) \\
&= R_{L'}(x,g)A(xg)
\end{aligned}$$

holds everywhere. Since all σ-finite quasi-invariant measures on $H\backslash G$ are equivalent, the result follows by Theorem V.6. $\square$

COROLLARY V.7. *Suppose R is a unitary representation of the G space $H\backslash G$ and μ is a σ-finite G quasi-invariant measure on $H\backslash G$. Then $L(h) = R(He, h)$ is a unitary representation of H and $\operatorname{ind}_H^G L \cong \operatorname{ind} R$.*

PROOF. Let σ be a Borel cross section for $H\backslash G$. Define A by

$$A(x) = R(x, \sigma(x)^{-1}).$$

Then

$$\begin{aligned}
A(x)R_L(x,g) &= R(x,\sigma(x)^{-1})L(\sigma(x)g\sigma(xg)^{-1}) \\
&= R(x,\sigma(x)^{-1})R(H,\sigma(x)g\sigma(xg)^{-1}) \\
&= R(x,g\sigma(xg)^{-1}) \\
&= R(x,g)R(xg,\sigma(xg)^{-1}) \\
&= R(x,g)A(xg).
\end{aligned}$$

□

Assume H and K are closed subgroups of G. Then $H \times K$ acts on G by $g \cdot (h,k) = h^{-1}gk$. This action is continuous and the orbits are the $H : K$ double cosets HgK. Denote the space of orbits with the quotient Borel structure by $H\backslash G/K$.

THEOREM V.7 (MACKEY'S SUBGROUP THEOREM).
Suppose $H\backslash G/K$ is a standard Borel space and μ is a G quasi-invariant σ-finite measure on $H\backslash G$. Let p be the natural projection of $H\backslash G$ onto $H\backslash G/K$ and let ν be a σ-finite measure on $H\backslash G/K$ equivalent to $p_*\mu$. If one disintegrates μ into $\int_{H\backslash G/K} \mu_y d\nu(y)$ where each μ_y is a K quasi-invariant σ-finite measure on $p^{-1}(y) = HgK$ where $g \in y$, then in the direct integral decomposition of $ind_H^G L|_K = ind_H^G(L,\sigma,\mu)|_K = ind_{H\backslash G\times G}^G R_L|_K = \int_{H\backslash G/K}^{\oplus} ind_{p^{-1}(y)\times K}^K R_L^y \, d\nu(y)$, one has

$$ind_{p^{-1}(y)\times K}^K R_L^y \cong ind_{g^{-1}Hg\cap K}^K L^g \quad a.e. \ y$$

where L^g is defined on $g^{-1}Hg \cap K$ by $L^g(k) = L(gkg^{-1})$. Hence

$$ind_H^G L|_K \cong \int_{H\backslash G/K}^{\oplus} ind_{g^{-1}Hg\cap K}^K L^g \, d\nu(HgK).$$

PROOF. We let $y = HgK$. We may assume μ_y is nonzero. We note that the K orbit of Hg in $H\backslash G$ is HgK and that the stabilizer of Hg is $K_y = g^{-1}Hg \cap K$ and thus the mapping $\psi : Hgk \mapsto (g^{-1}Hg \cap K)k$ is a Borel space K equivariant isomorphism of this K orbit onto the K space $K_y\backslash K$. In particular, $\mu_y' = \psi_*\mu_y$ is a K quasi-invariant σ-finite measure on $K_y\backslash K$. Now the representation R_L^y on $p^{-1}(y) \times K$ is defined by

$$R_L^y(Hgk',k) = L(\sigma(Hgk')k\sigma(Hgk'k)^{-1}).$$

Since ψ is a K equivariant Borel isomorphism from HgK onto $K_y\backslash K$ carrying the measure μ_y to the measure μ_y', ψ yields a unitary isomorphism U from $L^2(HgK,\mu_y,\mathbb{H})$ onto $L^2(K_y\backslash K,\mu_y',\mathbb{H})$ defined by

$Uf(K_yk) = f(Hgk)$. Under this isomorphism the representation $\mathrm{ind}_{p^{-1}(y)\times H}^K R_L^y$ is equivalent to the representation W^y defined by

$$W_y(k)f(K_yk') = (\frac{d\mu_y' \cdot k^{-1}}{d\mu_y'})^{\frac{1}{2}}(K_yk')L(\sigma(Hgk')k\sigma(Hgk'k)^{-1})f(K_yk'k).$$

Hence W_y is equivalent to the representation induced by the unitary representation $R_L'^y$ of K space $K_y\backslash K$ defined by $R_L'^y(K_yk',k) = L(\sigma(Hgk')k\sigma(Hgk'k)^{-1})$. By Corollary V.7, we see W_y is unitarily equivalent to the representation induced from K_y by the representation $L'^y(k) = L(\sigma(Hg)k\sigma(Hg)^{-1})$. Since this representation has form $k \mapsto L(h)L(gkg^{-1})L(h)^{-1}$, we see by Corollary V.6 that W^y is unitarily equivalent to $\mathrm{ind}_{g^{-1}Hg\cap K}^K L^g$. This completes the proof. $\square$

8. Multiplier Representations

Let G be a locally compact second countable group. Then a Borel mapping $x \mapsto U_x$ from G into $\mathcal{U}$, the unitary group of some separable Hilbert space, is said to be a **multiplier** representation or a **projective** representation if $U_e = I$ and

$$U_{xy} = \sigma(x,y)U_xU_y$$

for some function $\sigma : G \times G \to T$ where $T = \{z \in \mathbb{C} : |z| = 1\}$. The function σ is called the multiplier of the representation U.

PROPOSITION V.8. *Suppose σ is a multiplier of a projective representation of G. Then σ is a Borel function which satisfies:*

    (a) *$\sigma(e,x) = \sigma(x,e) = 1$ for all x and*
    (b) *$\sigma(xy,z)\sigma(x,y) = \sigma(x,yz)\sigma(y,z)$ for all x, y, z.*

PROOF. σ is Borel since U is Borel and $\mathcal{U}$ is a Polish group. Since $U_x = U_{ex} = \sigma(e,x)U_eU_x = \sigma(e,x)U_x$, we see $\sigma(e,x) = 1$. Similarly $\sigma(x,e) = 1$. Moreover, since $(xy)z = x(yz)$, we see

$$U_{(xy)z} = \sigma(xy,z)\sigma(x,y)U_xU_yU_z = \sigma(x,yz)\sigma(y,z)U_xU_yU_z = U_{x(yz)}.$$

This gives (b). $\square$

THEOREM V.8. *Suppose σ is a Borel function of $G \times G$ into T satisfying (a) and (b). Then $G^\sigma = T \times G$ with multiplication defined by*

$$(z_1,x)(z_2,y) = (z_1z_2\sigma(x,y),xy)$$

is a standard Borel group. Moreover, if λ is Lebesgue measure on T and m is a right Haar measure on G, then $\lambda \times m$ is a right invariant σ-finite measure on G^σ; and thus G^σ has a unique second countable *locally compact topology for which G^σ is a topological group. Further-more, the correspondence $U \mapsto U|_G$ is a one-to-one correspondence between the unitary representations of G^σ satisfying $U_{(z,e)} = \bar{z}I$ and the multiplier representations of G having multiplier σ.*

PROOF. That G^σ is a group follows immediately from (a) and (b). Moreover, G^σ with the product Borel structure is standard and the functions $(z_1, x), (z_2, y) \mapsto (z_1 z_2 \sigma(x, y), xy)$ and $(z, x) \mapsto (z, x)^{-1} = (\bar{z}\bar{\sigma}(x, x^{-1}), x^{-1})$ are Borel. Hence G^σ is a standard Borel group. Moreover, $m^\sigma = \lambda \times m$ is a right invariant measure; and hence by Theorem IV.6, we know there is a locally compact second countable topology on G^σ making G^σ into a topological group. An easy calculation shows that $g \mapsto U(1, g)$ is a σ representation of G if U is a unitary representation of G^σ satisfying $U(z, e) = \bar{z}I$. Next note that if W is a multiplier representation of G with multiplier σ, then $U(z, g) = \bar{z}W_g$ is a Borel homomorphism of G^σ into $\mathcal{U}(\mathbb{H})$ and thus is a unitary representation of G^σ. □

Suppose σ is a Borel function on $G \times G$ satisfying (a) and (b). To see that σ is a multiplier, we exhibit a multiplier representation of G with multiplier σ. Let U be the operator valued function defined by $U_x f(y) = \bar{\sigma}(y, x) f(yx)$ for $f \in L^2(G)$. Then

$$
\begin{aligned}
U_{xy} f(z) &= \bar{\sigma}(z, xy) f(zxy) \\
&= \sigma(x, y)(\bar{\sigma}(z, xy)\bar{\sigma}(x, y)) f(zxy) \\
&= \sigma(x, y)(\bar{\sigma}(zx, y)\bar{\sigma}(z, x)) f(zxy) \\
&= \sigma(x, y)\bar{\sigma}(z, x) U_y f(zx) \\
&= \sigma(x, y) U_x U_y f(z).
\end{aligned}
$$

Thus the multipliers on a group G are precisely the Borel functions on $G \times G$ into T satisfying (a) and (b) of Proposition V.6.

DEFINITION. Two multipliers σ and σ' on $G \times G$ are said to be **equivalent** if there is a Borel function $b : G \to T$ with

$$
\sigma'(x, y) = b(xy)\bar{b}(x)\bar{b}(y)\sigma(x, y).
$$

REMARK. $x \mapsto U_x$ is a multiplier representation with multiplier σ iff $x \mapsto b(x)U_x$ is a multiplier representation with multiplier $(db)(\sigma)$ where $db(x, y) = b(xy)\bar{b}(x)\bar{b}(y)$. Moreover, a multiplier σ is said to be trivial if $\sigma = db$ for some b.

9. Extensions of Representations to Multiplier Representations

THEOREM V.9. *Suppose U is an irreducible unitary representation of the closed normal subgroup N of G such that for each $g \in G$ the representations $n \mapsto U_n$ and $n \mapsto U_{gng^{-1}}$ are unitarily equivalent. Then there is a multiplier representation $\tilde{U}$ of G such that $\tilde{U}|_N = U$ and $\tilde{U}_g U_n \tilde{U}_g^{-1} = U_{gng^{-1}}$ for $n \in N$ and $g \in G$. In particular, U extends to a multiplier representation of G.*

PROOF. Let $X = \{(g, A) : A \in \mathcal{U}(\mathbb{H}) \text{ and } AU_n A^{-1} = U_{gng^{-1}} \text{ for all } n\}$. We first note that X is a Borel subset of $G \times \mathcal{U}(\mathbb{H})$. Indeed, if n_j form a dense countable subset of N, then (g, A) belongs to X iff $AU_{n_j}A^{-1} = U_{gn_jg^{-1}}$ for all j. Moreover, the projection p sending (g, A) to g is onto and Borel. Furthermore, by Schur's Lemma, $p^{-1}(g) = \{g\} \times TA_g$ where A_g is any unitary operator satisfying $A_g U_n A_g^{-1} = U_{gng^{-1}}$ for all n. In particular, $p^{-1}(g)$ is compact for each g. Thus by Theorem I.22 (one could also use Theorem I.14), there is a Borel function $W : G \to \mathcal{U}(\mathbb{H})$ satisfying $W_g U_n W_g^{-1} = U_{gng^{-1}}$ for all n and g. Define $\tilde{U}_g = W_g$ for $g \notin N$ and $\tilde{U}_n = U_n$ otherwise. Clearly $\tilde{U}$ is a Borel function. To see that it is a multiplier representation, note that the operator $\tilde{U}_{g_1 g_2} \tilde{U}_{g_2}^{-1} \tilde{U}_{g_1}^{-1}$ commutes with irreducible representation U. By Schur's Lemma, there is a scalar $\sigma(g_1, g_2)$ such that $\tilde{U}_{g_1 g_2} \tilde{U}_{g_2}^{-1} \tilde{U}_{g_1}^{-1} = \sigma(g_1, g_2)I$. □

Note if W is a projective representation of G with $W_g U_n W_g^{-1} = U_{gng^{-1}}$ for all n and g, then $W_g \tilde{U}_g^{-1}$ centralizes U; and thus by Schur's Lemma, there is a function $b : G \to T$ such that $W_g = b(g)\tilde{U}_g$. In particular, the multiplier of W is equivalent to the multiplier of $\tilde{U}$. Hence the collection of all multipliers obtained this way forms an equivalence class of multipliers on G, and this equivalence class is known as the Mackey obstruction to extending the representation U on N to G. In particular one has:

COROLLARY V.8. *Let U be an irreducible unitary representation of N where N is a closed normal subgroup of G. Suppose the representations $U^g(n) \equiv U(gng^{-1})$ are all unitarily equivalent to $U|_N$. Then U extends to a unitary representation of G iff the Mackey obstruction is trivial (i.e., contains the identically 1 multiplier).*

COROLLARY V.9. *Assume U is a multiplier representation of G and N is a closed normal subgroup on which U is a unitary representation. Furthermore, suppose $U_g U_n U_g^{-1} = U_{gng^{-1}}$ for $n \in N$ and*

$g \in G$. Then there is a multiplier σ on G/N such the Mackey obstruction for $U|_N$ is the equivalence class of the multiplier $(x, y) \mapsto \sigma(xN, yN)$.

PROOF. Proposition IV.2 implies there is a Borel subset B of G containing e meeting each coset xN exactly once. Define $\tilde{U}_{nb} = U_n U_b$. Note $\tilde{U}_n = U_n$. Then since the mapping $(n, b) \mapsto nb$ is one-to-one and Borel from the standard Borel space $N \times B$ onto G, the function $\tilde{U}$ is Borel. Moreover, it has the property that $\tilde{U}_g U_n \tilde{U}_g^{-1} = U_{gng^{-1}}$ and thus is a multiplier representation of G. Moreover, if $g = nb$, then $\tilde{U}_{n'g} = \tilde{U}_{n'nb} = U_{n'n} U_b = U_{n'} U_n U_b = U_{n'} \tilde{U}_g$. Thus if σ is the multiplier of $\tilde{U}$, we have $\sigma(n', g) = 1$. Similarly

$$\tilde{U}_{gn'} = \tilde{U}_{nbn'} = \tilde{U}_n \tilde{U}_{bn'} = \tilde{U}_n \tilde{U}_b \tilde{U}_b^{-1} \tilde{U}_{bn'b^{-1}b}$$
$$= \tilde{U}_g \tilde{U}_b^{-1} \tilde{U}_{bn'b^{-1}} \tilde{U}_b = \tilde{U}_g U_{n'}$$

and thus $\sigma(g, n') = 1$. Hence, if $n_3 = g_2^{-1} n_1 g_2 n_2$, then

$$\sigma(g_1 n_1, g_2 n_2) = \sigma(g_1 n_1, g_2 n_2)\sigma(g_1, n_1) = \sigma(g_1, n_1 g_2 n_2)\sigma(n_1, g_2 n_2)$$
$$= \sigma(g_1, g_2 n_3) = \sigma(g_1, g_2 n_3)\sigma(g_2, n_3) = \sigma(g_1 g_2, n_3)\sigma(g_1, g_2)$$
$$= \sigma(g_1, g_2).$$

Thus $\sigma(g_1 N, g_2 N) \equiv \sigma(g_1, g_2)$ is well defined. This function on $G/N \times G/N$ is easily seen to be a multiplier. $\square$

10. The Action of a Group on the Dual of a Normal Subgroup

Let G be a locally compact second countable group and let N be a closed normal subgroup. Let m be a right Haar measure on N and $n \in \{\infty, 1, 2, 3, \dots\}$. Then the set $\mathrm{Hom}(N, \mathcal{U}_n)$ of unitary representations of N on the Hilbert space $\mathbb{H}_n$ is a closed subset of $\mathcal{M}(N, m, \mathcal{U}(\mathbb{H}_n))$ and hence is a Polish space. Moreover, the mapping $g \mapsto \phi_g$ where $\phi_g(n) = gng^{-1}$ is a Borel group homomorphism of G into $\mathcal{I}(N, m)$. Indeed, to see this mapping is Borel, it suffices by Proposition I.17 to note the mapping $g \mapsto \int g(n)h(gng^{-1})\, dm(n)$ is Borel if $g \in L^1(N, m)$ and h is a bounded Borel function on N. This, however, follows from Fubini's Theorem. Hence by Theorem IV.16 and Proposition V.2, we may conclude:

PROPOSITION V.9. Under the action $(f, g) \mapsto f \cdot g$ where $f \cdot g(n) = f(gng^{-1})$, the spaces $\mathcal{M}(N, m, \mathcal{U}_n)$, $\mathrm{Hom}(N, \mathcal{U}_n)$, and N_n^c are standard Borel G spaces.

COROLLARY V.10. *For each $g \in G$, the mapping $\gamma \mapsto \gamma \cdot g$ is a Borel mapping of $\hat{N}$.*

PROOF. Let $\hat{E}$ be a Borel subset of $\hat{N}$. Then its preimage under the mapping $\gamma \mapsto \gamma \cdot g$ is $\hat{E} \cdot g^{-1}$. But this set is Borel since the union of its elements is the Borel set in N^c consisting of those L such that $L \cdot g$ is in the Borel set obtained by taking the union of the elements in $\hat{E}$. $\square$

COROLLARY V.11. *Suppose $\hat{N}$ is standard. Then the mapping $([L], g) \mapsto [L \cdot g]$ is Borel; in particular $\hat{N}$ is a standard Borel G space.*

PROOF. Clearly this mapping defines an action. Let $\hat{E}$ be a Borel subset of $\hat{N}$. Hence $E \equiv \cup_{\gamma \in \hat{E}} \gamma$ is a Borel subset of N^c. Define a mapping of $N^c \times G$ into $\hat{N} \times G$ by $W(L, g) = ([L \cdot g], g^{-1})$. This mapping is Borel and onto. Hence $W(E \times G)$ is an analytic set. But if $\tilde{E}$ is the complement of E, $W(\tilde{E} \times G)$ is the complement to $W(E \times G)$. Since this set is also analytic, we see by Corollary I.2 that the set $W(E \times G)$ must be a Borel subset of $\hat{N} \times G$. But the $\{(\gamma, g) \in \hat{N} \times G : \gamma \cdot g \in \hat{E}\} = W(E \times G)$. Thus the mapping $(\gamma, g) \mapsto \gamma \cdot g$ is Borel from $\hat{N} \times G$ into $\hat{N}$. $\square$

11. Restrictions of Representations to Normal Subgroups

Suppose U is a primary unitary representation of G. Then the von Neumann algebra $\mathcal{A}_U$ has trivial center. Suppose in addition $U|_N$ is a type I representation of N. Thus the von Neumann algebra $\mathcal{A}_N = U(N)''$ is type I. Let P_k be the largest central projection in $\mathcal{A}_N$ which is homogeneous of degree k, and let Q_k be the largest projection of multiplicity k in the Boolean algebra of central projections in $\mathcal{A}_N$. Note that the von Neumann algebra $\mathcal{A}_N$ is normalized by each unitary operator U_g since $U_g U_n U_g^{-1} = U_{gng^{-1}}$. By Corollary III.22, $U_g P_k U_g^{-1} = P_k$ and $U_g Q_k U_g^{-1} = Q_k$ for all g. This implies there exist a unique n and a unique m with $P_n = I$ and $Q_m = I$. Hence, by Corollary III.22 and Theorem III.24, we may assume the Hilbert space for U is $L^2(S, \mu, n\mathbb{H})$ for some standard Borel space S and σ-finite measure μ and $\mathcal{A}_N = \int_S^{\oplus} n\mathcal{B}(\mathbb{H}) \, d\mu(s)$ where $n \dim(\mathbb{H}) = m$. Theorem IV.17 implies there is a weakly Borel function $s \mapsto L_s$ of S into the unitary group $\mathcal{U}(\mathbb{H})$ such that $U(n) = \int_S^{\oplus} nL_s(n) \, d\mu(s)$. By Theorem III.16, $U(N)'' = \int_S^{\oplus} nL_s(N)'' \, d\mu(s)$; and hence $L_s(N)'' = \mathcal{B}(\mathbb{H})$ for a.e. s. In particular, by Proposition IV.10, L_s is an irreducible representation of N on $\mathbb{H}$ for a.e. s. The representation

$L \equiv \int_S^{\oplus} L_s \, d\mu(s)$ is locally simple. Indeed $\mathcal{A}_L = L^{\infty}(S, \mu, \mathbb{C}I)$ and thus is abelian.

In order to insure L has standard support in the dual of N, we now assume $\hat{N}$ is a standard Borel space. By Theorem IV.21, by removing a set of measure 0 from S if necessary, we may assume the map $s \mapsto [L_s]$ from S into $\hat{N}$ is one-to-one. By Lemma IV.7, its range is a Borel subset of $\hat{N}$ forming a standard Borel space, for it is Borel isomorphic to S. Hence the mapping $\Phi[L_s] = L_s$ is a Borel selection for the measure μ_* defined on $\hat{N}$ by $\mu_*(E) = \mu(\Phi(E \cap [L_S]))$. Moreover, $L = \int^{\oplus} \Phi(\gamma) \, d\mu_*(\gamma)$. Hence we may conclude:

PROPOSITION V.10. *Let N be a closed normal subgroup of G with $\hat{N}$ standard. Suppose U is a primary unitary representation of G such that the representation $U|_N$ is type I. Then there are unique nonnegative integers n and m and a σ-finite measure μ on $\hat{N}$ such that if $\mathbb{H}$ is a Hilbert space of dimension n and Φ is a Borel function from $\hat{N}$ into $\mathrm{Hom}(G, \mathcal{U}(\mathbb{H}))$ satisfying $[\Phi(\gamma)] = \gamma$ for μ a.e. γ, then $U|_N$ is unitarily equivalent to $nL[\mu, \Phi]$.*

We now suppose $U|_N = nL[\mu, \Phi]$ where μ is a σ-finite measure on $\hat{N}$ and Φ is a cross section from $\hat{N}$ into $\mathrm{Hom}(G, \mathcal{U}(\mathbb{H}))$. Thus the Hilbert space of U is $L^2(\hat{N}, \mu, n\mathbb{H})$. Let P be the canonical projection valued measure on $L^2(\hat{N}, \mu, n\mathbb{H})$.

The following theorems assume that $\hat{N}$ is a standard Borel space and that the representation $U|_N$ is type I. In the next chapter, we show that $\hat{N}$ is a standard Borel space iff every unitary representation of N is type I. This was first proven by Glimm using C^* algebras. A simplified proof based on the behavior of the action of the unitary group on the dual space was later given by Effros. Hence the hypotheses in the following results can all be modified to reflect this equivalence.

THEOREM V.10. *Let U be a unitary representation of G, and let N be a closed normal subgroup. Assume $\hat{N}$ is a standard Borel space. Suppose $U|_N = nL[\mu, \Phi]$ where μ is a type I measure on $\hat{N}$. Then if P is the canonical projection valued measure on $L^2(\hat{N}, \mu, n\mathbb{H})$, the pair (U, P) forms a system of imprimitivity on the G space $\hat{N}$. In particular, μ is a quasi-invariant measure on $\hat{N}$.*

PROOF. Let $L = U|_N$. Then $L \cdot g(x) = L(gxg^{-1})$ is a unitary representation of N and $U_g L(x) U_g^{-1} = L(gxg^{-1})$. But L is the unitary representation on $\int_{\hat{N}}^{\oplus} n\mathcal{H}(\gamma) \, d\mu(\gamma) = L^2(\hat{N}, \mu, n\mathbb{H})$ defined by $L = \int^{\oplus} n\Phi(\gamma) \, d\mu(\gamma)$. Thus $L \cdot g(x) = \int^{\oplus} n\Phi(\gamma)(gxg^{-1}) \, d\mu(\gamma)$. Define $\Phi'(\gamma) = \Phi(\gamma \cdot g^{-1}) \cdot g$. Then Φ' is a Borel function from $\hat{N}$ into $\mathrm{Hom}(N, \mathcal{U}(\mathbb{H}))$ satisfying $[\Phi'(\gamma)] = \gamma$. Thus $L \cdot g(x) = \int^{\oplus} n\Phi'(\gamma \cdot g)(x) \, d\mu(\gamma)$. Let $\mu \cdot g$ be the measure on $\hat{N}$ defined by $\mu \cdot g(E) = \mu(E \cdot g^{-1})$. Define a unitary transformation W from $L^2(\hat{N}, \mu, n\mathbb{H}) = \int_{\hat{N}}^{\oplus} n\mathcal{H}_{\gamma \cdot g} \, d\mu(\gamma)$ onto $L^2(\hat{N}, \mu \cdot g, n\mathbb{H}) = \int_{\hat{N}}^{\oplus} n\mathcal{H}_{\gamma} \, d\mu \cdot g(\gamma)$ by

$$W f(\gamma) = f(\gamma \cdot g^{-1}).$$

Then $W L \cdot g(x) W^{-1} = \int^{\oplus} n\Phi'(\gamma)(x) \, d\mu \cdot g(\gamma) = nL[\mu \cdot g, \Phi'](x)$. Let P' be the canonical projection valued measure on $L^2(\hat{N}, \mu \cdot g, n\mathbb{H})$. We know that the unitary operator $W U_g$ is a unitary equivalence between the representations $nL[\mu, \Phi]$ and $nL[\mu \cdot g, \Phi']$. By Theorem IV.22, the measure $\mu \cdot g$ is equivalent to μ; and by Corollary IV.14, one has $W U_g P(E) U_g^{-1} W^{-1} = P'(E)$ for all Borel subsets E. Thus $U_g P(E) U_g^{-1} = W^{-1} P'(E) W$ for all E. But $W^{-1} P'(E) W = P(E \cdot g^{-1})$. Hence μ is quasi-invariant, and the pair (U, P) form a system of imprimitivity. $\square$

Before we continue the analysis of the representation U of G in terms of $U|_N$, we need to establish some relationships between the representations R of a G space X with a quasi-invariant measure μ and the representations $\mathrm{ind}_{X \times G}^G R$ of G. It is easiest to do this in terms of the von Neumann algebras they generate.

Let (X, μ) be a standard Borel G space where μ is a quasi-invariant σ-finite measure. Let R be a unitary representation of G space X on Hilbert space $\mathbb{H}$. Let $\mathcal{A}'_R$ be the set of all strongly Borel essentially bounded mappings $x \mapsto A(x)$ from X into $B(\mathbb{H})$ satisfying $A(x) R(x, g) = R(x, g) A(x \cdot g)$ for μ a.e. x for each g, identified if they are equal almost everywhere. Since bounded Borel fields of operators are in one-to-one correspondence with operators A on $L^2(X, \mu, \mathbb{H}) = \int^{\oplus} \mathbb{H} \, d\mu(x)$ which commute with the canonical projection valued measure for this direct integral decomposition, we may identify $\mathcal{A}'_R$ with the von Neumann algebra of all operators of form $\int^{\oplus} A(x) \, d\mu(x)$ satisfying $A(x) R(x, g) = R(x, g) A(x \cdot g)$ a.e. x for each g.

DEFINITION. Let R be a unitary representation of a standard Borel G space X equipped with a σ-finite quasi-invariant measure μ.

Then the von Neumann algebra $\mathcal{A}_R$ generated by R is defined to be the commutant of the von Neumann algebra $\mathcal{A}_R'$. The representation R is said to be type I, primary, irreducible, or locally simple if $\mathcal{A}_R$ is type I, a factor, $B(L^2(X, \mu, \mathbb{H}))$, or has abelian commutant, respectively.

An easy consequence of the definition is that R is irreducible iff the only Borel fields in $\mathcal{A}_R'$ are the essentially constant scalar fields. Indeed, suppose R is irreducible. Thus $\mathcal{A}_R' = \mathcal{A}_R''' = B(L^2(X, \mu, \mathbb{H}))'$ is the collection of all scalar operators. Hence if $A(x)R(x, g) = R(x, g)A(x \cdot g)$ a.e. x for each g, then $A(x) = cI$ a.e. x for some scalar c.

DEFINITION. Two representations R and R' of a standard Borel G space (X, μ) are said to be unitarily equivalent if there is a strongly Borel field $U(x)$ of unitary operators between $\mathbb{H}$ and $\mathbb{H}'$ satisfying $U(x)R(x, g) = R'(x, g)U(x \cdot g)$ for μ a.e. x for each g.

Note that if R and R' are equivalent unitary representations of G space (X, μ), then R is type I, irreducible, locally simple or primary iff R' is type I, irreducible, locally simple or primary, respectively. Indeed, their von Neumann algebras are isomorphic under the spatial isometry $(Uf)(x) = U(x)f(x)$.

PROPOSITION V.11. *Suppose $X = H \backslash G$ and μ is a σ-finite quasi-invariant measure on $H \backslash G$. Let $\sigma : H \backslash G \to G$ be a Borel cross section. For each unitary representation L of H, let R_L be the unitary representation of the G space $H \backslash G$ defined by $R_L(x, g) = L(\sigma(x)g\sigma(x \cdot g)^{-1})$. Then every unitary representation of $H \backslash G \times G$ is unitarily equivalent to R_L for some L. Furthermore, R_L is type I, primary, irreducible or locally simple iff L is type I, primary, irreducible or locally simple, respectively. In addition, R_{L_1} is unitarily equivalent to R_{L_2} iff L_1 and L_2 are unitarily equivalent.*

PROOF. The argument in the proof of Corollary V.7 shows that if R is a unitary representation of $H \backslash G \times G$, then R is unitarily equivalent to R_L where $L(h) = R(H, h)$.

Next we show there is a natural isomorphism of $\mathcal{A}_L'$ onto $\mathcal{A}_R'$. Indeed, define $\Phi(A)(x) = A$ for $A \in \mathcal{A}_L'$. A straightforward calculation shows $\Phi(A)(x)R_L(x, g) = R_L(x, g)\Phi(A)(x \cdot g)$ for all x and g. This map is clearly one-to-one and into. Next suppose $A(x)R_L(x, g) = R_L(x, g)A(x \cdot g)$ a.e. x for each g. We now argue as in the proof of Lemma V.1. Let $M = \{(x, g) : A(x)R_L(x, g) = R_L(x, g)A(x \cdot g)\}$.

Then if $(x, g) \in M$ and $(x \cdot g, g') \in M$, then $(x, gg') \in M$; and as seen in Lemma V.1's proof, there is a g_0 with $A(Hg_0)R_L(Hg_0, g_0^{-1}hg_0) = R_L(Hg_0, g_0^{-1}hg_0)A(Hg_0)$ for all $h \in H$. Furthermore, this g_0 may be chosen so that $A(Hg_0)R_L(Hg_0, g) = R_L(Hg_0, g)A(Hg_0g)$ for a.e. g. Thus $A(Hg_0)L_h = L_hA(Hg_0)$ for all h; and if $x_0 = Hg_0$, then $A(x_0 \cdot g) = R_L(x_0, g)^{-1}A(x_0)R_L(x_0, g) = A(x_0)$ a.e. g. Thus $\Phi(A(x_0))(x) = A(x)$ for μ a.e. x. This shows Φ is onto. Hence R_L is type I, locally simple or irreducible iff L is. To see that L is primary iff R_L is primary, we show Φ restricted to $\mathcal{A}_L \cap \mathcal{A}'_L$ is onto $\mathcal{A}_R \cap \mathcal{A}'_R$. Indeed, suppose $x \mapsto A(x)$ is in $\mathcal{A}_R \cap \mathcal{A}'_R$. We know $A(x) = A$ a.e. x for some A in $\mathcal{A}'_L$. But $\int^\oplus A(x)\,d\mu(x)$ commutes with all the operators $\int^\oplus B(x)\,d\mu(x)$ where $B(x) = B$ for all x and $B \in \mathcal{A}'_L$. Thus $AB = BA$ for all $B \in \mathcal{A}'_L$ and $A \in \mathcal{A}_L \cap \mathcal{A}'_L$. Thus L is primary iff R_L is primary.

Finally, suppose R_{L_1} is unitarily equivalent to R_{L_2}. Thus there is a Borel field of unitary operators $U(x)$ from $\mathbb{H}$ to $\mathbb{H}'$ satisfying $U(x)R_{L_1}(x, g) = R_{L_2}(x, g)U(x \cdot g)$ a.e. x for each g. The same argument as above shows there is an $x_0 = Hg_0$ satisfying

$$U(x_0)R_{L_1}(x_0, g_0^{-1}hg_0) = R_{L_2}(x, g_0^{-1}hg_0)U(x_0)$$

for all h. Thus $U(x_0)L_1(h) = L_2(h)U(x_0)$ for all h, and L_1 and L_2 are unitarily equivalent. The converse is immediate. $\square$

THEOREM V.11. *Suppose $\hat{N}$ is a standard Borel space and μ is a type I quasi-invariant measure on $\hat{N}$. Let Φ be a Borel mapping from $\hat{N}$ into $\mathrm{Hom}(G, \mathcal{U}(\mathbb{H}))$ satisfying $[\Phi(\gamma)] = \gamma$ for μ a.e. γ. Identify representations R of $\hat{N} \times G$ if they are equal $\mu \times m_G$ a.e. (γ, g). Then the mapping $R \mapsto \mathrm{ind}R$ is a one-to-one correspondence between the representations R of $\hat{N} \times G$ such that $R(\gamma, x) = n\Phi(\gamma)(x)$ for all $x \in N$ for μ a.e. γ and the unitary representations U of G such that $U|_N = nL[\mu, \Phi]$. Moreover, R is type I, locally simple, irreducible or primary iff $\mathrm{ind}R$ is type I, locally simple, irreducible or primary. Furthermore, if $\mathrm{ind}R$ is primary, the measure μ is ergodic.*

PROOF. Let $U = \mathrm{ind}R$. Then $U(g)f(\gamma) = \frac{d\mu \cdot g^{-1}}{d\mu}(\gamma)R(\gamma, g)f(\gamma \cdot g)$ for $f \in L^2(\hat{N}, \mu, n\mathbb{H})$. Hence $U|_N = nL[\mu, \Phi]$ iff $R(\gamma, x) = n\Phi(\gamma)(x)$ a.e. γ for each $x \in N$. By Fubini's Theorem, this occurs if $R(\gamma, x) = n\Phi(\gamma)(x)$ a.e. x for μ a.e. γ. But since each function $x \mapsto R(\gamma, x)$ is a unitary representation of N, we would then have $R(\gamma, x) = n\Phi(\gamma)(x)$ for all x a.e. γ. Note the mapping $R \mapsto U = \mathrm{ind}R$ is one-to-one. We show it is onto. Suppose $U|_N = nL[\mu, \Phi]$. Let P be the canonical projection valued measure on $L^2(\hat{N}, \mu, n\mathbb{H})$. Then since μ is a type I measure, Proposition IV.27 implies the range of P is the Boolean algebra of central projections in the type I von Neumann algebra $\mathcal{A}_N = U(N)''$. Moreover, by Theorem V.10 and Corollary V.5, there is a unitary cocycle representation R of $\hat{N} \times G$ such that $U = \text{ind}R$. This shows the map is onto.

To prove the remaining statements, we first note that $\mathcal{A}_U = \mathcal{A}_R$ and $\mathcal{A}'_U = \mathcal{A}'_R$. Indeed, since each projection $P(E)$ belongs to $\mathcal{A}_N$, it belongs to $\mathcal{A}_U$. By Theorem III.14, if $T \in \mathcal{A}'_U$, then $T = \int^\oplus T(\gamma)\, d\mu(\gamma)$ where $\gamma \mapsto T(\gamma)$ is an essentially bounded Borel function from $\hat{N}$ into $B(n\mathbb{H})$. A direct calculation shows $TU(g) = U(g)T$ for all g iff $T(\gamma)R(\gamma, g) = R(\gamma, g)T(\gamma \cdot g)$ for μ a.e. γ for each g. Hence $\mathcal{A}'_U = \mathcal{A}'_R$. The double commutant Theorem then implies $\mathcal{A}_U = \mathcal{A}''_U = \mathcal{A}''_R = \mathcal{A}_R$. Thus U is type I, locally simple, irreducible or primary iff R is type I, locally simple, irreducible, or primary, respectively.

Finally, note that if U is a primary representation and E is an almost invariant Borel subset of $\hat{N}$, then $U(g)^{-1}P(E)U(g) = P(Eg) = P(E)$ for all g. Thus $P(E) \in \mathcal{A}_U \cap \mathcal{A}'_U = \mathbb{C}I$. Hence $P(E) = 0$ or $P(E) = I$. This implies $\mu(E) = 0$ or $\mu(\hat{N} - E) = 0$, and μ is ergodic. $\square$

We summarize these results in the following Theorem.

THEOREM V.12. *Suppose $\hat{N}$ is a standard Borel space. Let U be a primary unitary representation of G such that $U|_N$ is type I. Then there are a type I quasi-invariant ergodic measure μ and an $n \in \{\infty, 1, 2, \cdots\}$ such that for any selection $\Phi : \hat{N} \to N^c$ with $[\Phi(\gamma)] = \gamma$ for μ a.e. γ, there exists a representation R of $\hat{N} \times G$ on $n\mathbb{H}$ satisfying $R(\gamma, x) = n\Phi(\gamma)(x)$ for all x in N for μ a.e. γ for which U is unitarily equivalent to $\text{ind}R$. Moreover, if μ' is a σ-finite measure on $\hat{N}$ and R' is a representation satisfying $R'(\gamma, x) = n'\Phi'(\gamma)(x)$ for all $x \in N$ for μ' a.e. γ for some selection Φ' and U is unitarily equivalent to $\text{ind}R'$, then μ' is equivalent to μ, $n' = n$, and the representations R and R' are unitarily equivalent. Furthermore, the representation U is irreducible or type I iff R is irreducible or type I.*

PROOF. This result follows from Theorem V.10, Theorem IV.22, Theorem V.6 and Theorem V.11. $\square$

12. Transitive Quasi-orbits and Mackey's Subgroup Method

DEFINITION. Let N be a closed normal subgroup of G. Suppose U is a primary unitary representation of G and $U|_N$ is type I. The unique ergodic measure class $[\mu]$ for which there is an n such that $U|_N \cong nL[\mu, \Phi]$ is called the **quasi-orbit** of the representation U.

Since we are assuming $\hat{N}$ is standard, $\hat{N}$ is a standard Borel G space. In particular, Theorem IV.8 implies the stabilizer $H = \{g : \gamma \cdot g = \gamma\}$ is a closed subgroup of G containing N. Moreover, by Kuratowski's Theorem, since $H \backslash G$ is standard and the mapping $Hg \mapsto \gamma \cdot g$ is one-to-one and Borel, the orbit γG is a Borel subset of $\hat{N}$ which is Borel isomorphic to $H \backslash G$. Furthermore, by Proposition IV.7, any two quasi-invariant σ-finite measures based on γG are equivalent. Hence there is exactly one quasi-orbit based on γG. These quasi-orbits are called the transitive quasi-orbits and as G spaces with quasi-invariant measures $H \backslash G$ and γG are isomorphic.

With these preliminaries we conclude with the following Theorem.

THEOREM V.13. *Let $\hat{N}$ be standard. Let $\gamma \in \hat{N}$. Assume the quasi-orbit γG is type I. Let $\sigma : H \backslash G \to G$ be a Borel cross section for the right cosets of the closed subgroup $H = \{g : \gamma \cdot g = \gamma\}$, and let μ be a σ-finite quasi-invariant measure on $H \backslash G$. Then the mapping $L \mapsto ind_H^G(L, \sigma, \mu)$ is a correspondence between the unitary representations of H whose restriction to N is unitarily equivalent to a multiple of the representation γ and representations U of G whose restrictions to N are type I and have quasi-orbit γG. Moreover, any primary representation U of G whose restriction to N is type I and which has quasi-orbit γG is unitarily equivalent to $ind_H^G(L, \sigma, \mu)$ for some such L. Furthermore, $ind_H^G(L, \sigma, \mu)$ is type I or irreducible precisely when L is type I or irreducible; and $ind_H^G(L, \sigma, \mu)$ is unitarily equivalent to $ind_H^G(L', \sigma, \mu)$ iff L is unitarily equivalent to L'.*

PROOF. Let U be a primary representation of G whose restriction to N is type I with quasi-orbit γG. Let $\varphi : H \backslash G \to \gamma G$ be the isomorphism $\varphi(Hg) = \gamma g$. Let $\Gamma \in \gamma$ and define Φ on $\hat{N}$ by $\Phi(\gamma g)(x) = \Gamma(\sigma(Hg)x\sigma(Hg)^{-1})$ and $\Phi(\gamma') = \Gamma$ for $\gamma' \notin \gamma G$. Then $[\Phi(\gamma')] = \gamma'$ for $\varphi_* \mu$ a.e. γ'. By Theorem V.12, we know there is a unique n and a primary representation R' of $\hat{N} \times G$ satisfying $R'(\gamma g, x) = n\Gamma(\sigma(Hg)x\sigma(Hg)^{-1})$ for all $x \in N$ for a.e. g for which $U \cong indR'$. Let R be the representation on $H \backslash G \times G$ defined

by $R(Hg, g') = R'(\gamma g, g')$. Then $R(Hg, x) = n\Gamma(\sigma(Hg)x\sigma(Hg)^{-1})$
for all x in N for a.e. g. By Proposition V.11, there is a unitary
representation L of H such that R_L is unitarily equivalent to R.
Hence the representation R'_L on $\gamma G \times G$ defined by $R'_L(\gamma g, g') =$
$R_L(Hg, g')$ is unitarily equivalent to the representation $(\gamma, g) \mapsto$
$R'(\gamma g, g')$. In particular, U is unitarily equivalent to $\text{ind}R'_L$. An
easy argument, using the Borel measure preserving equivariant iso-
morphism φ from $H\backslash G$ onto γG, shows $\text{ind}R'_L$ is unitarily equivalent to $\text{ind}L = \text{ind}R_L$. Thus U is unitarily equivalent to $\text{ind}L$.
Since R_L is unitarily equivalent to R, there is a Borel map A from
$H\backslash G$ into $\mathcal{U}(n\mathbb{H})$ satisfying $A(Hg)R(Hg, g')A(Hgg')^{-1} = R_L(Hg, g')$
a.e. g' for μ a.e. Hg. By Lemma V.1, there is a g_0 such that
$A(Hg_0)R(Hg_0, g_0^{-1}hg_0)A(Hg_0)^{-1} = R_L(Hg_0, g_0^{-1}hg_0)$ for all h. Thus
if $x \in N$, we see

$$A(Hg_0)(n\Gamma(\sigma(Hg_0)x\sigma(Hg_0)^{-1}))A(Hg_0)^{-1} = L(\sigma(Hg_0)x\sigma(Hg_0)^{-1});$$

and thus L restricted to N is unitarily equivalent to $n\Gamma$. Moreover, by
Theorem V.11, U is type I or irreducible iff R' is type I or irreducible.
But R' is type I or irreducible iff R'_L is type I or irreducible which
occurs iff R_L is type I or irreducible. Thus by Proposition V.11, U
is type I or irreducible iff L is type I or irreducible, respectively.

Suppose $L|_N = n\Gamma$ where $\Gamma \in \gamma$. A easy argument shows $\text{ind}L$
is unitarily equivalent to the representation $\text{ind}R'_L$ where R'_L is the
representation defined $\varphi_*\mu \times m_G$ a.e. on $\hat{N} \times G$ by $R'_L(\gamma g, g') =$
$L(\sigma(Hg)g'\sigma(Hgg')^{-1})$. In particular,

$$R'_L(\gamma g, x) = n\Gamma(\sigma(Hg)x\sigma(Hg)^{-1}) = n\Phi(\gamma g)(x)$$

for $x \in N$. Hence $\text{ind}L|_N$ is unitarily equivalent to $nL[\varphi_*\mu, \Phi]$ and
hence is type I, and $\text{ind}L$ has quasi-orbit γG.

Finally suppose that the representations $\text{ind}L$ and $\text{ind}L'$ are unitarily equivalent. By Theorem V.12, R'_L and $R'_{L'}$ are unitarily equivalent. Hence R_L and $R_{L'}$ are unitarily equivalent representations of
$H\backslash G \times G$. By Proposition V.11, we conclude that L and L' are unitarily equivalent. If L and L' are unitarily equivalent, then by Proposition V.11, the representations R_L and $R_{L'}$ are unitarily equivalent.
By Theorem V.6, the representations $\text{ind}L$ and $\text{ind}L'$ are unitarily
equivalent. $\square$

Suppose $\hat{N}$ is standard and $\Gamma \in \gamma$ where $\gamma \in \hat{N}$. Corollary V.9 and
Theorem V.9 imply there are a multiplier σ on H/N and a multiplier

representation $\tilde{\Gamma}$ on H with multiplier $(h_1, h_2) \mapsto \sigma(Nh_1, Nh_2)$ such that $\tilde{\Gamma}(n) = \Gamma(n)$ for $n \in N$.

THEOREM V.14. *Suppose Γ is an irreducible unitary representation of N where $\hat{N}$ is standard. Let H be the stabilizer of $[\Gamma]$. Then each unitary representation L of H whose restriction to N is unitarily equivalent to a multiple of Γ is unitarily equivalent to a representation of form $U \otimes \tilde{\Gamma}$ where U is a multiplier representation of H/N with multiplier $\bar{\sigma}$ and $U \otimes \tilde{\Gamma}$ is the representation on the tensor product of the Hilbert space of U with the Hilbert space of Γ defined by $(U \otimes \tilde{\Gamma})(h) = U(Nh) \otimes \tilde{\Gamma}(h)$. Moreover, $U \otimes \tilde{\Gamma}$ is irreducible iff U is irreducible, type I iff U is type I, primary iff U is primary, and locally simple iff U is locally simple. Finally $U_1 \otimes \tilde{\Gamma}$ is unitarily equivalent to $U_2 \otimes \tilde{\Gamma}$ iff U_1 and U_2 are unitarily equivalent.*

PROOF. Suppose $L|_N = m\Gamma$. Hence the Hilbert space for L is $m\mathbb{H}$ where $\mathbb{H}$ is the Hilbert space for Γ. There is a unitary isomorphism T from Hilbert space $m\mathbb{H}$ onto the Hilbert space $\mathbb{K} \otimes \mathbb{H}$ where $\mathbb{K}$ is a Hilbert space of dimension m satisfying $T(mA)T^{-1} = I \otimes A$ for any bounded linear operator A on $\mathbb{H}$. In particular, $T(mB(\mathbb{H}))T^{-1} = I \otimes B(\mathbb{H})$ and $T((mB(\mathbb{H}))')T^{-1} = B(\mathbb{K}) \otimes I$. Define $W(h) = L(h)(m\tilde{\Gamma}(h))^{-1}$. Then since $L(h)L(n)L(h)^{-1} = L(hnh^{-1})$ and $\tilde{\Gamma}(h)\Gamma(n)\tilde{\Gamma}(h)^{-1} = \Gamma(hnh^{-1})$ for $n \in N$, we see $W(h)$ commutes with $m\Gamma(n)$ for all n in N. Since the von Neumann algebra generated by the operators $m\Gamma(n)$ for $n \in N$ is $mB(\mathbb{H})$, one has $W(h)$ is in $(mB(\mathbb{H}))'$; and we see $TW(h)T^{-1} = \tilde{U}(h) \otimes I$ for some operator $\tilde{U}(h)$. This implies $TL(h)T^{-1} = \tilde{U}(h) \otimes \tilde{\Gamma}(h)$ for each $h \in H$. Hence we may assume $L(h) = \tilde{U}(h) \otimes \tilde{\Gamma}(h)$ for $h \in H$. Since $L(n) = I \otimes \Gamma(n)$, we see $\tilde{U}(n) = I$ for $n \in N$. Moreover, since L is a unitary representation having trivial multiplier and $\tilde{\Gamma}$ is a multiplier representation with multiplier σ, one has

$$\tilde{U}(h_1 h_2) \otimes \tilde{\Gamma}(h_1 h_2) = (\tilde{U}(h_1) \otimes \tilde{\Gamma}(h_1))(\tilde{U}(h_2) \otimes \tilde{\Gamma}(h_2))$$
$$= \tilde{U}(h_1)\tilde{U}(h_2) \otimes \tilde{\Gamma}(h_1)\tilde{\Gamma}(h_2)$$
$$= \bar{\sigma}(Nh_1, Nh_2)\tilde{U}(h_1)\tilde{U}(h_2) \otimes \tilde{\Gamma}(h_1 h_2).$$

Thus $\tilde{U}(h_1 h_2) = \bar{\sigma}(Nh_1, Nh_2)\tilde{U}(h_1)\tilde{U}(h_2)$. Moreover $\tilde{U}(n) = I$ since $L(n) = I \otimes \Gamma(n)$ for $n \in N$. Hence $U(Nh) = \tilde{U}(h)$ is a unitary multiplier representation of H with multiplier $\bar{\sigma}$. But since the commutant of the set of operators $L(N) = (I \otimes \Gamma)(N)$ is $B(\mathbb{K}) \otimes I$, the commutant of the set of operators $L(H)$ is $\tilde{U}(H)' \otimes I$; thus L is primary or locally simple precisely when U is primary or locally simple.

Moreover, L is irreducible iff $\tilde{U}(H)' = \mathbb{C}I$ iff U is irreducible. In addition, the von Neumann algebra generated by L is type I iff $L(H)'$ is type I iff $\tilde{U}(H)'$ is type I iff the representation U is type I.

Finally suppose $L_1 = U_1 \otimes \tilde{\Gamma}$ and $L_2 = U_2 \otimes \tilde{\Gamma}$ are unitarily equivalent. We clearly may assume that U_1 and U_2 are representations on the same Hilbert space $\mathbb{K}$. If T is a unitary equivalence between L_1 and L_2, we see $TL_1(n)T^{-1} = L_2(n)$; and thus T is in the commutant of $I \otimes \Gamma(N)$ which is $B(\mathbb{K}) \otimes I$. Hence $T = S \otimes I$ where S is a unitary operator on $\mathbb{K}$. From this it follows that $SU_1(Nh)S^{-1} = U_2(Nh)$, and the representations U_1 and U_2 are unitarily equivalent. $\square$

As stated earlier, we shall show in the next chapter that the Borel structure on $\hat{N}$ is standard iff every unitary representation of N is type I. Hence again, every σ-finite measure on $\hat{N}$ is type I and the type I conditions in the following theorems are superfluous.

THEOREM V.15. *Let N be a closed normal subgroup of G. Suppose $\hat{N}$ is a standard Borel space and the quasi-orbit γG in $\hat{N}$ is type I. Then the mapping $U \mapsto \operatorname{ind}_H^G(U \otimes \tilde{\Gamma})$ is a correspondence between the primary $\bar{\sigma}$ representations of H/N and primary representations of G having quasi-orbit γG. Moreover, $\operatorname{ind}_H^G(U \otimes \tilde{\Gamma})$ is type I or irreducible iff U is type I or irreducible, respectively. Furthermore, every primary representation of G with quasi-orbit γG is unitarily equivalent to $\operatorname{ind}_H^G(U \otimes \tilde{\Gamma})$ for some primary $\bar{\sigma}$ representation of H/N; and this correspondence is one-to-one in that $\operatorname{ind}_H^G(U_1 \otimes \tilde{\Gamma})$ is unitarily equivalent to $\operatorname{ind}_H^G(U_2 \otimes \tilde{\Gamma})$ iff U_1 and U_2 are unitarily equivalent.*

Finally we have the following celebrated Theorem of G. Mackey.

THEOREM V.16. *Suppose every representation of N is type I, $\hat{N}$ is a standard Borel space, and the quotient Borel orbit space $\hat{N}^G$ is countably separated. Then if U is an irreducible unitary representation of G, there exist a unique transitive quasi-orbit γG and an irreducible $\bar{\sigma}$ representation W of H/N, where H is the stabilizer of γ in G, such that U is unitarily equivalent to $\operatorname{ind}_H^G(W \otimes \tilde{\Gamma})$. Moreover, W is uniquely determined up to unitary equivalence.*

PROOF. This result follows from the previous Theorem, Proposition IV.6, Theorem V.12 and Theorem V.13. $\square$

CHAPTER VI

DUAL TOPOLOGIES

1. The Primitive Ideal Space

Let $\mathcal{A}$ be a C^* algebra. The primitive ideal space of $\mathcal{A}$ is the collection $\mathrm{Prim}(\mathcal{A})$ consisting of the ideals $\ker \pi$ where π is an irreducible $*$ representation of $\mathcal{A}$. The elements of $\mathrm{Prim}(\mathcal{A})$ are the primitive ideals; they are closed two sided $*$ ideals. Let E be a subset of $\mathrm{Prim}(\mathcal{A})$. The kernel, $\ker E$, is the closed $*$ ideal of $\mathcal{A}$ defined by $\ker E = \cap_{I \in E} I$.

LEMMA VI.1. *Let $\mathcal{I}$ be a closed two sided ideal. Then $\mathcal{I} = \ker E$ for some subset E of $\mathrm{Prim}(\mathcal{A})$. Moreover, if $J \in \mathrm{Prim}(\mathcal{A})$ and if I_1 and I_2 are two sided ideals in $\mathcal{A}$ and $I_1 \cap I_2 \subseteq J$, then $I_1 \subseteq J$ or $I_2 \subseteq J$. Thus primitive ideals are prime.*

PROOF. By Corollary II.5, $\mathcal{A}/\mathcal{I}$ is a C^* algebra. By Theorem II.10 and Proposition II.16, for each $x \notin \mathcal{I}$ there is an irreducible representation π' of $\mathcal{A}/\mathcal{I}$ with $\pi'(x + \mathcal{I}) \neq 0$. Set $\pi(a) = \pi'(a + \mathcal{I})$. Then π is an irreducible representation of $\mathcal{A}$ with $\ker \pi \supseteq \mathcal{I}$ and $x \notin \ker \pi$. Thus $\mathcal{I} = \ker E$ where $E = \{\ker \pi : \pi(\mathcal{I}) = 0\}$.

Let $J = \ker \pi$ where π is an irreducible $*$ representation of $\mathcal{A}$. Then if I_1 and I_2 are ideals with $I_1 \cap I_2 \subseteq J$, then $\pi(I_1)\pi(I_2) \subseteq \pi(I_1 \cap I_2) = \{0\}$. But if $\pi(I_2)$ is nonzero, $\pi(I_2)\mathbb{H}$ is dense in $\mathbb{H}$; for π is irreducible and $\pi(\mathcal{A})\pi(I_2)\mathbb{H} \subset \pi(I_2)\mathbb{H}$. Hence either $\pi(I_2) = 0$ or $\pi(I_1) = 0$ on the dense subspace $\pi(I_2)\mathbb{H}$. Thus $I_2 \subset J$ or $I_1 \subseteq J$. $\square$

LEMMA VI.2. *Let $\mathcal{B}$ be a nonzero C^* subalgebra of C^* algebra $\mathcal{A}$. Every continuous positive linear functional ω on $\mathcal{B}$ has an extension to a positive linear functional ω' on $\mathcal{A}$ with the same norm.*

PROOF. We may assume $\omega \neq 0$. We add an identity to $\mathcal{A}$ if $\mathcal{A}$ has no identity. Let e be the identity of $\mathcal{A}$. Now e may or may not be in $\mathcal{B}$. If $e \in \mathcal{B}$, then by Corollary II.8, $\omega(e) = ||\omega||$. Otherwise, we consider the subalgebra $\mathcal{B}_e = \mathbb{C}e + \mathcal{B}$. It is a C^* subalgebra. To see this, note if $\lambda_n e + b_n$ converges in $\mathcal{A}$, then λ_n is bounded. Indeed, if not, then $e - \lambda_{n_k}^{-1} b_{n_k} \to 0$ for some subsequence, and then $e \in \mathcal{B}$. Replacing $\lambda_n e + b_n$ by a subsequence, we may assume λ_n converges. This implies b_n converges in $\mathcal{B}$. Thus the limit of $\lambda_n e + b_n$ is in $\mathcal{B}_e$.

By Proposition II.7, $\mathcal{B}$ has an approximate identity. Hence by Corollary II.9, ω satisfies $\omega(b)^* = \overline{\omega(b)}$ and $|\omega(b)|^2 \leq ||\omega|| \, \omega(b^*b)$. Proposition II.13 implies ω' defined on $\mathcal{B}_e$ by $\omega'(\lambda e + b) = \lambda ||\omega|| + \omega(b)$ is a positive linear functional on $\mathcal{B}_e$ extending ω with norm $\omega'(e)$. Thus we have returned to the first case where $\mathcal{B}$ contains the identity of $\mathcal{A}$.

Let $\mathcal{A}_h$ be the Hermitian elements in $\mathcal{A}$. This is a real Banach space. Let P be the collection of all x with $x \geq 0$. By Corollary II.3, P is normed closed. Moreover, (a) of Proposition II.5 implies the elements $x \in \mathcal{A}_h$ with $||(||x||e - x)|| < ||x||$ are interior to P and every element in P is a limit of such x. Proposition II.5 shows the interior $\text{Int}(P)$ is an open convex set. Furthermore, $\text{Int}(P) \cap \ker \omega = \emptyset$. By the Hahn–Banach Theorem, there is a closed hyperplane H containing $\ker \omega$ with $H \cap \text{Int}(P) = \emptyset$. Now choose a positive element $p \in \mathcal{B}_h$ with $\omega(p) = 1$ and $||(||p||e - p)|| < 1$. Define $\omega'(h + \lambda p) = \lambda$ if $h \in H$ and $\lambda \in \mathbb{C}$. Note ω' extends ω and ω' is continuous. Suppose $x \in \text{Int}(P)$ and $x = \lambda p + h$. If $\lambda < 0$, then $h = x - \lambda p \in \text{Int}(P)$ which is a contradiction. Thus $\omega'(x) \geq 0$ if $x \in \text{Int}(P)$. Hence $\omega'(x) \geq 0$ for $x \in P$.

Define $\omega'(a_1 + ia_2) = \omega'(a_1) + i\omega'(a_2)$ for a_1 and a_2 in $\mathcal{A}_h$. Then ω' is a positive linear functional on $\mathcal{A}$ which extends ω. Since $\omega'(e) = \omega(e)$, $||\omega'|| = ||\omega||$. $\square$

COROLLARY VI.1. *Let $\mathcal{B}$ be a C^* subalgebra of C^* algebra $\mathcal{A}$. Then every pure state ω on $\mathcal{B}$ has an extension to a pure state ω' on $\mathcal{A}$.*

PROOF. Let W be the set of all states on $\mathcal{A}$ which extend ω. Then W is a closed convex compact set in the weak $*$ topology. Thus by the Krein–Milman Theorem, W has an extreme point ω'. We claim ω' is an extreme point in the closed convex subset of positive linear functionals in the unit ball of the dual of $\mathcal{A}$. Indeed, suppose $\omega' = r\omega_1' + s\omega_2'$ where $r + s = 1$; $r, s \geq 0$. Then $\omega = r\omega_1 + s\omega_2$ where $\omega_j = \omega_j'|_{\mathcal{B}}$. Clearly ω_1 and ω_2 are in positive linear functionals in the unit ball of the dual of $\mathcal{B}$. Since ω is pure, Lemma II.6 implies $r = 0$ or $s = 0$. □

Recall by Corollary II.5 that two sided closed ideals in a C^* algebra are $*$ ideals.

PROPOSITION VI.1. *Let $\mathbb{H}$ be a Hilbert space. Suppose $\mathcal{A}$ is a C^* algebra and $\mathcal{I}$ is a closed two sided ideal in $\mathcal{A}$.*

(a) *Then the mapping $\pi \mapsto \tilde{\pi}$ from the irreducible $*$ representations of $\mathcal{A}$ on $\mathbb{H}$ with $\ker \pi \supseteq \mathcal{I}$ into the collection of irreducible $*$ representations of $\mathcal{A}/\mathcal{I}$ on $\mathbb{H}$ defined by*

$$\tilde{\pi}(a + \mathcal{I}) = \pi(a)$$

is one-to-one and onto.

(b) *The mapping $\pi \mapsto \pi|_{\mathcal{I}}$ defined on the set of irreducible $*$ representations of $\mathcal{A}$ on $\mathbb{H}$ with $\pi|_{\mathcal{I}} \neq 0$ is one-to-one and onto the set of irreducible $*$ representations of $\mathcal{I}$ on $\mathbb{H}$.*

PROOF. Note (a) is obvious. For (b) we first note it is one-to-one. Indeed, suppose π and π' are irreducible $*$ representations of $\mathcal{A}$ agreeing and nonzero on $\mathcal{I}$. Then since $\mathcal{I}$ is an ideal, we see

$$\pi(a)\pi(b)\xi = \pi(ab)\xi = \pi'(ab)\xi = \pi'(a)\pi'(b)\xi = \pi'(a)\pi(b)\xi$$

if $a \in \mathcal{A}$, $b \in \mathcal{I}$. Thus $\pi(a) = \pi'(a)$ on $\pi(\mathcal{I})\mathbb{H}$. Moreover, $\pi(\mathcal{I})\mathbb{H}$ is invariant under both π and π'. Since these are irreducible, $\pi(\mathcal{I})\mathbb{H}$ is dense in $\mathbb{H}$ and $\pi(a) = \pi'(a)$ for all a. Also note if $\mathbb{H}_0$ is a closed nonzero invariant subspace for $\pi|_{\mathcal{I}}$, then $\pi(\mathcal{I})\mathbb{H}_0 \supseteq \pi(\mathcal{I}\mathcal{A})\mathbb{H}_0 = \pi(\mathcal{I})\pi(\mathcal{A})\mathbb{H}_0 \neq 0$ since $\pi(\mathcal{A})\mathbb{H}_0$ is dense in $\mathbb{H}$ and $\pi(\mathcal{I}) \neq \{0\}$. But then $\mathbb{H}_0 \supset \pi(\mathcal{A})\pi(\mathcal{I})\mathbb{H}_0$ and $\pi(\mathcal{A})\pi(\mathcal{I})\mathbb{H}_0$ is dense since it is invariant under π. Hence $\mathbb{H}_0 = \mathbb{H}$. Thus the mapping is one-to-one into the set of irreducible $*$ representations of $\mathcal{I}$.

To see this mapping is onto, let ρ be an irreducible $*$ representation of $\mathcal{I}$ on $\mathbb{H}$. Let ξ be a unit vector in $\mathbb{H}$ and set $\omega(b) = \langle \rho(b)\xi, \xi \rangle$. Then by Theorem II.7 and Proposition II.14, ω is a pure state on the closed ideal $\mathcal{I}$ of $\mathcal{A}$. But $\mathcal{I}$ is a C^* subalgebra of $\mathcal{A}$. Thus by Corollary VI.1, ω has an extension to a pure state ω' of $\mathcal{A}$. Do the GNS construction to form a Hilbert space $\mathbb{H}'$ and an irreducible representation π' of $\mathcal{A}$ on $\mathbb{H}'$ having cyclic vector ξ' of length one so that $\omega'(a) = \langle \pi'(a)\xi', \xi' \rangle$. Thus $\omega(b) = \langle \pi'(b)\xi', \xi' \rangle$. In particular $\pi'|_{\mathcal{I}}$ is nonzero and thus irreducible and $\pi'(\mathcal{I})\xi$ is dense in $\mathbb{H}'$. Hence by the uniqueness of the GNS construction given in Theorem II.7, we may assume $\mathbb{H} = \mathbb{H}'$ and $\xi' = \xi$. This gives $\langle \pi'(b)\xi, \xi \rangle = \langle \rho(b)\xi, \xi \rangle$ for all $b \in \mathcal{I}$, and thus $\pi'(b) = \rho(b)$ for all $b \in \mathcal{I}$. $\square$

PROPOSITION VI.2. *Let $\mathcal{I}$ be a closed two sided ideal of the C^* algebra $\mathcal{A}$. Then*

(a) *the mapping $J \mapsto J + \mathcal{I}$ defined on $\{J \in Prim(\mathcal{A}) : J \supset \mathcal{I}\}$ is a bijection onto $Prim(\mathcal{A}/\mathcal{I})$ and*

(b) *the mapping $J \mapsto J \cap \mathcal{I}$ defined on $\{J \in Prim(\mathcal{A}) : J \cap \mathcal{I} \neq \mathcal{I}\}$ is a bijection onto $Prim(\mathcal{I})$.*

PROOF. Again (a) is clear. For (b) we know by Proposition VI.1 that if J' is in $Prim(\mathcal{I})$, then there is an irreducible representation π of $\mathcal{A}$ such that $\ker \pi \cap \mathcal{I} = J'$. Thus the mapping is onto.

To see it is one-to-one, suppose $J_1 \cap \mathcal{I} = J_2 \cap \mathcal{I} \neq \mathcal{I}$. But $J_1 \supseteq J_2 \cap \mathcal{I}$. Since J_1 is prime, either $J_2 \subseteq J_1$ or $\mathcal{I} \subseteq J_1$. But $\mathcal{I}$ is not a subset of J_1. Hence $J_2 \subseteq J_1$. By symmetry, $J_1 \subseteq J_2$. $\square$

2. The Hull-Kernel Topology

The dual space $\hat{\mathcal{A}}$ of a C^* algebra $\mathcal{A}$ is the collection of all classes of unitarily equivalent irreducible $*$ representations of $\mathcal{A}$. More specifically for each cardinal number n, let $\mathbb{H}_n$ be a Hilbert space of dimension n and let $\mathcal{A}_n^i$ be the set of all irreducible $*$ representations of $\mathcal{A}$ on $\mathbb{H}_n$. Then $\hat{\mathcal{A}}_n = \{[\pi] : \pi \in \mathcal{A}_n^i\}$ where $[\pi]$ is the collection of all $\pi' \in \mathcal{A}_n^i$ unitarily equivalent to π. Then $\hat{\mathcal{A}} = \cup_{n \leq |\mathcal{A}|} \hat{\mathcal{A}}_n$.

There is a natural action of $\mathcal{U}(\mathbb{H}_n)$ on $\mathcal{A}_n^i$, namely $\pi \cdot g(a) = g^{-1}\pi(a)g$. Thus we see $\hat{\mathcal{A}}_n = \mathcal{A}_n^i/\mathcal{U}(\mathbb{H}_n)$ is the space of orbits under this action. Since equivalent representations of $\mathcal{A}$ have the same kernels, the mapping $[\pi] \mapsto \ker \pi$ is a well defined mapping of $\hat{\mathcal{A}}$ onto $Prim(\mathcal{A})$.

DEFINITION. Let I be a closed two sided $*$ ideal of $\mathcal{A}$. The **hull**, hull(I), of I is the collection of all $J \in Prim(\mathcal{A})$ such that $J \supseteq I$.

DEFINITION. Let E be a subset of $Prim(\mathcal{A})$. Then the **hull-kernel** closure $\overline{E}$ of E is the set hull(ker(E)).

LEMMA VI.3. $\overline{\emptyset} = \emptyset$, $E \subseteq \overline{E}$, $\overline{\overline{E}} = \overline{E}$, and $\overline{E_1 \cup E_2} = \overline{E}_1 \cup \overline{E}_2$.

PROOF. Note $\ker \phi = \mathcal{A}$ and hull($\mathcal{A}$) $= \emptyset$ and thus $\overline{\emptyset} = \emptyset$. Next note if $J \in E$, then $J \supseteq \ker E$ and thus $J \in \overline{E}$. Hence $E \subseteq \overline{E}$. Also note $\ker E = \ker \overline{E}$ and thus $\overline{E} = \text{hull}(\ker E) = \text{hull}(\ker(\overline{E})) = \overline{\overline{E}}$.

Next note $\ker E_i \supseteq \ker(E_1 \cup E_2)$ and thus $\overline{E_1} \cup \overline{E_2} \subseteq \overline{E_1 \cup E_2}$. Also note if $J \in \overline{E_1 \cup E_2}$, then $J \supseteq \ker E_1 \cap \ker E_2$. But since J is prime, either $J \supseteq \ker E_1$ or $J \supseteq \ker E_2$. Thus $J \in \overline{E_1} \cup \overline{E_2}$. $\square$

Note that $\cap_i \overline{E}_i \subseteq \overline{E}_i$ and thus $\overline{\cap_i \overline{E}_i} \subseteq \overline{\overline{E}}_i = \overline{E}_i$ for each i. Thus $\overline{\cap_i \overline{E}_i} \subseteq \cap_i \overline{E}_i \subseteq \overline{\cap_i \overline{E}_i}$. Hence $\cap_i \overline{E}_i = \overline{\cap_i \overline{E}_i}$. From this and Lemma VI.3, it is easy to see the subsets $\mathrm{Prim}(\mathcal{A}) - \overline{E}$ where $E \subseteq \mathrm{Prim}(\mathcal{A})$ form a topology on $\mathrm{Prim}(\mathcal{A})$. It is called the **hull-kernel** topology.

PROPOSITION VI.3. *The hull-kernel topology on* $\mathrm{Prim}(\mathcal{A})$ *is* T_0. *Moreover, if* $J \in \mathrm{Prim}(\mathcal{A})$, *then* $\{J\}$ *is closed iff* J *is a maximal closed ideal.*

PROOF. Suppose $I_1 \neq I_2$ in $\mathrm{Prim}(\mathcal{A})$. Then either I_2 doesn't contain I_1 or I_1 doesn't contain I_2. Hence $I_2 \notin \mathrm{hull}(I_1) = \mathrm{hull}(\ker(\{I_1\}))$ which equals $\overline{\{I_1\}}$ or $I_1 \notin \mathrm{hull}(I_2)$ which equals $\overline{\{I_2\}}$. Hence $\mathrm{Prim}(\mathcal{A})$ is T_0.

Note $\{J\}$ is closed iff the only primitive ideal containing J is J iff J is primitive and J is a maximal closed ideal (since by Lemma VI.1, every proper closed ideal is contained in a primitive ideal). $\square$

PROPOSITION VI.4. *Suppose* $\mathcal{A}$ *has an identity* e. *Then* $\mathrm{Prim}(\mathcal{A})$ *is compact.*

PROOF. Suppose $\mathcal{A}$ has identity e. Let $\{F_i\}$ be a family of closed subsets of $\mathrm{Prim}(\mathcal{A})$ having an empty intersection. This implies the algebraic sum $\sum \ker F_i = \mathcal{A}$, for if not this would be a proper ideal not containing e. Since the interior of the unit ball about e consists of invertible elements, it would follow that the closure of $\sum \ker F_i$ would be a proper closed ideal and thus would be contained in a primitive ideal J. But then $J \in \mathrm{hull}(\ker F_i) = F_i$ for all i and $\cap F_i \neq \emptyset$. Hence there is a finite set S of indices i for which $e = \sum_{i \in S} e_i$ where $e_i \in \ker F_i$. Thus $\mathcal{A} = \sum_{i \in S} \ker F_i$. This implies $\cap_{i \in S} F_i = \emptyset$, for there is no primitive ideal containing $\mathcal{A}$. $\square$

DEFINITION. The hull-kernel topology on $\hat{\mathcal{A}}$ is the smallest topology on $\hat{\mathcal{A}}$ for which the mapping $[\pi] \mapsto \ker \pi$ is continuous; i.e., the open sets of $\hat{\mathcal{A}}$ are the sets of form $\{[\pi] : \ker \pi \in V\}$ where V is an open set of $\mathrm{Prim}(\mathcal{A})$.

LEMMA VI.4. *Let* $\mathcal{A}$ *be a* C^* *algebra without identity. Then if* f *is a continuous function with compact support on* $\mathbb{R}$ *and* $f(0) = 0$, *there is for each self adjoint element* $a \in \mathcal{A}$ *an element* $f(a)$ *in* $\mathcal{A}$ *such that for every representation* π *of* $\mathcal{A}$, *one has* $\pi(f(a)) = f(\pi(a))$.

PROOF. Consider $\mathcal{A}_e$, the extension of $\mathcal{A}$ to a C^* algebra with identity. Then $f(a)$ exists in $\mathcal{A}_e$. Moreover, by the Stone–Weierstrass Theorem, there is a sequence of polynomials $p_n(\lambda)$ with $p_n(0) = 0$ so that p_n converges uniformly to f. It follows that $p_n(a)$ converges in norm to $f(a)$. But $p_n(a) \in \mathcal{A}$. Thus $f(a) \in \mathcal{A}$ since $\mathcal{A}$ is norm closed in $\mathcal{A}_e$. $\square$

LEMMA VI.5. *Let $\mathcal{A}$ be a C^* algebra and suppose a is an Hermitian element in $\mathcal{A}$. Let F be a closed subset of $\mathbb{R}$. Then*

    (a) *if $\mathcal{A}$ has an identity, $\{[\pi] \in \hat{\mathcal{A}} : \sigma(\pi(a)) \subseteq F\}$ is closed in the hull-kernel topology;*

    (b) *if $\mathcal{A}$ has no identity, then $\{[\pi] \in \hat{\mathcal{A}} : \sigma(\pi(a)) \subseteq F \cup \{0\}\}$ is closed in the hull-kernel topology.*

PROOF. We do case (b). The argument for (a) is similar. Let $E = \{[\pi] \in \hat{\mathcal{A}} : \sigma(\pi(a)) \subseteq F \cup \{0\}\}$ and suppose $[\rho] \in \overline{E}$. Then $\ker \rho \supseteq \ker E = \cap_{[\pi] \in E} \ker \pi$. Let $\lambda \in \sigma(\rho(a))$. We claim $\lambda \in F \cup \{0\}$ and consequentially $[\rho] \in E$. If not, there is a continuous function f with compact support such that $f = 0$ on $F \cup \{0\}$ and $f(\lambda) \neq 0$. In particular $f(\rho(a)) \neq 0$ and thus $\rho(f(a)) \neq 0$. However, if $[\pi] \in E$, we have $f(\pi(a)) = 0$. Thus $\pi(f(a)) = 0$ for all $[\pi] \in E$ and thus $f(a) \in \ker E$. Hence $\rho(f(a)) = 0$. But $\rho(f(a)) \neq 0$ which is a contradiction. $\square$

PROPOSITION VI.5. *The map $[\pi] \mapsto ||\pi(x)||$ is lower semicontinuous on $\hat{\mathcal{A}}$ in the hull-kernel topology.*

PROOF. It suffices to show $||\pi(x^*x)||$ is lower semicontinuous. Let $\alpha \geq 0$. Then $||\pi(x^*x)|| \leq \alpha$ iff $\sigma(\pi(x^*x)) \subseteq [0, \alpha]$. Hence by Lemma VI.5, the set $\{[\pi] : ||\pi(x^*x)|| \leq \alpha\}$ is closed. $\square$

PROPOSITION VI.6. *Let $\mathcal{A}$ be a separable C^* algebra. Then the hull-kernel topology on $\hat{\mathcal{A}}$ is second countable.*

PROOF. Let x_i be a countable dense subset of $\mathcal{A}$. For each i, let $V_i = \{[\pi] : ||\pi(x_i)|| > 1\}$. These are open sets. Let U be an open set and let $[\pi] \in U$. Thus π is not in the hull of the kernel of $\hat{\mathcal{A}} - U$. Thus there is an a with $\pi(a) \neq 0$ and $\pi'(a) = 0$ for all $[\pi'] \in \hat{\mathcal{A}} - U$. We may assume $||\pi(a)|| = 2$. Choose x_i with $||x_i - a|| < 1$. Then $||\pi(x_i)|| > 1$ and thus $\pi \in V_i$. Moreover, if $[\pi'] \notin U$, then $\pi'(a) = 0$ and thus $||\pi'(x_i)|| = ||\pi'(x_i - a)|| < 1$. Thus $[\pi'] \notin V_i$. Thus $V_i \subseteq U$. $\square$

LEMMA VI.6. *Suppose for each $i \in I$, A_i is a bounded invertible operator on $\mathbb{H}_i$ and $\sup \|A_i\| < \infty$. Then $\oplus A_i$ has a bounded inverse operator on $\oplus \mathbb{H}_i$.*

PROOF. Let $B_i = A_i^{-1}$. Define a densely defined operator on $\oplus \mathbb{H}_i$ by $\mathrm{dom}(B) = \{(v_i) \in \oplus \mathbb{H}_i : (B_i v_i) \in \oplus \mathbb{H}_i\}$ and set $B(v_i) = (B_i v_i)$. We claim B is a closed operator. Indeed, suppose $v^j \to v$ and $B(v^j) \to w$ in $\oplus \mathbb{H}_i$. Then $v_i^j \to v_i$ for each i and $B_i(v_i^j) \to w_i$ for each i. Hence $A_i B_i(v_i^j) \to A_i(w_i)$ for each i and we see $v_i^j \to A_i(w_i)$. Hence $A_i(w_i) = v_i$ and this implies $B_i A_i(w_i) = w_i = B_i v_i$ for all i. Thus the operator B is closed. By the Closed Graph Theorem, B is continuous. Clearly $B = A^{-1}$. $\square$

PROPOSITION VI.7. *Let $\mathcal{A}$ be a C^* algebra and let $x \in \mathcal{A}$. Then there is a $\pi \in \hat{\mathcal{A}}$ with $\|\pi(x)\| = \|x\|$.*

PROOF. By considering $x^* x$ instead of x, we may assume x is Hermitian. Moreover, by Corollary II.13, there is a collection π_i of irreducible representations π_i of $\mathcal{A}$ such that $\rho = \oplus \pi_i$ is a faithful representation of $\mathcal{A}$. Thus $\|\rho(x)\| = \|x\|$. In particular $\|x\| \in \sigma(\rho(x))$. However, if $\|x\| \notin \sigma(\pi_i(x))$ for all i, then by Lemma VI.6, $\|x\| \notin \sigma(\oplus \pi_i(x))$ and thus $\|x\| \notin \sigma(\rho(x))$. Hence $\|x\| \in \sigma(\pi_i(x))$ for some i. Thus $\|x\| \leq \|\pi_i(x)\| \leq \|x\|$. $\square$

THEOREM VI.1. *The hull-kernel topology on $\hat{\mathcal{A}}$ is locally compact.*

PROOF. By Proposition VI.5, it suffices to show that if $x \in \mathcal{A}$ and $r > 0$, then the set $K = \{[\pi] \in \hat{\mathcal{A}} : \|\pi(x)\| \geq r\}$ is compact in its relative topology. Let $\mathcal{F}$ be a decreasing filter of relatively closed nonempty subsets of K. For each $F \in \mathcal{F}$, let $J_F = \ker F$. Then $J_{F'} \subset J_F$ if $F \subset F'$. Thus the set $\{J_F : F \in \mathcal{F}\}$ is an increasing filter of closed $*$ ideals in $\mathcal{A}$. Set $J = \overline{\cup J_F}$. Then J is a closed two sided $*$ ideal in $\mathcal{A}$. But since $\|x + J_F\| \geq \|\pi(x)\|$ for $\pi \in F$, we see $\|x + J_F\| \geq r$ for all F and thus $\|x + J\| \geq r$. By Propositions VI.2 and VI.7, there is an irreducible representation π of $\mathcal{A}$ such that $\pi(J) = 0$ and $\|\pi(x)\| = \|x + J\| \geq r$. Hence $[\pi] \in K$ and since $\pi(J_F) = 0$, $[\pi] \in \mathrm{hull}(J_F) = \overline{F}$ for all F. Thus $\pi \in \cap_{F \in \mathcal{F}}(\overline{F} \cap K) = \cap_{F \in \mathcal{F}} F$ and K is compact. $\square$

COROLLARY VI.2. *If $\hat{\mathcal{A}}$ is Hausdorff, then for each $x \in \mathcal{A}$, the mapping $[\pi] \mapsto \|\pi(x)\|$ is continuous.*

PROOF. For each $r > 0$, the set $\{[\pi] : ||\pi(x)|| \geq r\}$ is compact and thus is a closed set. Hence for each $r > 0$, $\{[\pi] : ||\pi(x)|| < r\}$ is open. Since the mapping is already known to be lower semicontinuous, we also know for each $r \geq 0$, the set $\{[\pi] : ||\pi(x)|| > r\}$ is open. $\square$

COROLLARY VI.3. *Prim($\mathcal{A}$) is locally compact with the hull-kernel topology.*

PROOF. Let $I \in \mathrm{Prim}(\mathcal{A})$. Choose an irreducible $*$ representation π with $\ker \pi = I$. Since $\hat{\mathcal{A}}$ is locally compact, there is a compact neighborhood V of $[\pi]$. Since the open sets in $\hat{\mathcal{A}}$ are the sets $\{[\pi'] : \ker \pi' \in W\}$ where W is an open set in $\mathrm{Prim}(\mathcal{A})$, we see that $\ker'(V) \equiv \{J \in \mathrm{Prim}(\mathcal{A}) : J = \ker \pi', [\pi'] \in V\}$ is a neighborhood of I. Moreover, it is compact. Indeed, if $\ker' V \subset \cup_\alpha U_\alpha$ where U_α are open sets in $\mathrm{Prim}(\mathcal{A})$, then

$$V \subset \cup_\alpha \{[\pi'] : \ker \pi' \in U_\alpha\};$$

and since V is compact, there are $\alpha_1, \alpha_2, \cdots, \alpha_n$ with V covered by the sets $\{[\pi'] : \ker \pi' \in U_{\alpha_i}\}$. Hence $\ker' V \subset U_{\alpha_1} \cup U_{\alpha_2} \cup \cdots \cup U_{\alpha_n}$. $\square$

Suppose $\mathcal{A}$ is a C^* algebra with no identity. We have seen in Proposition II.2 how to extend $\mathcal{A}$ to a C^* algebra $\mathcal{A}_e$ with identity. Moreover, it is easy to see that if 0 is the 0 representation of $\mathcal{A}$ on a one dimensional Hilbert space, $0_e(a + re) = r$ is an irreducible $*$ representation of $\mathcal{A}_e$.

PROPOSITION VI.8. *Suppose $\mathcal{A}$ is a C^* algebra with no identity. Then $\mathrm{Prim}(\mathcal{A}_e) = \mathrm{Prim}(\mathcal{A}) \cup \{\mathcal{A}\}$ and the relative topology of $\mathrm{Prim}(\mathcal{A}_e)$ on $\mathrm{Prim}(\mathcal{A})$ is the hull-kernel topology. Moreover, $\mathrm{Prim}(\mathcal{A})$ is an open subset of $\mathrm{Prim}(\mathcal{A}_e)$.*

PROOF. Note the mapping $\pi \mapsto \pi_e$ defined by $\pi_e(a + re) = \pi(a) + rI$ is a one-to-one correspondence between the irreducible representations of $\mathcal{A}$ and the irreducible representations π' of $\mathcal{A}_e$ whose restrictions to $\mathcal{A}$ are nonzero. Since all irreducible representations of $\mathcal{A}_e$ whose restrictions to $\mathcal{A}$ are 0 are all equivalent to 0_e and since $\ker \pi_e = \ker \pi$, we see $\mathrm{Prim}(\mathcal{A}_e) = \mathrm{Prim}(\mathcal{A}) \cup \{\mathcal{A}\}$. It follows from the definition of hull-kernel topology that $\mathrm{Prim}(\mathcal{A})$ with the hull-kernel topology is the relative topology of the hull-kernel topology on $\mathrm{Prim}(\mathcal{A}_e)$. Since $\{\mathcal{A}\} = \mathrm{hull}(\ker(\{\mathcal{A}\}))$, we see $\{\mathcal{A}\}$ is closed in $\mathrm{Prim}(\mathcal{A}_e)$; and thus $\mathrm{Prim}(\mathcal{A})$ is open in $\mathrm{Prim}(\mathcal{A}_e)$. $\square$

COROLLARY VI.4. *The mapping* $[\pi] \mapsto [\pi]_e$ *defined by* $[\pi] \mapsto [\pi_e]$ *from* $\hat{A}$ *into* $(\hat{A_e})$ *is a homeomorphism of* $\hat{A}$ *onto the open subset* $(\hat{A})_e$ *of* $(\hat{A_e})$.

3. Finite Dimensional Representations

LEMMA VI.7. *For each* $n \geq 1$, *there is a smallest positive integer* r *such that if* V_n *is an* n-*dimensional vector space and* $T_1, T_2, \ldots, T_r$ *are linear transformations on* V_n, *then*

$$\sum_{\sigma \in S_r} \epsilon_\sigma T_{\sigma 1} T_{\sigma 2} \cdots T_{\sigma r} = 0.$$

Moreover, if $r(n)$ *denotes this* r, *then* $r(1) < r(2) < \cdots$.

PROOF. First we note that if $\lambda \in V_n^*$ and $v \in V_n$, then

$$(T_1, T_2, \ldots, T_{n^2+1}) \mapsto \lambda(\sum_{\sigma \in S_{n^2+1}} \epsilon_\sigma T_{\sigma 1} T_{\sigma 2} \cdots T_{\sigma(n^2+1)} v)$$

is an alternating $n^2 + 1$ form on an n^2 dimensional space and thus is identically 0 for all λ and v. Thus

$$\sum_{\sigma \in S_{n^2+1}} \epsilon_\sigma T_{\sigma 1} T_{\sigma 2} \cdots T_{\sigma(n^2+1)} = 0$$

for all $T_1, T_2, \ldots, T_{n^2+1}$. Hence r exists and $r(n) \leq n^2 + 1$.

Now suppose there is $T_1, T_2, \cdots, T_r \in L(V_n, V_n)$ such that

$$T = \sum_{\sigma \in S_r} \epsilon_\sigma T_{\sigma 1} T_{\sigma 2} \cdots T_{\sigma r} \neq 0.$$

Let $V_{n+1} = V_n \oplus \mathbb{C}e_{n+1}$. Set $T_i' = T_i \oplus 0$ for $i = 1, 2, \ldots, r$ and $T' = T \oplus 0$. Set T_{r+1}' to be an operator which is 0 on V_n and sends e_{n+1} to a vector v_n in V_n where $Tv_n \neq 0$ and set $T_{r+2}' = 0 \oplus I$. Consider the sum

$$\sum_{\tau \in S_{r+2}} \epsilon_\tau T_{\tau 1}' T_{\tau 2}' \cdots T_{\tau r}' T_{\tau(r+1)}' T_{\tau(r+2)}'.$$

Note that each term in the sum where $T_{\tau(r+2)}' \neq T_{r+2}'$ is 0, for every other transformation has range in V_n and T_{r+2}' vanishes on V_n. But since $T_i' T_{r+2}' = 0$ for $i \leq r$, we see all terms are 0 except those for which $\tau(r+1) = r+1$ and $\tau(r+2) = r+2$. Thus

$$\sum_{\tau \in S_{r+2}} \epsilon_\tau T_{\tau 1}' T_{\tau 2}' \cdots T_{\tau r}' T_{\tau(r+1)}' T_{\tau(r+2)}' = \sum_{\tau \in S_r} \epsilon_\tau T_{\tau 1}' T_{\tau 2}' \cdots T_{\tau r}' T_{r+1}' T_{r+2}'.$$

Hence $\sum_{\tau \in S_{r+2}} \epsilon_\tau T'_{\tau 1} T'_{\tau 2} \cdots T'_{\tau r} T'_{\tau(r+1)} T'_{\tau(r+2)} = T' T'_{r+1} T'_{r+2}$. Applying this to $0 \oplus e_{n+1}$, we obtain $T' T'_{r+1} T'_{r+2} e_{n+1} = T' T'_{r+1} e_{n+1} = T v_n \neq 0$. Thus

$$\sum_{\tau \in S_{r+2}} \epsilon_\tau T'_{\tau 1} T'_{\tau 2} \cdots T'_{\tau r} T'_{\tau(r+1)} T'_{\tau(r+2)} \neq 0.$$

Hence $r(n+1) \geq r(n) + 2$. $\square$

THEOREM VI.2. *Let $\mathcal{A}$ be a C^* algebra and let $\hat{\mathcal{A}}$ have the hull-kernel topology. Let n be a positive integer. Set $\hat{\mathcal{A}}_{\leq n} = \{\pi \in \hat{\mathcal{A}} : \dim \pi \leq n\}$ and set $\hat{\mathcal{A}}_n = \{\pi \in \hat{\mathcal{A}} : \dim \pi = n\}$. Then*

(a) *$\hat{\mathcal{A}}_{\leq n}$ is a closed subset of $\hat{\mathcal{A}}$, and $\hat{\mathcal{A}}_n$ is an open subset of $\hat{\mathcal{A}}_{\leq n}$.*

(b) *Let $I_n = \ker(\hat{\mathcal{A}}_{\leq n})$. Then $\hat{\mathcal{A}}_{\leq n}$ is canonically homeomorphic to $(\mathcal{A}/I_n\hat{)}$, and $\hat{\mathcal{A}}_n$ is canonically homeomorphic to $(I_{n-1}/I_n\hat{)}$.*

PROOF. We prove (a). We show $\mathrm{hull}(\ker(\hat{\mathcal{A}}_{\leq n})) = \hat{\mathcal{A}}_{\leq n}$. Indeed, suppose $\pi \in \hat{\mathcal{A}}$ and $\pi(\ker(\hat{\mathcal{A}}_{\leq n})) = 0$. Suppose $\dim \pi > n$. By taking a subspace of $\mathbb{H}(\pi)$ of dimension n and using Lemma VI.7, we can obtain bounded transformations $T_1, T_2, \ldots, T_{r(n)}$ in $B(\mathbb{H}(\pi))$ such that

$$\sum_{\sigma \in S_{r(n)}} \epsilon_\sigma T_{\sigma 1} T_{\sigma 2} \cdots T_{\sigma r(n)} \neq 0.$$

Since $\pi(\mathcal{A})$ is strongly dense in $B(\mathbb{H}(\pi))$, we may assume $T_i = \pi(a_i)$ for some $a_i \in \mathcal{A}$. Thus if $x = \sum_{\sigma \in S_{r(n)}} \epsilon_\sigma a_{\sigma 1} a_{\sigma 2} \cdots a_{\sigma r(n)}$, then $\pi(x) \neq 0$. But by Lemma VI.7, $\pi'(x) = 0$ for all $\pi' \in \hat{\mathcal{A}}_{\leq n}$. Thus $\pi \notin \mathrm{hull}(\ker(\hat{\mathcal{A}}_{\leq n}))$ and $\hat{\mathcal{A}}_{\leq n}$ is closed. Since $\hat{\mathcal{A}}_{\leq n-1}$ is closed, $\hat{\mathcal{A}}_n = \hat{\mathcal{A}}_{\leq n} - \hat{\mathcal{A}}_{\leq n-1}$ is open in $\hat{\mathcal{A}}_{\leq n}$.

Note (b) follows from Proposition VI.1. That the topologies correspond can be shown using the fact that the topologies are defined in terms of kernels and hulls of representations and these are algebraic definitions. $\square$

4. Dual Pairings

Let E and F be vector spaces over $\mathbb{R}$ or $\mathbb{C}$, and let $\langle \cdot, \cdot \rangle$ be a nondegenerate bilinear form on $E \times F$. Then (E, F) form a **dual pairing**. The weak topology on E determined by $\langle \cdot, \cdot \rangle$ is the topology defined by the pseudonorms $p_y(x) = |\langle x, y \rangle|$ where $y \in F$. It is a locally convex topology and is the smallest for which the linear functions $x \mapsto \langle x, y \rangle$ are continuous. It is called the $\sigma(E, F)$ topology or the weak topology determined by the pairing. Analogously, the $\sigma(F, E)$ topology is the topology defined on F by the pseudonorms

$p_x(y) = |\langle x, y \rangle|$ where $x \in E$. We recall that a set A is balanced if $rA \subset A$ whenever $|r| \leq 1$. A set A is absolutely convex if it is balanced and convex.

PROPOSITION VI.9. *Let ϕ be a linear functional on E. Then ϕ is $\sigma(E, F)$ continuous iff there is a $y \in F$ such that $\phi(x) = \langle x, y \rangle$.*

PROOF. Suppose ϕ is $\sigma(E, F)$ continuous. Then there exist y_i, $i = 1, 2, \cdots, n$, such that

$$|\phi(x)| \leq 1 \text{ if } |\langle x, y_i \rangle| \leq 1 \text{ for } i = 1, 2, \cdots, n.$$

In particular, $\phi(x) = 0$ if $\langle x, y_i \rangle = 0$ for $i = 1, 2, \cdots, n$. Hence the mapping $(\langle x, y_1 \rangle, \langle x, y_2 \rangle, \ldots, \langle x, y_n \rangle) \mapsto \phi(x)$ is a well defined linear functional which has an extension to a linear function L on n dimensional space. But then $L(c_1, c_2, \ldots, c_n) = \sum_{i=1}^{n} a_i c_i$. This gives

$$\phi(x) = L(\langle x, y_1 \rangle, \langle x, y_2 \rangle, \ldots, \langle x, y_n \rangle) = \langle x, a_1 y_1 + a_2 y_2 + \cdots + a_n y_n \rangle.$$

□

We recall the following fact about locally convex topological vector spaces (this is a form of the Hahn–Banach Theorem).

THEOREM VI.3. *Let A be a closed convex subset in a real locally convex topological vector space X. If $0 \in A$ and $x \notin A$, then there is a continuous linear functional ϕ on X such that $\phi(y) \leq 1$ for $y \in A$ and $\phi(x) > 1$. If X is real or complex and A is absolutely convex, then there is a continuous linear functional ϕ such that $|\phi(y)| \leq 1$ on A and $|\phi(x)| > 1$.*

DEFINITION (POLAR SETS). Let (E, F) be a dual pairing. Let $A \subset E$. Then the absolute polar set A^0 to A is $\{y : \sup_{x \in A} |\langle x, y \rangle| \leq 1\}$ while the polar set to A is $A^0 = \{y : \langle x, y \rangle \leq 1 \text{ for } x \in A\}$. Similarly one has the notion of polar and absolute polar for subsets of F.

PROPOSITION VI.10. *Let A^0 be the (absolute) polar of the set A.*

(1) *A^0 is (absolutely) convex and $\sigma(F, E)$ closed in F.*
(2) *$A \subset B$ implies $B^0 \subset A^0$.*
(3) *If $\lambda \neq 0$ then $(\lambda A)^0 = \frac{1}{\lambda} A^0$.*
(4) *$(\cup A_i)^0 = \cap A_i^0$.*

(5) *The absolute bipolar A^{00} of the set A is the smallest $\sigma(E, F)$ closed absolutely convex set containing A in E. The bipolar of a set A is the smallest $\sigma(E, F)$ closed convex set containing $A \cup \{0\}$.*

(6) *Let A_i be a family of $\sigma(E, F)$ closed (absolutely) convex sets each containing 0. Then $(\cap A_i)^0$ is the smallest $\sigma(F, E)$ closed (absolutely) convex set containing A_i^0 for all i.*

PROOF. We only do the polar case; the absolute polar case is argued similarly. Note $0 \in A^0$. Since $A^0 = \cap_{x \in A}\{y : \langle x, y \rangle \leq 1\}$ is an intersection of $\sigma(F, E)$ closed convex sets, (1) holds. Straightforward arguments give (2), (3) and (4). For (5), let B be the smallest $\sigma(E, F)$ convex set containing A and 0. Then $B \subset A^{00}$. Suppose $x \in A^{00} - B$. Then by Theorem VI.3, there is a continuous linear functional ϕ such that $\phi(x) > 1$ and $\phi(x') \leq 1$ for $x' \in B$. By Proposition VI.9, $\phi(\cdot) = \langle \cdot, y \rangle$ for some y. Hence $y \in B^0$ and thus $y \in A^0$. But then $\langle x, y \rangle = \phi(x) \leq 1$ which is a contradiction. Hence $B = A^{00}$.

For (6), suppose B is a closed convex set containing A_i^0 for all i. Then $B^0 \subset A_i^{00}$ for all i gives $B^0 \subset \cap(A_i^{00})$. Hence $B^{00} = B \supset (\cap(A_i^{00}))^0 = (\cap A_i)^0$. $\square$

5. Weak Containment and Hull-Kernel Closures

Let X be a real separated locally convex vector space, and let X' be its dual space, i.e., the space of all continuous linear functionals on X. Set $\langle x, x' \rangle = x'(x)$ for $x \in X$ and $x' \in X'$. Theorem VI.3 implies $(x, x') \mapsto \langle x, x' \rangle$ is a dual pairing between X and X'. In particular, if A is a subset of X (or of X'), then A^{00} is the smallest weakly closed convex subset of X (or X') containing $A \cup \{0\}$.

LEMMA VI.8. *Let $\mathcal{A}$ be a C^* subalgebra of $B(\mathbb{H})$. Let ω be a state on $\mathcal{A}$. Then there is a net ω_i of states on $\mathcal{A}$ each having form $\omega_i(a) = \sum_{k=1}^{n}\langle a\xi_i, \xi_i \rangle$ which converges to ω in the weak $*$ topology; i.e., $\lim \omega_i(a) = \omega(a)$ for all a.*

PROOF. By Lemma VI.2, we may assume $\mathcal{A}$ contains the identity I of $B(\mathbb{H})$. Let $E = \{\phi_\xi : ||\xi|| = 1\}$ where $\phi_\xi(a) = \langle a\xi, \xi \rangle$. We show the bipolar E^{00} of E contains all states. Note

$$a \in E^0 \text{ iff}$$
$$\phi_\xi(a) \le 1 \text{ for all } ||\xi|| = 1 \text{ iff}$$
$$\langle a\xi, \xi \rangle \le 1 \text{ for all } ||\xi|| = 1 \text{ iff}$$
$$a \le I \text{ iff}$$
$$0 \le I - a \text{ in } B(\mathbb{H}) \text{ iff}$$
$$0 \le I - a \text{ in } \mathcal{A} \text{ iff}$$
$$a \le I.$$

Hence $E^{00} = \{\omega \in \mathcal{A}' : \omega(a) \le 1 \text{ for } a \le I\}$. By Corollary II.8, E^{00} contains all the states of $\mathcal{A}$. In particular, if ω is a state on $\mathcal{A}$, there exists a net $\omega_i = \sum_{j=1}^{n_i} r_{i,j} \phi_{\xi_{i,j}}$ where $r_{i,j} \ge 0$, $\sum r_{i,j} \le 1$ such that $\omega_i(a) \to \omega(a)$ for each a. Thus $\omega_i \ge 0$ for each i and $\omega_i(I) \to \omega(I) = 1$. Thus $\frac{1}{\omega_i(I)}\omega_i$ are states having the desired property. $\square$

THEOREM VI.4. *Let $\mathcal{A}$ be a C^* algebra, and let E be a set of representations. Let π be a $*$ representation of $\mathcal{A}$. Then the following are equivalent:*

(1) $\ker \pi \supseteq \ker E \equiv \cap_{\pi' \in E} \ker \pi'$
(2) *Given $\xi \in \mathbb{H}$. Then there is a net ω_i of positive linear functionals on $\mathcal{A}$ having form $\omega_i(a) = \sum_{k=0}^{n_i} \langle \pi_{i,k}(a)\xi_{i,k}, \xi_{i,k} \rangle$ where $\pi_{i,k} \in E$ and $\xi_{i,k} \in \mathbb{H}(\pi_{i,k})$ satisfying $\lim_i \omega_i(a) = \langle \pi(a)\xi, \xi \rangle$ for each a. Moreover, this net may be chosen with $||\omega_i|| \le ||\xi||^2$ for all i.*

PROOF. We show (2) implies (1). To see this suppose $a \in \ker E$. Then $\pi_{i,k}(a) = 0$ for all i and k. Thus $\omega_i(a) = 0$ for all i. Hence $\langle \pi(a)\xi, \xi \rangle = 0$ for all ξ. This implies $\pi(a) = 0$, and thus $a \in \ker \pi$.

To do the converse, we may suppose $||\xi|| = 1$. Hence $\omega(a) = \langle \pi(a)\xi, \xi \rangle$ is a state on $\mathcal{A}$. Now suppose $\ker E \subseteq \ker \pi$. Let $\mathcal{I} = \ker E$. Then by Corollary II.5, $\mathcal{A}/\mathcal{I}$ is a C^* algebra; and π factors to a representation $\tilde{\pi}$ of $\mathcal{A}/\mathcal{I}$. Moreover, the mapping $a + \mathcal{I} \mapsto \sum_{\pi' \in E} \pi'(a)$ is a faithful representation of $\mathcal{A}/\mathcal{I}$ on $\oplus \mathbb{H}(\pi')$. Since this is an isometric isomorphism of $\mathcal{A}/\mathcal{I}$ with a C^* subalgebra $\tilde{\mathcal{A}}$ of $B(\oplus \mathbb{H}(\pi'))$, Lemma VI.8 implies there is a net ω_i of states of $\mathcal{A}$ of form $\omega_i(a) = \sum_{k=1}^{n} \langle \tilde{\pi}(a)\xi_k^\infty, \xi_k^\infty \rangle$ where $\xi_k^\infty \in \oplus_{\pi' \in E} \mathbb{H}(\pi')$ with $\omega_i(a) \to \omega(a)$ for all a. Now $\xi_k^\infty = (\xi_{k,\pi'})_{\pi' \in E}$. For each i, there is a finite subset F_i of E such that if $\omega_i'(a) = \sum_{k=1}^{n} \sum_{\pi' \in F_i} \langle \pi'(a)\xi_{k,\pi'}^\infty, \xi_{k,\pi'}^\infty \rangle$, then $||\omega_i - \omega_i'|| \to 0$. Since $\omega_i - \omega_i' \ge 0$ and $\omega_i = (\omega_i - \omega_i') + \omega_i'$, we see that $||\omega_i'|| \le ||\omega_i|| = 1 = ||\xi||^2$. Indeed, if $\omega_1 = \omega_2 + \omega_3$ and ω_2 and ω_3 are

positive and continuous, then $||\omega_1|| = ||\omega_2|| + ||\omega_3||$; this follows by
Corollary II.8, Proposition II.2, and Proposition II.13. $\square$

DEFINITION. Let E be a collection of $*$ representations on a C^*
algebra $\mathcal{A}$. Let π be a $*$ representation of $\mathcal{A}$. We say π is weakly
contained in E if either (1) or (2) of Theorem VI.4 hold.

Note if π is irreducible and E is a subset of $\hat{A}$, then π is weakly
contained in E iff π is in the hull-kernel closure of E.

LEMMA VI.9. *Let X be a normed linear space, and let K be a weak
$*$ compact subspace of the dual space X^*. Let C be the smallest weak
$*$ closed convex set containing K. Then the extreme points of C are
in K.*

PROOF. Give X^* the weak $*$ topology. Let $\mathcal{B}(K)$ be the collection
of all regular probability measures on K. For each $\mu \in \mathcal{B}(K)$, define
an element ϕ_μ in X^* by $\phi_\mu(x) = \int_K \phi(x) \, d\mu(\phi)$. Note $\mathcal{B}(K)$ is a
compact subset of $C(K)^*$ when equipped with the weak $*$ topology.
Moreover, $\mu \mapsto \phi_\mu$ is continuous in the weak $*$ topology of $C(K)^*$ to
the weak $*$ topology of X^*. Thus $\phi_{\mathcal{B}(K)}$ is a weak $*$ compact subset
of X^*. Clearly it is convex. It also contains K. Indeed, if $\phi \in K$
and ϵ_ϕ is the point mass at ϕ, then $\phi_{\epsilon_\phi} = \phi$. Hence $\phi_{\mathcal{B}(K)} \supset C$.
We claim they are equal. Hence, we need to show each ϕ_μ is a
weak $*$ limit of some net of points in the convex hull of K. Let $\mathcal{N}$ be
the collection of all tuples $(E_1, E_2, \cdots, E_k, \phi_1, \phi_2, \cdots, \phi_k)$ where $P =
\{E_1, E_2, \cdots, E_k\}$ is a partition of K and $\phi_i \in E_i$ for $i = 1, 2, \cdots, k$.
Order the pairs by $(P, \phi_i) > (P', \phi_i')$ if P is a refinement of P'. Define
$\phi_{(P,\phi_i)} = \sum \mu(E_i)\phi_i$. This defines a net in the algebraic convex hull of
K. We claim it converges in the weak $*$ topology to ϕ_μ. Indeed, note
if $x \in X$, then $\phi_\mu(x) = \int_K \phi(x) \, d\mu(\phi)$ is the integral of a continuous
function over a compact subset of X^*. Using uniform continuity, for
each $\epsilon > 0$ there is a neighborhood V of 0 in X^* such that if ϕ_1 and
ϕ_2 are in K and $\phi_1 - \phi_2 \in V$, then $|\phi_1(x) - \phi_2(x)| < \epsilon$. Hence take a
partition $P = \{E_1, E_2, \cdots, E_k\}$ where $E_i - E_i \subset V$ for each i. Then
$|\phi_{(P',\phi')}(x) - \phi_\mu(x)| \le \epsilon$ if P' is a refinement of P. Thus $\phi_{\mathcal{B}(K)} = C$.

Let ϕ_μ be an extreme point of C. We claim $\phi_\mu \in K$. Indeed,
let ϕ be in the support of the measure μ. Then for any open set
U of K containing ϕ, we have $\mu(U) > 0$. If $\mu(U) < 1$, then $\mu =
\mu(U)\frac{\mu|_U}{\mu(U)} + \mu(K - U)\frac{\mu|_{K-U}}{\mu(K-U)}$; and thus $\phi_\mu = \mu(U)\phi_\nu + \mu(K - U)\phi_{\nu'}$
where $\nu = \frac{\mu|_U}{\mu(U)} \in \mathcal{B}(K)$ and $\nu' = \frac{\mu|_{K-U}}{\mu(K-U)} \in \mathcal{B}(K)$. But since ϕ_μ is
an extreme point, $\phi_\mu = \phi_\nu = \phi_{\nu'}$. Now ϕ_ν converges to ϕ_{ϵ_ϕ} in the

weak $*$ topology as U converges to $\{\phi\}$. Hence $\phi_\mu = \phi \in K$. If no open subset U of K containing ϕ satisfies $\mu(U) < 1$, then regularity implies $\mu\{\phi\} = 1$ and $\phi_\mu = \phi$. $\quad\square$

THEOREM VI.5. *Let π be an irreducible $*$ representation of a C^* algebra $\mathcal{A}$ on $\mathbb{H}$. Let $E \subset \hat{\mathcal{A}}$. Then the following are equivalent.*

(a) *π is in the hull-kernel closure of E.*
(b) *There is a nonzero $\xi \in \mathbb{H}$ for which there is a net ω_i of positive linear functionals on $\mathcal{A}$ with form $\omega_i(a) = \sum_{k=0}^{n_i} \langle \pi_{i,k}(a)\xi_{i,k}, \xi_{i,k} \rangle$ where $\pi_{i,k} \in E$ and $\xi_{i,k} \in \mathbb{H}(\pi_{i,k})$ satisfying $\lim_i \omega_i(a) = \langle \pi(a)\xi, \xi \rangle$ for each a.*
(c) *For any $\xi \in \mathbb{H}$, there is a net (π_i, ξ_i) with $\pi_i \in E$ and $\xi_i \in \mathbb{H}_i$ where $\|\xi_i\| \le \|\xi\|$ such that $\langle \pi_i(a)\xi_i, \xi_i \rangle \to \langle \pi(a)\xi, \xi \rangle$ for all a.*

PROOF. By Theorem VI.4, (a) implies (b). Suppose (b). Hence there is a nonzero ξ with $\langle \pi(a)\xi, \xi \rangle = \lim \omega_i(a)$ where

$$\omega_i(a) = \sum_{k=1}^{n_i} \langle \pi_i(a)\xi_{i,k}, \xi_{i,k} \rangle$$

and $\pi_{i,k} \in E$. Suppose $a \in \ker E$. Since $\ker E$ is a two sided $*$ ideal, $b^* a^* ab \in \ker E$ for all $b \in \mathcal{A}$. Thus $\omega_i(b^* a^* ab) = 0$ for all i. Hence $\langle \pi(b^* a^* ab)\xi, \xi \rangle = 0$, and we see $\pi(a)\pi(b)\xi = 0$ for all $b \in \mathcal{A}$. Since π is irreducible, the nonzero invariant collection $\{\pi(b)\xi : b \in \mathcal{A}\}$ is dense in $\mathbb{H}$. Consequently $\pi(a) = 0$; i.e., $a \in \ker \pi$. Thus (b) implies (a).

Clearly (c) implies (b). We show (a) implies (c). By Proposition VI.8, we may assume $\mathcal{A}$ has an identity. Note it is sufficient to show (c) for unit vectors ξ. Hence assume $\|\xi\| = 1$. Let $\omega(a) = \langle \pi(a)\xi, \xi \rangle$. Then ω is a positive linear functional on $\mathcal{A}$ with norm 1. Set K to be the weak $*$ closure of the set of all functionals ω' where $\omega'(a) = \langle \pi'(a)\xi', \xi' \rangle$ with $\|\xi'\| \le 1$ and $\pi' \in E$. Then K is weak $*$ compact. Moreover, by Theorem VI.4, there is a net ω_i of form

$$\omega_i(a) = \sum_{k=1}^{n_i} \langle \pi_i(a)\xi_{i,k}, \xi_{i,k} \rangle$$

satisfying $\|\omega_i\| \le 1$ for all i and $\lim_i \omega_i(a) = \omega(a)$ for all a. But then $\|\omega_i\| = \omega_i(e) = \sum_k \|\xi_{i,k}\|^2 \le 1$ for each i. This implies ω_i is in the algebraic convex hull of K. Indeed, set $r^2 = \|\omega_i\|$. Then $r^2 \le 1$ and

$$\omega_i(a) = \sum_k \frac{\|\xi_{i,k}\|^2}{r^2} \left\langle \pi_i(a) \frac{r}{\|\xi_{i,k}\|}\xi_{i,k}, \frac{r}{\|\xi_{i,k}\|}\xi_{i,k} \right\rangle.$$

Since $\omega_i(a) \to \omega(a)$ for all a, ω is in the weak $*$ closed convex hull C of K. Since π is irreducible, Theorem II.7 and Proposition II.14 imply ω is a pure state. By Lemma II.6, ω is an extreme point of C. Lemma VI.9 implies ω is in K. This proves (c). $\square$

6. The Simple-Strong Topology on $\mathrm{Hom}(\mathcal{A}, B(\mathbb{H}))$

Let $\mathcal{A}$ be a separable C^* algebra, and let $\mathbb{H}$ be a fixed separable Hilbert space. Let $\mathrm{Hom}(\mathcal{A}, B(\mathbb{H}))$ be the collection of all $*$ representations π of $\mathcal{A}$ on $\mathbb{H}$. Define a topology on $\mathrm{Hom}(\mathcal{A}, B(\mathbb{H}))$ by π_n converges to π iff $\pi_n(a)$ converges weakly to $\pi(a)$ for each $a \in \mathcal{A}$.

REMARK. $\pi_n(a)$ converges weakly to $\pi(a)$ for all $a \in \mathcal{A}$ iff $\pi_n(a)$ converges strongly to $\pi(a)$ for all a.

PROOF. Let $\pi_n(a)$ converge weakly to $\pi(a)$ for all a. Then

$$\|\pi_n(a)v - \pi(a)v\|^2 = \langle \pi_n(a)v - \pi(a)v, \pi_n(a)v - \pi(a)v\rangle$$
$$= \langle \pi_n(a^*a)v, v\rangle - \langle \pi_n(a)v, \pi(a)v\rangle - \langle \pi(a)v, \pi_n(a)v\rangle + \langle \pi(a)v, \pi(a)v\rangle$$
$$\to \langle \pi(a^*a)v, v\rangle - \langle \pi(a)v, \pi(a)v\rangle - \langle \pi(a)v, \pi(a)v\rangle + \langle \pi(a)v, \pi(a)v\rangle$$
$$= 0 \text{ as } n \to \infty.$$

$\square$

PROPOSITION VI.11. $Hom(\mathcal{A}, B(\mathbb{H}))$ *is a complete separable metric space.*

PROOF. Let $\{\phi_n\}$ be a complete orthonormal basis of $\mathbb{H}$. Let $\{a_m\}$ be a countable dense subset of the unit ball of $\mathcal{A}$. Let $\pi_1, \pi_2 \in \mathrm{Hom}(\mathcal{A}, B(\mathbb{H}))$. Define $d(\pi_1, \pi_2) = \sum_{m,n} 2^{-n-m} \|\pi_1(a_m)\phi_n - \pi_2(a_m)\phi_n\|$. Then d is a metric defining the topology of pointwise strong convergence on $\mathrm{Hom}(\mathcal{A}, B(\mathbb{H}))$. It is complete, for if $\pi_n \in \mathrm{Hom}(\mathcal{A}, B(\mathbb{H}))$ and $\pi_n(a) \to \pi(a)$ strongly for each a, then $\pi \in \mathrm{Hom}(\mathcal{A}, B(\mathbb{H}))$. Indeed, since $\|\pi_n(a)\| \le \|a\|$, we see

$$\|\pi_n(ab)v - \pi(a)\pi(b)v\| \le$$
$$\|\pi_n(a)\pi_n(b)v - \pi_n(a)\pi(b)v\| + \|\pi_n(a)\pi(b)v - \pi(a)\pi(b)v\|$$
$$\le \|a\| \, \|\pi_n(b)v - \pi(b)v\| + \|\pi_n(a)\pi(b)v - \pi(a)\pi(b)v\| \to 0 \text{ as } n \to \infty.$$

Thus $\pi(ab) = \pi(a)\pi(b)$. A similar argument works for sums and scalar products. For adjoints note

$$\|\pi_n(a^*)v - \pi(a)^*v\|^2 = \langle(\pi_n(a) - \pi(a))^*v, (\pi_n(a) - \pi(a))^*v\rangle$$
$$= \langle v, (\pi_n(a) - \pi(a))(\pi_n(a) - \pi(a))^*v\rangle$$
$$= \langle v, \pi_n(aa^*)v\rangle - \langle v, \pi_n(a)\pi(a)^*v\rangle - \langle v, \pi(a)\pi_n(a^*)v\rangle + \langle v, \pi(a)\pi(a)^*v\rangle$$
$$\to \langle v, \pi(a)\pi(a^*)v\rangle - \langle v, \pi(a)\pi(a)^*v\rangle - \langle v, \pi(a)\pi(a^*)v\rangle + \langle v, \pi(a)\pi(a)^*v\rangle = 0$$

as $n \to 0$.

With the product topology, $\mathbb{H}^{\mathbb{N} \times \mathbb{N}}$ is a separable metric space, and the mapping $\pi \mapsto \{\pi(a_m)\phi_n\}$ is a bicontinuous mapping from $\text{Hom}(\mathcal{A}, B(\mathbb{H}))$ into $\mathbb{H}^{\mathbb{N} \times \mathbb{N}}$. Thus $\text{Hom}(\mathcal{A}, B(\mathbb{H}))$ is a complete separable metric space. $\square$

We know by Theorem III.3 that the space $B_*(\mathbb{H})$ of σ weakly continuous linear functionals on $B(\mathbb{H})$ is a predual of $B(\mathbb{H})$. In fact, if $\omega \in B_*(\mathbb{H})$, then $\omega = \sum \omega_{v_i, w_i}$ where $v_i, w_i \in \mathbb{H}$, $\omega_{v_i, w_i}(T) = \langle Tv_i, w_i \rangle$ and $\sum (||v_i||^2 + ||w_i||^2) < \infty$.

Define a function F from $\text{Hom}(\mathcal{A}, B(\mathbb{H})) \times B(\mathbb{H})$ into $l_\infty(\mathbb{N}^3)$ by

$$F(\pi, T) = \{\langle (\pi(a_i)T - T\pi(a_i))\phi_j, \phi_k \rangle\}_{i,j,k}.$$

Note that F is continuous in the product topology and is a linear mapping in T for fixed π.

LEMMA VI.10. *There exists a continuous mapping F_* from $\text{Hom}(\mathcal{A}, B(\mathbb{H})) \times l_1(\mathbb{N}^3)$ into $B_*(\mathbb{H})$ such that $F_*(\pi, \cdot)$ is a linear transformation on $l_1(\mathbb{N}^3)$ for each π and $(F_*(\pi, \cdot))^* = F(\pi, \cdot)$.*

PROOF. Define F_* by

$$F_*(\pi, \{\lambda_{i,j,k}\}) = \sum \lambda_{i,j,k}(\omega_{\phi_j, \pi(a_i^*)\phi_k} - \omega_{\pi(a_i)\phi_j, \phi_k}).$$

F_* is continuous and linear in $\lambda = \{\lambda_{i,j,k}\} \in l_1(\mathbb{N}^3)$. Fix $\pi \in \text{Hom}(\mathcal{A}, B(\mathbb{H}))$ and set $G(\lambda) = F_*(\pi, \lambda)$. We calculate $G^*(T)$ for $T \in B(\mathbb{H})$.

$$\begin{aligned}
G^*(T)(\lambda) &= \langle G^*(T), \lambda \rangle \\
&= \langle T, G(\lambda) \rangle \\
&= \sum \lambda_{i,j,k}(\omega_{\phi_j, \pi(a_i^*)\phi_k}(T) - \omega_{\pi(a_i)\phi_j, \phi_k}(T)) \\
&= \sum \lambda_{i,j,k}(\langle \pi(a_i)T\phi_j, \phi_k \rangle - \langle T\pi(a_i)\phi_j, \phi_k \rangle) \\
&= \sum \lambda_{i,j,k}F(\pi, T)_{i,j,k}.
\end{aligned}$$

Hence $G^*(T) = F(\pi, T)$. $\square$

THEOREM VI.6. *The set $\text{Hom}(\mathcal{A}, B(\mathbb{H}))^i$ of irreducible representations π in $\text{Hom}(\mathcal{A}, B(\mathbb{H}))$ is a G_δ set in $\text{Hom}(\mathcal{A}, B(\mathbb{H}))$ and hence is a Polish space in the relative topology.*

PROOF. Note the set of T with $F(\pi, T) = 0$ is exactly the von Neumann algebra of all operators commuting with the representation π. But $\ker F(\pi, \cdot) = \operatorname{range}(F_*(\pi, \cdot))^{\perp}$. Hence $\dim(\ker F(\pi, \cdot)) = \dim(B_*(\mathbb{H})/M(\pi))$ where $M(\pi) = F_*(\pi, l_1(\mathbb{N}^3))$. Let $\omega_1, \omega_2, \cdots, \omega_k \in B_*(\mathbb{H})$. Let $r_1, r_2, \cdots, r_k$ be an arbitrary k sequence of complex numbers with rational real and imaginary parts. Let $\{\lambda_n\}$ be a countable dense subset of $l_1(\mathbb{N}^3)$. Then $F = \{\omega_1, \omega_2, \cdots, \omega_k\}$ is a linearly dependent subset of $B_*(\mathbb{H})$ modulo $M(\pi)$ iff $\pi \in S_F$ where S_F is the set

$$\cap_{m=1}^{\infty} \cup_{\frac{1}{2} \leq \sum |r_i|^2 \leq 1} \cup_{n=1}^{\infty} \{\pi : ||r_1\omega_1 + r_2\omega_2 + \cdots + r_k\omega_k - F_*(\pi, \lambda_n)|| < \frac{1}{m}\}.$$

Let C be a countable dense subset of $B_*(\mathbb{H})$. Then $\dim(\ker F(\pi, \cdot)) \leq n$ iff $\pi \in \cap_{\infty > |F| > n, F \subset C} S_F$. Since each S_F is a G_δ, we see the set of all representations of $\mathcal{A}$ with $\dim(\ker F(\pi, \cdot)) \leq 1$ is a G_δ set and thus by Proposition I.1 is a Polish space in the relative topology. But this space is the space of irreducible representations of $\mathcal{A}$ on $\mathbb{H}$. □

7. The Dual Space Topology on $\hat{\mathcal{A}}$

We let $\mathcal{A}_{\mathbb{H}}^i = \operatorname{Hom}(\mathcal{A}, B(\mathbb{H}))^i$ be the space of irreducible $*$ representations of the C^* algebra $\mathcal{A}$ on $\mathbb{H}$. Equipped with the topology of pointwise convergence in the weak or equivalently strong operator topology, this is a Polish space. Let $G = \mathcal{U}(\mathbb{H})$ be the Polish group of unitary operators on $\mathbb{H}$. (See Proposition IV.1.) Define an action of G on $\mathcal{A}_{\mathbb{H}}^i$ by

$$\pi \cdot g(a) = g^{-1}\pi(a)g.$$

LEMMA VI.11. *The mapping $(\pi, g) \mapsto \pi \cdot g$ is a continuous from $\mathcal{A}_{\mathbb{H}}^i \times \mathcal{U}(\mathbb{H})$ into $\mathcal{A}_{\mathbb{H}}^i$, and thus $(\mathcal{A}_{\mathbb{H}}^i, \mathcal{U}(\mathbb{H}))$ is a Polish transformation space.*

PROOF. Let g_n converge to $g *$ strongly and $\pi_n(a)$ converge to $\pi(a)$ strongly for each a. Since $g_n^{-1} = g_n^*$ and $g^{-1} = g^*$, g_n^{-1} converges strongly to g^{-1}. Lemma IV.1 implies $g_n^{-1}\pi_n(a)g_n$ converges strongly to $g^{-1}\pi(a)g$, for $||\pi_n(a)|| \leq ||a||$ for all n and $||g_n|| = 1$ for all n. □

DEFINITION. Let $n = \dim \mathbb{H}$. Then $n \in \{\infty, 1, 2, 3, \cdots\}$. Set $\hat{\mathcal{A}}_n = \mathcal{A}_{\mathbb{H}}^i/\mathcal{U}(\mathbb{H})$. Equip each $\hat{\mathcal{A}}_n$ with the quotient topology. Set $\hat{\mathcal{A}} = \hat{\mathcal{A}}_\infty \cup \hat{\mathcal{A}}_1 \cup \hat{\mathcal{A}}_2 \cup \cdots$ and give $\hat{\mathcal{A}}$ the disjoint union topology. Then $\hat{\mathcal{A}}$ is called the dual space of $\mathcal{A}$, and the topology on $\hat{\mathcal{A}}$ is called the dual space topology.

THEOREM VI.7. *The relative hull-kernel topology is identical to the dual space topology on $\hat{\mathcal{A}}_n$.*

PROOF. Let F be a closed subset of $\hat{A}_n$ in the relative hull-kernel topology. Let $[\pi]$ be in the closure of F in the dual space topology. Thus π is the limit in $\mathrm{Hom}(\mathcal{A}, \mathbb{H})^i$ of a sequence of representations π_k where $[\pi_k] \in F$. In particular, if ψ is a nonzero vector in $\mathbb{H}$, then $\langle \pi_k(a)\psi, \psi \rangle \to \langle \pi(a)\psi, \psi \rangle$ for all $a \in \mathcal{A}$. By Theorem VI.5, $[\pi]$ is in the relative hull-kernel closure of F. Thus $[\pi] \in F$, and F is closed in the dual space topology.

Let F be closed in the dual space topology. Let π be in the hull-kernel closure of F. By Corollary VI.4, we see π_e is in the hull-kernel closure of F_e if $\mathcal{A}$ has no unit. We know by Theorem VI.5 that if $\xi \in \mathbb{H}$ and $||\xi|| = 1$, there exist nets $\xi_\gamma \in \mathbb{H}$ and $\pi_\gamma \in F$ with $||\xi_\gamma|| \leq 1$ such that $\langle \pi_\gamma(a)\xi_\gamma, \xi_\gamma \rangle \to \langle \pi(a)\xi, \xi \rangle$ for $a \in \mathcal{A}$. If $\mathcal{A}$ has no unit, we may thus assume $\langle \pi_{\gamma,e}(e)\xi_\gamma, \xi_\gamma \rangle \to \langle \xi, \xi \rangle = 1$. In any case, we see we may assume $\mathcal{A}$ has a unit and $\langle \pi_\gamma(a)\xi_\gamma, \xi_\gamma \rangle \to \langle \pi(a)\xi, \xi \rangle$ for all a in $\mathcal{A}$.

We next show that if S is a finite subset of $\mathcal{A}$ containing e, then for large γ there is a unitary operator U^γ on $\mathbb{H}$ such that $||\pi_\gamma(a)\xi_\gamma - U^\gamma \pi(a)\xi|| \to 0$ for $a \in S$ as γ increases. First, let $\mathbb{H}_0$ be the finite dimensional subspace of $\mathbb{H}$ spanned by the vectors $\pi(a)\xi$ for $a \in S$. Let $e_1, e_2, \cdots, e_m$ be an orthonormal basis of $\mathbb{H}_0$. Now for each $a \in S$, we have $\pi(a)\xi = \sum_{j=1}^m \lambda_j(a)e_j$ where $\lambda_j(a) = \langle \pi(a)\xi, e_j \rangle$. Moreover, we can choose $\tau_j(a) \in \mathbb{C}$ so that $e_j = \sum_{a \in S} \tau_j(a)\pi(a)\xi$. Define $a_{\tau,j} = \sum_{a \in S} \tau_j(a)a$. Note $\pi(a_{\tau,j})\xi = e_j$ and thus

$$\begin{aligned}
\langle \pi_\gamma(a_{\tau,j})\xi_\gamma, \pi_\gamma(a_{\tau,k})\xi_\gamma \rangle &= \langle \pi_\gamma(a_{\tau,k}^* a_{\tau,j})\xi_\gamma, \xi_\gamma \rangle \\
&\to \langle \pi(a_{\tau,k}^*)\pi(a_{\tau,j})\xi, \xi \rangle \\
&= \langle \pi(a_{\tau,j})\xi, \pi(a_{\tau,k})\xi \rangle \\
&= \langle e_j, e_k \rangle
\end{aligned}$$

as γ increases. Hence, by taking γ large, we see the vectors $\pi_\gamma(a_{\tau,j})\xi_\gamma$ for $j = 1, 2, \cdots, m$ are linearly independent. In fact, for γ large, using Gramm-Schmidt orthonormalization, we can form an orthonormal basis e_j^γ, $j = 1, 2, \cdots, m$, of the linear span of $\pi_\gamma(a_{\tau,j})\xi_\gamma$ such that $e_j^\gamma - \pi_\gamma(a_{\tau,j})\xi_\gamma \to 0$ as γ increases. Take U^γ to be a unitary operator on $\mathbb{H}$ satisfying $U^\gamma e_j = e_j^\gamma$. Then for $a \in S$, we see, since $||\xi_\gamma|| \leq 1$, that

$$||\pi_\gamma(a)\xi_\gamma - U^\gamma \pi(a)\xi||^2 =$$
$$\langle \pi_\gamma(a)\xi_\gamma, \pi_\gamma(a)\xi_\gamma \rangle - 2\mathrm{Re}\langle \pi_\gamma(a)\xi_\gamma, U^\gamma \pi(a)\xi \rangle + \langle U^\gamma \pi(a)\xi, U^\gamma \pi(a)\xi \rangle$$
$$= \langle \pi_\gamma(a^*a)\xi_\gamma, \xi_\gamma \rangle - 2\mathrm{Re}\langle \pi_\gamma(a)\xi_\gamma, \sum_{j=1}^m \lambda_j(a)U^\gamma e_j \rangle + \langle \pi(a^*a)\xi, \xi \rangle.$$

Hence

$$\|\pi_\gamma(a)\xi_\gamma - U^\gamma\pi(a)\xi\|^2$$

$$= \langle\pi_\gamma(a^*a)\xi_\gamma,\xi_\gamma\rangle - 2\mathrm{Re}\langle\pi_\gamma(a)\xi_\gamma, \sum_{j=1}^m \lambda_j(a)e_j^\gamma\rangle + \langle\pi(a^*a)\xi,\xi\rangle$$

$$= \langle\pi_\gamma(a^*a)\xi_\gamma,\xi_\gamma\rangle - 2\mathrm{Re}\langle\pi_\gamma(a)\xi_\gamma, \sum_{j=1}^m \lambda_j(a)\pi_\gamma(a_{\tau,j})\xi_\gamma\rangle + \langle\pi(a^*a)\xi,\xi\rangle$$

$$-2\mathrm{Re}\langle\pi_\gamma(a)\xi_\gamma, \sum_{j=1}^m \lambda_j(a)(e_j^\gamma - \pi_\gamma(a_{\tau,j})\xi_\gamma)\rangle$$

$$= \langle\pi_\gamma(a^*a)\xi_\gamma,\xi_\gamma\rangle - 2\sum_{j=1}^m \bar\lambda_j(a)\langle\pi_\gamma(a_{\tau,j}^*a)\xi_\gamma,\xi_\gamma\rangle + \langle\pi(a^*a)\xi,\xi\rangle$$

$$-2\mathrm{Re}\langle\pi_\gamma(a)\xi_\gamma, \sum_{j=1}^m \lambda_j(a)(e_j^\gamma - \pi_\gamma(a_{\tau,j})\xi_\gamma)\rangle$$

$$\to 2\langle\pi(a^*a)\xi,\xi\rangle - 2\sum_{j=1}^m \bar\lambda_j(a)\langle\pi(a_{\tau,j}^*a)\xi,\xi\rangle$$

$$= 2\langle\pi(a^*a)\xi,\xi\rangle - 2\langle\pi(a)\xi, \sum_{j=1}^m \lambda_j(a)\pi(a_{\tau,j})\xi\rangle$$

$$= 2\langle\pi(a^*a)\xi,\xi\rangle - 2\langle\pi(a)\xi, \sum_{j=1}^m \lambda_j(a)e_j\rangle$$

$$= 2\langle\pi(a)\xi,\pi(a)\xi\rangle - 2\langle\pi(a)\xi,\pi(a)\xi\rangle = 0$$

as γ increases. In particular, since

$$\|\pi_\gamma(e)\xi_\gamma - U^\gamma\pi(e)\xi\| = \|\xi_\gamma - U^\gamma\xi\| \to 0,$$

one has

$$\|\pi_\gamma(a)U^\gamma\xi - U^\gamma\pi(a)\xi\| \le$$
$$\|\pi_\gamma(a)U^\gamma\xi - \pi_\gamma(a)\cdot\xi_\gamma\| + \|\pi_\gamma(a)\cdot\xi_\gamma - U^\gamma\pi(a)\xi\|$$
$$\le \|a\|\,\|U^\gamma\xi - \xi_\gamma\| + \|\pi_\gamma(a)\xi_\gamma - U^\gamma\pi(a)\xi\| \to 0.$$

Now let $\epsilon > 0$ and vectors $\xi_1,\xi_2,\cdots,\xi_r$ and $a_1,a_2,\cdots,a_r$ be given. Since π is irreducible, $\pi(\mathcal{A})\xi$ is dense in $\mathbb{H}$. Hence there exist $b_k \in \mathcal{A}$ such that $\|\pi(b_k)\xi - \xi_k\| < \frac{\epsilon}{2\max_j \|a_j\|}$ for $k = 1,2,\cdots,r$. Take

$$S = \{e,b_1,b_2,\cdots,b_r,a_1b_1,a_2b_2,\cdots,a_rb_r\}.$$

Then

$$||\pi_\gamma(a_k)U^\gamma\xi_k - U^\gamma\pi(a_k)\xi_k|| \le ||\pi_\gamma(a_k)U^\gamma\xi_k - \pi_\gamma(a_k)U^\gamma\pi(b_k)\xi|| +$$
$$||\pi_\gamma(a_k)U^\gamma\pi(b_k)\xi - U^\gamma\pi(a_k)\pi(b_k)\xi|| +$$
$$||U^\gamma\pi(a_k)\pi(b_k)\xi - U^\gamma\pi(a_k)\xi_k||$$
$$\le 2||a_k|| \, ||\xi_k - \pi(b_k)\xi|| + ||\pi_\gamma(a_k)U^\gamma\pi(b_k)\xi - \pi_\gamma(a_k)\pi_\gamma(b_k)U^\gamma\xi|| +$$
$$||\pi_\gamma(a_k)\pi_\gamma(b_k)U^\gamma\xi - U^\gamma\pi(a_k)\pi(b_k)\xi||$$
$$\le 2||a_k|| \, ||\xi_k - \pi(b_k)\xi|| + ||a_k|| \, ||U^\gamma\pi(b_k)\xi - \pi_\gamma(b_k)U^\gamma\xi|| +$$
$$||\pi_\gamma(a_kb_k)U^\gamma\xi - U^\gamma\pi(a_kb_k)\xi||$$
$$< \epsilon$$

for γ large. Thus for γ large,

$$||(U^\gamma)^{-1}\pi_\gamma(a_k)U^\gamma\xi_k - \pi(a_k)\xi_k|| < \epsilon,$$

for $k = 1, 2, \cdots, r$. Thus $[\pi]$ is in the closure of F in the dual space topology. $\square$

COROLLARY VI.5. *Suppose the dual space topology on $\hat{A}_n$ is T_0. Then two irreducible representations of A on $\mathbb{H}_n$ are unitarily equivalent iff they have the same kernels.*

PROOF. Suppose π_1 and π_2 have the same kernel. Then the hull-kernel closures of the sets $\{[\pi_1]\}$ and $\{[\pi_2]\}$ are equal. Since the dual topology and the hull-kernel topology agree on $\hat{A}_n$ and $\hat{A}_n$ is T_0, $[\pi_1] = [\pi_2]$. $\square$

COROLLARY VI.6. *The hull-kernel topology and the dual space topology on $\hat{A}$ generate the same Borel sets.*

PROOF. From Theorem VI.2, the sets $\hat{A}_n$ are Borel sets in the hull-kernel topology. Hence the Borel structures are the same if the relative Borel structures on the sets $\hat{A}_n$ are the same. But this follows from Theorem VI.7. $\square$

8. Ergodic Measures on $\mathrm{Hom}(\mathcal{A}, B(\mathbb{H}))$

We have seen $\mathrm{Hom}(\mathcal{A}, \mathbb{H})^i$ is a Polish G space where G is the unitary group of $\mathbb{H}$ and G acts on $\mathrm{Hom}(\mathcal{A}, \mathbb{H})^i$ by $\pi \cdot g(a) = g^{-1}\pi(a)g$. In section 4, we defined the notion of ergodicity for quasi-invariant measure. Here we need the more general notion of an ergodic measure which may not be quasi-invariant. Specifically, a Borel measure μ on $\mathrm{Hom}(\mathcal{A}, \mathbb{H})^i$ is **ergodic** under the action of the unitary group if every invariant set measurable with respect to the completion of μ is null or conull. (Note any nonzero measure on a transitive Borel G space S would be ergodic.)

We first remark that orbits are measurable relative to the completion of μ. Indeed, since $\text{Hom}(\mathcal{A}, \mathbb{H})^i$ is a Polish G space, an orbit $\gamma \cdot G$ is the image of a Borel set under a continuous mapping $g \mapsto \gamma \cdot g$ and thus is an analytic set. But Theorem I.12 states analytic sets are universally measurable.

LEMMA VI.12. *Let $s \mapsto \pi_s$ be a Borel field of representations of a separable C^* algebra $\mathcal{A}$ over the Hilbert bundle $\mathcal{H}$. If π_s are μ a.e. equivalent to π_0, then $\pi = \int_S^{\oplus} \pi_s \, d\mu(s)$ is unitarily equivalent to $n\pi_0$ where $n = \dim L^2(S, \mu)$.*

PROOF. By Theorem III.12, we may assume the Hilbert bundle is trivial. Hence $\mathcal{H}_s = \{s\} \times \mathbb{H}$ for all s. Moreover, by removing a null set from S, we may assume $\pi_s \cong \pi_0$ for all s. Let $\mathcal{U}$ be the unitary group of $\mathbb{H}$. Set $E = \{(s, U) : \pi_0 \cdot U = \pi_s\}$. Since $\mathcal{A}$ is separable, the set E is a Borel subset of $S \times \mathcal{U}$. Moreover, each section $E_s = \{U : (s, U) \in E\}$ is nonempty. Theorem I.14 implies there is a conull Borel subset S_0 of S and a Borel function $s \mapsto U_s \in \mathcal{U}$ on S_0 satisfying $\pi_0 \cdot U_s = \pi_s$. Define a unitary mapping T on $L^2(S_0, \mu, \mathbb{H})$ by $Tf(s) = U_s^{-1} f(s)$. Then

$$
\begin{aligned}
T^{-1}\pi(a)Tf(s) &= U_s \pi(a) T f(s) \\
&= U_s \pi_s(a) T f(s) \\
&= U_s \pi_s(a) U_s^{-1} f(s) \\
&= \left(\pi_s(a) \cdot U_s^{-1}\right) f(s) \\
&= \pi_0(a) f(s) \text{ for } s \in S_0.
\end{aligned}
$$

Thus $T^{-1}\pi(a)Tf(s) = (\pi_0(a) \otimes I) f(s)$ where I is the identity operator on $L^2(S, \mu, \mathbb{H})$. Hence π is unitarily equivalent to $n\pi_0$. $\square$

THEOREM VI.8. *Suppose μ is an ergodic Borel σ-finite measure on $\text{Hom}(\mathcal{A}, \mathbb{H})^i$. Let $\pi = \int_{\text{Hom}(\mathcal{A}, \mathbb{H})^i} \gamma \, d\mu(\gamma)$. Then π is a primary representation of $\mathcal{A}$. Moreover, π is Type I iff the completion of μ vanishes off some orbit.*

PROOF. Set $X = \text{Hom}(\mathcal{A}, \mathbb{H})^i$. One may assume the measure μ on X is finite. Indeed, if ν is a finite measure equivalent to μ, then the representations $\pi_\mu = \int^{\oplus} \gamma \, d\mu(\gamma)$ and $\pi_\nu = \int^{\oplus} \gamma \, d\nu(\gamma)$ are unitarily equivalent.

Now note the strong closure $\mathcal{B}$ of $\pi(\mathcal{A})$ is a von Neumann algebra. Indeed, Corollary III.7 shows that the closure is a $*$ closed subalgebra of $B(\mathbb{H})$. Thus we need only show $\mathcal{B}$ contains the identity operator. By Theorem III.9, the projection Q onto the closure of the subspace $\mathcal{B}\mathbb{H}$ is in $\mathcal{B}' \cap \mathcal{B}''$. Since in the direct integral $\int^{\oplus} \gamma \, d\mu(\gamma)$, each γ is irreducible, it follows by Corollary IV.12 that the range $\mathcal{M}$ of the canonical projection valued measure on $\int^{\oplus} \mathbb{H} \, d\mu(\gamma) = L^2(X, \mu, \mathbb{H})$ is a maximal Boolean algebra of projections in $\pi(\mathcal{A})'$. Thus there is a Borel subset F of X such that $Qf = 1_F f$ for all $f \in L^2(X, \mu, \mathbb{H})$. But then $Q\pi(a)f(\gamma) = \gamma(a)f(\gamma) = 1_F(\gamma)\gamma(a)f(\gamma)$ for μ a.e. γ for each a. Thus for each $a \in \mathcal{A}$, one has $\gamma(a) = 0$ for μ a.e. γ not in F. Let a_i be a countable dense subset of $\mathcal{A}$. Then $\gamma(a_i) = 0$ for all i for μ a.e. $\gamma \notin F$. Since each γ is continuous, we see $\gamma = 0$ for μ a.e. $\gamma \notin F$. But no irreducible representation of $\mathcal{A}$ on $\mathbb{H}$ is 0. Thus $\mu(X - F) = 0$ and $Q = I$, and $\mathcal{B}$ contains the identity operator.

Now let P be a nonzero orthogonal projection in the center of the von Neumann algebra $\mathcal{B}$ generated by $\pi(\mathcal{A})$. To show π is primary, we need to show $P = I$. Again, since the range of the canonical projection value measure on X is maximal, there is a Borel subset E of X having positive μ measure such that $Pf = 1_E f$ for $f \in L^2(X, \mu, \mathbb{H})$.

Now the operator P is in the unit ball of $\mathcal{B}$. By the Kaplansky density theorem P is in the strong closure of the unit ball of $\pi(\mathcal{A})$. But $L^2(X, \mathbb{H})$ is separable. Thus by Lemma III.3, the unit ball of bounded operators on $L^2(X, \mathbb{H})$ is Polish with the strong operator topology. This implies there exists a sequence a_n with $||\pi(a_n)|| \leq 1$ such that $\pi(a_n) \to P$ strongly as $n \to \infty$. Now $\pi(a_n)f(\gamma) = \gamma(a_n)f(\gamma)$ for all $f \in L^2(X, \mathbb{H})$. Hence $\int ||\gamma(a_n)f(\gamma) - 1_E(\gamma)f(\gamma)||^2 \, d\mu(\gamma) \to 0$ for all f. Taking $f = v$ gives $\int ||\gamma(a_n)v - 1_E(\gamma)v||^2 \, d\mu(\gamma) \to 0$ for each vector $v \in \mathbb{H}$. Thus for any $v \in \mathbb{H}$, $||\gamma(a_n)v - 1_E(\gamma)v||$ converges in measure to 0. Hence if $\{v_i\}_{i=1}^{\infty}$ is a countable dense subset of $\mathbb{H}$, there is a subsequence a_{n_k} of a_n such that $\gamma(a_{n_k})v_i - 1_E(\gamma)v_i \to 0$ for all i for μ a.e. γ. But $||\gamma(a_n)|| \leq 1$ a.e. γ since $||\pi(a_n)|| \leq 1$. Thus $\gamma(a_{n_k})v - 1_E(\gamma)v \to 0$ for all v for μ a.e. γ. Hence except for a null set N of γ's, $\gamma(a_{n_k})$ converges strongly to $1_E(\gamma)I$. Now $(E - N) \cdot G \subset E \cup N$. Indeed, if $\gamma \in E - N$, then $\gamma(a_{n_k}) \to I$ as $k \to \infty$. Hence $\gamma \cdot g(a_{n_k}) = g^{-1}\gamma(a_{n_k})g \to I$ as $k \to \infty$. Hence if $g \cdot \gamma \notin N$, $\gamma \cdot g \in E$ for $\gamma \cdot g(a_{n_k}) \to 1_E(\gamma \cdot g)I$ if $\gamma \cdot g \notin N$. Hence E is conull if $(E - N) \cdot G$ is conull. But $(E - N) \cdot G$ is a measurable invariant set which contains the set $E - N$ which has positive measure. Since μ is ergodic, $(E - N) \cdot G$ is conull, and thus $P = I$ and π is primary.

Suppose μ vanishes off the orbit $\gamma_0 \cdot G$. Then there is a Borel subset

E of $\gamma_0 \cdot G$ such that $\mu(X - E) = 0$. Thus $L^2(X, \mathbb{H}) = \mathbb{L}^2(E, \mathbb{H})$, and all the representations in E are unitarily equivalent. Lemma VI.12 implies π is unitarily equivalent to $n\gamma_0$ where $n = \dim L^2(E, \mu)$. Thus the von Neumann algebra generated by π is unitarily isomorphic to $n\gamma_0(\mathcal{A})'' = nB(\mathbb{H})$ which is type I.

Conversely, suppose the von Neumann algebra generated by π is type I. Then $\pi(\mathcal{A})'$ is a type I von Neumann algebra with no nontrivial central projections. It follows by Theorem III.22 that $\pi(\mathcal{A})'$ is a homogeneous Neumann algebra. Let n be the degree of homogeneity. By Theorem III.21, there is a unitary operator U from $L^2(X, \mu, \mathbb{H})$ onto $n\mathbb{H}'$ for some Hilbert space $\mathbb{H}'$ such that $U\pi(\mathcal{A})'U^{-1} = \mathcal{M}_n(\mathbb{C}\mathbb{I})$. Hence if $\mathcal{B} = \pi(\mathcal{A})''$, one has $U\mathcal{B}U^{-1} = nB(\mathbb{H}')$.

Realizing $n\mathbb{H}' = \mathbb{H}' \otimes \mathbb{H}''$ where $\mathbb{H}''$ is a Hilbert space of dimension n, we see $U\mathcal{B}U^{-1}$ consists of all operators on $\mathbb{H}' \otimes \mathbb{H}''$ having form $A \otimes I$ for some $A \in B(\mathbb{H}')$. In particular, $U\pi(a)U^{-1} = \rho(a) \otimes I$ for an irreducible representation ρ of $\mathcal{A}$ on $\mathbb{H}'$. Since the commutant of $\rho(\mathcal{A}) \otimes I$ is $I \otimes B(\mathbb{H}'')$, $U\mathcal{M}U^{-1} = I \otimes \mathcal{M}_0$ for some Boolean algebra of projections $\mathcal{M}_0$ on $L^2(X, \nu)$, and $I \otimes \mathcal{M}_0$ is a maximal Boolean algebra of projections in the commutant of $\rho(\mathcal{A}) \otimes I$. In particular, $\mathcal{M}_0$ is a maximal σ-algebra of projections on $\mathbb{H}''$ isomorphic to $\mathcal{M}$.

Since $\mathcal{M}_0$ is maximal, Corollary III.4 implies any separating vector is cyclic. In particular $\mathcal{M}_0$ has multiplicity one. By Theorems III.1 and III.8, we may assume $\mathcal{M}_0$ is the range of the canonical projection valued measure on $L^2(X, \nu)$ where ν is an appropriate finite measure on X.

In particular, we would now have $\rho(a) \otimes I = \int^\oplus \rho(a) \, d\nu(\gamma)$. This is a direct integral decomposition of the representation $\rho \otimes I$ into irreducible representations on $\mathbb{H}' \otimes \mathbb{H}'' = \mathbb{H}' \otimes L^2(X, \nu) = \int^\oplus \mathbb{H}' \, d\nu(\gamma)$, and the canonical projection valued measure for this decomposition has range isomorphic to $\mathcal{M}_0$. Moreover, U is a unitary operator from $L^2(X, \mu, \mathbb{H})$ onto $L^2(X, \nu, \mathbb{H}')$ satisfying $U\pi(a)U^{-1} = \rho(a) \otimes I$ for all $a \in \mathcal{A}$, and $U\mathcal{M}U^{-1}$ is the range of the canonical projection valued measure on $L^2(X, \nu, \mathbb{H}')$.

By Theorem III.15, there is a Borel isomorphism ϕ of X such that $\phi_*\mu \sim \nu$ and a strongly Borel mapping $\gamma \mapsto U(\gamma)$ from X into the bounded linear operators from $\mathbb{H}$ to $\mathbb{H}'$ such that $U(\gamma)$ is a unitary operator for μ a.e. γ and $Uf(\phi(\gamma)) = \frac{d\nu \circ \phi}{d\mu}(\gamma)^{1/2} U(\gamma) f(\gamma)$ for $f \in L^2(X, \mu, \mathbb{H})$. An easy calculation using $U\pi(a)f(\gamma) = (\rho(a) \otimes I)Uf(\gamma)$ shows $U(\gamma)\gamma(a)f(\gamma) = \rho(a)U(\gamma)f(\gamma)$ for μ a.e. γ for each $f \in L^2(X, \mu, \mathbb{H})$. Taking constant functions $f_i(\gamma) = v_i$ where v_i form a countable dense subset of $\mathbb{H}$ gives $U(\gamma)\gamma(a)v_i = \rho(a)U(\gamma)v_i$ for all i for μ a.e. γ. Hence $\gamma(a) = U(\gamma)^{-1}\rho(a)U(\gamma)$ for μ a.e. γ for each a. Using the separability of $\mathcal{A}$, one then sees by the continuity of γ and ρ that $\gamma(a) = U(\gamma)^{-1}\rho(a)U(\gamma)$ for all a for μ a.e γ. Fix γ_0 where $U(\gamma_0)$ is a unitary equivalence of γ_0 and ρ. Set $g(\gamma) = U(\gamma_0)^{-1}U(\gamma)$. We see $\gamma = \gamma_0 \cdot g(\gamma)$ for μ a.e. γ, and hence the measure μ vanishes off the orbit of γ_0. $\square$

9. Orbit Spaces of a Polish Action

THEOREM VI.9. *Let X be a Polish G space. Then the following are equivalent:*

(a) *For each x, the mapping $G_x g \mapsto x \cdot g$ is a topological isomorphism of $G_x \backslash G$ onto $x \cdot G$.*

(b) *Each orbit $x \cdot G$ is second category in itself.*

(c) *Each orbit is a G_δ set.*

(d) *X/G is a T_0 space.*

PROOF. Suppose $G_x g \mapsto x \cdot g$ is a topological isomorphism. To show (b), we need only show $G_x \backslash G$ is second category. Indeed, suppose F_i are a countable family of closed sets whose union is $G_x \backslash G$. Then these are closed sets in G which are left G_x invariant having union G. But since G is second category, F_i has interior in G for some i. But the projection mapping from G to $G_x \backslash G$ is an open mapping. Therefore, F_i has interior in $G_x \backslash G$ for some i. This implies $G_x \backslash G$ is second category. Hence (a) implies (b).

We claim (b) implies (a). Suppose we show for each x and each open set U of G containing e, the set $x \cdot U$ has interior in $x \cdot G$ where $x \cdot G$ has the relative topology. Then the mapping $g \mapsto x \cdot g$ is an open mapping. Indeed, let V be an open set in G and $g_0 \in V$. Choose W an open set containing e with $g_0 WW^{-1} \subset V$. Then $x \cdot g_0 W$ contains an nonempty open subset A. Since the mapping $y \mapsto y \cdot g$ is a homeomorphism for each g, $x \cdot g_0$ is an interior point in $x \cdot g_0 WW^{-1} \subset x \cdot V$. Thus $x \cdot V$ is open. This implies the natural mapping from $G_x \backslash G$ onto $x \cdot G$ is an open mapping. Since it is continuous, we could then conclude that this mapping is a homeomorphism.

Hence suppose U is an open subset of G containing e. Let V be an open subset of G containing e with $VV^{-1} \subset U$. By Proposition IV.2, there is a Borel mapping $\sigma : G_x \backslash G \to G$ such that $G_x \sigma(z) = z$. Thus $x \cdot V$ is the image of $G_x V$ under the Borel mapping $z \mapsto x \cdot \sigma(z)$. Since this mapping is one-to-one, the Kuratowski Theorem implies $x \cdot V$ is a Borel set. Moreover, the set $x \cdot V$ is second category in $x \cdot G$. Indeed, if not, we know there is a sequence g_i such that $G = \cup V g_i$ and if $x \cdot V$ were first category in $x \cdot G$, then $x \cdot G = \cup x \cdot V g_i$ would be first category in $x \cdot G$. By Proposition I.9, we know every Borel set is Baire. Hence there is an open set A in $x \cdot G$ whose symmetric difference with $x \cdot V$ is meager in $x \cdot G$. Since $x \cdot V$ is second category, A is nonempty. Moreover $AV^{-1} \subset x \cdot VV^{-1}$. Indeed, suppose $x \cdot g \notin x \cdot VV^{-1}$. Then $(x \cdot gV) \cap x \cdot V = \emptyset$. Thus $(x \cdot gV) \cap A$ is meager, for it is a subset of $A - x \cdot V$. Let W be the set of elements $g' \in G$ with $x \cdot g' \in A$. Then W is open and $x \cdot W = A$ and $x \cdot gV \cap A = x \cdot (gV \cap W)$. But since $gV \cap W$ is an open set, we know by the argument above that $x \cdot (gV \cap W)$ is either nonmeager in $x \cdot G$ or empty. Thus, since $x \cdot (gV \cap W)$ is meager, we have $x \cdot gV \cap A = \emptyset$; and we conclude that $x \cdot g \notin AV^{-1}$. Thus $x \cdot VV^{-1}$ has interior points. Thus for each x and any open set U containing e, the set $x \cdot U$ has interior in $x \cdot G$.

We show (b) implies (d). For each $x \in X$, let F_x be the closure in X/G of the singleton set $\{x \cdot G\}$. Let p be the natural projection of X onto X/G. Then $p^{-1}(F_x)$ is a closed subset of X containing $p^{-1}(x \cdot G)$. Thus the closure of $p^{-1}(x \cdot G)$ is contained in $p^{-1}(F_x)$. Moreover if $p(z) \in F_x$ and U is an open set containing z, then $p(U)$ is an open set containing $p(z)$ and hence $x \cdot G \in p(U)$. This implies $U \cap p^{-1}(x \cdot G) \neq \emptyset$, and thus z is in the closure of $p^{-1}(x \cdot G)$, and we see $p^{-1}(F_x)$ is the closure in X of the set $x \cdot G = p^{-1}(x \cdot G)$.

Suppose X/G is not T_0. Hence there are distinct orbits $x \cdot G$ and $y \cdot G$ with $F_x = F_y$. Since $x \cdot G$ is second category in $x \cdot G$ and $y \cdot G$ is second category in $y \cdot G$, they are second category in their closure $C = p^{-1}(F_x) = p^{-1}(F_y) = \overline{x \cdot G} = \overline{y \cdot G}$. But $x \cdot G$ is a Borel subset of C and hence is Baire. Thus there is an open set A in C with $(A - x \cdot G) \cup (x \cdot G - A)$ meager in C. Clearly A is nonempty; and since the topology on C is second countable, and the union A^* of all such sets is a countable union of such sets and is G invariant, we see that $(A^* - x \cdot G) \cup (x \cdot G - A^*)$ is meager in C. Finally, $p(A^*)$ is a nonempty open set in $p(C) = F_y$ and thus contains $p(y)$. This implies $y \cdot G$ is a subset of the meager subset $A^* - x \cdot G$ of C. In particular, $y \cdot G$ is meager in C, and this gives a contradiction.

To see (d) implies (c), we choose a countable base U_i for the topology on X and set $V_i = p(U_i)$. Then the V_i form a countable base for the topology on X/G. Let $q = x \cdot G \in X/G$. Set $B_i = U_i$ if $q \in V_i$ and $B_i = X - U_i$ if $q \notin V_i$. Each set $p^{-1}(B_i)$ is either open or closed, and thus each $p^{-1}(B_i)$ is a G_δ set in X. Thus $p^{-1}(q) = x \cdot G = \cap p^{-1}(B_i)$ is a G_δ set in X.

Lastly (c) implies (b) since by Proposition I.1, a subset in X with the relative topology is Polish iff it is a G_δ set and all Polish spaces are second category in themselves by the Baire Category Theorem. □

10. Condition C and the Orbit Space

Condition C: Suppose G is a Polish group and G acts topologically on a Polish space X. This action is said to satisfy Condition C if for each neighborhood N of e, there is a neighborhood M of e such that $\overline{xM} \subset xN$ for all $x \in X$.

REMARK. If G is locally compact and X is Hausdorff, then (X, G) satisfies condition C; for one can take M to be a compact neighborhood of e contained in N.

LEMMA VI.13. *Condition C implies orbits $x \cdot G$ are F_σ sets in X and thus are Borel subsets of X.*

PROOF. Suppose M is a neighborhood of e with $\overline{xM} \subset xG$ for all $x \in X$. Since G is second countable, there is a sequence g_i in G such that $G = \cup Mg_i$. Hence $xG \subset \cup xMg_i \subset \cup \overline{xM}g_i \subset xG$. Thus xG is an F_σ set. □

Another consequence of condition C is the following:

LEMMA VI.14. *Each neighborhood N of e in G contains a neighborhood M of e such that for any x and any countable decreasing neighborhood base $\{U_n\}$ at x, one has $\cap \overline{U_n M} \subset xN$.*

PROOF. Let M_1 and M be neighborhoods of e such that $\overline{xM_1} \subset xN$ for any x and $\overline{yM^{-1}} \subset yM_1^{-1}$ for any y. Suppose $y \in \cap \overline{U_n M}$. Then $y = u_n g_n$ where $u_n \in U_n$ and $g_n \in M$. Thus $u_n = yg_n^{-1}$ converges to x and thus $x \in \overline{yM^{-1}} \subset yM_1^{-1}$. This gives $y \in xM_1$. Thus $\cap \overline{U_n M} \subset \overline{xM_1} \subset xN$. □

DEFINITION. A subset E of a topological space is locally closed if for each $x \in E$, there is an open neighborhood N_x of x such that $E \cap N_x$ is relatively closed in N_x. Equivalently, E is the intersection of an open and closed set. Indeed, $E = (\cup_x N_x) \cap \overline{E}$ and, in particular, E is relatively open in its closure.

The quotient Borel structure on X/G consists of all subsets E of X/G for which $\cup_{\mathcal{O} \in E} \mathcal{O}$ is a Borel set in X. In general this is larger than the σ algebra generated by the open sets in X/G.

THEOREM VI.10. *Let (G, X) be a Polish transformation group satisfying condition C. Then (a)-(d) of Theorem VI.9 are equivalent to each of the following.*

(e) *Each orbit is locally closed.*

(f) *The quotient topology on X/G generates the quotient Borel structure.*

(g) *The quotient Borel structure on X/G is countably separated.*

(h) *X/G has no nontrivial atomic measures.*

(i) *Every ergodic probability measure on X is supported on an orbit.*

PROOF. We show (b) implies (e). Let $p \in X/G$. To see p is locally closed, it suffices to show $\pi^{-1}(p)$ is open in $\overline{\pi^{-1}(p)} = \pi^{-1}(\overline{\{p\}})$. Hence we may suppose $X/G = \overline{\{p\}}$ and $X = \pi^{-1}(p)$. Thus since $\pi^{-1}(p)$ is second category in itself, it is second category in X. But by Lemma VI.13, $\pi^{-1}(p)$ is an F_σ set. This implies $\pi^{-1}(p)$ contains an open set W of X. Thus $\pi^{-1}(p) = WG$ is an open set in X. Hence p is locally closed.

Note every locally closed set is the intersection of an open and closed set and thus is a G_δ set. Hence (e) implies (c).

Suppose (d) holds. Let V_n be a countable base of open sets for the quotient topology on X/G. Such a base was shown to exist in the proof of Theorem VI.9. Since X/G is T_0, they separate the points of X/G. Hence X/G is countably separated. Thus (d) implies (g).

We show (g) implies (h). Since X/G is countably separated, by Theorem I.1, we may thus suppose $X/G \subset [0, 1]$ with the relative Borel structure. Let μ be an atomic probability measure on X/G. Since $[0, 1]$ is countably separated, $\mu(\{p\}) = 1$ for some p. Thus every atom on X/G is a point mass and thus trivial.

Assume μ is an ergodic probability measure on X. Define $\hat{\mu}$ on X/G by $\hat{\mu}(E) = \mu(\pi^{-1}(E))$. Then $\hat{\mu}$ is an atom on X/G. Hence if (h) holds, $\hat{\mu}$ is a point mass. Thus μ is concentrated on an orbit, and thus (h) implies (i).

To complete the circle of equivalence of (d) with (g), (h), and (i), we show (i) implies (d). Suppose X/G is not T_0. Hence there are distinct orbits p and q in X/G with $q \in \overline{\{p\}}$ and $p \in \overline{\{q\}}$. Set $X_0 = \pi^{-1}(p)$. Then (X_0, G) is a Polish transformation group satisfying property C. To show (i) implies (d), it will suffice to construct an ergodic measure on X_0 whose support is not on an orbit. Thus we may assume $X = \pi^{-1}(p)$, and X contains two distinct orbits $\pi^{-1}(p)$ and $\pi^{-1}(q)$, each which have closure X. Let $\pi^{-1}(p) = xG$.

By Lemma VI.14, there is a symmetric open set M containing e in G such that $\cap Q_m M \subset yG$ for any decreasing base of open sets Q_m at y for any $y \in X$. We define inductively for each $w \in \cup_{n \geq 0}\{0,1\}^n$ an open set $P(w)$ and for each n an element $g(n) \in G$ such that

(1) $P_w M \cap P_{w'} = \emptyset$ if $w \neq w'$ and $w, w' \in \{0,1\}^n$;
(2) $x \in P_{O_n}$ where $O_n = (0, \cdots, 0) \in \{0,1\}^n$ for all n;
(3) $\overline{P_w} \subset P_{w|_{n-1}}$ if $w \in \{0,1\}^n$ and $w|_{n-1} = (i_1, i_2, \cdots, i_{n-1})$ where $w = (i_1, i_2, \cdots, i_n)$;
(4) diameter$(P_w) < \frac{1}{n}$ if $w \in \{0,1\}^n$; and
(5) $P_{(0,0,\cdots,0,i_{n+1},\cdots,i_k)} g(n) = P_{(0,0,\cdots,0,1,i_{n+1},\cdots,i_k)}$.

We construct P_w and $g(n)$ inductively on n. For $n = 0$, set $P_\emptyset = X$ and $g(0) = e$. Suppose P_w and $g(k)$ have been defined for $0 \leq k \leq n$. If $n = 0$, set $N = M$; else set

$$N = \cup_{1 \leq k_1 < k_2 < \cdots < k_r \leq n} g(k_r) \cdots g(k_1) M g(k_1)^{-1} \cdots g(k_r)^{-1}.$$

Thus $N = \cup_{i=1}^s M_i$ where $M_i = h_i M h_i^{-1}$ and the h_i list the elements $g(k_r) \cdots g(k_1)$.

Let R_m be a neighborhood base of open sets for the point x. Then $R_m h_i$ is a base of open sets for the point xh_i. Hence $\cap R_m h_i M \subset xh_i G = xG$. Also $\cap R_m h_i M h_i^{-1} \subset \cap R_m h_i M h_i^{-1} \subset xh_i G h_i^{-1} = xG$. Since $N = \cup_{i=1}^s M_i$ where $M_i = h_i M h_i^{-1}$, this implies $\cap_m R_m N = \cup_{k_1,k_2,\cdots} \cap_m R_m M_{k_j}$ where $1 \leq k_j \leq s$. Thus each such sequence k_j assumes some value i infinitely many times. Therefore, $\cup_{k_1,k_2,\cdots} \cap_m R_m M_{k_j} \subset \cup_{i=1}^s (\cap R_m M_i)$. This gives $\overline{\cap R_m N} \subset \cup_{i=1}^s (\cap_m R_m M_i) \subset \cup_{i=1}^s \cap_m R_m h_i M h_i^{-1} \subset xG$.

The open sets $R_m N$ cannot all be dense in P_{0_n} for otherwise their intersection would be dense in P_{0_n} by the Baire Category Theorem (P_{0_n} is an open subset of X and thus is Polish); and this would give $P_{0_n} = P_{0_n} \cap \overline{\cap_m R_m N} \subset xG$, and thus $xG = \pi^{-1}(p)$ would be an open set missing $\pi^{-1}(q)$. Hence $q \notin \overline{\{p\}}$, a contradiction.

Hence we can choose m with $P_{0_n} - \overline{R_m N} \neq \emptyset$. As xG is dense and the group operations are continuous, we can find a $g(n+1)$ with $x \cdot g(n+1) \in P_{0_n} - \overline{R_m N}$ and an open neighborhood $P_{0_{n+1}}$ of x such that

$$\overline{P_{0_{n+1}}} \subset P_{0_n} \cap R_m,$$
$$\overline{P_{0_{n+1}}} \cdot g(n+1) \subset P_{0_n} - \overline{R_m N}, \text{ and}$$
$$\text{diameter}(P_{0_{n+1}} g(k_1) \cdots g(k_r)) < \frac{1}{n+1} \text{ for any } 1 \leq k_1 < \cdots < k_r \leq n+1.$$

Define $P_{(i_1,i_2,\cdots,i_{n+1})} = P_{0_{n+1}}g(k_r)g(k_{r-1})\cdots g(k_1)$ where $k_1 < k_2 < \cdots < k_r$ are the positions of the 1's in $(i_1, i_2, \cdots, i_{n+1})$.

Clearly (2), (4), and (5) are satisfied. Note $\overline{P}_{0_{n+1}} \subset P_{0_n}$ and if $i = (i_1, i_2, \cdots, i_n)$ is given and $\omega = (i, 1)$, then since

$$P_i = P_{0_n}g(k_s)g(k_{s-1})\cdots g(k_1)$$

where $1 \leq k_1 < k_2 < \cdots < k_s \leq n$ are the positions of the 1's in i, we have $P_i = P_{0_n}g(k_s)\cdots g(k_1) \supset \overline{P}_{0_{n+1}}g(n+1)g(k_s)\cdots g(k_1) = \overline{P}_\omega$; and thus (3) holds.

To see (1) holds, suppose $i = (i_1, \cdots, i_{n+1}) \neq j = (j_1, \cdots, j_{n+1})$. If $i_k \neq j_k$ for some $k < n+1$, then $P_iM \cap P_j \subset P_{i|_k}M \cap P_{j|_k} = \emptyset$. Hence we may assume $i_k = j_k$ for $k < n+1$ and $i_{n+1} = 0$, $j_{n+1} = 1$. Let $1 \leq k_1 < k_2 < \cdots < k_s \leq n$ be the positions of the 1's in $i|_n$. Then

$$\begin{aligned}
P_iM &= P_{0_{n+1}}g(k_s)\cdots g(k_1)Mg(k_1)^{-1}\cdots g(k_s)^{-1}g(k_s)\cdots g(k_1) \\
&\subset P_{0_{n+1}}Ng(k_s)\cdots g(k_1) \text{ while} \\
P_j &= P_{0_{n+1}}g(n+1)g(k_s)\cdots g(k_1)
\end{aligned}$$

and these are disjoint, for $P_{0_{n+1}}g(n+1)$ and $P_{0_{n+1}}N$ are disjoint since $P_{0_{n+1}} \subset P_{0_n} \cap R_m$ and $P_{0_{n+1}}g(n+1) \subset P_{0_n} - R_mN$.

Define a mapping θ from $Z = \{0,1\}^{\mathbb{N}}$ into X by setting $\theta(i)$ to be the unique element in $\cap_n P_{i|_n}$. Then θ is a one-to-one mapping of Z into X. Moreover, $\theta(\{i : i_1 = k_1, \cdots, i_r = k_r\}) = \theta(Z) \cap P_{(k_1,k_2,\cdots,k_r)}$. Since $\text{diam}P_{(k_1,k_2,\cdots,k_r)} < \frac{1}{r}$, we see θ is continuous. Note if i and j are in $\{0,1\}^{\mathbb{N}}$ and disagree at only finitely many entries, then $\theta(i)$ and $\theta(j)$ are in the same orbit. This follows from $P_\omega = P_{0_n}g(k_s)g(k_{s-1})\cdots g(k_1)$ where $0 \leq k_1 < k_2 < \cdots < k_s \leq n$ are the positions of the 1's in $\omega \in \{0,1\}^n$.

Let $\nu = \prod \epsilon$ where ϵ is counting measure on $\{0,1\}$. Then ν is a probability measure on Z which satisfies $\nu(\{i : i|_n = (k_1, \cdots, k_n)\}) = 2^{-n}$. If $\{0,1\}$ is given addition mod 2 group structure, then Z is a compact group, and ν is a Haar measure on Z. Define $\mu = \theta_*\nu$. We show μ is an ergodic measure on X.

If $\omega \in \{0,1\}^n$, we let $\tilde{\omega}$ be the element in Z having form $\omega \times (0,0,\dots)$. Suppose E' is an invariant subset of X of positive measure. Set $E = \theta^{-1}(E')$. Let $0 < \epsilon < 1$. Choose an open set Q with $E \subset Q \subset Z$ and $\nu(E) > (1-\epsilon)\nu(Q)$. For each n and each $\omega \in \{0,1\}^n$, let C_ω be the cylinder set $C_\omega = \{i \in Z : i_j = \omega_j \text{ for } j = 1, 2, \cdots, n\}$. These are open sets and form a base of open closed sets for the topology on Z. Hence $Q = \cup_j C_{\omega_j}$ for some pairwise disjoint collection

C_{ω_j}. Since $\nu(E) > (1 - \epsilon)\nu(Q)$, there is a j with $\nu(E \cap C_{\omega_j}) > (1 - \epsilon)\nu(C_{\omega_j})$. Note $C_{\omega_j} = C_{0_n}\tilde{\omega}_j$.

Let ω be any other tuple in $\{0,1\}^n$. Note $E \cap C_\omega = (E \cap C_{\omega_j})\tilde{\omega}_j\tilde{\omega}$. Indeed, $i \in E \cap C_\omega$ iff $\theta(i) \in E'$ and $i|_n = \omega$ iff $\theta(i\tilde{\omega}\tilde{\omega}_j) \in E'$ and $(i\tilde{\omega}\tilde{\omega}_j)|_n = \omega_j$ iff $i = j\tilde{\omega}_j\tilde{\omega}$ where $j \in E \cap C_{\omega_j}$. Thus $\nu(E \cap C_\omega) = \nu((E \cap C_{\omega_j})\tilde{\omega}_j\omega) = \nu(E \cap C_{\omega_j}) \geq (1 - \epsilon)\nu(C_{\omega_j}) = (1 - \epsilon)\nu(C_\omega)$ for all $\omega \in \{0,1\}^n$. Since the C_ω's for $\omega \in \{0,1\}^n$ partition Z, we see $\nu(E) \geq 1 - \epsilon$. This being true for all $\epsilon > 0$ implies $\nu(E) = 1$. Thus $\mu(E') = 1$.

To finish we need to show μ is not concentrated on an orbit. First note points in Z have measure 0. Hence to show orbits have measure 0, it suffices to show orbits in X intersect $\theta(Z)$ in countably many points. Fix $x \in X$. Let M_1 be an open neighborhood of e with $M_1^{-1}M_1 \subset M$. As G is second countable, there are h_k, $k = 1, 2, \cdots$ with $G = \cup h_k M_1$. Hence $xG \cap \theta(Z) = \cup(xh_k M_1 \cap \theta(Z))$. Let $xh_k g$ and $xh_k g'$ be distinct elements of $xh_k M_1 \cap \theta(Z)$. Thus there is an n and ω, $\omega' \in \{0,1\}^n$ with $\omega \neq \omega'$ and $xh_k g \in P_\omega$ and $xh_k g' \in P_{\omega'}j$. Hence $xh_k g' = xh_k g(g^{-1}g') \in P_\omega(g^{-1}g') \cap P_{\omega'} \subset P_\omega M \cap P_{\omega'} = \emptyset$ which is a contradiction. Hence $xG \cap \theta(Z)$ is countable.

To see (g) implies (f), suppose X/G is countably separated. By Theorem I.10, X/G is an analytic space. In the proof of Theorem VI.9, we saw the topology on X/G has a countable base. Moreover, the topology on X/G is T_0 since (g) implies (d). Hence the Borel structure generated by the quotient topology is countably separated. By Corollary I.7, the quotient Borel structure is generated by any countable separating family of Borel sets. Hence the quotient topology generates the quotient Borel structure.

To finish we show (f) implies (d). If X/G were not T_0, then the quotient topology would not separate points. Hence the Borel structure it generates would not separate points. However, the quotient Borel structure separates points by Lemma VI.13 since orbits are Borel sets. $\square$

DEFINITION. A topological space Y is almost Hausdorff if each nonempty closed subset of Y contains a relatively open nonempty set that is Hausdorff in the relative topology.

REMARK. Y is almost Hausdorff iff there is an ordinal α_0 and a collection R_α of open sets, $\alpha \leq \alpha_0$, such that

(a) $\alpha < \beta \implies R_\alpha \subset R_\beta$, $R_\alpha \neq R_\beta$
(b) $\alpha < \alpha_0 \implies R_{\alpha+1} - R_\alpha$ is Hausdorff

(c) if $\lambda \leq \alpha_0$ is a limit ordinal, then $R_\lambda = \cup_{\alpha < \lambda} R_\alpha$
(d) $R_0 = \emptyset$ and $R_{\alpha_0} = Y$.

PROOF. Suppose Y is almost Hausdorff. Let α' be an ordinal such that the set of $\alpha < \alpha'$ has cardinality larger than the cardinality of the collection of all open subsets of Y. Define $R_0 = \emptyset$. Suppose R_α have been defined for all $\alpha < \lambda$. If λ is a successor ordinal, i.e., $\lambda = \alpha + 1$, then $Y - R_\alpha$ is closed. If $Y = R_\alpha$ we are done. So we may assume $Y - R_\alpha \neq \emptyset$. Thus there is an open set G such that $G \cap (Y - R_\alpha)$ is nonempty and Hausdorff. Set $R_\lambda = G \cup R_\alpha$. If λ is not a successor ordinal, set $R_\lambda = \cup_{\alpha < \lambda} R_\alpha$. Clearly by a cardinality argument, there is an $\alpha_0 < \alpha'$ with $R_{\alpha_0} = Y$.

Conversely, suppose such a collection R_α exists. Let S be any nonempty subset of Y. Take the smallest ordinal β with $R_\beta \cap S \neq \emptyset$. Clearly β is not a limit ordinal, for otherwise $R_\alpha \cap S \neq \emptyset$ for some $\alpha < \beta$. Thus $\beta = \alpha + 1$. Hence $S \cap R_\beta = S \cap (R_{\alpha+1} - R_\alpha)$ is Hausdorff. $\square$

We note that if Y is second countable and R_α, $\alpha \leq \alpha_0$, is a collection satisfying (a)-(d), then α_0 is countable. Indeed, let G_n be a base for the topology of Y. For each α, let $n(\alpha)$ be the first n for which $G_n \subset R_{\alpha+1}$ and $G_n \cap (R_{\alpha+1} - R_\alpha) \neq \emptyset$. Then $\alpha \mapsto n(\alpha)$ is one-to-one; for if $\beta < \alpha$ and $G_n \subset R_{\beta+1}$, then $G_n \cap (R_{\alpha+1} - R_\alpha) = \emptyset$.

LEMMA VI.15. *Let R be an equivalence relation on X. Give X/R the quotient topology. Then if X/R is Hausdorff, R is a closed subset of $X \times X$. Conversely, if the natural projection π from X onto X/R is an open mapping and if R is a closed set, then space X/R is Hausdorff.*

PROOF. Assume X/R is Hausdorff. Let $(x, y) \notin R$. Thus $\pi(x) \neq \pi(y)$. Hence there are open sets U, V in X/R with $U \cap V = \emptyset$ and $\pi(x) \in U$, $\pi(y) \in V$. Hence $\pi^{-1}(U) \times \pi^{-1}(V)$ is an open set in $X \times X$ which contains (x, y) but does not meet R. Thus R is closed.

Suppose π is an open mapping and R is closed in $X \times X$. Suppose $\pi(x) \neq \pi(y)$. Then $(x, y) \notin R$. Hence there exist open sets U and V containing x and y, respectively, such that $(U \times V) \cap R = \emptyset$. Then $\pi(U)$ and $\pi(V)$ are disjoint open sets in X/R with $\pi(x) \in \pi(U)$ and $\pi(y) \in \pi(V)$. Thus X/R is Hausdorff. $\square$

11. Condition D and the Orbit Space

DEFINITION. A topological transformation group (X, G) satisfies Condition D if X is first countable and each open neighborhood N of e contains an open neighborhood M of e such that

$$\cap_m \overline{Q_m M} \subset xN$$

for any x and any decreasing neighborhood base $\{Q_m\}_{m=1}^\infty$ at x consisting of open sets.

Note Condition D implies Condition C for $\overline{xM} \subset xN$.

THEOREM VI.11. *Let (X, G) be a Polish transformation group satisfying Condition D. Then (a)-(i) of Theorems VI.9 and VI.10 are equivalent to each of the following:*

(j) *X/G is almost Hausdorff.*

(k) *X/G is standard.*

(l) *The quotient mapping $\pi : X \to X/G$ has a Borel transversal; i.e., a Borel mapping $\phi : X/G \to X$ satisfying $\pi \circ \phi = \mathrm{id}$.*

(m) *The quotient mapping π has a Borel cross section.*

(n) *For each closed set F in X/G and each neighborhood N of e in G, there is a nonempty relatively open set U in $\pi^{-1}(F)$ such that for $x \in U$, $xG \cap U = xN \cap U$.*

PROOF. We show (a) implies (n). Hence suppose $G_x g \mapsto xg$ is a homeomorphism of $G_x \backslash G$ onto $x \cdot G$ for any x. Let F be a closed set in X/G. Let N be open neighborhood of e in G. Replacing X by $\pi^{-1}(F)$ which is closed in X, we need only show there is an open set U with $xG \cap U = xN \cap U$ for all $x \in U$.

Suppose such an open set U does not exist. Let M be an open neighborhood of e in G such that $\cap \overline{Q_m M} \subset xN$ for all x and any decreasing neighborhood base of open sets Q_m at x.

Set $P_0 = X$. Suppose $P_0, P_1, \cdots, P_n$ and $g_1, g_2, \cdots, g_n$ have been defined so that

(1) $\overline{P_{i+1}} \subset P_i$

(2) diameter$(P_i) < \frac{1}{i}$ for $i = 1, 2, \cdots, n$

(3) $P_{i+1} g_{i+1} \subset P_i - \overline{P_{i+1} M}$ for $i = 1, 2, \cdots, n-1$

(4) each P_i is an open set.

By our assumption $xG \cap P_n \neq xN \cap P_n$ for some $x \in P_n$. Let Q_m be an open neighborhood base at x. Since $\cap_m \overline{Q_m M} \subset xN$, there is an m such that $xG \cap P_n \neq \overline{Q_m M} \cap P_n$. Hence choose g_{n+1} with $xg_{n+1} \in P_n - \overline{Q_m M}$ and an open neighborhood P_{n+1} of x with diameter smaller than $\frac{1}{n+1}$ such that $P_{n+1}g_{n+1} \subset P_n - \overline{Q_m M}$ and $\overline{P}_{n+1} \subset P_n \cap Q_m$. Inductively, we see we may assume (1)-(4) for all n.

Since X is complete $\cap_m P_m$ contains exactly one point x_0. But $x_0 g_{n+1} \in P_n$ for all n. Thus $x_0 g_{n+1}$ converges to x_0. However, (3) implies $x_0 g_{n+1} \notin x_0 M$ for all n. Thus $x_0 M$ is not a neighborhood of x_0 in $x_0 G$ which is a contraction, for $g \mapsto G_x g$ is an open mapping.

To see (n) implies (j), suppose (X, G) satisfies (n). Let F be a closed subset of X/G. We need to find a nonempty relatively open subset of F such that the relative topology on this subset is Hausdorff. Replacing X by $\pi^{-1}(F)$, we may suppose $\pi^{-1}(F) = X$. (Condition D and (n) are still satisfied.) Hence we need only show X/G has a nonempty open Hausdorff subspace.

Let M be an open neighborhood of e in G such that $\cap_m \overline{Q_m M} \subset xG$ for any x and any decreasing neighborhood base Q_m at x consisting of open sets. Let U be a nonempty open set in X such that

$$xG \cap U = xM \cap U$$

for all $x \in U$. Let R be the relative equivalence relation defined by the action of G on the open set U. Since the mapping $\pi : U \to U/R$ is open, to show $\pi(U)$ is Hausdorff, it suffices by Lemma VI.15 to show R is a closed subset of $U \times U$. Suppose $(x_n, x_n g_n) \in R$ and converges to $(x, y) \in U \times U$. As $x_n G \cap U = x_n M \cap U$ for all n, we may assume $g_n \in M$ for each n. Let Q_m be a decreasing neighborhood base of open sets at x. Then $y \in \overline{Q_m M}$ for each m. Thus $y \in \cap_m \overline{Q_m M} \subset xG$ and thus $(x, y) \in R$.

We next show (j) implies (m). If X/G is almost Hausdorff, there exist open sets R_α, $\alpha \leq \alpha_0$, satisfying conditions (a)-(d). We know X/G has a countable base for its topology; and thus as remarked earlier, the ordinal α_0 is a countable. Set $F_\alpha = \pi^{-1}(R_{\alpha+1} - R_\alpha)$ for $\alpha < \alpha_0$. Since F_α is the intersection of an open and closed set, F_α is a G_δ set for all $\alpha < \alpha_0$. Thus by Proposition I.1 of Chapter 1, F_α is Polish in the relative topology. Moreover, since the sets $R_{\alpha+1} - R_\alpha$ are Hausdorff, the orbits in F_α are closed sets in F_α. Moreover, the saturation of any relatively open subset of F_α is relatively open in F_α. By Theorem I.15, there is a Borel set E_α meeting each orbit in F_α in exactly one point. Let $E = \cup E_\alpha$. Then E is a Borel cross section for X/G.

To see (m) implies (k), let E be a Borel cross-section for X/G. Then the mapping $\pi|_E$ is one-to-one, Borel and onto. Moreover, the saturation of any Borel subset W of E is analytic and coanalytic, for its complement is the saturation of $E - W$. By Corollary I.2, $\pi^{-1}(\pi(W)) = W \cdot G$ is Borel. Hence $\pi(W)$ is Borel in the quotient Borel structure. Thus E and X/G are Borel isomorphic, and hence X/G is standard.

Next we show (m) implies (l). Assuming (m), we know X/G is standard. Let E be a Borel cross-section for X/G and define $\phi :$ $X/G \to E$ by $\phi(\pi(x)) = y$ if $y \in E$ and $\pi(x) = \pi(y)$. Note ϕ is Borel for ϕ^{-1} is Borel since $\phi^{-1}(y) = \pi(y)$ is a Borel isomorphism from E onto X/G.

Note (l) implies (g), for if $\phi : X/G \to X$ is a Borel transversal, then since X is countably separated and ϕ is one-to-one, X/G is countably separated.

Finally (k) implies (g) for standard spaces are always countably generated. $\square$

PROPOSITION VI.12. *Let $\mathcal{A}$ be a separable C^* algebra, and let $\mathbb{H}$ be a separable Hilbert space. Then the action of $\mathcal{U}(\mathbb{H})$ on $\mathcal{A}^i_{\mathbb{H}}$ satisfies condition D.*

PROOF. We need to show for each open neighborhood N of I in $\mathcal{U}(\mathbb{H})$, there is an open neighborhood M of I such that for any decreasing neighborhood base Q_m of open sets at $\pi \in \mathcal{A}^i_{\mathbb{H}}$, one has $\cap \overline{Q_m M} \subseteq \pi \cdot N$. But the weak topology on $\mathcal{U}(\mathbb{H})$ is the same as the strong $*$ topology on $\mathcal{U}(\mathbb{H})$. Indeed, suppose $g_n \to g$ weakly as $n \to \infty$. Then

$$||g_n v - gv||^2 = (g_n v, g_n v) - (g_n v, gv) - (gv, g_n v) + (gv, gv)$$
$$= (v, v) - (g_n v, gv) - (gv, g_n v) + (v, v) \to 0$$

and

$$||g_n^* v - g^* v||^2 = (g_n^* v, g_n^* v) - (g_n^* v, g^* v) - (g^* v, g_n^* v) + (g^* v, g^* v)$$
$$= (v, v) - (v, g_n g^* v) - (g_n g^* v, v) + (v, v) \to 0$$

as $n \to \infty$. Hence we may suppose $N = N_\epsilon(v_1, \cdots, v_n, w_1, \cdots, w_n) = \{g \in \mathcal{U}(\mathbb{H}) : |\langle (g - I)v_i, w_i \rangle| < \epsilon \text{ for } i = 1, 2, \cdots, n\}$ where $\epsilon > 0$ and $||v_i|| = ||w_i|| = 1$ for $i = 1, 2, \cdots, n$. Set

$$M = N_{\frac{\epsilon}{3}}(v_1, \cdots, v_n, w_1, \cdots, w_n) \cap \{g : |\langle gv_1, v_1 \rangle > 1 - \frac{\epsilon}{3}\}.$$

Let $\pi' \in \cap \overline{Q_m M}$. Since $\mathcal{A}^i_{\mathbb{H}}$ is Polish, it is first countable; and thus we may select $\pi_m \in Q_m$ and $g_m \in M$ such that $\pi_m \cdot g_m \to \pi'$ in $\mathcal{A}^i_{\mathbb{H}}$. Since the unit ball in $B(\mathbb{H})$ is weakly compact, by taking a subsequence we may suppose that g_m converges weakly to a $T \in$

$B(\mathbb{H})_1$. We have $g_m^{-1}\pi_m(a)g_m$ converges strongly to $\pi'(a)$ for each $a \in \mathcal{A}$. Since the sequence g_m is bounded and converges weakly to T, one has

$$|\langle \pi_m(a)g_m v, w\rangle - \langle T\pi'(a)v, w\rangle| = |\langle g_m g_m^{-1}\pi_m(a)g_m v, w\rangle - \langle T\pi'(a)v, w\rangle|$$
$$\leq |\langle g_m g_m^{-1}\pi_m(a)g_m v, w\rangle - \langle g_m \pi'(a)v, w\rangle| + |\langle g_m \pi'(a)v, w\rangle - |\langle T\pi'(a)v, w\rangle|$$
$$\leq \|g_m\| \, \|(g_m^{-1}\pi_m(a)g_m - \pi'(a))v\| \, \|w\| + |\langle g_m \pi'(a)v, w\rangle - \langle T\pi'(a)v, w\rangle|$$
$$\to 0 \text{ as } n \to \infty.$$

Similarly, since $\pi_m(a)$ converges to $\pi(a)$ strongly and $\|g_m\| = 1$, $\pi_m(a)g_m$ converges to $\pi(a)T$ weakly. Hence

$$\pi(a)T = T\pi'(a)$$

for all a. Since π and π' are irreducible and $|\langle Tv_1, v_1\rangle| \geq 1 - \frac{\epsilon}{3}$, Proposition II.11 implies $T^*T = TT^* = c^2 I$ where $1 \geq c = \|T\| > 0$. Hence $T = cV$ where V is a unitary operator on $\mathbb{H}$; and since $|\langle Tv_1, v_1\rangle| \geq 1 - \frac{\epsilon}{3}$, we see $c \geq 1 - \frac{\epsilon}{3}$. This implies $V^{-1}\pi(a)V = \pi \cdot V(a) = \pi'(a)$. Since $g_m \in M$,

$$|\langle (V - I)v_i, w_i\rangle| \leq |\langle (V - T)v_i, w_i\rangle| + |\langle (T - I)v_i, w_i\rangle|$$
$$\leq \|V - T\| + \lim_{m \to \infty} |\langle (g_m - I)v_i, w_i\rangle|$$
$$\leq |1 - c| + \frac{\epsilon}{3} = \frac{2\epsilon}{3} < \epsilon \text{ for } i = 1, 2, \dots, n.$$

Thus $V \in N$ and $\pi' \in \pi \cdot N$. $\square$

12. The Equivalence of Smooth and Type I

THEOREM VI.12. *Suppose G is a second countable locally compact group. Then the following are equivalent:*

(a) $\hat{G}_\infty$ *is countably separated*

(b) $\hat{G}$ *is a standard Borel space*

(c) $C^*(G)$ *is a type I C^* algebra*

(d) $\hat{C}^*(G)$ *is T_0 in the dual space topology.*

PROOF. Let $\mathbb{H}$ be a separable Hilbert. Let $\mathcal{A} = C^*(G)$ and set $\mathcal{A}_\infty^i = \mathcal{A}_\mathbb{H}^i = \mathrm{Hom}(\mathcal{A}, B(\mathbb{H}))^i$ where $\mathcal{A}_\infty^i$ has the topology of pointwise convergence in the weak operator topology. By Corollary IV.9, the mapping $\pi \mapsto \pi'$ from G_∞^i into $\mathcal{A}_\infty^i$ defined by $\pi'(\tau(f)) = \int f(g)\,\pi(g)\,dg$ is a one-to-one onto mapping. Also the inverse mapping $\pi' \mapsto \pi$ satisfies $\pi' \mapsto \pi(f)$ is strongly Borel for each $f \in L^1(G)$. By Proposition IV.23, the inverse mapping is Borel. Kuratowski's Theorem implies the mapping $\pi \mapsto \pi'$ is a Borel isomorphism from

G_∞^i onto $\mathcal{A}_\infty^i$. This isomorphism preserves the action of the unitary group $\mathcal{U} = \mathcal{U}(\mathbb{H})$. Since these are both Borel actions, it follows that the quotient Borel structures on $\hat{G}_\infty = G_\infty^i/\mathcal{U}$ and $\hat{\mathcal{A}}_\infty = \mathcal{A}_\infty^i/\mathcal{U}$ are Borel isomorphic. For finite n, we know by the remarks after Theorem IV.16 that $\hat{G}_n$ is a standard Borel space. By Proposition VI.12 and Theorems VI.10 and VI.11, $\hat{\mathcal{A}}_\infty$ is countably separated iff $\hat{\mathcal{A}}_\infty$ is a standard Borel space. Thus $\hat{G}_\infty$ is a standard Borel space iff it is a countably separated Borel space. Hence (a) and (b) are equivalent.

To see (b) implies (d), first note by Corollary IV.7 that $\hat{\mathcal{A}}_n = \mathcal{A}_n^i/\mathcal{U}_n$ has a Borel cross section when n is finite, for the unitary group of a finite dimensional Hilbert space is compact. It follows by Theorems VI.9 and VI.11 that $\hat{\mathcal{A}}_n$ is T_0 for $n < \infty$. Hence $\hat{\mathcal{A}}$ is T_0 in the dual space topology iff $\hat{\mathcal{A}}_\infty$ is T_0. But (b) implies $\hat{G}_\infty$ is standard, and thus $\hat{\mathcal{A}}_\infty$ is standard and hence T_0 by Theorem VI.11.

To see (d) implies (c), it suffices by Proposition III.24 to show all primary representations π of $\mathcal{A}$ on a separable Hilbert space are type I. Let $\pi = \int^\oplus \pi_s \, d\mu(s)$ be a direct integral decomposition into irreducible representations. Using Theorem III.12 and Proposition III.22, we may assume the Hilbert bundle is trivial, and hence all π_s live on the same Hilbert space $\mathbb{H}$. By Proposition III.23, $\ker \pi_s = \ker \pi$ for μ a.e. s. Since $\hat{\mathcal{A}}$ is T_0, it follows from Corollary VI.5 that there is an irreducible representation π_0 with $\pi_s \cong \pi_0$ for μ a.e. s. By Lemma VI.12, π is unitarily equivalent to $n\pi_0$ where $n = \dim L^2(S, \mu)$. Thus π is type I.

Finally to see (c) implies (b), it suffices to show $\hat{\mathcal{A}}_\infty$ is countably separated. Suppose not. Then by Theorem VI.10, $\mathcal{A}_\infty^i$ has a nontrivial ergodic measure μ. Theorem VI.8 implies the representation $\int_{\mathcal{A}_\infty^i} \gamma \, d\mu(\gamma)$ is not type I. This is a contradiction. $\square$

THEOREM VI.13. *Let $\mathcal{A}$ be a separable C^* algebra. Then the following are equivalent:*

(a) $\hat{\mathcal{A}}_\infty$ *is countably separated*
(b) $\hat{\mathcal{A}}$ *is a standard Borel space*
(c) $\mathcal{A}$ *is a type I*
(d) $\hat{\mathcal{A}}$ *is T_0 with the dual space topology.*

13. Representations of Separable Type I C* Algebras

Let $\mathcal{A}$ be a separable type I C^* algebra. We note that the results on unitary representations of second countable locally compact groups and their direct integrals hold for nondegenerate $*$ representations of $\mathcal{A}$ on separable Hilbert spaces. The arguments are almost carbon copies of those presented in Sections 7 and 12 of Chapter IV. In particular, Theorems IV.10, IV.22 and IV.23 imply the following theorem.

THEOREM VI.14. *Let $\mathcal{A}$ be a separable C^* algebra which is type I, and let π be a nondegenerate $*$ representation of $\mathcal{A}$ on a separable Hilbert space $\mathbb{H}$. Then there exist a finite measure μ on $\hat{A}$, a Borel function $n : \hat{A} \to \{\infty, 1, 2, \ldots\}$, a Hilbert bundle $\gamma \mapsto \mathcal{H}_\gamma$ over $\hat{A}$, and a Borel field of irreducible $*$-representations $\gamma \mapsto \pi_\gamma$ with $\pi_\gamma \in \gamma$ for μ a.e. γ such that π is unitarily equivalent to*

$$\int^\oplus n(\gamma)\pi_\gamma \, d\mu(\gamma).$$

Moreover, the measure class of μ is uniquely determined by π, and the function $\gamma \mapsto n(\gamma)$ is unique a.e. γ.

CHAPTER VII

LEFT HILBERT ALGEBRAS

1. Hilbert–Schmidt Operators

DEFINITION. Let $\mathbb{H}$ and $\mathbb{K}$ be Hilbert spaces. A linear operator T from $\mathbb{H}$ into $\mathbb{K}$ is said to be **Hilbert–Schmidt** if there is an orthonormal basis e_α of $\mathbb{H}$ such that $\sum \|Te_\alpha\|^2 < \infty$.

PROPOSITION VII.1. *Let T be a Hilbert–Schmidt operator. Then $\sum \|Te_\alpha\|^2$ is independent of orthonormal basis. Moreover, if f_β is an orthonormal basis of $\mathbb{K}$, then $\sum \|Te_\alpha\|^2 = \sum_{\alpha,\beta} |\langle Te_\alpha, f_\beta\rangle|^2$.*

PROOF. Let f_β be an orthonormal basis of $\mathbb{K}$. Then

$$\sum \|Te_\alpha\|^2 = \sum_{\alpha,\beta} |\langle Te_\alpha, f_\beta\rangle|^2 = \sum_{\alpha,\beta} |\langle e_\alpha, T^*f_\beta\rangle|^2 = \sum_\beta \|T^*f_\beta\|^2.$$

$\square$

DEFINITION. Let T be a Hilbert–Schmidt operator from $\mathbb{H}$ into $\mathbb{K}$. Define $\|T\|_2^2 = \sum_\alpha \|Te_\alpha\|^2 = \sum_{\alpha,\beta} |\langle Te_\alpha, f_\beta\rangle|^2 = \sum_\beta \|T^*f_\beta\|^2$. Then $\|T\|_2$ is called the Hilbert–Schmidt norm of T.

COROLLARY VII.1. *Let T be a Hilbert–Schmidt operator from $\mathbb{H}$ to $\mathbb{K}$. Then T^* is a Hilbert–Schmidt operator from $\mathbb{K}$ to $\mathbb{H}$ and $\|T^*\|_2 = \|T\|_2$. Moreover, if U is a bounded linear operator on $\mathbb{K}$ and V is a bounded linear operator on $\mathbb{H}$, then UTV is Hilbert–Schmidt and $\|UTV\|_2 \le \|U\|\,\|T\|_2\|V\|$.*

PROOF. We only need to prove the last statement. First note

$$\|UT\|_2^2 = \sum \|UTe_\alpha\|^2 \le \|U\|^2 \sum \|Te_\alpha\|^2,$$

265

and thus $||UT||_2 \leq ||U|| \, ||T||_2$. Furthermore, $||TV||_2 = ||V^*T^*||_2 \leq ||V^*|| \, ||T^*||_2 = ||V|| \, ||T||_2$. Putting these together gives $||UTV||_2 \leq ||U|| \, ||T||_2 ||V||$. $\square$

THEOREM VII.1. *Let* $\mathcal{HS}(\mathbb{H}, \mathbb{K})$ *be the set of Hilbert–Schmidt operators from* $\mathbb{H}$ *into* $\mathbb{K}$. *Then* $\mathcal{HS}(\mathbb{H}, \mathbb{K})$ *is a Banach space under the Hilbert–Schmidt norm. Moreover,* $||T|| \leq ||T||_2$.

PROOF. $||T||_2 = 0$ iff $T = 0$. Also

$$\begin{aligned} ||T_1 + T_2||_2^2 &= \sum ||T_1 e_\alpha + T_2 e_\alpha||^2 \\ &\leq \sum \left(||T_1 e_\alpha||^2 + 2||T_1 e_\alpha|| \, ||T_2 e_\alpha|| + ||T_2 e_\alpha||^2 \right) \\ &\leq ||T_1||_2^2 + 2||T_1||_2 ||T_2||_2 + ||T_2||^2 \end{aligned}$$

by the Cauchy–Schwarz inequality. Thus $||T_1 + T_2||_2 \leq ||T_1||_2 + ||T_2||_2$, and $\mathcal{HS}(\mathbb{H}, \mathbb{K})$ is a normed linear space.

Let v be a vector in $\mathbb{H}$ with $||v|| = 1$. Then $||T||_2 \geq ||Tv||$. Hence $||T||_2 \geq ||T||$. In particular, if T_n is Cauchy in $\mathcal{HS}(\mathbb{H}, \mathbb{K})$, then there is a bounded operator T from $\mathbb{H}$ into $\mathbb{K}$ such that $T_n \to T$ in norm. We claim $T_n \to T$ in $\mathcal{HS}(\mathbb{H}, \mathbb{K})$. Indeed, let $\epsilon > 0$. Choose N such that $m, n > N$ imply $||T_m - T_n||_2^2 \leq \epsilon^2$. Let e_α be a orthonormal basis of $\mathbb{H}$. Let F be any finite subset of the α. Thus for $m, n > N$, we have

$$\sum_{\alpha \in F} ||T_m e_\alpha - T_n e_\alpha||^2 \leq \epsilon^2.$$

Let $n \to \infty$. Then for $m > N$ we see $\sum_{\alpha \in F} ||T_m e_\alpha - T e_\alpha||^2 \leq \epsilon^2$, and thus $\sum_\alpha ||T_m e_\alpha - T e_\alpha||^2 \leq \epsilon^2$. Hence for $m > N$, we see $T_m - T \in \mathcal{HS}(\mathbb{H}, \mathbb{K})$; and since $T_m \in \mathcal{HS}(\mathbb{H}, \mathbb{K})$, we have $T \in \mathcal{HS}(\mathbb{H}, \mathbb{K})$. Moreover, $||T_n - T||_2 \leq \epsilon$ for $n > N$. $\square$

LEMMA VII.1. *Let* B *be a Banach space. Suppose the norm satisfies*

$$||x + y||^2 + ||x - y||^2 = 2(||x||^2 + ||y||^2) \text{ (the parallelogram law).}$$

Then B *is a Hilbert space with inner product*

$$\langle x, y \rangle = \frac{1}{4} \sum_{j=0}^{3} i^j ||x + i^j y||^2.$$

PROOF. First note $\overline{\langle x, y \rangle} = \langle y, x \rangle$ for

$$\sum_{j=0}^{3} (-i)^j ||x + i^j y||^2 = \sum_{j=0}^{3} (-i)^j ||(-i)^j x + y||^2.$$

Also

$$||x + x' + y||^2 - ||x + x' - y||^2 =$$
$$\frac{1}{2}(||x + x' + y||^2 + ||x + y - x'||^2 + ||x + x' + y||^2 + ||x - x' - y||^2$$
$$- ||x + x' - y||^2 - ||x - x' - y||^2 - ||x + x' - y||^2 - ||x + y - x'||^2).$$

Hence using the parallelogram law, one sees

$$||x + x' + y||^2 - ||x + x' - y||^2 =$$
$$= ||x + y||^2 + ||x'||^2 + ||x' + y||^2 + ||x||^2$$
$$- ||x - y||^2 - ||x'||^2 - ||x' - y||^2 - ||x||^2$$
$$= ||x + y||^2 + ||x' + y||^2 - ||x - y||^2 - ||x' - y||^2.$$

Thus

$$4\langle x + x', y \rangle = \sum_{j=0}^{3} i^j ||x + x' + i^j y||^2$$
$$= (||x + x' + y||^2 - ||x + x' - y||^2) + i(||x + x' + iy||^2 - ||x + x' - iy||^2)$$
$$= (||x + y||^2 - ||x - y||^2) + i(||x + iy||^2 - ||x - iy||^2)$$
$$+ (||x' + y||^2 - ||x' - y||^2) + i(||x' + iy||^2 - ||x' - iy||^2)$$
$$= 4\langle x, y \rangle + 4\langle x', y \rangle.$$

Together these imply $\langle \cdot, \cdot \rangle$ is sesquilinear. Also from the second line in the above steps, one notes by taking $x' = 0$ and $y = x$ that

$$4\text{Re}\langle x, x \rangle = ||2x||^2 = 4||x||^2.$$

But $\langle x, x \rangle$ is real since $\overline{\langle x, x \rangle} = \langle x, x \rangle$. Thus $\langle x, x \rangle = ||x||^2$. □

THEOREM VII.2. $\mathcal{HS}(\mathbb{H}, \mathbb{K})$ *is a Hilbert space, and the inner product is defined by*

$$\langle T_1, T_2 \rangle = \sum \langle T_2^* T_1 e_\alpha, e_\alpha \rangle = \sum \langle T_1 T_2^* f_\beta, f_\beta \rangle.$$

PROOF. Apply Lemma VII.1 to each of the following forms for the Hilbert–Schmidt norm.

$$||T||_2^2 = \sum ||Te_\alpha||^2$$

and

$$||T||_2^2 = \sum ||T^* f_\beta||^2.$$

□

We recall a continuous linear operator T from $\mathbb{H}$ into $\mathbb{K}$ is finite rank if the range of T is a finite dimensional subspace of $\mathbb{K}$.

PROPOSITION VII.2. *The finite rank operators in* $\mathcal{HS}(\mathbb{H}, \mathbb{K})$ *form a dense subspace of* $\mathcal{HS}(\mathbb{H}, \mathbb{K})$.

PROOF. Let $\epsilon > 0$ and suppose $T \in \mathcal{HS}(\mathbb{H}, \mathbb{K})$. Let e_α be an orthonormal basis of $\mathbb{H}$. Choose a finite set F of α such that one has $\sum_{\alpha \notin F} ||Te_\alpha||^2 < \epsilon^2$. Define $T_0 = T$ on the span of the e_α for α in F, and set $T_0 = 0$ on the orthogonal complement of this span. Then $||T - T_0||_2 < \epsilon$. □

PROPOSITION VII.3. $T \in \mathcal{HS}(\mathbb{H}, \mathbb{K})$ *iff* $T^*T \in B_1(\mathbb{H})$ *iff* $TT^* \in B_1(\mathbb{K})$. *Moreover,* $||T||_2 = ||T^*T||_1 = ||TT^*||_1$.

PROOF. Suppose $T^*T \in B_1(\mathbb{H})$. Let $\{e_\alpha\}$ be a complete orthonormal basis of $\mathbb{H}$. By Proposition III.8, $\sum_\alpha \langle T^*Te_\alpha, e_\alpha \rangle < \infty$. Hence $\sum ||Te_\alpha||^2 < \infty$ and $T \in \mathcal{HS}(\mathbb{H}, \mathbb{K})$. Also $||T||_2^2 = \sum ||Te_\alpha||^2 = \sum_\alpha \langle T^*Te_\alpha, e_\alpha \rangle = \mathrm{Tr}(T^*T) = ||T^*T||_1$. Conversely, if $T \in \mathcal{HS}(\mathbb{H}, \mathbb{K})$, then $\sum_\alpha \langle T^*Te_\alpha, e_\alpha \rangle = \sum ||Te_\alpha||^2 < \infty$, and thus by Theorem III.6, T^*T is trace class.

Note $T \in \mathcal{HS}(\mathbb{H}, \mathbb{K})$ iff $T^* \in \mathcal{HS}(\mathbb{K}, \mathbb{H})$; and by what has just been demonstrated, $T^* \in \mathcal{HS}(\mathbb{K}, \mathbb{H})$ iff $TT^* \in B_1(\mathbb{K})$. Moreover, $||T||_2^2 = ||T^*||_2^2 = ||TT^*||_1$. □

Now suppose π is a unitary representation of G on $\mathbb{H}$. Let $\mathbb{H}^*$ be the dual space of $\mathbb{H}$. If $v \in \mathbb{H}$ and ϕ_v is the element in $\mathbb{H}^*$ defined by $\phi_v(w) = \langle w, v \rangle$, then $v \mapsto \phi_v$ is a conjugate linear isomorphism of $\mathbb{H}$ onto $\mathbb{H}^*$; and $\mathbb{H}^*$ equipped with inner product $\langle \phi_v, \phi_w \rangle = \langle w, v \rangle$ is the Hilbert space $\bar{\mathbb{H}}$ described in Section 2 of Chapter III. Define a representation $\bar{\pi}$ of G on the conjugate Hilbert space $\bar{\mathbb{H}}$ by $\bar{\pi}(g)\phi_v(w) = \phi_v(\pi(g^{-1})w)$. We see using the identification of $\bar{\mathbb{H}}$ with $\mathbb{H}$ under the isomorphism $\phi_v \mapsto v$ that $\bar{\pi}(g)v = \pi(g)v$. Indeed:

$$\bar{\pi}(g)\phi_v(w) = \phi_v(\pi(g^{-1})w)$$
$$= \langle \pi(g^{-1})w, v \rangle$$
$$= \langle w, \pi(g)v \rangle$$
$$= \phi_{\pi(g)v}(w).$$

$\bar{\pi}$ is called the conjugate representation of π.

2. Tensor Products

Let $\mathbb{H}$ and $\mathbb{K}$ be Hilbert spaces. Let $\mathcal{HS}(\bar{\mathbb{K}}, \mathbb{H})$ be the Hilbert space of Hilbert–Schmidt operators T from $\bar{\mathbb{K}}$ to $\mathbb{H}$. We denote this space by $\mathbb{H} \otimes \mathbb{K}$ and call it a tensor product of $\mathbb{H}$ and $\mathbb{K}$. By Proposition VII.2, the finite rank operators in $\mathcal{HS}(\bar{\mathbb{K}}, \mathbb{H})$ forms a dense subspace of $\mathbb{H} \otimes \mathbb{K}$. For $v \in \mathbb{H}$ and $w \in \mathbb{K}$, define finite rank operator $v \otimes w$ in $\mathbb{H} \otimes \mathbb{K}$ by $(v \otimes w)(w') = \langle w', w \rangle_{\bar{\mathbb{K}}} v = \langle w, w' \rangle_{\mathbb{K}} v$.

Suppose T is a Hilbert–Schmidt operator from $\mathbb{H}$ to $\mathbb{K}$ with finite dimensional range. Let $e_1, e_2, \cdots, e_n$ be an orthonormal set spanning the range of T. Then $Tw = \sum \langle Tw, e_i \rangle e_i = \sum \langle w, T^* e_i \rangle_{\bar{\mathbb{K}}} e_i = \sum_{i=1}^{n} e_i \otimes T^* e_i(w)$. Hence

$$T = \sum e_i \otimes T^* e_i$$

for any orthonormal set e_i spanning the range of T. In particular, the linear span of the elementary tensors $v \otimes w$ form a dense subspace of $\mathbb{H} \otimes \mathbb{K}$.

THEOREM VII.3. *The mapping $\otimes$ from $\mathbb{H} \times \mathbb{K}$ into $\mathbb{H} \otimes \mathbb{K}$ sending (v, w) to $v \otimes w$ is a continuous bilinear function such that the linear span of $\otimes(\mathbb{H} \times \mathbb{K})$ is dense in $\mathbb{H} \otimes \mathbb{K}$. Additionally, $\langle v \otimes w, v' \otimes w' \rangle = \langle v, v' \rangle \langle w, w' \rangle$. Moreover, if e_α is a complete orthonormal basis of $\mathbb{H}$ and f_β is a complete orthonormal basis of $\mathbb{K}$, then the vectors $e_\alpha \otimes f_\beta$ form a complete orthonormal basis of $\mathbb{H} \otimes \mathbb{K}$.*

PROOF. Let $v_1, v_2 \in \mathbb{H}$ and let $w_1, w_2 \in \mathbb{K}$. By Theorem VII.2,

$$\langle v_1 \otimes w_1, v_2 \otimes w_2 \rangle = \mathrm{Tr}((v_1 \otimes w_1)(v_2 \otimes w_2)^*).$$

But note the operator $T = (v_1 \otimes w_1)(v_2 \otimes w_2)^*$ is zero on the orthogonal complement of v_2 and has range in the linear span of v_1. In fact

$$\langle (v_2 \otimes w_2)w, v \rangle_{\mathbb{H}} = \langle \langle w_2, w \rangle_{\mathbb{K}} v_2, v \rangle_{\mathbb{H}}$$
$$= \langle \langle v_2, v \rangle_{\mathbb{H}} w_2, w \rangle_{\mathbb{K}}$$
$$= \langle w, \langle v_2, v \rangle_{\mathbb{H}} w_2 \rangle_{\bar{\mathbb{K}}}$$
$$= \langle w, w_2 \otimes v_2(v) \rangle_{\bar{\mathbb{K}}}.$$

Thus

$$Tv = (v_1 \otimes w_1)(\langle v_2, v \rangle_{\mathbb{H}} w_2) = \langle v, v_2 \rangle_{\mathbb{H}}(v_1 \otimes w_1)w_2 = \langle v, v_2 \rangle_{\mathbb{H}} \langle w_1, w_2 \rangle_{\mathbb{K}} v_1.$$

Let W be the linear span of v_1 and v_2. Then $Tv_1 = \langle v_1, v_2 \rangle \langle w_1, w_2 \rangle v_1$ and $Tv_2 = ||v_2||^2 \langle w_1, w_2 \rangle v_1$. Thus $\text{Tr}(T) = \langle v_1, v_2 \rangle \langle w_1, w_2 \rangle$, and this gives

$$\langle v_1 \otimes w_2, v_1 \otimes w_2 \rangle = \langle v_1, v_2 \rangle \langle w_1, w_2 \rangle.$$

In particular $||v \otimes w||^2 = ||v||^2 ||w||^2$, and the bilinear mapping $\otimes$ is continuous.

Since the linear span of the vectors $v \otimes w$ are the finite rank operators, the linear span of the range of $\otimes$ is dense. Moreover, if e_α form an orthonormal basis of $\mathbb{H}$ and f_β form an orthonormal basis of $\mathbb{K}$, the closure of the linear span of the vectors $e_\alpha \otimes f_\beta$ contains all vectors $v \otimes w$. But then the span of the vectors $e_\alpha \otimes f_\beta$ is dense in $\mathbb{H} \otimes \mathbb{K}$. Since these vectors are orthonormal, they form a complete orthonormal basis of $\mathbb{H} \otimes \mathbb{K}$. $\square$

COROLLARY VII.2. *If T is a bounded linear operator on $\mathbb{H}$ and S is a bounded linear operator on $\mathbb{K}$, then there is a unique bounded linear operator $T \otimes S$ on $\mathbb{H} \otimes \mathbb{K}$ satisfying $T \otimes S(v \otimes w) = Tv \otimes Sw$ for all v and w. It is defined by $T \otimes S(L) = TL\bar{S}^*$*

PROOF. Define $T \otimes S$ on $\mathcal{HS}(\bar{\mathbb{K}}, \mathbb{H})$ by $T \otimes S(U) = TU\bar{S}^*$ where $\bar{S}$ is the operator on $\bar{\mathbb{K}}$ defined by $\bar{S}w = w$. By Corollary VII.1, $||T \otimes S|| \leq ||T|| \, ||\bar{S}|| = ||T|| \, ||S||$. Thus it is bounded. Moreover,

$$
\begin{aligned}
(T \otimes S)(v \otimes w)(w') &= T(v \otimes w)\bar{S}^* w' \\
&= T(\langle \bar{S}^* w', w \rangle_{\bar{\mathbb{K}}} v) \\
&= \langle w', \bar{S}w \rangle_{\bar{\mathbb{K}}} Tv \\
&= \langle Sw, w' \rangle_{\mathbb{K}} Tv \\
&= (Tv \otimes Sw)(w').
\end{aligned}
$$

$\square$

REMARK. The space $\mathbb{H} \otimes \mathbb{K}$ is the set of all bounded continuous operators T from $\bar{\mathbb{K}}$ into $\mathbb{H}$ such that TT^* is a trace class operator on $\mathbb{H}$. $\mathbb{H} \otimes \bar{\mathbb{H}}$ is the set of Hilbert–Schmidt operators on $\mathbb{H}$.

DEFINITION. Let π and π' be unitary representations of G on Hilbert spaces $\mathbb{H}$ and $\mathbb{K}$ respectively. Define a unitary representation $\pi \otimes \pi'$ on $\mathbb{H} \otimes \mathbb{K}$ by $(\pi \otimes \pi')(g) = \pi(g) \otimes \pi'(g)$ for $g \in G$.

Note each operator $\pi \otimes \pi'(g)$ is unitary, for it preserves the inner product and has inverse $\pi \otimes \pi'(g^{-1})$. They are also strongly continuous since $g \mapsto \pi(g)v \otimes \pi'(g)v'$ is continuous for each $v \in \mathbb{H}$ and $v' \in \mathbb{K}'$. (The span of the $v \otimes v'$ forms a dense subspace.)

3. Left Hilbert Algebras and Examples

DEFINITION. Let $\mathbb{H}$ be a Hilbert space. Let $\mathcal{A} \subset \mathbb{H}$ be an associative $\#$ algebra over $\mathbb{C}$ which is a dense subspace of $\mathbb{H}$. Then $\mathcal{A}$ is a left Hilbert algebra if the following hold:

(a) $\xi \mapsto \eta\xi$ is a bounded linear operator on $\mathcal{A}$ for all $\eta \in \mathcal{A}$,
(b) $\langle \eta\xi, \xi' \rangle = \langle \xi, \eta^{\#}\xi' \rangle$ for all $\eta, \xi, \xi' \in \mathcal{A}$,
(c) $\mathcal{A}^2$, the linear span of the set $\{\eta\xi : \eta, \xi \in \mathcal{A}\}$, is $\mathbb{H}$ dense in $\mathcal{A}$, and
(d) $\xi \mapsto \xi^{\#}$ has a closed extension in $\mathbb{H}$.

EXAMPLE (THE HILBERT–SCHMIDT OPERATORS).
Let $\mathbb{H}$ be a Hilbert space. Let $\mathbb{H}_2$ be the Hilbert space of all Hilbert–Schmidt operators on $\mathbb{H}$. Set $\mathcal{A} = \mathbb{H}_2$ and for $T \in \mathbb{H}_2$, let $T^{\#} = T^*$. Take multiplication to be a composition of operators. By Corollary VII.1, $S \mapsto TS$ is bounded on $\mathbb{H}_2$ for each $T \in \mathbb{H}_2$ and $S \mapsto S^{\#}$ is an isometry. An easy computation using Theorem VII.2 shows $\langle ST, T' \rangle_{\mathcal{HS}(\mathbb{H})} = \langle T, S^*T' \rangle_{\mathcal{HS}(\mathbb{H})}$. Finally note $\mathcal{A}^2$ contains all finite rank operators. Indeed, suppose T is a finite rank operator and P is the orthogonal projection onto the range of T. Then $P, T \in \mathcal{A}$ and $T = PT \in \mathcal{A}^2$. Hence Proposition VII.2 shows $\mathcal{A}^2$ is dense in $\mathcal{A} = \mathbb{H}_2$. Concluding we see $\mathcal{HS}(\mathbb{H})$ is a left Hilbert algebra in $\mathcal{HS}(\mathbb{H})$.

EXAMPLE $(C_c(G))$. Let G be a locally compact group. Take $\mathbb{H}$ to be $L^2(G)$ where G has left Haar measure. Set $\mathcal{A} = C_c(G)$, and set $f^{\#}(x) = \overline{f(x^{-1})}\Delta(x^{-1})$. To see (a), let $f, g \in C_c(G)$. Then

$$\|f * g\|_2^2 = \int |f * g(x)|^2 \, dx$$

$$= \int \left| \int f(y)g(y^{-1}x) \, dy \right|^2 dx$$

$$= \int \left| \int f(y)g(x) \, dy \right|^2 dx$$

$$= \left| \int f(y) \, dy \right|^2 \int |g(x)|^2 \, dx$$

$$\leq \int |f(y)|^2 \, dy \int |g(x)|^2 \, dx$$

$$= \|f\|_2^2 \|g\|_2^2,$$

and hence $||f * g||_2 \leq ||f||_2\,||g||_2$. For (b) note

$$\langle f * g, h \rangle = \int f * g(x)\bar{h}(x)\, dx$$

$$= \iint f(y)g(y^{-1}x)\bar{h}(x)\, dy\, dx$$

$$= \iint f(y)g(x)\bar{h}(yx)\, dx\, dy$$

$$= \iint g(x)f(y^{-1})\Delta(y^{-1})\bar{h}(y^{-1}x)\, dy\, dx$$

$$= \iint g(x)\overline{\bar{f}(y^{-1})\Delta(y^{-1})}h(y^{-1}x)\, dy\, dx$$

$$= \int g(x)\overline{\int f^{\#}(y)h(y^{-1}x)\, dy}\, dx$$

$$= \langle g, f^{\#} * h \rangle.$$

Since $\mathcal{A}$ has approximate units, (c) is true. To see (d), define * to be the involution with domain $D = \{f \in L^2(G) : f^* \in L^2\}$ where $f^*(x) = \bar{f}(x^{-1})\Delta(x^{-1})$. Clearly * extends $^{\#}$; and if $f_i \to f$ in L^2 and $f_i^* \to g$ in L^2, then $f^* = g$. Thus * is closed.

EXAMPLE. Let $\mathbb{H}$ be a finite dimensional Hilbert space. Let $\mathcal{L}(\mathbb{H})$ be the space of linear transformations of $\mathbb{H}$. Set $\phi(A) = \operatorname{Tr}(A) = \sum a_{i,i}$ for $A \in \mathcal{L}(\mathbb{H})$. Note ϕ is a faithful finite trace for $A \geq 0$ and $\phi(A) = 0$ implies $A = 0$. For each positive invertible operator K in $\mathcal{L}(\mathbb{H})$, define $\phi_K(A) = \phi(KA)$. Then ϕ_K is a positive linear functional. Indeed, if $A \geq 0$, then $\phi_K(A) = \phi(K^{1/2}AK^{1/2}) \geq 0$. Note ϕ_K is not a trace; however it is faithful since $K^{1/2}AK^{1/2} \neq 0$ iff $A \neq 0$. Now let ψ be any positive faithful functional on $\mathcal{L}(\mathbb{H})$. Let π_ψ be the representation of $\mathcal{L}(\mathbb{H})$ obtained by the GNS construction, and let ξ_ψ be the corresponding cyclic vector satisfying $\psi(A) = \langle \pi_\psi(A)\xi_\psi, \xi_\psi \rangle$. Note π_ψ is finite dimensional and $N = \{X : \psi(X^*X) = 0\} = \{0\}$. Thus $\xi_\psi = I + \{0\}$ and $\mathbb{H}_\psi = \mathcal{L}(\mathbb{H})/\{0\} = \mathcal{L}(\mathbb{H})$. Hence $\dim \mathbb{H}_\psi = (\dim \mathbb{H})^2$. Hence π_ψ is an isomorphism. Indeed, $A \mapsto \pi_\psi(A)\xi_\psi$ is a linear onto mapping between vector spaces of the same dimension. Hence $A = 0$ iff $\pi_\psi(A)\xi_\psi = 0$. Thus $\pi_\psi(A) = 0$ implies $A = 0$, and ξ_ψ is a separating vector for $\pi_\psi(\mathcal{L}(\mathbb{H}))$. In particular $\mathcal{A} = \pi_\psi(\mathcal{L}(\mathbb{H}))\xi_\psi$ is a Hilbert space with algebra structure copied from $\mathcal{L}(\mathbb{H})$. Define

$(\pi(A)\xi_\psi)^\# = \pi(A^*)\xi_\psi$. (a) is obvious. For (b) note

$$\langle \pi(A)\xi_\psi \pi(B)\xi_\psi, \pi(C)\xi_\psi \rangle = \langle \pi(B)\xi_\psi, \pi(A^*)\pi(C)\xi_\psi \rangle$$
$$= \langle \pi(B)\xi_\psi, \pi(A^*)\xi_\psi \pi(C)\xi_\psi \rangle$$
$$= \langle \pi(B)\xi_\psi, (\pi(A)\xi_\psi)^\# \pi(C)\xi_\psi \rangle.$$

Clearly (c) holds since $\mathcal{A}^2 = \mathcal{A}$, and (d) follows by finite dimensionality. Hence every faithful positive linear functional ψ of $\mathcal{L}(\mathbb{H})$ where $\mathbb{H}$ is finite dimensional makes $\mathbb{H}_\psi$ into a left Hilbert algebra.

EXAMPLE. Let $\mathcal{M}$ be a nondegenerate von Neumann algebra on a Hilbert space $\mathbb{H}$. Let ξ_0 be a separating cyclic vector for $\mathcal{M}$. Set $\mathcal{A} = \{T\xi_0 : T \in \mathcal{M}\}$. Define multiplication by

$$T\xi_0\, S\xi_0 = TS\xi_0 \text{ and } \# \text{ by}$$

$$(T\xi_0)^\# = T^*\xi_0.$$

These are well defined for ξ_0 is a separating vector. Since $\|T\xi_0\, S\xi_0\| = \|TS\xi_0\| \le \|T\|\,\|S\xi_0\|$, (a) holds. Note (b) is clear, and (c) follows from $I \in \mathcal{M}$. For (d), it suffices to show the adjoint of $\#$ is densely defined. By Lemma III.8, since ξ_0 is a separating vector for $\mathcal{M}$, it is a cyclic vector for $\mathcal{M}'$. Hence $\mathcal{M}'\xi_0$ is dense in $\mathbb{H}$. Define $^\flat$ on $\mathcal{M}'\xi_0$ by $(T'\xi_0)^\flat = T'^*\xi_0$. Then the adjoint of $\#$ extends $^\flat$. Indeed,

$$\langle (T\xi_0)^\#, S'\xi_0 \rangle = \langle T^*\xi_0, S'\xi_0 \rangle$$
$$= \langle \xi_0, TS'\xi_0 \rangle$$
$$= \langle \xi_0, S'T\xi_0 \rangle$$
$$= \langle S'^*\xi_0, T\xi_0 \rangle.$$

Thus $\mathcal{A}$ is a left Hilbert algebra.

EXAMPLE. Let G be a discrete group. Set $\mathbb{H} = L^2(G)$ and $\mathcal{M} = L(G)''$ where L is the left regular representation. Set $\xi_0 = 1_{\{e\}}$. Then $\mathcal{M}' = R(G)''$ where R is the right regular representation. We claim ξ_0 is a separating cyclic vector for the von Neumann algebra $\mathcal{M}$. First note $f * \xi_0 = f$ for $f \in L^1(G)$. Thus ξ_0 is cyclic for $\mathcal{M}$. Similarly $\xi_0 * f = f$, and thus ξ_0 is cyclic for $R(G)''$. Thus ξ_0 is separating for $\mathcal{M}$.

DEFINITION. A weight ϕ on a von Neumann algebra $\mathcal{M}$ is a function $\phi : \mathcal{M}^+ \to [0, \infty]$ satisfying

$$\phi(S + T) = \phi(S) + \phi(T)$$
$$\phi(\lambda S) = \lambda\phi(S) \text{ for } \lambda \ge 0.$$

The weight ϕ is faithful if $\phi(A) = 0$ implies $A = 0$. It is semifinite if the linear span of all $A > 0$ with $\phi(A) < \infty$ is weakly dense in $\mathcal{M}$.

One can do the Gelfand-Naimark-Segal construction with weights on von Neumann algebras. Namely, set $\mathcal{M}_\phi = \{X : \phi(X^*X) < \infty\}$ and $\mathcal{N}_\phi = \{X : \phi(X^*X) = 0\}$. Then $\mathcal{M}_\phi/\mathcal{N}_\phi$ is a pre-Hilbert space, and one can define a representation π_ϕ on this space by $\pi_\phi(X)(Y + \mathcal{N}_\phi) = XY + \mathcal{N}_\phi$. Define a multiplication and an adjoint by $(X + \mathcal{N}_\phi)(Y + \mathcal{N}_\phi) = XY + \mathcal{N}_\phi$ and $(X + \mathcal{N}_\phi)^{\#} = X^* + \mathcal{N}_\phi$. We shall see this will give a left Hilbert algebra. If ϕ is faithful, one will have $\mathcal{M}$ as the von Neumann algebra generated by this left Hilbert algebra.

4. Properties of Left Hilbert Algebras

DEFINITION. Let $\mathcal{A}$ be a left Hilbert algebra. For each ξ in $\mathcal{A}$, let $\pi(\xi)$ be the bounded operator on $\mathbb{H}$ extending $\eta \mapsto \xi\eta$. Since $\mathcal{A}^2$ is dense in $\mathbb{H}$, we know $\pi(\mathcal{A})\mathbb{H}$ is dense in $\mathbb{H}$; i.e., π is nondegenerate. The left ring $\mathcal{L}(\mathcal{A})$ of $\mathcal{A}$ is the von Neumann algebra generated by $\pi(\mathcal{A})$; i.e., $\overline{\pi(\mathcal{A})}^w = \pi(\mathcal{A})''$.

PROPOSITION VII.4. π is a faithful nondegenerate $*$-representation of $\mathcal{A}$.

PROOF. We note π is linear. To see $\pi(a^{\#}) = \pi(a)^*$, note

$$\langle \pi(a)b, c \rangle = \langle ab, c \rangle$$
$$= \langle b, a^{\#}c \rangle$$
$$= \langle b, \pi(a^{\#})c \rangle$$

for $a, b, c \in \mathcal{A}$. Since $\mathcal{A}$ is dense in $\mathbb{H}$ and $\pi(a)$ is a bounded linear operator, $\pi(a)^* = \pi(a^{\#})$. Finally to see π is faithful, suppose $\pi(a) = 0$. Then $ab = 0$ for all $b \in \mathcal{A}$. Taking adjoints implies $b^{\#}a^{\#} = 0$ for all b. Hence $\langle a^{\#}, bc \rangle = \langle b^{\#}a^{\#}, c \rangle = 0$ for all $b, c \in \mathcal{A}$. Since $\mathcal{A}^2$ is dense in $\mathbb{H}$, $a = (a^{\#})^{\#} = 0$. $\square$

DEFINITION. Let S_0 denote the map $\xi \mapsto \xi^{\#}$ on $\mathcal{A}$. Let $S_{00} = S_0|_{\mathcal{A}^2}$. (Recall $\mathcal{A}^2$ is the linear span of all products $\xi\eta$ where $\xi, \eta \in \mathcal{A}$.) Let S be the closure of S_{00}. It exists since S_0 has closure.

REMARK. It is not clear that $\mathcal{A}$ is a subset of the domain of S. This will be established later.

Let T be a conjugate linear transformation of a Hilbert space $\mathbb{H}$. Then T^* is defined by $\eta \in \text{Dom}\, T^*$ if $\xi \mapsto \langle \eta, T\xi \rangle$ is continuous (linear). Moreover, $T^*\eta$ is the vector satisfying $\langle \eta, T\xi \rangle = \langle \xi, T^*\eta \rangle$. Note T^* is conjugate linear.

LEMMA VII.2. *Define $F = S^*$. (F is for "flat" and S is for "sharp".) Set $\Delta = S^*S = FS$. (Δ is the modular operator.) Then Δ is a positive self adjoint linear (generally unbounded) operator. Furthermore*

(i) $F^2 = I$, $S^2 = I$; *i.e.*, $\xi \in \text{Dom}\, F \implies F\xi \in \text{Dom}\, F$ *and* $FF\xi = \xi$.

(ii) $\langle F\eta, \xi \rangle = \langle S\xi, \eta \rangle$ *if $\eta \in \text{Dom}\, F$ and $\xi \in \text{Dom}\, S$.*

PROOF. (ii) is immediate. We show (i). Let $\eta \in \text{Dom}\, F$. Since S is closed, $S^{**} = S$; and thus $F^* = S$. Now if $\xi \in \mathcal{A}^2$, then

$$\langle F\eta, S\xi \rangle = \langle S^2\xi, \eta \rangle = \langle \xi, \eta \rangle$$

since $S\xi \in \mathcal{A}^2$. Hence $F\eta \in \text{Dom}\, F$ and $F^2\eta = \eta$.

Also by (ii), we have $\langle F\eta, S\xi \rangle = \langle \xi, F^2\eta \rangle$ if $\eta \in \text{Dom}\, F$ and $\xi \in \text{Dom}\, S$. Thus $\langle F\eta, S\xi \rangle = \langle \xi, \eta \rangle$. Hence $S\xi \in \text{Dom}\, F^*$ and $\langle F^*S\xi, \eta \rangle = \langle \xi, \eta \rangle$. Hence $F^*S\xi = \xi$. This gives $S^2\xi = \xi$.

Finally recall S^*S is self adjoint and positive. $\square$

COROLLARY VII.3. *0 is not in the point spectrum of Δ, and $\Delta^{-1} = SF$.*

PROOF. $\Delta\xi = 0$ implies $S^*S\xi = 0$. Hence $FS\xi = 0$. Thus $FFS\xi = 0$. Hence $S\xi = 0$. Thus $\xi = SS\xi = 0$. Let $\xi \in \text{Dom}\, \Delta$. Then $\Delta\xi = FS\xi$. Thus $FFS\xi = S\xi \in \text{Dom}\, S$. Therefore $SFFS\xi = \xi$. Thus $SF\Delta\xi = \xi$. Let $\xi \in \text{Dom}\, SF$. Then $SF\xi \in \text{Dom}\, \Delta$ for $SF\xi \in \text{Dom}\, S$. Also $\Delta SF\xi = FSSF\xi = FF\xi = \xi$. $\square$

THEOREM VII.4. *There exists a conjugate linear unitary operator J on $\mathbb{H}$ such that*

(i) $J^2 = I$

(ii) $J\Delta^{1/2} = S = \Delta^{-1/2}J$

(iii) $J\Delta^{-1/2} = F = \Delta^{1/2}J$

(iv) $J\Delta^\alpha J = \Delta^{-\bar{\alpha}}$ *for $\alpha \in \mathbb{C}$*

(v) $JSJ = F$.

PROOF. $\Delta = S^*S$ is a closed positive operator on $\mathbb{H}$; and since S is closed, there is a unique partial isometry $J : \overline{\text{Rang}S^*} \to \overline{\text{Rang}S}$ such that $J\Delta^{1/2} = S$. (Note in this case since S is antilinear, so is

J.) Since Rang$S$ and Rang$F$ are dense in $\mathbb{H}$ and $F = S^*$, we see J is unitary. Since $S = S^{-1}$, we see $S = (J\Delta^{1/2})^{-1} = \Delta^{-1/2}J^{-1} = \Delta^{-1/2}J^*$. Thus $J\Delta^{1/2} = \Delta^{-1/2}J^*$. Hence $J\Delta^{1/2}J = \Delta^{-1/2}$. Now $JJJ^*\Delta^{1/2}J = J\Delta^{1/2}J = \Delta^{-1/2}$. Since the polar decomposition of $\Delta^{-1/2}$ is unique, $J^2 = I$ and $J^*\Delta^{1/2}J = \Delta^{-1/2}$. Thus $\Delta^{-1/2} = J\Delta^{1/2}J$. Thus $\Delta^{-1} = J\Delta J$.

Let E_λ be the resolution of the identity for Δ. Thus $\Delta = \int \lambda dE_\lambda$. Now $J\Delta J = \Delta^{-1} = \int_0^\infty \lambda^{-1} dE_\lambda = \int_0^\infty \lambda dE_{\lambda^{-1}}$. But since $J\Delta J = \int \bar{\lambda} d(JE_\lambda J)$, one has $dE_{\lambda^{-1}} = JdE_\lambda J$. Now $\Delta^\alpha = \int \lambda^\alpha dE_\lambda$. Therefore

$$J\Delta^\alpha J = \int \overline{\lambda^\alpha} JdE_\lambda J = \int \lambda^{\bar{\alpha}} dE_{\lambda^{-1}} = \int \lambda^{-\bar{\alpha}} dE_\lambda = \Delta^{-\bar{\alpha}}.$$

Finally, note $S = J\Delta^{1/2} = \Delta^{-1/2}J$ implies $F = S^* = \Delta^{1/2}J = J\Delta^{-1/2}$. Also $JSJ = J^2\Delta^{1/2}J = J\Delta^{-1/2}J^2 = FJ^2 = F$. $\square$

EXAMPLE ($C_c(G)$ CONTINUED). We have $\mathcal{A} = C_c(G)$ and

$$\xi^\#(x) = \Delta(x^{-1})\bar{\xi}(x^{-1}).$$

Set $\tilde{\xi}(x) = \bar{\xi}(x^{-1})$. Then $\text{Dom}(S) = \{f \in L^2(G) : \Delta^{-1}\tilde{f} \in L^2(G)\}$ and $Sf = \Delta^{-1}\tilde{f}$. Now $\langle Ff, g \rangle = \langle Sg, f \rangle$ for $g \in \text{Dom}(S)$ and $f \in \text{Dom}(F)$. Hence

$$\int Ff(x)\bar{g}(x)\,dx = \int \Delta(x^{-1})\tilde{g}(x)\bar{f}(x)\,dx$$

$$= \int \bar{f}(x^{-1})\tilde{g}(x^{-1})\,dx$$

$$= \int \bar{f}(x^{-1})\bar{g}(x)\,dx.$$

Thus $Ff(x) = \bar{f}(x^{-1})$. Moreover, $\text{Dom}(F) = \{f : \tilde{f} \in L^2(G)\}$. Now since $\Delta = FS$, we see $\Delta f(x) = \Delta(x)f(x)$. Finally for J, note $J\Delta^{1/2} = S$. Hence $J(\Delta^{1/2}f)(x) = \Delta(x^{-1})\bar{f}(x^{-1}) = \Delta(x)^{-1/2}(\Delta^{1/2}f)\tilde{}(x)$. Thus

$$Jh(x) = \Delta(x)^{-1/2}\tilde{h}(x).$$

REMARK. π is a * representation of the $^\#$ algebra $\mathcal{A}$.

PROOF. First note $\pi(\xi\eta)\xi' = \xi\eta\xi' = \pi(\xi)\pi(\eta)\xi'$ and

$$\langle \pi(\xi)\eta, \xi' \rangle = \langle \xi\eta, \xi' \rangle$$

$$= \langle \eta, \xi^\#\xi' \rangle$$

$$= \langle \eta, \pi(\xi^\#)\xi' \rangle.$$

Thus $\pi(\xi)^* = \pi(\xi^\#)$. $\square$

5. The Right Hilbert Algebra

DEFINITION. $\mathcal{A}'$, the right Hilbert algebra of $\mathcal{A}$, is defined as $\{\eta \in \text{Dom}(F) : \xi \mapsto \pi(\xi)\eta, \xi \in \mathcal{A}$ is bounded$\}$.

PROPOSITION VII.5. η is in $\mathcal{A}'$ iff there is a bounded operator $A \in \mathcal{L}(\mathbb{H})$ and an element $\eta' \in \mathbb{H}$ such that $A\xi = \pi(\xi)\eta$ and $A^*\xi = \pi(\xi)\eta'$ for $\xi \in \mathcal{A}$. If this is the case, then $F\eta = \eta'$.

PROOF. Assume $\eta \in \mathcal{A}'$. Set $\eta' = F\eta$. Let A be the extension of $\xi \mapsto \pi(\xi)\eta$. Thus $A\xi = \pi(\xi)\eta$ for $\xi \in \mathcal{A}$. Now for $\xi, \xi' \in \mathcal{A}$

$$
\begin{aligned}
\langle A\xi, \xi' \rangle &= \langle \pi(\xi)\eta, \xi' \rangle \\
&= \langle \eta, \pi(\xi^\#)\xi' \rangle \\
&= \langle \eta, \xi^\# \xi' \rangle \\
&= \langle \eta, S(\xi'^\# \xi) \rangle \\
&= \langle \eta, S_{00}(\xi'^\# \xi) \rangle \\
&= \langle \xi'^\# \xi, F\eta \rangle \\
&= \langle \xi, \pi(\xi')F\eta \rangle.
\end{aligned}
$$

Thus $A^*\xi' = \pi(\xi')F\eta$.

Now assume $\eta \in \mathbb{H}$ and $A \in \mathcal{L}(\mathbb{H})$ and $A\xi = \pi(\xi)\eta$, $A^*\xi = \pi(\xi)\eta'$. We claim $\eta \in \text{Dom}(F)$ and $F\eta = \eta'$. Recall $F = S^* = S_{00}^*$. Thus

$$
\begin{aligned}
\langle \eta, \xi_1^\# \xi_2 \rangle &= \langle \eta, \pi(\xi_1)^* \xi_2 \rangle \\
&= \langle \pi(\xi_1)\eta, \xi_2 \rangle \\
&= \langle A\xi_1, \xi_2 \rangle \\
&= \langle \xi_1, A^*\xi_2 \rangle \\
&= \langle \xi_1, \pi(\xi_2)\eta' \rangle \\
&= \langle \xi_2^\# \xi_1, \eta' \rangle.
\end{aligned}
$$

Therefore, $\langle \eta, S(\xi_2^\# \xi_1) \rangle = \langle \xi_2^\# \xi_1, \eta' \rangle$. Hence $\eta \in \text{Dom}(S_{00}^*) = \text{Dom}(S^*)$ and $F\eta = S^*\eta = \eta'$. $\square$

COROLLARY VII.4. Let η_i be a net in $\mathcal{A}'$ such that

$$\sup \|\eta_i\| < \infty \text{ and } \sup \|F\eta_i\| < \infty.$$

Then if $\pi'(\eta_i)$ converges weakly to a bounded operator A, there is a $\eta \in \mathcal{A}'$ such that $\|\eta\| \leq \sup_i \|\eta_i\|$, $\pi'(\eta) = A$ and $\pi'(F\eta) = A^*$.

PROOF. We claim $\lim_i \langle \xi, \eta_i \rangle$ exists for all $\xi \in \mathbb{H}$. Indeed, since η_i is bounded, it suffices to show this limit exists for $\xi \in \mathcal{A}^2$. Let $a, b \in \mathcal{A}$. Then

$$
\begin{aligned}
\lim_i \langle a^\# b, \eta_i \rangle &= \lim_i \langle \pi(a^\#) b, \eta_i \rangle \\
&= \lim_i \langle b, \pi(a) \eta_i \rangle \\
&= \lim_i \langle b, \pi'(\eta_i) a \rangle \\
&= \langle b, Aa \rangle.
\end{aligned}
$$

Hence $\lim_i \langle \xi, \eta_i \rangle$ exists. Similarly, $\lim_i \langle \xi, F\eta_i \rangle$ exists for all ξ, and

$$
\lim_i \langle a^\# b, F\eta_i \rangle = \langle Ab, a \rangle.
$$

Since these are bounded linear functionals, there are η and η' with $\langle a^\# b, \eta \rangle = \langle b, Aa \rangle$ and $\langle a^\# b, \eta' \rangle = \langle Ab, a \rangle$ for all $a, b \in \mathcal{A}$. Thus $\langle b, \pi(a) \eta \rangle = \langle b, Aa \rangle$ and $\langle b, \pi(a) \eta' \rangle = \langle b, A^* a \rangle$. This implies $Aa = \pi(a) \eta$ and $A^* a = \pi(a) \eta'$. Hence $A = \pi'(\eta)$ and $F\eta = \eta'$. Clearly $\|\eta\| \leq \sup_i \|\eta_i\|$. $\square$

DEFINITION. For $\eta \in \mathcal{A}'$, define $\pi'(\eta)$ to be the unique bounded linear operator extending $\xi \mapsto \pi(\xi)\eta$ to all of $\mathbb{H}$.

COROLLARY VII.5. F maps $\mathcal{A}'$ onto $\mathcal{A}'$ and $\pi'(F\eta) = \pi'(\eta)^*$.

PROOF. Let $\eta \in \mathcal{A}'$. By Proposition VII.5, the operators $\pi'(\eta)$ and $\pi'(\eta)^*$ satisfy $\pi'(\eta)\xi = \pi(\xi)\eta$ and $\pi'(\eta)^*\xi = \pi(\xi)F\eta$ for $\xi \in \mathcal{A}$. Proposition VII.5 yields $F\eta \in \mathcal{A}'$ and $\pi'(F\eta) = \pi'(\eta)^*$. $\square$

We will turn $\mathcal{A}'$ into a left Hilbert algebra with F as the involution. All we need to do is define multiplication.

LEMMA VII.3. *Let $\mathcal{A}$ be a left Hilbert algebra.*

(i) *For $\eta \in \mathcal{A}'$, $\pi'(\eta) \in \mathcal{L}(\mathcal{A})' = \pi(\mathcal{A})'$ where $\mathcal{L}(\mathcal{A})$ is the left ring of $\mathcal{A}$.*

(ii) *If $\eta_1, \eta_2 \in \mathcal{A}'$, then $\pi'(\eta_1)\eta_2 \in \mathcal{A}'$ and $F(\pi'(\eta_1)\eta_2) = \pi'(F\eta_2)F\eta_1$.*

(iii) *For $\eta_1, \eta_2 \in \mathcal{A}'$, $\pi'(\pi'(\eta_1)\eta_2) = \pi'(\eta_1)\pi'(\eta_2)$.*

PROOF. For (i) note

$$
\pi'(\eta)\pi(\xi)\xi' = \pi'(\eta)\xi\xi' = \pi(\xi\xi')\eta = \pi(\xi)\pi(\xi')\eta = \pi(\xi)\pi'(\eta)\xi'
$$

if $\xi, \xi' \in \mathcal{A}$. Thus $\pi'(\eta)\pi(\xi) = \pi(\xi)\pi'(\eta)$ for all ξ. Thus $\pi'(\eta) \in \pi(\mathcal{A})'$.

To see (ii), we apply Proposition VII.5. Set $A = \pi'(\eta_1)\pi'(\eta_2)$, $\eta_3 = \pi'(\eta_1)\eta_2$, and $\eta_3' = \pi'(F\eta_2)F\eta_1$. We show $A\xi = \pi(\xi)\eta_3$ and $A^*\xi = \pi(\xi)\eta_3'$. First

$$\pi'(\eta_1)\pi'(\eta_2)\xi = \pi'(\eta_1)\pi(\xi)\eta_2 = \pi(\xi)\pi'(\eta_1)\eta_2 = \pi(\xi)\eta_3$$

and second

$$\pi'(\eta_2)^*\pi'(\eta_1)^*\xi = \pi'(F\eta_2)\pi'(F\eta_1)\xi = \pi'(F\eta_2)\pi(\xi)F\eta_1$$
$$= \pi(\xi)\pi'(F\eta_2)F\eta_1 = \pi(\xi)\eta_3'.$$

This establishes (ii). It also shows $\pi'(\eta_3)\xi = \pi'(\eta_1)\pi'(\eta_2)\xi$ and hence (iii). $\square$

REMARK. We define a multiplication on $\mathcal{A}'$ by $\eta_1\eta_2 = \pi'(\eta_1)\eta_2$. Then $\mathcal{A}'$ is an F algebra. We shall show later that $\mathcal{A}'$ is a left Hilbert algebra —"the right Hilbert algebra of $\mathcal{A}$". One can form $\mathcal{A}''$. It will be shown that $\mathcal{A}'' \supset \mathcal{A}$ and $\mathcal{A}''' = \mathcal{A}'$. Note $\mathcal{A}'' = \{\xi \in \mathrm{Dom}(S) : \eta \mapsto \pi'(\eta)\xi \text{ is bounded}\}$.

EXAMPLE ($C_c(G)$ CONTINUED). Here $\mathcal{A} = C_c(G)$ and

$$\mathcal{A}' = \{f \in \mathrm{Dom}(F) : h \mapsto h * f \text{ is bounded}\}$$
$$= \{f : \tilde{f} \in L^2(G), h \mapsto h * f \text{ is bounded}\}.$$

Note $\mathcal{A}' \supseteq C_c(G)$. Therefore, $\mathcal{A}''$ consists of all $f \in L^2(G)$ such that $f \in \mathrm{Dom}(S)$ and $h \mapsto f*h$ is bounded on $C_c(G)$. Recall $f \in \mathrm{Dom}(S)$ iff $\Delta^{-1}\tilde{f} \in L^2$.

6. Operators Affiliated with a von Neumann Algebra

DEFINITION. Let $\mathcal{M}$ be a von Neumann algebra, T an operator on $\mathbb{H}$, possibly unbounded. Then T is affiliated with $\mathcal{M}$ iff for every $A \in \mathcal{M}'$, $TA \geq AT$ (i.e., $A\mathrm{Dom}\,(T) \subset \mathrm{Dom}(T)$ and $TAx = ATx$ for all $x \in \mathrm{Dom}(T)$).

PROPOSITION VII.6. *Let T be a closed operator on $\mathbb{H}$. Set $|T| = (T^*T)^{1/2}$, and let $T = U|T|$ be its left polar decomposition. (Here U is a partial isometry with initial domain $\overline{\mathrm{Rang}(T^*)}$ and final domain $\overline{\mathrm{Rang}(T)}$.) Let E be the resolution of the identity for $|T|$. Then T is affiliated with $\mathcal{M}$ iff $U \in \mathcal{M}$ and $E(S) \in \mathcal{M}$ for every Borel subset S of $\mathbb{R}^+ \cup \{0\}$.*

PROOF. Assume T is affiliated with $\mathcal{M}$. Then $TA = AT$ for every unitary operator $A \in \mathcal{M}'$. To see this suppose Ax is in the domain of T. We need to show x is in the domain of T. But $TA^{-1} \geq A^{-1}T$, and thus Ax is in the domain of TA^{-1}. Hence x is in the domain of T. Thus if A is unitary, $AU|T| = U|T|A = UAA^*|T|A$. Setting $T' = AT$, we have two polar decompositions of T'; namely

$AU|T|$ and $UA(A^*|T|A)$. Thus $AU = UA$ and $|T| = A^*|T|A$ for all unitary $A \in \mathcal{M}'$. By uniqueness of the resolution of the identity, $A^*E(S)A = E(S)$ for all Borel sets S and all unitary A in $\mathcal{M}'$. This gives $U \in \mathcal{M}$ and $E(S) \in \mathcal{M}$ by the double commutant theorem (unitaries generate $\mathcal{M}$ weakly).

Conversely, assume $U \in \mathcal{M}$ and $E(S) \in \mathcal{M}$ for all Borel subsets S of $\mathbb{R}^+ \cup \{0\}$. Let $A \in \mathcal{M}'$. We show $TA \geq AT$. Suppose x is in the domain of T. Then $\int \lambda^2 \, d\langle E_\lambda x, x \rangle < \infty$ and $Tx = U \lim_{n \to \infty} \int_0^n \lambda \, dE_\lambda x = \lim_n U \int_0^n \lambda \, dE_\lambda x$. Now

$$\int \lambda^2 \langle dE_\lambda Ax, Ax \rangle \leq ||A||^2 \int \lambda^2 \langle dE_\lambda x, x \rangle$$
$$< \infty,$$

for

$$\langle E(S)Ax, Ax \rangle = \langle E(S)Ax, E(S)Ax \rangle = \langle AE(S)x, AE(S)x \rangle$$
$$\leq ||A||^2 \langle E(S)x, x \rangle.$$

Thus $Ax \in \mathrm{Dom}(T)$ and

$$TAx = \lim_n U \int_0^n \lambda \, dE_\lambda Ax = \lim_n AU \int_0^n \lambda \, dE_\lambda x = AU|T|x = ATx.$$

Thus $TA \geq AT$. $\square$

LEMMA VII.4. *Let* $\eta \in Dom(F)$. *Then* $\xi \mapsto \pi(\xi)\eta$ *is defined on* $\mathcal{A}$ *and has closure as an operator on* $\mathbb{H}$. *Denote the closure by* $\pi'(\eta)$. *Then* $\pi'(\eta)$ *is affiliated with* $L(\mathcal{A})'$ *and* $\pi'(\eta)^* \geq \pi'(F\eta)$.

PROOF. Let $\eta \in \mathrm{Dom}(F)$. Set $A\xi = \pi(\xi)\eta$ and $B\xi = \pi(\xi)F\eta$ for $\xi \in \mathcal{A}$. To show A has closure, we show A^* is densely defined. We do this by showing $A^* \geq B$. (Note by symmetry, $B^* \geq A$; and thus $\pi'(F\eta)^* \geq \pi'(\eta)$. Taking adjoints gives $\pi'(\eta)^* \geq \pi'(F\eta)$.) Indeed, for ξ_1 and ξ_2 in $\mathcal{A}$

$$\langle A\xi_1, \xi_2 \rangle = \langle \pi(\xi_1)\eta, \xi_2 \rangle = \langle \eta, \xi_1^\# \xi_2 \rangle = \langle \eta, S(\xi_2^\# \xi_1) \rangle = \langle \xi_2^\# \xi_1, F\eta \rangle$$
$$= \langle \xi_1, \pi(\xi_2)F\eta \rangle = \langle \xi_1, B\xi_2 \rangle.$$

Thus $A^* \geq B$. We now show $\pi'(\eta)$ is associated to $\mathcal{L}(\mathcal{A})'$. We need to show $\pi'(\eta)A \geq A\pi'(\eta)$ for all $A \in \mathcal{L}(\mathcal{A})$. Now $\mathcal{L}(\mathcal{A})$ is the strong closure of $\pi(\mathcal{A})$. Let $\xi \in \mathcal{A}$ and $\eta_1 \in \mathrm{Dom}(\pi'(\eta))$. Then there is a sequence $\xi_n \to \eta_1$ with $\xi_n \in \mathcal{A}$ and $\pi'(\eta)\xi_n = \pi(\xi_n)\eta \to \pi'(\eta)\eta_1$. Thus $\pi(\xi)\xi_n = \xi\xi_n \to \pi(\xi)\eta_1$ and $\pi(\xi)\pi(\xi_n)\eta \to \pi(\xi)\pi'(\eta)\eta_1$. Hence

$$\pi(\xi)\eta_1 \in \mathrm{Dom}(\pi'(\eta)) \text{ and } \pi'(\eta)\pi(\xi)\eta_1 = \pi(\xi)\pi'(\eta)\eta_1.$$

This shows $\pi'(\eta)A \geq A\pi'(\eta)$ for all $A \in \pi(\mathcal{A})$. Now let A be the strong limit of a net $\pi(\xi_i)$. Let $\eta_1 \in \mathrm{Dom}(\pi'(\eta))$. Then $\pi(\xi_i)\eta_1 \in \mathrm{Dom}(\pi'(\eta))$ and $\pi'(\eta)\pi(\xi_i)\eta_1 = \pi(\xi_i)\pi'(\eta)\eta_1$. Now $\pi(\xi_i)\eta_1 \to A\eta_1$ and $\pi'(\eta)\pi(\xi_i)\eta_1 \to A\pi'(\eta)\eta_1$. Thus one has $A\eta_1 \in \mathrm{Dom}(\pi'(\eta))$ and $\pi'(\eta)A\eta_1 = A\pi'(\eta)\eta_1$. Hence $\pi'(\eta)A \geq A\pi'(\eta)$ for all $A \in \mathcal{L}(\mathcal{A})$.   □

7. The Resolvent of the Modular Function Δ

We know Δ is a positive self adjoint unbounded operator on $\mathbb{H}$. Hence for $\lambda \notin [0, \infty)$, $(\Delta - \lambda)^{-1}$ is a bounded operator on $\mathbb{H}$. The same is true of $(\Delta^{-1} - \lambda)^{-1}$.

EXAMPLE $(C_c(G)$ CONTINUED$)$.
Here $\mathcal{A} = C_c(G)$, $\Delta f(x) = \Delta(x)f(x)$, and

$$\mathcal{A}' = \{f \in L^2(G) : \tilde{f} \in L^2 \text{ and } h \mapsto h * f \text{ is bounded on } L^2\}.$$

Now let $g \in C_c(G) = \mathcal{A}$. Let $r > 0$. Then $(\Delta^{-1}+r)^{-1}g(x) = \frac{g(x)}{r+\Delta^{-1}(x)}$ which is in $C_c(G) \subseteq \mathcal{A}'$.

THEOREM VII.5. *Let $r > 0$. Then $(\Delta^{-1} + r)^{-1}\mathcal{A} \subseteq \mathcal{A}'$.*

PROOF. Let $\eta = (\Delta^{-1}+r)^{-1}\xi$ where $\xi \in \mathcal{A}$. Then $\eta \in \mathrm{Dom}(\Delta^{-1/2})$ for

$$\Delta^{-1/2}(\Delta^{-1}+r)^{-1} = \int \lambda^{-1/2}(\lambda^{-1}+r)^{-1}\,dE_\lambda = \int \frac{1}{\lambda^{-1/2}+r\lambda^{1/2}}\,dE_\lambda$$

which is a bounded operator. Since, by Theorem VII.4, $F = J\Delta^{-1/2}$, $\eta \in \mathrm{Dom}(F)$. We thus need only show $\pi'(\eta)$ is bounded. Now $\pi'(\eta)$ is a closed operator and has a polar decomposition UK where $K \geq 0$. We need to show K is bounded. Let $K = \int_0^\infty \lambda\,dE_\lambda$. Set $P = E[n, m]$ where $0 \leq n < m < \infty$. Let $\xi \in \mathcal{A}$. Then

$$\|\pi(\xi)\|^2\|PF\eta\|^2 \geq \langle \pi(\xi)PF\eta, \pi(\xi)PF\eta \rangle = \langle P\pi(\xi)F\eta, P\pi(\xi)F\eta \rangle$$

since $\pi'(\eta)$ is affiliated to $\mathcal{L}(\mathcal{A})'$. Now $\pi(\xi)F\eta = \pi'(F\eta)\xi$. Therefore by Lemma VII.4,

$$\|\pi(\xi)\|^2\|PF\eta\|^2 \geq \langle P\pi'(F\eta)\xi, P\pi'(F\eta)\xi \rangle = \langle P\pi'(\eta)^*\xi, P\pi'(\eta)^*\xi \rangle.$$

Since $\pi'(\eta) = UK$, one has $\pi'(\eta)^* = KU^*$; and thus

$$\|\pi(\xi)\|^2\|PF\eta\|^2 \geq \langle PKU^*\xi, PKU^*\xi \rangle.$$

Now $\xi = (\Delta^{-1} + r)\eta$. Therefore,

$$
\begin{aligned}
\|\pi(\xi)\|^2 \|PF\eta\|^2 &\geq \langle PKU^*(\Delta^{-1} + r)\eta, PKU^*(\Delta^{-1} + r)\eta \rangle \\
&= \|PKU^*(\Delta^{-1} + r)\eta\|^2 \\
&\geq \|PKU^*(\Delta^{-1} + r)\eta\|^2 - \|PKU^*(\Delta^{-1} - r)\eta\|^2 \\
&= 4r\mathrm{Re}\langle PKU^*\Delta^{-1}\eta, PKU^*\eta \rangle \\
&= 4r\mathrm{Re}\langle \Delta^{-1}\eta, UKPKU^*\eta \rangle \\
&= 4r\mathrm{Re}\langle \Delta^{-1}\eta, UPK^2U^*\eta \rangle.
\end{aligned}
$$

We will show $UPK^2U^*\eta \in \mathcal{A}'$ and $F(UPK^2U^*\eta) = PK^2F\eta$. To do this we use Proposition VII.5. First note

$$
\begin{aligned}
\pi(\xi)UPK^2U^*\eta &= UPK^2U^*\pi(\xi)\eta = UPK^2U^*\pi'(\eta)\xi \\
&= UPK^2U^*UK\xi \\
&= UPK^3\xi \\
&= A\xi
\end{aligned}
$$

where $A = UPK^3$ is a bounded operator since $P = E[n, m]$. Now by Lemma VII.4,

$$
\begin{aligned}
A^*\xi &= K^3PU^*\xi = PK^2KU^*\xi = PK^2\pi'(\eta)^*\xi \\
&= PK^2\pi'(F\eta)\xi = PK^2\pi(\xi)F\eta = \pi(\xi)(PK^2F\eta).
\end{aligned}
$$

Hence by Proposition VII.5, $UPK^2U^*\eta \in \mathcal{A}'$ and $F(UPK^2U^*\eta) = PK^2F\eta$. Therefore

$$
\begin{aligned}
\|\pi(\xi)\|^2\|PF\eta\|^2 &\geq 4r\mathrm{Re}\langle \Delta^{-1}\eta, UPK^2U^*\eta \rangle \\
&= 4r\mathrm{Re}\langle SF\eta, UPK^2U^*\eta \rangle \\
&= 4r\mathrm{Re}\langle F\eta, FUPK^2U^*\eta \rangle \\
&= 4r\mathrm{Re}\langle F\eta, PK^2F\eta \rangle \\
&= 4r\|PKF\eta\|^2.
\end{aligned}
$$

Now $PKF\eta = \int_n^m \lambda\, dE_\lambda F\eta$ and thus $\|PKF\eta\| \geq n\|PF\eta\|$. This gives

$$
\|\pi(\xi)\|^2\|PF\eta\|^2 \geq 4rn\|PF\eta\|^2
$$

for all $P = E[n, m]$. Hence if $E[n, \infty)F\eta \neq 0$, then $\|\pi(\xi)\| \geq 4rn$. Since $\pi(\xi)$ is bounded, we have $E[n, \infty)F\eta = 0$ for some n. This implies for each $\xi \in \mathcal{A}$,

$$
\begin{aligned}
0 &= E[n, \infty) \pi(\xi) F \eta \\
&= E[n, \infty) \pi'(F \eta) \xi \\
&= E[n, \infty) \pi'(\eta)^* \xi \\
&= E[n, \infty) K U^* \xi
\end{aligned}
$$

since $\pi'(\eta)^* \geq \pi'(F\eta)$. Since U is a partial isometry from $\overline{\mathrm{Rang} K}$ onto $\overline{\mathrm{Rang} \pi'(\eta)}$, U^* is a partial isometry onto $\overline{\mathrm{Rang} K} = \ker K^\perp$ since K is self adjoint. Hence $E[n, \infty) K = 0$. Thus $K = E[0, n) K$, and K is bounded. $\square$

LEMMA VII.5. *Let* $\eta_1, \eta_2 \in Dom(\Delta^{1/2}) \cap Dom(\Delta^{-1/2})$. *Then for* $\xi \in \mathcal{A}$, $r > 0$ *and* $\xi_1 = (\Delta^{-1} + r)^{-1}\xi \in \mathcal{A}'$, *one has* $\langle \pi(\xi)\eta_1, \eta_2 \rangle = \langle J\pi'(\xi_1)^* J\Delta^{1/2}\eta_1, \Delta^{-1/2}\eta_2 \rangle + r\langle J\pi'(\xi_1)^* J\Delta^{-1/2}\eta_1, \Delta^{1/2}\eta_2 \rangle$.

REMARK. Let $\Delta = \int_0^\infty \lambda \, dE_\lambda$. Since Δ^{-1} exists, the point spectrum does not contain 0. Hence if $P_n = E[\frac{1}{n}, n]$, then $\cup P_n \mathbb{H}$ is dense in $\mathbb{H}$. Now $P_n \mathbb{H} \subset \mathrm{Dom}(\Delta^s)$ for all real s. Thus $\cap_{s \in \mathbb{R}} \mathrm{Dom}(\Delta^s)$ is dense in $\mathbb{H}$. Hence $\mathrm{Dom}(\Delta^{1/2}) \cap \mathrm{Dom}(\Delta^{-1/2})$ is dense in $\mathbb{H}$.

PROOF. Let $\eta_1, \eta_2 \in \mathcal{A}'$. Now $\xi_1 \in \mathcal{A}'$ by Theorem VII.5. Hence using Lemma VII.2 and Lemma VII.3, one has

$$
\begin{aligned}
\langle \pi(\xi)\eta_1, \eta_2 \rangle &= \langle \pi'(\eta_1)\xi, \eta_2 \rangle = \langle \xi, \pi'(\eta_1)^* \eta_2 \rangle \\
&= \langle (\Delta^{-1} + r)\xi_1, \pi'(\eta_1)^* \eta_2 \rangle \\
&= \langle SF\xi_1, \pi'(F\eta_1)\eta_2 \rangle + r\langle \xi_1, \pi'(\eta_1)^* \eta_2 \rangle \\
&= \langle F\pi'(F\eta_1)\eta_2, F\xi_1 \rangle + r\langle \xi_1, \pi'(\eta_1)^* \eta_2 \rangle \\
&= \langle \pi'(F\eta_2)\eta_1, F\xi_1 \rangle + r\langle \pi'(\eta_1)\xi_1, \eta_2 \rangle \\
&= \langle \eta_1, \pi'(\eta_2)F\xi_1 \rangle + r\langle \pi'(\eta_1)\xi_1, \eta_2 \rangle \\
&= \langle \eta_1, F(\pi'(\xi_1)F\eta_2) \rangle + r\langle F(\pi'(F\xi_1)F\eta_1), \eta_2 \rangle.
\end{aligned}
$$

Now take $\eta_1, \eta_2 \in (\Delta^{-1} + 1)^{-1}\mathcal{A}^2 \subseteq \mathcal{A}'$. Since $\mathcal{A}^2 \subseteq \mathrm{Dom}(S)$ and $S = J\Delta^{1/2}$, $\mathcal{A}^2 \subseteq \mathrm{Dom}(\Delta^{1/2})$. Hence $\eta_1, \eta_2 \in (\Delta^{-1}+1)^{-1}\mathrm{Dom}(\Delta^{1/2})$. This implies η_1, η_2 are in $\mathrm{Dom}(\Delta^{1/2})$. Indeed, one has

$$
\Delta^{1/2}(\Delta^{-1} + 1)^{-1}\gamma = (\Delta^{-1} + 1)^{-1}\Delta^{1/2}\gamma
$$

for $\gamma \in \mathrm{Dom}(\Delta^{1/2})$. Also $\eta_1, \eta_2 \in \mathrm{Dom}(\Delta^{-1/2})$ for $\eta_i \in \mathcal{A}' \subseteq \mathrm{Dom}(F)$. Hence $\eta_1, \eta_2 \in \mathrm{Dom}(\Delta^{1/2}) \cap \mathrm{Dom}(\Delta^{-1/2})$. Therefore, continuing from above, and using $F = J\Delta^{-1/2} = \Delta^{1/2}J$, we see

$$\langle \pi(\xi)\eta_1, \eta_2 \rangle = \langle \eta_1, \Delta^{1/2}J\pi'(\xi_1)J\Delta^{-1/2}\eta_2 \rangle + r\langle \Delta^{1/2}J\pi'(\xi_1)^*J\Delta^{-1/2}\eta_1, \eta_2 \rangle$$
$$= \langle J\pi'(\xi_1)^*J\Delta^{1/2}\eta_1, \Delta^{-1/2}\eta_2 \rangle + r\langle J\pi'(\xi_1)^*J\Delta^{-1/2}\eta_1, \Delta^{1/2}\eta_2 \rangle.$$

Hence we have shown the result for $\eta_1, \eta_2 \in (\Delta^{-1} + 1)\mathcal{A}^2 \subseteq \mathcal{A}' \cap \text{Dom}(\Delta^{1/2}) \cap \text{Dom}(\Delta^{-1/2})$.

To finish the proof we show if $\eta \in \text{Dom}(\Delta^{1/2}) \cap \text{Dom}(\Delta^{-1/2})$, there is a sequence $\xi'_n \in (\Delta^{-1} + 1)^{-1}\mathcal{A}^2$ such that $\lim \xi'_n = \eta$, $\lim \Delta^{1/2}\xi'_n = \Delta^{1/2}\eta$ and $\lim \Delta^{-1/2}\xi'_n = \Delta^{-1/2}\eta$. Now $\Delta^{1/2}\mathcal{A}^2 = JS\mathcal{A}^2 = J\mathcal{A}^2$ is dense in $\mathbb{H}$. Hence there is a sequence $\xi_n \in \mathcal{A}^2$ such that $\Delta^{1/2}\eta + \Delta^{-1/2}\eta = \lim_n \Delta^{1/2}\xi_n$. Set $\xi'_n = (\Delta^{-1} + 1)^{-1}\xi_n$. Now $(\Delta^{-1} + 1)^{-1}$, $\Delta^{-1/2}(\Delta^{-1}+1)^{-1}$ and $\Delta^{-1}(\Delta^{-1}+1)^{-1}$ are bounded operators. Hence

$$\xi'_n = (\Delta^{-1} + 1)^{-1}\xi_n$$
$$= \Delta^{-1/2}(\Delta^{-1} + 1)^{-1}\Delta^{1/2}\xi_n$$
$$\longrightarrow \Delta^{-1/2}(\Delta^{-1} + 1)^{-1}(\Delta^{1/2} + \Delta^{-1/2})\eta$$
$$= \eta,$$

$$\Delta^{1/2}\xi'_n = \Delta^{1/2}(\Delta^{-1} + 1)^{-1}\xi_n$$
$$= (\Delta^{-1} + 1)^{-1}\Delta^{1/2}\xi_n$$
$$\longrightarrow (\Delta^{-1} + 1)^{-1}(\Delta^{1/2} + \Delta^{-1/2})\eta$$
$$= \Delta^{1/2}\eta$$

and

$$\Delta^{-1/2}\xi'_n = \Delta^{-1/2}(\Delta^{-1} + 1)^{-1}\xi_n$$
$$= \Delta^{-1}(\Delta^{-1} + 1)^{-1}\Delta^{1/2}\xi_n$$
$$\longrightarrow \Delta^{-1}(\Delta^{-1} + 1)^{-1}(\Delta^{1/2} + \Delta^{-1/2})\eta$$
$$= \Delta^{-1/2}\eta$$

as $n \longrightarrow \infty$. $\square$

8. A Conjugate Isomorphism of $\mathcal{A}$ and $\mathcal{A}'$

DEFINITION. Let Δ be a closed, positive, nonsingular operator on $\mathbb{H}$. Let $X \in B(\mathbb{H})$. Define $\psi_r(X)$ by

$$\psi_r(X) = \int_{-\infty}^{\infty} \frac{r^{it-1/2}}{e^{\pi t} + e^{-\pi t}} \Delta^{it} X \Delta^{-it} \, dt.$$

LEMMA VII.6. *Let $r > 0$. Then if $X, Y \in B(\mathbb{H})$ satisfy*

$$\langle X\eta_1, \eta_2 \rangle = \langle Y\Delta^{1/2}\eta_1, \Delta^{-1/2}\eta_2 \rangle + r\langle Y\Delta^{-1/2}\eta_1, \Delta^{1/2}\eta_2 \rangle$$

for all $\eta_1, \eta_2 \in Dom(\Delta^{1/2}) \cap Dom(\Delta^{-1/2})$, then $Y = \psi_r(X)$.

PROOF. Let E be the resolution of the identity for Δ. Let $P = E[\frac{1}{n}, n]$. Set $f(z) = \frac{-ir^{iz}}{e^{\pi z} - e^{-\pi z}}\Delta^{iz}PX\Delta^{-iz}P$. Then f is analytic except at $z = ki$ where $k \in \mathbb{Z}$. Set $z = t + i\nu$. Then if $t \neq 0$,

$$\left|e^{\pi t} - e^{-\pi t}\right| \leq \left|e^{\pi t} - e^{-\pi t}e^{-2\pi i\nu}\right| = \left|e^{\pi t + i\pi\nu} - e^{-\pi t - i\pi\nu}\right|$$

for rotating $e^{-\pi t}$ results in something further from $e^{\pi t}$. Thus

$$\|f(z)\| \leq \left|\frac{r^{it-\nu}}{e^{\pi(t+i\nu)} - e^{-\pi(t+i\nu)}}\right| \|\Delta^{it-\nu}PX\Delta^{-it+\nu}P\|$$

$$\leq \frac{r^{-\nu}}{\left|e^{\pi t} - e^{-\pi t}\right|}\|P\Delta^{-\nu}\|\,\|X\|\,\|\Delta^{\nu}P\| \leq \frac{r^{-\nu}}{\left|e^{\pi t} - e^{-\pi t}\right|}n^{2\nu}\|X\|.$$

Hence if $-1/2 \leq \nu \leq 1/2$, then $f(z)$ converges uniformly to 0 as $t \to \infty$. Therefore

$$\int_{-\infty}^{\infty} f\left(t - \frac{1}{2}i\right)dt - \int_{-\infty}^{\infty} f\left(t + \frac{1}{2}i\right)dt = 2\pi i \sum_{-1/2 < \mathrm{Im}\, z < 1/2} \mathrm{Res}\, f = 2\pi i \mathrm{Res}_{z=0} f.$$

Now

$$\mathrm{Res}_{z=0} f = \lim_{z \to 0} zf(z) = \lim_{z \to 0} \frac{-izr^{iz}}{e^{\pi z} - e^{-\pi z}}PXP$$

$$= -\frac{i}{2\pi} \lim_{z \to 0}\left(\frac{e^{\pi z} - e^{-\pi z}}{2\pi z}\right)^{-1} PXP = -\frac{i}{2\pi}PXP.$$

Hence $\int_{-\infty}^{\infty} f(t - \frac{1}{2}i)\,dt - \int_{-\infty}^{\infty} f(t + \frac{1}{2}i)\,dt = PXP$. This yields

$$PXP = \int_{-\infty}^{\infty} \frac{-ir^{it+1/2}}{e^{\pi t - \pi i/2} - e^{-\pi t + \pi i/2}}\Delta^{it+1/2}PX\Delta^{-it-1/2}P\,dt$$

$$- \int_{-\infty}^{\infty} \frac{-ir^{it-1/2}}{e^{\pi t + \pi i/2} - e^{-\pi t - \pi i/2}}\Delta^{it-1/2}PX\Delta^{-it+1/2}P\,dt$$

$$= P\int_{-\infty}^{\infty} \frac{r^{it+1/2}}{e^{\pi t} + e^{-\pi t}}\Delta^{1/2}\Delta^{it}X\Delta^{-it}\Delta^{-1/2}\,dtP$$

$$+ P\int_{-\infty}^{\infty} \frac{r^{it-1/2}}{e^{\pi t} + e^{-\pi t}}\Delta^{-1/2}\Delta^{it}X\Delta^{-it}\Delta^{1/2}\,dtP$$

$$= rP\Delta^{1/2}\psi_r(X)\Delta^{-1/2}P + P\Delta^{-1/2}\psi_r(X)\Delta^{1/2}P.$$

Thus $PXP = P\Delta^{-1/2}\psi_r(X)\Delta^{1/2}P + rP\Delta^{1/2}\psi_r(X)\Delta^{-1/2}P$. From the hypothesis, one can see

$$PXP = P\Delta^{-1/2}Y\Delta^{1/2}P + rP\Delta^{1/2}Y\Delta^{-1/2}P$$

for $P\mathbb{H} \subset \text{Dom}(\Delta^{1/2}) \cap \text{Dom}(\Delta^{-1/2})$ and $P\Delta^{\pm 1/2}$ are bounded operators. Set $A = \psi_r(X) - Y$. Thus $P\Delta^{-1/2}A\Delta^{1/2}P = -rP\Delta^{1/2}A\Delta^{-1/2}P$. This gives

$$PA^*\Delta^{1/2}P(P\Delta^{-1/2}A\Delta^{1/2}P) = -rPA^*\Delta^{1/2}P(\Delta^{1/2}A\Delta^{-1/2}P)$$
$$= -rPA^*\Delta PA\Delta^{-1/2}P.$$

Therefore, $PA^*PA\Delta^{1/2}P = -rPA^*\Delta PA\Delta^{-1/2}P$. Hence

$$PA^*PA\Delta P = -rPA^*P\Delta PAP \le 0.$$

But $PA^*PA(\Delta P) = (PA^*PPAP)P\Delta$ is a product of two positive operators, and this product is self adjoint for it is a negative operator. Thus they commute, and $PA^*PA\Delta P \ge 0$. This gives $PA^*PA\Delta P = 0$. Thus $PA^*PAP = 0$. This gives $PAP = 0$. Letting $n \to \infty$ gives $A = 0$, and thus $Y = \psi_r(X)$. $\square$

LEMMA VII.7. *If $r > 0$, then*

$$\Delta^{-1/2}(\Delta^{-1} + r)^{-1} = \int_{-\infty}^{\infty} \frac{r^{it-1/2}}{e^{\pi t} + e^{-\pi t}}\Delta^{it}\,dt.$$

PROOF. Set $f(z) = \frac{-ir^{iz}}{e^{\pi z} - e^{-\pi z}}\Delta^{iz}P$ where $P = E[\frac{1}{n}, n]$. The same argument as in the proof of Lemma VII.6 shows

$$P = P\int_{-\infty}^{\infty} \frac{-ir^{it+1/2}}{e^{\pi t - \pi i/2} - e^{-\pi t + \pi i/2}}\Delta^{it+1/2}\,dtP$$
$$- P\int_{-\infty}^{\infty} \frac{-ir^{it-1/2}}{e^{\pi t + \pi i/2} - e^{-\pi t - \pi i/2}}\Delta^{it-1/2}\,dtP.$$

Therefore, one sees $P = rP\Delta^{1/2}R_rP + P\Delta^{-1/2}R_rP$ where $R_r = \int_{-\infty}^{\infty} \frac{r^{it-1/2}}{e^{\pi t} + e^{-\pi t}}\Delta^{it}\,dt$. Hence $r\Delta^{1/2}R_rP + \Delta^{-1/2}R_rP = P$ and so $(\Delta^{-1} + r)\Delta^{1/2}R_rP = P$. Letting $n \to \infty$, we see $(\Delta^{-1} + r)\Delta^{1/2}R_r = I$. Thus $R_r = \Delta^{-1/2}(\Delta^{-1} + r)^{-1}$. $\square$

COROLLARY VII.6. *For $\xi \in \mathcal{A}$, one has*

$$\psi_r(\pi(\xi)) = J\pi'((\Delta^{-1} + r)^{-1}\xi)^*J.$$

PROOF. This follows from Lemmas VII.5 and VII.6. $\square$

REMARK. The following were some of the motivating ideas in the development of the theory; namely

(1) $\mathcal{L}(\mathcal{A}) = \pi(\mathcal{A})''$
(2) $J\mathcal{A} \subseteq \mathcal{A}'$
(3) $J\mathcal{A}' = \mathcal{A}''$
(4) $J\mathcal{L}(\mathcal{A})J = \mathcal{L}(\mathcal{A})'$.

Now knowing $\psi_r(\pi(\xi)) = J\pi'((\Delta^{-1}+r)^{-1}\xi)^*J$, one would suspect, since $J\pi'((\Delta^{-1}+r)^{-1}\xi)^*J \in \mathcal{L}(\mathcal{A})$, that it is a function of $\pi(\xi)$.

EXAMPLE ($C_c(G)$ CONTINUED). Here $\mathcal{A} = C_c(G)$. Set

$$\rho = (\Delta^{-1} + r)^{-1}\xi.$$

Then

$$J\pi'(\rho)^* J\eta = \Delta^{-1/2}(\pi'(\rho)^* \widetilde{J\eta})$$
$$= \Delta^{-1/2}(\pi'(F\rho)\widetilde{J\eta})$$
$$= \Delta^{-1/2}(J\eta * \widetilde{F\rho})$$
$$= \Delta^{-1/2}((\Delta^{-1/2}\tilde{\eta}) * \widetilde{\rho})$$
$$= \Delta^{-1/2}(\rho * (\Delta^{1/2}\eta))$$

for

$$\overline{(\Delta^{-1/2}\tilde{\eta} * \tilde{\rho}(x^{-1}))} = \overline{\int \Delta^{-1/2}\tilde{\eta}(y)\tilde{\rho}(y^{-1}x^{-1})\,dy}$$
$$= \int \Delta^{-1/2}(y)\eta(y^{-1})\rho(xy)\,dy$$
$$= \int \rho(y)\Delta^{-1/2}(x^{-1}y)\eta(y^{-1}x)\,dy$$
$$= \int \rho(y)\Delta^{1/2}(y^{-1}x)\eta(y^{-1}x)\,dy$$
$$= \rho * (\Delta^{1/2}\eta)(x).$$

Therefore $J\pi'(\rho)^* J = \Delta^{-1/2}(\rho * (\Delta^{1/2}\eta))$. Now

$$\Delta^{-1/2}(\rho * (\Delta^{1/2}\eta))(x) = \int \rho(y)\Delta^{1/2}(y^{-1})\eta(y^{-1}x)\,dy = (\Delta^{-1/2}\rho) * \eta(x).$$

Hence $J\pi'(\rho)^* J\eta = (\Delta^{-1/2}\rho) * \eta$. This suggests

$$J\pi'((\Delta^{-1} + r)^{-1}\xi)^* J = \pi(\Delta^{-1/2}(\Delta^{-1} + r)^{-1}\xi).$$

As seen earlier it was useful to determine a formula for $\Delta^{-1/2}(\Delta^{-1} + r)^{-1}$ in terms of Δ^{it}. This approach can be motivated. Consider

$$\lambda^{-1/2}(\lambda^{-1} + r)^{-1} = \int_{-\infty}^{\infty} f_r(t)\lambda^{it}\, dt = \int_{-\infty}^{\infty} f_r(t)e^{it\ln\lambda}\, dt = \hat{f}_r(\ln\lambda).$$

Then $f_r = F^{-1}(e^{-\lambda/2}(e^{-\lambda}+r))$ where F is the Fourier transform. The main ideas were to use contour integrals and residues to evaluate this integral.

THEOREM VII.6. *Let $t \in \mathbb{R}$ and $\xi \in \mathcal{A}$. Then $J\Delta^{it}\xi \in \mathcal{A}'$ and $\pi'(J\Delta^{it}\xi) = J\Delta^{it}\pi(\xi)\Delta^{-it}J$ and $F(J\Delta^{it}\xi) = J\Delta^{it}\xi^{\#}$.*

PROOF. By Corollary VII.6, $\psi_r(\pi(\xi)) = J\pi'((\Delta^{-1}+r)^{-1}\xi)^*J$. By Theorems VII.4 and VII.5, if $\eta \in \mathcal{A}$, then

$$J\psi_r(\pi(\xi))J\eta = \pi'((\Delta^{-1} + r)^{-1}\xi)^*\eta = \pi'(F(\Delta^{-1} + r)^{-1}\xi)\eta$$
$$= \pi(\eta)F(\Delta^{-1} + r)^{-1}\xi = \pi(\eta)J\Delta^{-1/2}(\Delta^{-1} + r)^{-1}\xi.$$

Using Lemma VII.7, the definition of $\psi_r(\pi(\xi))$, and the conjugate linearity of J, we see

$$r^{-1/2}\int_{-\infty}^{\infty}\frac{r^{-it}}{e^{\pi t} + e^{-\pi t}}J\Delta^{it}\pi(\xi)\Delta^{-it}J\eta\, dt = r^{-1/2}\int_{-\infty}^{\infty}\frac{r^{-it}}{e^{\pi t} + e^{-\pi t}}\pi(\eta)J\Delta^{it}\xi\, dt$$

for all $r > 0$. Taking $r = e^s$ where $s \in \mathbb{R}$ gives

$$\int_{-\infty}^{\infty}\frac{e^{-ist}}{e^{\pi t} + e^{-\pi t}}(J\Delta^{it}\pi(\xi)\Delta^{-it}J\eta - \pi(\eta)J\Delta^{it}\xi)\, dt = 0.$$

Since the Fourier transform is one-to-one, we see

$$J\Delta^{it}\pi(\xi)\Delta^{-it}J\eta = \pi(\eta)J\Delta^{it}\xi$$

for all $\xi, \eta \in \mathcal{A}$. But this also holds for $\xi^{\#}$. Hence we have both

$$\pi(\eta)J\Delta^{it}\xi = (J\Delta^{it}\pi(\xi)\Delta^{-it}J)\eta \text{ and}$$
$$\pi(\eta)(J\Delta^{it}\xi^{\#}) = (J\Delta^{it}\pi(\xi^{\#})\Delta^{-it}J)\eta = (J\Delta^{it}\pi(\xi)\Delta^{-it}J)^*\eta.$$

By Proposition VII.5, $J\Delta^{it}\xi \in \mathcal{A}'$ and $F(J\Delta^{it}\xi) = J\Delta^{it}\xi^{\#}$. Also $\pi'(J\Delta^{it}\xi) = J\Delta^{it}\pi(\xi)\Delta^{-it}J$. □

COROLLARY VII.7. *$J\mathcal{A} \subseteq \mathcal{A}'$, $\pi'(J\xi) = J\pi(\xi)J$, and $F(J\xi) = J\xi^{\#}$ for $\xi \in \mathcal{A}$.*

COROLLARY VII.8. *$\mathcal{A} \subseteq Dom(S)$ and S is the closure of $^{\#}$.*

PROOF. By Theorem VII.4, $S = JFJ$. Since $J\mathcal{A} \subseteq \mathcal{A}' \subseteq Dom(F)$, we see $\mathcal{A} \subseteq Dom(S)$. □

Using Lemma VII.3 and $\pi'(\eta) \in \mathcal{L}(\mathcal{A})'$ for $\eta \in \mathcal{A}'$, one obtains the following corollary.

COROLLARY VII.9. $J\mathcal{L}(\mathcal{A})J \subseteq \mathcal{L}(\mathcal{A})'$.

LEMMA VII.8. Let $\xi', \eta' \in \mathcal{A}'$. Then $J\pi'(\xi')J\eta' = \pi'(\eta')J\xi'$.

PROOF. Let $\xi \in \mathcal{A}$. Then by Corollary VII.7,

$$\langle \pi'(\eta')J\xi', \xi^\# \rangle = \langle J\xi', \pi'(\eta')^*\xi^\# \rangle = \langle J\xi', \pi'(F\eta')\xi^\# \rangle$$
$$= \langle J\xi', \pi(\xi^\#)F\eta' \rangle = \langle \pi(\xi)J\xi', F\eta' \rangle$$
$$= \langle JJ\pi(\xi)J\xi', F\eta' \rangle$$
$$= \langle J\pi'(J\xi)\xi', F\eta' \rangle.$$

Using Lemma VII.3 and Theorem VII.4, we obtain

$$\langle J\pi'(J\xi)\xi', F\eta' \rangle = \langle JF\pi'(F\xi')FJ\xi, F\eta' \rangle$$
$$= \langle JF\pi'(\xi')^*FJ\xi, F\eta' \rangle = \langle SJ\pi'(\xi')^*JS\xi, F\eta' \rangle$$
$$= \langle FF\eta', J\pi'(\xi')^*J\xi^\# \rangle = \langle \eta', J\pi'(\xi')^*J\xi^\# \rangle$$
$$= \langle J\pi'(\xi')J\eta', \xi^\# \rangle.$$

Thus $\langle \pi'(\eta')J\xi', \xi^\# \rangle = \langle J\pi'(\xi')J\eta', \xi^\# \rangle$ for all $\xi \in \mathcal{A}$. From this $\pi'(\eta')J\xi' = J\pi'(\xi')J\eta'$ follows. □

9. $\mathcal{A}'$ is a Left Hilbert Algebra

By Lemma VII.3, we define $\eta_1\eta_2$ by $\eta_1\eta_2 = \pi'(\eta_1)\eta_2 \in \mathcal{A}'$ for $\eta_1, \eta_2 \in \mathcal{A}'$.

PROPOSITION VII.7. $\mathcal{A}'$ is a left Hilbert algebra. The adjoint in $\mathcal{A}'$ is F.

PROOF. By Theorem VII.5, $(\Delta^{-1} + r)^{-1}\mathcal{A} \subseteq \mathcal{A}'$, and thus $\mathcal{A}'$ is dense in $\mathbb{H}$. We check properties (a), (b), (c), and (d) in the definition of a left Hilbert algebra. Indeed, clearly (a) holds. For (b), note

$$\langle \eta\xi, \xi' \rangle = \langle \pi'(\eta)\xi, \xi' \rangle = \langle \xi, \pi'(F\eta)\xi' \rangle = \langle \xi, (F\eta)\xi' \rangle$$

for $\eta, \xi, \xi' \in \mathcal{A}'$. To see (c), note if ξ and η are in $\mathcal{A}$, then

$$(J\xi)(J\eta) = \pi'(J\xi)J\eta = J\pi(\xi)\eta.$$

Hence $(\mathcal{A}')^2 \supseteq (J\mathcal{A})^2 = J(\mathcal{A}^2)$. Hence $\mathcal{A}'^2$ is dense in $\mathbb{H}$ and consequentially dense in $\mathcal{A}'$. Clearly (d) holds for $F|_{\mathcal{A}'}$ has closure. □

THEOREM VII.7. If $\mathcal{A}$ is a left Hilbert algebra, then

(i) $J\mathcal{L}(\mathcal{A})J = \mathcal{L}(\mathcal{A}') = \mathcal{L}(\mathcal{A})'$ and

(ii) $\Delta^{it}\mathcal{L}(\mathcal{A})\Delta^{-it} = \mathcal{L}(\mathcal{A})$ *for all* $t \in \mathbb{R}$.

PROOF. We know $J\mathcal{L}(\mathcal{A})J \subseteq \mathcal{L}(\mathcal{A})'$ and $\mathcal{L}(\mathcal{A}') = \pi'(\mathcal{A}')'' \subseteq \mathcal{L}(\mathcal{A})'$. Next note if ξ', ξ'_1 and η' are in $\mathcal{A}'$, then using Lemmas VII.8 and VII.3, one has

$$
\begin{aligned}
J\pi'(\xi')J\pi'(\eta')\xi'_1 &= \pi'(\pi'(\eta')\xi'_1)J\xi' \\
&= \pi'(\eta'\xi'_1)J\xi' \\
&= \pi'(\eta')\pi'(\xi'_1)J\xi' \\
&= \pi'(\eta')J\pi'(\xi')J\xi'_1.
\end{aligned}
$$

Thus $J\mathcal{L}(\mathcal{A}')J \subseteq \mathcal{L}(\mathcal{A}')'$. We now show $\mathcal{L}(\mathcal{A}') = \mathcal{L}(\mathcal{A})'$. Let $X' \in \mathcal{L}(\mathcal{A})'$. Then for $\xi', \eta' \in \mathcal{A}'$, we have for $\xi \in \mathcal{A}$,

$$
\begin{aligned}
\pi'(\xi')X'\pi'(\eta')\xi &= \pi'(\xi')X'\pi(\xi)\eta' = \pi'(\xi')\pi(\xi)X'\eta' \\
&= \pi(\xi)\pi'(\xi')X'\eta' \text{ and} \\
(\pi'(\xi')X'\pi'(\eta'))^*\xi &= \pi'(F\eta')X'^*\pi'(F\xi')\xi \\
&= \pi'(F\eta')X'^*\pi(\xi)F\xi' \\
&= \pi(\xi)\pi'(F\eta')X'^*F\xi'.
\end{aligned}
$$

By Proposition VII.5, one has $\pi'(\xi')X'\eta' \in \mathcal{A}'$ and $F(\pi'(\xi')X'\eta') = \pi'(F\eta')X'^*F\xi'$. Also $\pi'(\xi')X'\pi'(\eta') = \pi'(\pi'(\xi')X'\eta') \in \pi'(\mathcal{A}')$. Hence $\pi'(\xi')\mathcal{L}(\mathcal{A})'\pi'(\eta') \subseteq \mathcal{L}(\mathcal{A}')$ for all $\xi', \eta' \in \mathcal{A}'$. Now $\pi'(\mathcal{A}')$ is dense in the von Neumann algebra $\mathcal{L}(\mathcal{A}')$ and $I \in \mathcal{L}(\mathcal{A}')$. By the Kaplansky density theorem III.7, there is a net $\pi'(\xi'_i)$ in the unit ball of $\pi'(\mathcal{A}')$ converging strongly to I. Hence $\pi'(\xi'_i)X'\pi'(\xi'_i)$ converges strongly to X' for $X' \in \mathcal{L}(\mathcal{A})'$. Hence $X' \in \mathcal{L}(\mathcal{A}')$. Thus $\mathcal{L}(\mathcal{A}') = \mathcal{L}(\mathcal{A})'$.

Now $J\mathcal{L}(\mathcal{A}')J \subseteq \mathcal{L}(\mathcal{A}')'$ by Corollary VII.9. So $\mathcal{L}(\mathcal{A}') \subseteq J\mathcal{L}(\mathcal{A}')'J = J\mathcal{L}(\mathcal{A})J$. But $J\mathcal{L}(\mathcal{A})J \subseteq \mathcal{L}(\mathcal{A}')$ by Corollary VII.7. Hence $\mathcal{L}(\mathcal{A}') = J\mathcal{L}(\mathcal{A})J$. This finishes (i).

For (ii), note by Theorem VII.6 that $J\Delta^{it}\pi(\mathcal{A})\Delta^{-it}J \subseteq \mathcal{L}(\mathcal{A}')$. Hence

$$
\Delta^{it}\pi(\mathcal{A})\Delta^{-it} \subseteq J\mathcal{L}(\mathcal{A}')J = \mathcal{L}(\mathcal{A}).
$$

This gives $\Delta^{it}\mathcal{L}(\mathcal{A})\Delta^{-it} = \mathcal{L}(\mathcal{A})$. $\square$

EXAMPLE ($C_c(G)$ CONTINUED). Let λ be the left regular representation on G and λ' be the right regular representation. Hence for $\xi \in L^2(G)$ and $g \in G$, $\lambda(g)\xi(x) = \xi(g^{-1}x)$ and $\lambda'(g)\xi(x) = \Delta(g)^{1/2}\xi(xg)$. Let $\mathcal{M}(\lambda)$ and $\mathcal{M}(\lambda')$ be the von Neumann algebras generated by $\lambda(G)$ and $\lambda'(G)$, respectively. Then $\mathcal{M}(\lambda)$ and $\mathcal{M}(\lambda')$

are commutants. Indeed, note for $f \in C_c(G)$, $\pi(f)\xi = f * \xi$. Thus $\mathcal{M}(\lambda) = \mathcal{L}(C_c(G))$. Now if $f \in C_c(G)$,

$$\lambda'(f)\xi(x) = \int f(g)\Delta(g)^{1/2}\xi(xg)\,dg = \int \xi(g)f(x^{-1}g)\Delta(x^{-1}g)^{1/2}\,dg$$

$$= \int \xi(g)\tilde{\bar{f}}(g^{-1}x)\Delta^{-1/2}(g^{-1}x)\,dg$$

$$= \xi * (\Delta^{-1/2}\tilde{\bar{f}}) = \xi * J(\bar{f})$$

$$= \pi'(J\bar{f})\xi.$$

This implies, since $J\bar{f} = \Delta^{-1/2}\tilde{\bar{f}} \in C_c(G)$ iff $f \in C_c(G)$, that $\lambda'(C_c(G)) = \pi'(JC_c(G)) = J\pi(C_c(G))J$, and thus

$$\mathcal{M}(\lambda') = \lambda'(C_c(G))''$$
$$= (J\pi(C_c(G))J)''$$
$$= J\pi(C_c(G))''J$$
$$= J\mathcal{L}(C_c(G))J$$
$$= \mathcal{L}(C_c(G))'$$
$$= \mathcal{M}(\lambda)'.$$

10. The Double Left Hilbert Algebra

Since $\mathcal{A}'$ is a left Hilbert algebra, one may form $\mathcal{A}''$. Now $\mathcal{A}'' = \{\xi \in \mathrm{Dom}(\overline{F|_{\mathcal{A}'}})^* : \eta \mapsto \pi'(\eta)\xi$ is continuous$\}$. We now show $\overline{F|_{\mathcal{A}'}} = F$ and consequentially $(\overline{F|_{\mathcal{A}'}})^* = F^* = S$. Since $\overline{F|_{\mathcal{A}'}} \leq F$, it suffices to show $\mathrm{Dom}(\overline{F|_{\mathcal{A}'}}) \supseteq \mathrm{Dom}(F)$. Now $S = JFJ$. Hence if $\eta \in \mathrm{Dom}(F)$, $\eta = J\xi$ where $\xi \in \mathrm{Dom}(S)$. Now S is the closure of $S|_{\mathcal{A}}$. Hence there is a sequence $\xi_n \to \xi$ with $S\xi_n \to S\xi$, $\xi_n \in \mathcal{A}$. Since $S\xi_n \in \mathcal{A}$, $J\xi_n$ and $JS\xi_n$ are in $\mathcal{A}'$ and $J\xi_n \to J\xi$ and $JS\xi_n \to JS\xi$. Now $F(J\xi_n) = JSJJ\xi_n \to JS\xi = JSJJ\xi = FJ\xi$. Thus $\eta \in \mathrm{Dom}(\overline{F|_{\mathcal{A}'}})$. Hence $F = \overline{(F|_{\mathcal{A}'})}$. Therefore, $\mathcal{A}'' = \{\xi \in \mathrm{Dom}(S) : \eta \mapsto \pi'(\eta)\xi, \eta \in \mathcal{A}'$ is continuous$\}$.

DEFINITION. For $\xi \in \mathcal{A}''$, set $\pi(\xi)$ to be the natural extension of $\eta \mapsto \pi'(\eta)\xi$ to all of $\mathbb{H}$.

Note $\mathcal{A} \subseteq \mathcal{A}''$ and for $\xi \in \mathcal{A}$, $\pi(\xi)$ is the same as the original operator. Indeed, if $\xi \in \mathcal{A}$, then $\xi \in \mathrm{Dom}(S)$ and $\pi'(\eta)\xi = \pi(\xi)\eta$ for $\eta \in \mathcal{A}'$.

THEOREM VII.8. *One has*

(i) $\mathcal{A}''$ *is a left Hilbert algebra with involution* S *and product* $(\xi, \eta) \mapsto \pi(\xi)\eta$ *and left von Neumann algebra* $\mathcal{L}(\mathcal{A})$.

(ii) $(\mathcal{A}'')' = \mathcal{A}'$, $(\mathcal{A}'')'' = \mathcal{A}''$.

PROOF. We show the left von Neumann algebra of $\mathcal{A}''$ is $\mathcal{L}(\mathcal{A})$. Since $\mathcal{A}'' \supseteq \mathcal{A}$, $\mathcal{L}(\mathcal{A}) \subseteq \mathcal{L}(\mathcal{A}'')$. Now $\mathcal{L}(\mathcal{A}) = J\mathcal{L}(\mathcal{A}')J = \mathcal{L}(\mathcal{A}'')$ by Theorem VII.7. This proves (i).

We show $(\mathcal{A}'')' = \mathcal{A}'$. Let $\xi' \in (\mathcal{A}'')'$. Hence $\xi' \in \text{Dom}(F)$ and the mapping $\eta \mapsto \pi(\eta)\xi'$, $\eta \in \mathcal{A}''$, is continuous. Since $\mathcal{A} \subseteq \mathcal{A}''$, we see $\xi' \in \mathcal{A}'$. Conversely, if $\xi \in \mathcal{A}'$, then $J\xi \in \mathcal{A}''$ and therefore $\xi = JJ\xi \in J\mathcal{A}'' \subseteq (\mathcal{A}'')'$. Hence $(\mathcal{A}'')' = \mathcal{A}'$. From this follows $(\mathcal{A}'')'' = \mathcal{A}''$. This completes (ii). □

11. Achieved Left Hilbert Algebras

DEFINITION. A left Hilbert algebra $\mathcal{A}$ is achieved (fulfilled) if $\mathcal{A}'' = \mathcal{A}$.

EXAMPLE $(C_c(G)$ CONTINUED$)$. The achievement of $\mathcal{A} = C_c(G)$ consists of all $f \in L^2(G)$ satisfying $Sf = \Delta^{-1}\tilde{f} \in L^2$ and $g \mapsto f * g$ is continuous on L^2.

THEOREM VII.9. *Let* $\mathcal{A}$ *be an achieved left Hilbert algebra. Then*

(i) $J\mathcal{A} = \mathcal{A}'$, $\pi'(J\xi) = J\pi(\xi)J$ *for* $\xi \in \mathcal{A}$

(ii) $\Delta^{it}\mathcal{A} = \mathcal{A}$ *and* $\pi(\Delta^{it}\xi) = \Delta^{it}\pi(\xi)\Delta^{-it}$ *for* $\xi \in \mathcal{A}$, $t \in \mathbb{R}$.

PROOF. By Corollary VII.7, $J\mathcal{A} \subseteq \mathcal{A}'$ and $J\mathcal{A}' \subseteq \mathcal{A}'' = \mathcal{A}$. Therefore $J\mathcal{A} = \mathcal{A}'$. Theorem VII.6 implies $J(\Delta^{it}\xi) \in \mathcal{A}'$ for $\xi \in \mathcal{A}$. Hence $\Delta^{it}\xi \in J\mathcal{A}' = \mathcal{A}$ for all $t \in \mathbb{R}$ and for $\xi \in \mathcal{A}$. So $\Delta^{it}\mathcal{A} = \mathcal{A}$ for all t. Again by Theorem VII.6, $\pi'(J\Delta^{it}\xi) = J\Delta^{it}\pi(\xi)\Delta^{-it}J$. Since $\Delta^{it}\xi \in \mathcal{A}$ for $\xi \in \mathcal{A}$, we see $J\pi(\Delta^{it}\xi)J = J\Delta^{it}\pi(\xi)\Delta^{-it}J$. Hence $\pi(\Delta^{it}\xi) = \Delta^{it}\pi(\xi)\Delta^{-it}$. □

12. The Subalgebra $\mathcal{L}(\mathcal{A})_0$

Let $z \mapsto A(z)$ be a mapping from $\mathbb{C}$ into $B(\mathbb{H})$. Recall A is strongly entire if for each $v \in \mathbb{H}$, $z \mapsto A(z)v$ is an entire function into $\mathbb{H}$. We note that a strongly continuous function $A(z)$ is strongly entire iff it is weakly entire.

DEFINITION. Let $\mathcal{A}$ be a left Hilbert algebra in a Hilbert space $\mathbb{H}$. Define $\mathcal{L}(\mathcal{A})_0$ to be the collection of all operators $A \in \mathcal{L}(\mathcal{A})$ such that $it \mapsto \Delta^{it}A\Delta^{-it}$ and $it \mapsto \Delta^{it}A^*\Delta^{-it}$ both have strongly entire extensions which are uniformly bounded on all vertical strips $-\delta \leq \text{Re}z \leq \delta$.

Suppose $F(z)$ is an entire extension of $it \mapsto \Delta^{it} A \Delta^{-it}$ where $A \in \mathcal{L}(\mathcal{A})$. Then $F(z) \in \mathcal{L}(\mathcal{A})$ for all z. Indeed, if T is in the commutant of $\mathcal{L}(\mathcal{A})$ and $v \in \mathbb{H}$, then by Theorem VII.7, the entire functions $F(z)Tv$ and $TF(z)v$ agree on $i\mathbb{R}$. Hence $F(z)Tv = TF(z)v$ for all v and z. This gives $F(z) \in \mathcal{L}(\mathcal{A})'' = \mathcal{L}(\mathcal{A})$.

LEMMA VII.9. *Suppose $F(z)$ and $F(\bar{z})^*$ are strongly entire $B(\mathbb{H})$ valued functions which are uniformly bounded on all vertical strips $-\delta \leq \text{Re}z \leq \delta$. If $z \mapsto G(z)$ is a continuous bounded $\mathbb{H}$ valued function on a strip $\alpha \leq \text{Re}z \leq \beta$ which is holomorphic on $\alpha < \text{Re}z < \beta$, then $z \mapsto F(z)G(z)$ is continuous and bounded on the closed strip $\alpha \leq \text{Re}z \leq \beta$ and holomorphic on $\alpha < \text{Re}z < \beta$.*

PROOF. Set $H(z) = F(z)G(z)$. Then $H(z)$ is defined on the strip $\alpha \leq \text{Re}z \leq \beta$ and has values in $\mathbb{H}$. Moreover, since both $F(z)$ and $G(z)$ are bounded on the strip $\alpha \leq \text{Re}z \leq \beta$, one sees $H(z)$ is bounded on this strip. Moreover, $H(z)$ is continuous. Indeed,

$$\|H(z) - H(z_0)\| = \|F(z)G(z) - F(z_0)G(z_0)\|$$
$$\leq \|F(z)G(z) - F(z)G(z_0)\| + \|F(z)G(z_0) - F(z_0)G(z_0)\|$$
$$\leq \|F(z)\| \, \|G(z) - G(z_0)\| + \|F(z)G(z_0) - F(z_0)G(z_0)\|$$
$$\to 0$$

as $z \to z_0$ in the strip $\alpha \leq \text{Re}z \leq \beta$, for $F(z)$ is uniformly bounded there, $G(z)$ is continuous on this strip, and $F(z)G(z_0)$ is holomorphic. Thus, to see $H(z)$ is holomorphic in the open strip $\alpha < \text{Re}z < \beta$, it suffices to show it is weakly holomorphic there. But

$$\langle H(z), w \rangle = \langle F(z)G(z), w \rangle$$
$$= \langle G(z), F(z)^* w \rangle.$$

Since $G(z)$ is holomorphic inside the strip and $z \mapsto F(z)^* w$ is anti-holomorphic inside the strip, $H(z)$ is weakly holomorphic and hence strongly holomorphic on $\alpha < \text{Re}z < \beta$. $\square$

LEMMA VII.10. *Let $A \in \mathcal{L}(\mathcal{A})_0$. Let $F(z)$ be the strongly entire extension of $it \mapsto \Delta^{it} A \Delta^{-it}$. Then $z \mapsto F(-\bar{z})^*$ is the strongly entire extension of $it \mapsto \Delta^{it} A^* \Delta^{-it}$.*

PROOF. Let $G(z)$ be the strongly entire extension of the mapping $it \mapsto \Delta^{it}A^*\Delta^{-it}$. Then $G_{v,w}(z) = \langle G(z)v, w \rangle$ is entire, and $G_{v,w}(it) = \langle \Delta^{it}A^*\Delta^{-it}v, w \rangle = \langle v, \Delta^{it}A\Delta^{-it}w \rangle = \overline{F_{w,v}(it)}$ where $F_{w,v}$ is defined by $F_{w,v}(z) = \langle F(z)w, v \rangle$. Since $\overline{F_{w,v}(-\bar{z})}$ is entire, we have

$$G_{v,w}(z) = \overline{F_{w,v}(-\bar{z})}$$

for all z. Hence $\langle G(z)v, w \rangle = \overline{\langle F(-\bar{z})w, v \rangle} = \langle v, F(-\bar{z})w \rangle$. Hence $G(z) = F(-\bar{z})^*$. $\square$

PROPOSITION VII.8. $\mathcal{L}(\mathcal{A})_0$ is a $*$ subalgebra of $\mathcal{L}(\mathcal{A})$.

PROOF. It is clearly closed under addition and scalar multiplication and adjoints.

We show $\mathcal{L}(\mathcal{A})_0$ is closed under multiplication. Suppose A and B are in $\mathcal{L}(\mathcal{A})_0$ and $F(z)$ and $G(z)$ are the strongly entire extensions of $it \mapsto \Delta^{it}A\Delta^{-it}$ and $it \mapsto \Delta^{it}B\Delta^{-it}$ to $\mathbb{C}$ which are bounded on vertical strips. Set $H(z) = F(z)G(z)$. Note $H(z)$ is strongly entire for if $v \in \mathbb{H}$, then $H(z)v = F(z)G(z)v$; and thus since $G(z)$ is bounded on vertical strips, Lemma VII.9 implies $z \mapsto H(z)v$ is strongly holomorphic on all vertical strips. Thus $z \mapsto H(z)$ is strongly entire and extends $H(it) = \Delta^{it}AB\Delta^{-it}$. A similar argument shows $it \mapsto \Delta^{it}B^*A^*\Delta^{-it}$ has a strongly entire extension that is bounded on vertical strips. $\square$

LEMMA VII.11. Let Δ be a positive unbounded self adjoint operator on a Hilbert space $\mathbb{H}$. Then $\xi \in Dom(\Delta^\delta)$ iff the function $it \mapsto \Delta^{it}\xi$ has a continuous bounded extension to the vertical strip $0 \le Re\, z \le \delta$ such that this extension is holomorphic within the strip. When this is the case, $z \mapsto \Delta^z \xi$ is the extension. Moreover, $Dom(\Delta^z) \subseteq Dom(\ln \Delta\, \Delta^z)$ if $0 < Re\, z < \delta$ and

$$\frac{d}{dz}\Delta^z\xi = (\ln \Delta)\Delta^z\xi.$$

PROOF. Let $\xi \in Dom(\Delta^\delta)$ and let P be the resolution of the identity for Δ. Hence $\int_0^\infty \lambda^{2\delta}\, d\langle P_\lambda \xi, \xi \rangle < \infty$. Thus since $\lambda^{2s} \le \lambda^{2\delta}$ if $1 \le \lambda$ and $0 \le s \le \delta$, we see $\int_0^\infty |\lambda^{s+it}|^2\, d\langle P_\lambda \xi, \xi \rangle = \int_0^\infty \lambda^{2s}\, d\langle P_\lambda \xi, \xi \rangle < \infty$; and thus $\xi \in Dom(\Delta^{s+it})$. Set $F(z) = \Delta^z\xi$ for $0 \le Re\, z \le \delta$. Note $\|F(z)\|^2 = \int \lambda^{2Re\, z}\, d\langle P_\lambda \xi, \xi \rangle \le \|P[0,1]\xi\|^2 + \int_1^\infty \lambda^{2\delta}\, d\langle P_\lambda \xi, \xi \rangle$. Thus $F(z)$ is uniformly bounded on the strip. To see $F(z)$ is continuous, note if $z_n \to z$ in the strip, then since

$$|\lambda^{z_n} - \lambda^z|^2 \le \left(|\lambda|^{Re\, z_n} + |\lambda|^{Re\, z}\right)^2 \le 4\chi_{[0,1]}(\lambda) + 4\lambda^{2\delta} \in L^1(d\langle P_\lambda \xi, \xi \rangle),$$

the Lebesgue dominated convergence theorem implies

$$\|\Delta^{z_n}\xi - \Delta^z\xi\|^2 = \left\| \int (\lambda^{z_n} - \lambda^z)\, d\langle P_\lambda \xi, \xi\rangle \right\|^2$$

$$= \int |\lambda^{z_n} - \lambda^z|^2\, d\langle P_\lambda \xi, \xi\rangle$$

$$\to 0 \text{ as } n \to \infty.$$

Hence $F(z)$ is continuous. We next show $F(z)$ is holomorphic within the strip. First let $z = s + it$ where $0 < s < \delta$. Note that $(\ln \lambda)^2 \lambda^{2s} \to 0$ as $\lambda \to 0+$. Thus $\xi \in \text{Dom}(\ln \Delta \, \Delta^z)$ iff

$$\int_1^\infty (\ln \lambda)^2 \lambda^{2s}\, d\langle P_\lambda \xi, \xi\rangle < \infty.$$

But $(\ln \lambda)^2 \lambda^{2s-2\delta} \to 0$ as $\lambda \to \infty$. Hence

$$\int_1^\infty (\ln \lambda)^2 \lambda^{2s}\, d\langle P_\lambda \xi, \xi\rangle = \int_1^\infty (\ln \lambda)^2 \lambda^{2s-2\delta} \lambda^{2\delta}\, d\langle P_\lambda \xi, \xi\rangle < \infty.$$

Next for each n, set $F_n(z) = P[0,n]F(z) = \Delta^z P[0,n]\xi$. Then $F_n(z) = \int_{[0,n]} \lambda^z\, dP_\lambda \xi$ for $\text{Re}\, z \geq 0$ and $F_n'(z) = \int_{[0,n]} \ln \lambda\, \lambda^z\, dP_\lambda \xi$ for $\text{Re}\, z > 0$. In particular, each $F_n(z)$ is holomorphic on the open right half plane $\text{Re}\, z > 0$. We claim $F_n(z)$ converges uniformly to $F(z)$ on the strip $0 \leq \text{Re}\, z \leq \delta$. Indeed, note

$$\|F_n(z) - F(z)\|^2 = \int_{(n,\infty)} \lambda^{2\text{Re}\, z}\, d\|P_\lambda \xi\|^2$$

$$\leq \int_{(n,\infty)} \lambda^{2\delta}\, d\|P_\lambda \xi\|^2 \to 0$$

as $n \to \infty$. This implies $F(z)$ is holomorphic on the open strip $0 < \text{Re}\, z < \delta$ and $F'(z) = \lim_n F_n'(z) = \int \ln \lambda\, \lambda^z\, dP_\lambda \xi$. Hence $\xi \in \text{Dom}(\ln \Delta \, \Delta^z)$ and $F'(z) = \ln \Delta \, \Delta^z\xi$.

Conversely, suppose $it \mapsto \Delta^{it}\xi$ has an extension $F(z)$ which is continuous and bounded on the closed strip $0 \leq \text{Re}\, z \leq \delta$ and holomorphic on the open strip $0 < \text{Re}\, z < \delta$. Let $\eta \in \text{Dom}(\Delta^\delta)$. By what has already been shown, $z \mapsto \Delta^z \eta$ is defined, continuous and bounded on the closed strip $0 \leq \text{Re}\, z \leq \delta$ and holomorphic within. Hence the function $z \mapsto \langle \xi, \Delta^{\bar{z}}\eta\rangle$ is continuous and bounded on this closed strip and holomorphic within the strip. We note that for $z = it$, this function gives $\langle \xi, \Delta^{-it}\eta\rangle = \langle \Delta^{it}\xi, \eta\rangle$. But $\langle F(z), \eta\rangle$ is continuous and bounded on the strip $0 \leq \text{Re}\, z \leq \delta$ and holomorphic on its interior. It also gives $\langle \Delta^{it}\xi, \eta\rangle$ for $z = it$. By uniqueness of holomorphic extensions, we see $\langle F(z), \eta \rangle = \langle \xi, \Delta^{\bar{z}} \eta \rangle$ for $0 \le \operatorname{Re} z \le \delta$. Taking $z = \delta$, we obtain

$$\langle F(\delta), \eta \rangle = \langle \xi, \Delta^{\delta} \eta \rangle.$$

Since this holds for all $\eta \in \Delta^{\delta}$ and Δ^{δ} is self adjoint, $\xi \in \operatorname{Dom}(\Delta^{\delta})$. $\square$

PROPOSITION VII.9. *Let $\mathcal{A}$ be an achieved left Hilbert algebra. Suppose $a \in \mathcal{A}$ and $A \in \mathcal{L}(\mathcal{A})_0$. Then $Aa \in \mathcal{A}$ and $\pi(Aa) = A\pi(a)$.*

PROOF. We first note that $\eta \mapsto \pi'(\eta)Aa = A\pi'(\eta)a$ is bounded on $\mathcal{A}'$. Indeed, it is the operator $A\pi(a)$ restricted to $\mathcal{A}'$. Hence to show $Aa \in \mathcal{A}''$, we need only show $Aa \in \operatorname{Dom}(S)$. Since $S = J\Delta^{1/2}$, this is equivalent to showing $Aa \in \operatorname{Dom}(\Delta^{1/2})$. Let $F(z)$ be the strongly entire extension of $\Delta(it)A\Delta(-it)$ which is bounded on vertical strips. Since $a \in \operatorname{Dom}(S) = \operatorname{Dom}(\Delta^{1/2})$, Lemma VII.11 implies there is bounded continuous function $G(z)$ on $0 \le \operatorname{Re} z \le \frac{1}{2}$ such that $G(it) = \Delta^{it}a$ and $G(z)$ is holomorphic on $0 < \operatorname{Re} z < \frac{1}{2}$. Set $H(z) = F(z)G(z)$. Lemma VII.9 implies $H(z)$ is bounded and continuous on the strip $0 \le \operatorname{Re} z \le \frac{1}{2}$ and is holomorphic in the open strip $0 < \operatorname{Re} z < \frac{1}{2}$. Moreover, $H(it) = (\Delta^{it}A\Delta^{-it})\Delta^{it}a = \Delta^{it}Aa$. Thus using Lemma VII.11 again, we see $Aa \in \operatorname{Dom}(\Delta^{1/2})$. Thus $Aa \in \operatorname{Dom}(S)$ and $\eta \mapsto \pi'(\eta)Aa = A\pi(a)\eta$ is bounded. This gives $Aa \in \mathcal{A}'' = \mathcal{A}$ and $\pi(Aa) = A\pi(a)$. $\square$

13. Tomita Algebras

DEFINITION. A Tomita algebra $\mathcal{A}_0$ is a left Hilbert algebra such that $\Delta(\alpha) = \Delta^{\alpha}$, $\alpha \in \mathbb{C}$, is a one parameter group of operators on $\mathcal{A}_0$ such that

$$\Delta(\alpha)(\xi\eta) = (\Delta(\alpha)\xi)(\Delta(\alpha)\eta) \text{ and}$$

(a) $\alpha \mapsto \Delta(\alpha)\xi$ is entire for each $\xi \in \mathcal{A}_0$
(b) $(\Delta(\alpha)\xi)^{\#} = \Delta(-\bar{\alpha})\xi^{\#}$
(c) $\langle \Delta(\alpha)\xi, \eta \rangle = \langle \xi, \Delta(\bar{\alpha})\eta \rangle$ for $\xi, \eta \in \mathcal{A}_0$
(d) $\langle \Delta(1)\xi^{\#}, \eta^{\#} \rangle = \langle \eta, \xi \rangle$ for $\xi, \eta \in \mathcal{A}_0$
(e) $(1 + \Delta(t))\mathcal{A}_0$ is dense in $\mathcal{A}_0$ for each $t \in \mathbb{R}$.

DEFINITION. Two left Hilbert algebras $\mathcal{A}_1$ and $\mathcal{A}_2$ are said to be equivalent if $\mathcal{L}(\mathcal{A}_1) = \mathcal{L}(\mathcal{A}_2)$.

For each achieved left Hilbert algebra $\mathcal{A}$, it will be shown there exists an equivalent Tomita subalgebra of $\mathcal{A}$.

EXAMPLE ($C_c(G)$ CONTINUED). Let $\mathcal{A}_0 = C_c(G)$. Then $\mathcal{A}_0$ is a Tomita algebra, for $\Delta^\alpha f(x) = \Delta(x)^\alpha f(x)$ is entire in α. Easy calculations show $\Delta(\alpha)(f * g) = (\Delta(\alpha)f) * (\Delta(\alpha)g)$ and $(\Delta(\alpha)f)^\# = \Delta(-\bar{\alpha})f^\#$. For (d) note

$$\langle \Delta(1)f^\#, g^\# \rangle = \int \Delta(x) f^\#(x) \overline{g^\#(x)} \, dx$$

$$= \int \Delta(x)\Delta(x^{-1})\bar{f}(x^{-1})\Delta(x^{-1})g(x^{-1}) \, dx$$

$$= \int \bar{f}(x)g(x) \, dx$$

$$= \langle g, f \rangle.$$

The rest follows easily.

REMARK. Let $\mathcal{A}$ be a left Hilbert algebra. Then

$$\cap_{\alpha \in \mathbb{C}} \mathrm{Dom}(\Delta^\alpha) = \cap_{t \in \mathbb{R}} \mathrm{Dom}(\Delta^t) = \cap_{n \in \mathbb{Z}} \mathrm{Dom}(\Delta^n).$$

PROOF. Note $\Delta^{t+is} = \Delta^{is}\Delta^t$ and Δ^{is} is unitary. Also for $n > 0$, $\xi \in \mathrm{Dom}(\Delta^n)$ iff $\int_1^\infty \lambda^{2n} \, d\langle E_\lambda \xi, \xi \rangle < \infty$ iff $\int_1^\infty \lambda^{2t} \, d\langle E_\lambda \xi, \xi \rangle < \infty$ for $0 \le t \le n$ iff $\xi \in \mathrm{Dom}(\Delta^t)$ for $0 \le t \le n$; and for $n < 0$, $\xi \in \mathrm{Dom}(\Delta^n)$ iff $\int_0^1 \lambda^{2n} \, d\langle E_\lambda \xi, \xi \rangle < \infty$ iff $\int_0^1 \lambda^{2t} \, d\langle E_\lambda \xi, \xi \rangle < \infty$ for $n \le t \le 0$ iff $\xi \in \mathrm{Dom}(\Delta^t)$ for $n \le t \le 0$. Thus $\cap_t \mathrm{Dom}\, \Delta^t = \cap_n \mathrm{Dom}\, \Delta^n$. $\square$

DEFINITION. Let $\mathcal{A}$ be a left Hilbert algebra. Define $\mathcal{A}_0$ to be the space of $\xi \in \mathcal{A}$ such that $\xi \in \mathrm{Dom}(\Delta^\alpha)$ and $\Delta^\alpha \xi \in \mathcal{A}$ for all $\alpha \in \mathbb{C}$.

PROPOSITION VII.10. If $\xi \in \mathcal{A}_0$, then $\alpha \mapsto \Delta(\alpha)\xi$ is an $\mathbb{H}$ valued entire function.

PROOF. $\xi \in \mathrm{Dom}(\Delta^\delta)$ for all $\delta \in \mathbb{R}$. Hence by Lemma VII.11 applied to Δ and Δ^{-1}, the result is obtained. More specifically, $\Delta^z \xi$ and $\Delta^{-z}\xi$ are holomorphic on $0 < \mathrm{Re}\, z < \infty$ and are continuous and bounded on the strips $0 \le \mathrm{Re}\, z < r$ for $r > 0$. Hence $f(z) = \Delta^z \xi$ is holomorphic except possibly on the line $\mathrm{Re}\, z = 0$. But $\oint_\Gamma f(\gamma) \, d\gamma$ for every closed piecewise smooth curve Γ in $\mathbb{C}$. Therefore, f is entire. $\square$

LEMMA VII.12. Let $\mathcal{A}$ be an achieved left Hilbert algebra. Then $\Delta^{it}(\xi\eta) = \Delta^{it}(\xi)\Delta^{it}(\eta)$ for all $\xi, \eta \in \mathcal{A}$.

PROOF. By Theorem VII.9, $\Delta^{it}\mathcal{A} \subseteq \mathcal{A}$ and

$$\pi(\Delta^{it}(\xi\eta)) = \Delta^{it}\pi(\xi\eta)\Delta^{-it} = \Delta^{it}\pi(\xi)\Delta^{-it}\Delta^{it}\pi(\eta)\Delta^{-it}$$

$$= \pi(\Delta^{it}\xi)\pi(\Delta^{it}\eta) = \pi(\Delta^{it}\xi\,\Delta^{it}\eta).$$

By Proposition VII.4, π is faithful. Hence $\Delta^{it}(\xi\eta) = (\Delta^{it}\xi)(\Delta^{it}\eta)$. □

We next show $\mathcal{A}_0$ is dense in $\mathcal{A}$.

LEMMA VII.13. *Let f be a continuous positive definite function on $\mathbb{R}$. Then if $\mathcal{A}$ is an achieved left Hilbert algebra, $f(\ln\Delta)$ is a bounded operator, $f(\ln\Delta)\mathcal{A} \subseteq \mathcal{A}$, and if $\xi \in \mathcal{A}$, $(f(\ln\Delta)\xi)^{\#} = f(\ln\Delta)\xi^{\#}$.*

PROOF. Since f is positive definite, there is by Bochner's theorem a finite positive Borel measure μ on $\mathbb{R}$ such that

$$f(\lambda) = \int e^{i\lambda t}\, d\mu(t).$$

Hence $f(\ln\lambda) = \int_{-\infty}^{\infty} \lambda^{it}\, d\mu(t)$ for $\lambda > 0$. Thus one has $f(\ln\Delta) = \int_{-\infty}^{\infty} \Delta^{it} d\mu(t)$. Hence if $\xi \in \mathcal{A}$, $f(\ln\Delta)\xi = \int_{-\infty}^{\infty} \Delta^{it}\xi\, d\mu(t)$. Since $\mathcal{A}$ is achieved, to show $f(\ln\Delta)\xi \in \mathcal{A}$, it suffices to show $\eta' \mapsto \pi'(\eta')f(\ln\Delta)\xi$ is continuous on $\mathcal{A}'$ and $f(\ln\Delta)\xi \in \text{Dom}(S)$. First note by Theorem VII.9 that

$$||\pi'(\eta')f(\ln\Delta)\xi|| = ||\pi'(\eta')\int_{-\infty}^{\infty} \Delta^{it}\xi\, d\mu(t)|| = ||\int_{-\infty}^{\infty} \pi'(\eta')\Delta^{it}\xi\, d\mu(t)||$$

$$= ||\int \pi(\Delta^{it}\xi)\eta'\, d\mu(t)|| = ||\int \Delta^{it}\pi(\xi)\Delta^{-it}\eta'\, d\mu(t)||$$

$$\leq ||\pi(\xi)||\,||\eta'||\mu(\mathbb{R}).$$

Since $S = J\Delta^{1/2}$, $f(\ln\Delta)\xi$ is the domain of S if $\Delta^{1/2}f(\ln\Delta)\xi$ is defined. But this is $f(\ln\Delta)\Delta^{1/2}\xi$. Hence $f(\ln\Delta)\xi \in \mathcal{A}$. Finally using $J\Delta^{it}J = \Delta^{it}$ from Theorem VII.4, we see

$$S(f(\ln\Delta)\xi) = J\Delta^{1/2}f(\ln\Delta)\xi$$

$$= Jf(\ln\Delta)\Delta^{1/2}\xi$$

$$= J\int \Delta^{it}\, d\mu(t)J\, J\Delta^{1/2}\xi$$

$$= \int J\Delta^{it}J\, d\mu(t)\, S\xi$$

$$= \int \Delta^{it}\, d\mu(t)\, S\xi$$

$$= f(\ln\Delta)\xi^{\#}.$$

□

COROLLARY VII.10. *Let A be an achieved left Hilbert algebra. For $n \in \mathbb{N}$, set*

$$f_n = \frac{1}{2n} \chi_{(-n,n)} * \chi_{(-n,n)}.$$

Then f_n is a continuous positive definite function and

$$f_n(\ln \Delta)A \subseteq A_0.$$

PROOF. Note

$$f_n(x) = \frac{1}{2n} \int 1_{(-n,n)}(y) 1_{(-n,n)}(x - y) \, dy = \frac{1}{2n} \langle \lambda_x 1_{(-n,n)}, 1_{(-n,n)} \rangle$$

where λ is the left regular representation on $\mathbb{R}$. By Theorem IV.13, f_n is continuous and positive definite. To finish we need to show for $\xi \in A$, $\Delta^\alpha f_n(\ln \Delta)\xi \in A$ for all $\alpha \in \mathbb{C}$. Let $\alpha = s + it$ where s and t are real. By Theorem VII.9, since $\Delta^{it}A \subseteq A$, it suffices to show $\Delta^s f_n(\ln \Delta) \subseteq A$. Set $e_s(x) = e^{sx}$. Then $(e_s \cdot f_n)(x) = \frac{1}{2n}\langle \lambda_x(e_s 1_{(-n,n)}), e_s 1_{(-n,n)} \rangle$ and thus is continuous and positive definite. By Lemma VII.13, $e_s(\ln \Delta)f_n(\ln \Delta)\xi \in A$. Hence $\Delta^\alpha f_n(\ln \Delta)\xi \in A$ for all α and $f_n(\ln \Delta)\xi \in A_0$. $\square$

PROPOSITION VII.11. *Suppose A is an achieved left Hilbert algebra. Then A_0 is dense in A.*

PROOF. We calculate f_n. Namely $f_n(y) = \frac{1}{2n}\int_{-n}^n 1_{(-n,n)}(y-x)\,dx$. Thus

$$f_n(y) = \begin{cases} \frac{1}{2n}(2n - |y|) & \text{if } |y| \leq 2n \\ 0 & \text{if } |y| > 2n. \end{cases}$$

Hence as $n \to \infty$, $f_n \to 1$ uniformly on compact sets. Since each $f_n \leq 1$, $f_n(\ln \Delta)$ converges strongly to I. Indeed, if $g_n(\lambda) = f_n(\ln \lambda)$, then

$$\|f_n(\ln \Delta)\xi - \xi\|^2 = \|g_n(\Delta)\xi - \xi\|^2 = \left\| \int g_n(\lambda) dE_\lambda \xi - \xi \right\|^2$$

$$\leq 2\|E[0,\epsilon)\xi\|^2 + \left\| \int_{[\epsilon,k]} g_n(\lambda) dE_\lambda \xi - E[0,k]\xi \right\|^2 + 2\|E(k,\infty)\xi\|^2 \to 0$$

for $n \to \infty$ as $k \to \infty$ and $\epsilon \to 0+$. $\square$

LEMMA VII.14. $A_0 \subseteq A \cap A'$ *if A is achieved.*

PROOF. By the definition of A_0, $\Delta^\alpha A_0 \subseteq A_0$ for all α. Since $SA_0 \subseteq A$, $J\Delta^{1/2}A_0 \subseteq A$. Hence $J^2\Delta^{1/2}A_0 \subseteq JA = A'$, by Theorem VII.9. Hence $\Delta^{1/2}A_0 \subseteq A'$. But then $A_0 = \Delta^{1/2}\Delta^{-1/2}A_0 \subseteq \Delta^{1/2}A_0 \subseteq A'$. $\square$

COROLLARY VII.11. *If $\mathcal{A}$ is achieved, $J\mathcal{A}_0 = \mathcal{A}_0$.*

PROOF. Let $\xi \in \mathcal{A}_0$. Then $\Delta^\alpha J\xi = J\Delta^{-\bar\alpha}\xi \in J\mathcal{A}_0 \subseteq J\mathcal{A}' = \mathcal{A}$ by Theorems VII.4 and VII.9. $\square$

COROLLARY VII.12. *Suppose $\mathcal{A}$ is achieved. Then $F\mathcal{A}_0 = \mathcal{A}_0$ and $S\mathcal{A}_0 = \mathcal{A}_0$.*

PROOF. These follow from $F = J\Delta^{-1/2} = \Delta^{1/2}J$ and $S = J\Delta^{1/2} = \Delta^{-1/2}J$. $\square$

PROPOSITION VII.12. *Let $\mathcal{A}$ be an achieved left Hilbert algebra. Then $\pi(\mathcal{A}_0) \subseteq \mathcal{L}(\mathcal{A})_0$. In particular, $\mathcal{L}(\mathcal{A})_0$ is a strongly operator dense $*$ subalgebra of $\mathcal{L}(\mathcal{A})$.*

PROOF. Let $a_0 \in \mathcal{A}_0$. By Proposition VII.10, $z \mapsto \Delta(z)a_0$ is entire. Set $F(z) = \pi(\Delta(z)a_0)$; and for $\xi, \eta \in \mathcal{A}_0$, set $F_{\xi,\eta}(z) = \langle\pi(\Delta(z)a_0)\xi, \eta\rangle$. We first note $F_{\xi,\eta}$ is an entire function. Indeed, by Lemma VII.14, $\mathcal{A}_0 \subseteq \mathcal{A} \cap \mathcal{A}'$, and thus

$$\begin{aligned} F_{\xi,\eta}(z) &= \langle(\Delta(z)a_0)\xi, \eta\rangle \\ &= \langle\pi'(\xi)(\Delta(z)a_0), \eta\rangle \\ &= \langle\Delta(z)a_0, \pi'(\xi)^*\eta\rangle \end{aligned}$$

is an entire function. Moreover, since $\Delta(s + it)a_0 = \Delta(it)\Delta(s)a_0$, one has

$$|F_{\xi,\eta}(s + it)| \le ||\Delta(s)a_0||\,||\pi'(\xi)^*\eta||;$$

and thus $F_{\xi,\eta}(z)$ is bounded on any vertical strip $a \le \mathrm{Re}\,z \le b$.

Set $M(s) = \sup_{t\in\mathbb{R}}|F_\xi(s + it)|$. By the Phragmen–Lindelof Theorem, $M(s) \le M(a)^{\frac{b-s}{b-a}}M(b)^{\frac{s-a}{b-a}}$ for $a \le s \le b$. Now since $F_{\xi,\eta}(s + it) = \langle\pi(\Delta(s + it)a_0)\xi, \eta\rangle = \langle\Delta(it)\pi(\Delta(s)a_0)\Delta(-it)\xi, \eta\rangle$, we see

$$M(s) \le ||\pi(\Delta(s)a_0||\,||\xi||\,||\eta||.$$

Hence

$$|F_{\xi,\eta}(s+it)| \le (||\pi(\Delta(a)a_0)||\,||\xi||\,||\eta||)^{\frac{b-s}{b-a}}(||\pi(\Delta(b)a_0)||\,||\xi||\,||\eta||)^{\frac{s-a}{b-a}}.$$

This implies there is a constant K depending only on a and b such that

$$|F_{\xi,\eta}(z)| \le K||\xi||\,||\eta||$$

for z in the strip $a \le \mathrm{Re}\,z \le b$. Thus $||\pi(\Delta(z)a_0)|| \le K$ when $a \le \mathrm{Re}\,z \le b$. Hence if ξ_i and η_i are sequences in $\mathcal{A}_0$ converging to ξ and η in $\mathbb{H}$, then F_{ξ_i,η_i} converges uniformly on vertical strips to $z \mapsto \langle\pi(\Delta(z)a_0)\xi, \eta\rangle$. Hence the functions $z \mapsto \langle\pi(\Delta(z)a_0)\xi, \eta\rangle$ are entire for all ξ and η and $z \mapsto \pi(\Delta(z)a_0)$ is bounded on vertical strips $a \leq \mathrm{Re}z \leq b$. Moreover, if $\eta \in \mathcal{A}'$, $\pi(\Delta(z)a_0)\eta = \pi'(\eta)(\Delta(z)a_0)$ is continuous in z. Thus $z \mapsto \pi(\Delta(z)a_0)\eta$ is continuous for all η in a dense subspace of $\mathbb{H}$. Since $z \mapsto \pi(\Delta(z)a_0)$ is uniformly bounded on compact sets, $z \mapsto \pi(\Delta(z)a_0)\xi$ is continuous everywhere for all $\xi \in \mathbb{H}$. Hence $z \mapsto \pi(\Delta(z)a_0)$ is both strongly continuous and weakly holomorphic. Thus $z \mapsto \pi(\Delta(z)a_0)$ is strongly holomorphic. It clearly extends $it \mapsto \pi(\Delta(it)a_0) = \Delta^{it}\pi(a_0)\Delta^{-it}$. The same argument gives $z \mapsto \pi(\Delta(z)a_0^{\#})$ is strongly entire and bounded vertical strips. Also $\pi(\Delta(it)a_0^{\#}) = \Delta^{it}\pi(a_0^{\#})\Delta^{-it} = \Delta^{it}\pi(a_0)^*\Delta^{-it}$. Hence $\pi(a_0) \in \mathcal{L}(\mathcal{A})_0$. $\square$

COROLLARY VII.13. *If $a \in \mathcal{A}_0$, then the strongly entire extension of $it \mapsto \Delta^{it}\pi(a_0)\Delta^{-it}$ is $z \mapsto \pi(\Delta^z a_0)$.*

LEMMA VII.15. *If $\xi, \eta \in \mathcal{A}_0$, where $\mathcal{A}$ is achieved, then $\xi\eta \in \mathcal{A}_0$ and $\Delta(\alpha)(\xi\eta) = (\Delta(\alpha)\xi)(\Delta(\alpha)\eta)$.*

PROOF. Set $F(z) = \pi(\Delta^z\xi)$. By Corollary VII.13 and Proposition VII.10, $F(z)$ is strongly entire and uniformly bounded on vertical strips $-\delta \leq \mathrm{Re}z \leq \delta$. Moreover, by Lemma VII.10, since $\pi(\xi) \in \mathcal{L}(\mathcal{A})_0$, $F(\bar{z})^*$ is strongly entire. Set $G(z) = \Delta^z\eta$. $G(z)$ is continuous and bounded on strips $-\delta \leq \mathrm{Re}z \leq \delta$ and and is entire. By Lemma VII.9, $H(z) = F(z)G(z)$ is entire everywhere and is bounded on vertical strips $-\delta \leq \mathrm{Re}z \leq \delta$. Using Lemma VII.11, we see, since $H(it) = \pi(\Delta^{it}\xi)\Delta^{it}\eta = \Delta^{it}\xi\Delta^{it}\eta = \Delta^{it}(\xi\eta)$ has an entire extension which is bounded and continuous on vertical strips $-\delta \leq \mathrm{Re}z \leq \delta$, that $\xi\eta \in \mathrm{Dom}(\Delta^{\delta})$ for all $\delta \in \mathbb{R}$. Also $\Delta^z(\xi\eta) = F(z)G(z) = \pi(\Delta^z\xi)\Delta^z\eta = \Delta^z\xi\Delta^z\eta \in \mathcal{A}$ for all z. Hence $\xi\eta \in \mathcal{A}_0$. $\square$

LEMMA VII.16. *Let $\mathcal{A}$ be an achieved left Hilbert algebra. Then*

$$(\Delta(\alpha)\xi)^{\#} = \Delta(-\bar{\alpha})\xi^{\#} \text{ for } \xi \in \mathcal{A}_0.$$

PROOF. By Theorem VII.4,

$$(\Delta(\alpha)\xi)^{\#} = S\Delta(\alpha)\xi = J\Delta^{1/2}\Delta(\alpha)\xi = J\Delta(\alpha)JJ\Delta^{1/2}\xi = \Delta(-\bar{\alpha})S\xi.$$

$\square$

LEMMA VII.17. *Let $\mathcal{A}$ be achieved. Then $\langle\Delta(\alpha)\xi, \eta\rangle = \langle\xi, \Delta(\bar{\alpha})\eta\rangle$ for $\xi, \eta \in \mathcal{A}_0$.*

PROOF. Let $\xi = s + it$. Then $\Delta^{s+it} = \Delta^s \Delta^{it}$. Hence

$$\langle \Delta(s+it)\xi, \eta \rangle = \langle \Delta^s \Delta^{it}\xi, \eta \rangle = \langle \xi, \Delta^{-it}\Delta^s \eta \rangle = \langle \xi, \Delta(s-it)\eta \rangle$$

for $\xi, \eta \in \mathcal{A}_0$. $\square$

Let T be a closed operator on $\mathbb{H}$. Define

$$\langle \xi, \eta \rangle_T = \langle \xi, \eta \rangle + \langle T\xi, T\eta \rangle.$$

This turns $\operatorname{Dom} T$ into a Hilbert space. Let $T = UH$, $H = (T^*T)^{1/2}$ be the left polar decomposition of T. Then $\operatorname{Dom} T = \operatorname{Dom} H$ as Hilbert spaces. Note a sequence ξ_n is Cauchy in $\operatorname{Dom}(T)$ iff the sequences ξ_n and $T\xi_n$ are both Cauchy in $\mathbb{H}$.

LEMMA VII.18. *A subset $\mathcal{M}$ of $\operatorname{Dom}(T)$ is dense in $\operatorname{Dom}(T)$ iff $(1+H)\mathcal{M}$ is dense in $\mathbb{H}$.*

PROOF. Set $K = (1+H^2)^{1/2}(1+H)^{-1}$. K is a bounded invertible self adjoint operator on $\mathbb{H}$. For $\xi \in \operatorname{Dom}(T)$, set $\eta = (1+H)\xi$. Since $\|T\xi\| = \|H\xi\|$, one has

$$\begin{aligned}
\|\xi\|_T^2 &= \|\xi\|^2 + \|T\xi\|^2 \\
&= \|\xi\|^2 + \|H\xi\|^2 \\
&= \|(1+H)^{-1}\eta\|^2 + \|H(1+H)^{-1}\eta\|^2 \\
&= \langle (1+H^2)(1+H)^{-2}\eta, \eta \rangle \\
&= \|K\eta\|^2 = \|K(1+H)\xi\|^2.
\end{aligned}$$

Thus $\xi \mapsto K(1+H)\xi$ is an isometry of $\operatorname{Dom}(T)$ onto $\mathbb{H}$. (It is onto $\mathbb{H}$ since $(1+H)\operatorname{Dom}(T)$ is dense in $\mathbb{H}$.) Therefore $\mathcal{M}$ is T dense in $\operatorname{Dom}(T)$ iff $(1+H)\mathcal{M}$ is dense in $\mathbb{H}$. $\square$

LEMMA VII.19. *Let $\mathcal{A}$ be an achieved left Hilbert algebra. Then $(1+\Delta(s))\mathcal{A}_0$ is dense in $\mathcal{A}_0$ for every real s and the closure of $\Delta(\alpha)|_{\mathcal{A}_0}$ is $\Delta(\alpha)$ for all $\alpha \in \mathbb{C}$.*

PROOF. We first show $(1 + \Delta(s))\mathcal{A}_0$ is dense in $\mathcal{A}_0$ for all real s. Clearly $(1 + \Delta(s))\mathcal{A}_0 \subseteq \mathcal{A}_0$. Take $\{f_n\}$ to be the sequence of functions defined in Corollary VII.10. Then $f_n(\ln \Delta)\mathcal{A}_0 \subseteq \mathcal{A}_0$ and $f_n(\ln \Delta)$ is bounded and $(1 + \Delta(s))\mathcal{A}_0 \supseteq (1 + \Delta(s))f_n(\ln \Delta)\mathcal{A}_0$. Thus if $\xi \perp (1 + \Delta(s))\mathcal{A}_0$, then $\langle (1 + \Delta(s))f_n(\ln \Delta)\mathcal{A}_0, \xi \rangle = 0$. But since f_n has compact support, $(1+\Delta(s))f_n(\ln \Delta)$ is bounded. Hence $\langle \mathcal{A}_0, (1+\Delta(s))f_n(\ln \Delta)\xi \rangle = 0$. Hence $(1+\Delta(s))f_n(\ln \Delta)\xi = 0$. This gives $f_n(\ln \Delta)\xi = 0$ for all n. By the proof of Proposition VII.11, $f_n(\ln \Delta) \to I$ strongly. Thus $\xi = 0$.

We claim the closure of $\Delta(\alpha)|_{\mathcal{A}_0}$ is $\Delta(\alpha)$. First note $\Delta(\alpha) = \Delta(s)\Delta(it)$ where $\alpha = s + it$ is the polar decomposition of $\Delta(\alpha)$. Since $(1 + \Delta(s))\mathcal{A}_0$ is dense in $\mathbb{H}$, Lemma VII.18 implies $\mathcal{A}_0$ is dense in the Hilbert space $\mathrm{Dom}(\Delta(\alpha))$. Hence if $\xi \in \mathrm{Dom}(\Delta(\alpha))$, there is a sequence ξ_n in $\mathcal{A}_0$ with $\xi_n \to \xi$ and $\Delta(\alpha)\xi_n \to \Delta(\alpha)\xi$. Thus the closure of $\Delta(\alpha)|_{\mathcal{A}_0}$ is $\Delta(\alpha)$. $\square$

14. The Maximum Tomita Subalgebra

THEOREM VII.10. *Let $\mathcal{A}$ be an achieved left Hilbert algebra. Then $\mathcal{A}_0$, i.e., $\{\xi : \Delta(\alpha)\xi \in \mathcal{A} \text{ for all } \alpha \in \mathbb{C}\}$, is the maximum Tomita subalgebra of $\mathcal{A}$. Also $\mathcal{A}_0$ is equivalent to $\mathcal{A}$, $\mathcal{A}_0' = \mathcal{A}'$, and $\mathcal{A}_0'' = \mathcal{A}'' = \mathcal{A}$.*

PROOF. Proposition VII.10, Lemmas VII.15, VII.16, VII.17, and VII.19, and

$$\langle \Delta(1)\xi, \eta \rangle = \langle FS\xi, \eta \rangle = \langle F^*\eta, S\xi \rangle = \langle S\eta, S\xi \rangle = \langle \eta^\#, \xi^\# \rangle$$

show $\mathcal{A}_0$ satisfies a, b, c, d, and e in the definition of a Tomita algebra. To finish showing $\mathcal{A}_0$ is a Tomita algebra, we need to insure $\mathcal{A}_0$ is a left Hilbert algebra. This will be done if $\mathcal{A}_0^2$ is dense in $\mathcal{A}_0$. We first show $\mathcal{A}\mathcal{A}_0$ is dense in $\mathbb{H}$. By Proposition VII.11, $\mathcal{A}_0$ is dense in $\mathcal{A}$. Choose $\xi_n \in \mathcal{A}_0$ converging to ξ where $\xi \in \mathcal{A}$. Then $\eta\xi_n \to \eta\xi$ for all $\eta \in \mathcal{A}$. Since $\mathcal{A}^2$ is dense, it follows that $\mathcal{A}\mathcal{A}_0$ is dense in $\mathbb{H}$.

Now we see $\mathcal{A}_0^2$ is dense in $\mathbb{H}$. Let $\xi_n \in \mathcal{A}_0$ with $\xi_n \to \xi$ where $\xi \in \mathcal{A}$. Thus $\xi_n\eta \to \xi\eta$ for every $\eta \in \mathcal{A}_0$ since $\mathcal{A}_0 \subseteq \mathcal{A}'$. Since $\mathcal{A}\mathcal{A}_0$ is dense and everything in $\mathcal{A}$ can be approximated in $\mathcal{A}_0$, we see $\mathcal{A}_0^2$ is dense.

Clearly $\mathcal{A}_0$ is the maximum Tomita subalgebra. To finish the proof, it suffices to show $\mathcal{A}_0' = \mathcal{A}'$. This will follow if $\overline{S}|_{\mathcal{A}_0} = S$. But

$$\overline{S}|_{\mathcal{A}_0} = \overline{J\Delta(1/2)}|_{\mathcal{A}_0} = J\overline{\Delta(1/2)}|_{\mathcal{A}_0} = J\Delta(1/2) = S$$

by Lemma VII.19. $\square$

15. Continuous Linear Functionals on Various Topologies

There is a natural pairing between a Banach space X and its dual X^* defined by $\langle x, x^* \rangle = x^*(x)$. By the Hahn–Banach Theorem, this pairing is nondegenerate. The $\sigma(X^*, X)$ topology on X^* is the weak $*$ topology on X^*.

There is a stronger locally convex topology on X^* called the bounded X topology which is denoted by BX. A set $E \subset X^*$ will be BX

open iff $E \cap X_r^*$ is relatively weak $*$ open in $X_r^* = \{x^* : ||x^*|| \leq r\}$ for $r > 0$.

Let X be a Banach space. For each sequence $s = \{x_k\}_{k=1}^\infty$ in X which converges to 0, define a seminorm $|| \cdot ||_s$ on X^* by

$$||x^*||_s = \sup_k |x^*(x_k)| = \sup_k |\langle x_k, x^* \rangle|.$$

DEFINITION. The locally convex topology defined on X^* by all the seminorms $|| \cdot ||_s$ where s is a sequence in X converging to 0 is called the BX topology or the bounded X topology on X^*.

Note the bounded X topology on X^* is stronger than the weak $*$ topology. Also note that if $s_1, s_2, \ldots, s_k$ are sequences in X converging to 0, then there is a sequence s in X converging to 0 whose range is the union of the ranges of the s_i. In particular,

$$||x^*||_s = \max\{||x^*||_{s_1}, \ldots, ||x^*||_{s_k}\}.$$

THEOREM VII.11. *A subset K of X^* is BX closed iff $K \cap X_r^*$ is closed in the weak $*$ topology for all $r > 0$.*

PROOF. We know $X_r^* = \{x^* : ||x^*|| \leq r\}$ is compact in the weak $*$ topology for $r > 0$. Assume K is BX closed. Let $x^* \in X_r^* - K \cap X_r^*$. Then there is a sequence s in X converging to 0 such that if $||y^* - x^*||_s < 1$, then $y^* \notin K$. Let $s = \{x_n\}_{n=1}^\infty$. Choose N such that $n > N$ implies $||x_n|| < \frac{1}{2r}$. Suppose $|(y^* - x^*)(x_k)| < 1$ for $k = 1, 2, \ldots, N$. Since $|\langle x_k, y^* - x^* \rangle| \leq |y^*(x_k)| + |x^*(x_k)| \leq ||y^*|| \, ||x_k|| + ||x^*|| \, ||x_k|| < 1$ if $||y^*|| \leq r$ and $k > N$, $\{y^* \in X_r^* : |\langle x_k, y^* - x^* \rangle| < 1$ for $k = 1, 2, \ldots, N\}$ has trivial intersection with $K \cap X_r^*$. Thus x^* is not a weak $*$ limit point of $K \cap X_r^*$. Hence $K \cap X_r^*$ is weak $*$ closed in X_r^*.

Conversely, suppose $K \cap X_r^*$ is weakly $*$ closed for all $r > 0$. For each finite subset F in X, let $F^0 = \{x^* : |x^*(x)| \leq 1 \text{ for } x \in F\}$. Suppose $x^* \notin K$ and a finite subset F_n has been obtained satisfying

$$(x^* + F_n^0 \cap X_n^*) \cap K = \emptyset.$$

We claim there is a finite set F in X satisfying $||x|| \leq \frac{1}{n}$ for $x \in F$ and $(x^* + (F_n \cup F)^0 \cap X_{n+1}^*) \cap K = \emptyset$. Suppose not. Then the sets $(x^* + (F_n \cup F)^0 \cap X_{n+1}^*) \cap K$ with $||x|| \leq \frac{1}{n}$ for $x \in F$ have the finite intersection property. Since these sets are weakly $*$ compact, there is a y^* in their intersection. Hence $y^* \in K$ and $|\langle x, y^* - x^* \rangle| \leq 1$ for all x satisfying either $||x|| \leq \frac{1}{n}$ or $x \in F_n$. Hence $||y^* - x^*|| \leq n$ and $y^* - x^* \in F_n^0$. Thus $y^* \in (x^* + F_n^0 \cap X_n^*) \cap K$, a contradiction.

Set $F_{n+1} = F_n \cup F$, and repeat. Enumerating the union of the F_n gives a sequence s in X converging to 0 satisfying

$$(x^* + \{y^* : ||y^*||_s \leq 1\}) \cap K = \emptyset.$$

Hence K is BX bounded. $\square$

THEOREM VII.12. *A linear functional on X^* is weak $*$ continuous iff it is BX continuous.*

PROOF. Since the BX topology is stronger than the weak $*$ topology on X^*, it suffices to show every BX continuous linear functional is weak $*$ continuous.

Suppose f is BX continuous. Then there is a sequence $s = \{x_k\}_{k=1}^{\infty}$ in X converging to 0 satisfying

$$|f(x^*)| \leq 1 \text{ if } ||x^*||_s \leq 1.$$

This implies $f(x^*) = 0$ if $||x^*||_s = 0$. Let c_0 be the Banach space of sequences converging to 0. Define $\Phi : X^* \to c_0$ by $\Phi(x^*) = \{x^*(x_k)\}_{k=1}^{\infty}$. Define F on $\Phi(X^*)$ by $F(\Phi(x^*)) = f(x^*)$. Note F is well defined. Moreover, F is continuous. Indeed, if $\Phi(x_n^*) \to 0$, then $||x_n^*||_s \to 0$ as $n \to \infty$ and consequently $f(x_n^*) \to 0$. By the Hahn–Banach Theorem, F has a continuous extension to c_0. But $c_0^* = l_1$, the space of summable sequences. Thus there exists a sequence $\{\lambda_k\}_{k=1}^{\infty}$ with $\sum |\lambda_k| < \infty$ satisfying

$$F(\Phi(x^*)) = \sum \lambda_k x^*(x_k).$$

This implies

$$f(x^*) = x^*\left(\sum \lambda_k x_k\right).$$

Thus f is weak $*$ continuous. $\square$

COROLLARY VII.14. *A convex subset of X^* is weak $*$ closed iff it is BX closed.*

PROOF. By the Hahn–Banach Theorem on locally convex spaces, a point x not in the closure of a convex set can be separated from the convex set by a continuous linear functional. $\square$

LEMMA VII.20. *Let M be a σ-weakly closed subspace of $B(\mathbb{H})$. Let M_* be the space of σ-weakly continuous linear functionals on M. Then M_* is isometricly isomorphic to $B_*(\mathbb{H})/M^{\perp}$ where $M^{\perp} = \{f \in B_*(\mathbb{H}) : f(M) = 0\}$. In particular, M_* is closed in M^* in the norm topology.*

PROOF. By Theorem III.3, $B_*(\mathbb{H})$ is a normed closed subspace in $B^*(\mathbb{H})$, and the mapping $A \mapsto L_A$ defined by $L_A(f) = f(A)$ is an isometry of $B(\mathbb{H})$ onto $B_*(\mathbb{H})^*$.

Define a mapping ϕ from $B_*(\mathbb{H})$ into M_* by $\phi(f) = f|_M$. By the Hahn–Banach Theorem, each $h \in M_*$ has a σ-weakly continuous extension to $B(\mathbb{H})$. Thus the mapping ϕ is onto. Moreover, $\|\phi(f)\| \leq \|f\|$. Hence if $M^\perp = \{f \in B_*(\mathbb{H}) : f(M) = 0\}$, the mapping $f + M^\perp \mapsto \phi(f)$ is an isomorphism from $B(\mathbb{H})_*/M^\perp$ onto M_*. Call it Φ. We show Φ is an isometry. To do this, we note $(B(\mathbb{H})_*/M^\perp)^* = M$. Indeed, if L is a norm continuous linear functional on $B(\mathbb{H})_*/M^\perp$, then $L(f + M^\perp) = f(A)$ for a unique $A \in B(\mathbb{H})$. Since $f(A) = 0$ for all $f \in M^\perp$, $A \in M$; for otherwise the Hahn–Banach Theorem implies there is a σ-weakly continuous linear functional f on $B(\mathbb{H})$ vanishing on M with $f(A) \neq 0$. Moreover, this is an isometry, for $\|L\| = \sup_{\|f+M^\perp\|<1} |f(A)| = \sup_{\|f\|<1} |f(A)| = \|A\|$. Hence $\|f + M^\perp\| = \sup_{A \in M, \|A\| \leq 1} |f(A)| = \|\phi(f)\| = \|\Phi(f + M^\perp)\|$.

Since $B_*(\mathbb{H})$ is norm closed in $B(\mathbb{H})^*$, $M_* = B_*(\mathbb{H})/M^\perp$ is a Banach space, and thus must be norm closed in M^*. □

COROLLARY VII.15. *Let M be a σ-weakly closed subspace of $B(\mathbb{H})$; and let M_* be the space of σ-weakly continuous linear functionals on M. Then M_* is a Banach subspace of $B(\mathbb{H})^*$, and the mapping $A \mapsto L_A$ from M to $(M_*)^*$ defined by $L_A(f) = f(A)$ is an isometry from M onto $(M_*)^*$.*

THEOREM VII.13. *Let M be a σ-weakly closed subspace of $B(\mathbb{H})$. Let $M_\sim$ be the weakly continuous linear functionals on M, and M_* be the σ-weakly continuous linear functionals on M. Then*

(i) *The following are equivalent for a linear functional f on M:*

    (i1) *f is weakly continuous*

    (i2) *f is strongly continuous*

    (i3) *f is strongly $*$ continuous*

    (i4) *$f = \sum_{i=1}^n \omega_{x_i, y_i}$.*

(ii) *The following are equivalent:*

    (ii1) *f is σ-weakly continuous*

    (ii2) *f is σ-strongly continuous*

    (ii3) *f is σ-strongly $*$ continuous*

    (ii4) *$f = \sum \omega_{x_i, y_i}$ where $\sum \|x_i\|^2 + \|y_i\|^2 < \infty$*

    (ii5) *$f|_{M_1}$ is σ-weakly continuous (equivalently weakly continuous)*

    (ii6) *$f|_{M_1}$ is σ-strongly continuous (or strongly continuous)* (ii7) $f|_{M_1}$ is σ-strongly $*$ continuous (or strongly $*$ continuous)

(iii) The norm closure of $M_\sim$ is M_*, and M is the dual of M_* under the natural pairing on $M_* \times M$. (This is an isometric isomorphism.)

(iv) Let K be a convex set in M. The following are equivalent:

   (iv1) K is σ-weakly closed

   (iv2) K is σ-strongly closed

   (iv3) K is σ-strongly $*$ closed

   (iv4) $K \cap M_r$ is σ-weakly closed for $r > 0$ (equivalently weakly closed)

   (iv5) $K \cap M_r$ is σ-strongly closed for $r > 0$ (equivalently strongly closed)

   (iv6) $K \cap M_r$ is σ-strongly $*$ closed for $r > 0$ (equivalently strongly $*$ closed).

PROOF. For (i), note (i4) implies (i1) implies (i2) implies (i3). We claim (i3) implies (i4). Let f be strongly $*$ continuous on M. By the Hahn–Banach Theorem, f has a strongly $*$ continuous extension to $B(\mathbb{H})$. Then Proposition III.6 implies (i4).

The equivalence of (ii1), (ii2), (ii3) and (ii4) is proved exactly the same way using Corollary III.5 instead of Proposition III.6. We finish (ii) later.

For (iii), we first show $\overline{M_\sim} = M_*$. Note $M_\sim$ is norm dense in M_* by (ii1)-(ii4) and the fact that $\sum_{i=1}^{n} \omega_{x_i, y_i}$ converges in norm to $\sum \omega_{x_i, y_i}$ if $\sum \|x_i\|^2 < \infty$ and $\sum \|y_i\|^2 < \infty$. Now by Lemma VII.20, M_* is norm closed in M^*. Thus $\overline{M_\sim} = M_*$. The rest of (iii) follows by Corollary VII.15.

The equivalence of (ii1), (ii2) and (ii3) imply the equivalence of (iv1), (iv2) and (iv3) and the equivalence of (iv4), (iv5) and (iv6).

To finish (iv), it suffices to show the equivalence of (iv1) and (iv4). Note $(M_*)^* = M$ under the natural dual pairing between M_* and M defined by $\langle f, A \rangle = f(A)$. Proposition VI.9 implies the $\sigma(M, M_*)$ topology on M is the σ-weak topology on M. Since M is the Banach space dual of M_*, this is the weak $*$ topology on M. By Corollary VII.14, a convex subset of M is weak $*$ closed iff it is bounded M_* closed. Hence a convex set K is closed in the σ-weak topology iff $K \cap M_r$ is σ-weakly closed for all $r > 0$.

To finish the proof, we complete (ii). Note any one of the equivalent statements (ii1), (ii2), (ii3) imply (ii5), (ii6), (ii7) all hold. Thus (ii) is finished if we show (ii5) implies (ii1). Suppose f is a linear functional on M such that $f|_{M_1}$ is σ-weakly continuous. Since f is linear, $f|_{M_r}$ is σ-weakly continuous for all $r > 0$. Hence $f^{-1}(0) \cap M_r$ is σ-weakly closed for all $r > 0$. By the equivalence of (iv1) and (iv4), $f^{-1}(0)$ is σ-weakly closed. Thus f is σ-weakly continuous. $\square$

16. Positive Linear Mappings Between von Neumann Algebras

DEFINITION. Let $\mathcal{A}$ and $\mathcal{B}$ be von Neumann algebras. Let $\Phi :$ $\mathcal{A} \to \mathcal{B}$ be a linear mapping. Then Φ is positive if $\Phi(\mathcal{A}^+) \subset \mathcal{B}^+$. A positive Φ is called normal if for each upper bounded directed subset $\mathcal{F}$ of $\mathcal{A}^+$, one has $\Phi(\sup \mathcal{F}) = \sup \Phi(\mathcal{F})$.

In particular, a positive linear functional f on a von Neumann algebra $\mathcal{A}$ is normal if for any increasing net T_i in $\mathcal{A}$ which converges strongly to T, then $f(T_i)$ converges to $f(T)$.

PROPOSITION VII.13. *Let f be a positive linear functional on a von Neumann algebra $\mathcal{A}$. Then the following are equivalent:*

(a) *f is σ-strongly $*$ continuous.*
(b) *f is σ-weakly continuous.*
(c) *There is a sequence v_i in $\mathbb{H}$ with $\sum \|v_i\|^2 < \infty$ such that $f = \sum_i \omega_{v_i, v_i}$.*
(d) *f is normal.*

PROOF. The equivalence of (a) and (b) follows from Proposition III.4 and the Hahn–Banach Theorem (extend the linear functional f to a continuous linear functional on $B(\mathbb{H})$). Also (c) implies (b) implies (d).

We establish (b) implies (c). By Corollary III.5, there exist w_i and w_i' such that $f(T) = \sum_i \langle Tw_i, w_i' \rangle$ where $\sum \|w_i\|^2 < \infty$ and $\sum \|w_i'\|^2 < \infty$. Hence $\omega(T) \equiv \sum \langle T(w_i + w_i'), w_i + w_i' \rangle$ is a positive linear functional on $\mathcal{A}$ and $0 \le f \le \omega$. Consider the representation π of $\mathcal{A}$ on $l_2 \otimes \mathbb{H}$ defined by $\pi(T)(\xi_i)_{i=1}^\infty = (T\xi_i)_{i=1}^\infty$. Then $\omega(T) = \langle \pi(T)(w + w'), w + w' \rangle$. It follows by Corollary II.10 that there is a positive linear operator S on $l_2 \otimes \mathbb{H}$ such that $f(T) = \langle \pi(T)S(w + w'), S(w + w') \rangle$ for $T \in \mathcal{A}$. Set $v_i = S(w + w')_i$. Then $\sum \|v_i\|^2 < \infty$ and $f(T) = \sum_i \langle Tv_i, v_i \rangle$. Hence $f = \sum_i \omega_{v_i, v_i}$.

To finish the proof, we show (d) implies (b). Suppose f is normal. Let $\mathcal{C}$ be the collection of all positive operators P in $\mathcal{A}$ with $P \le I$ having the property that f_P defined by $f_P(T) = f(TP)$ is σ-weakly continuous. We claim $\mathcal{C}$ has a maximal member. Indeed, let $\{P_\alpha\}$ be a linearly ordered subset of $\mathcal{C}$.

Note $\langle P_\alpha v, v \rangle$ converges for each v. Thus

$$\lim_\alpha \langle P_\alpha v, w \rangle = \lim_\alpha \frac{1}{4} \sum_{k=0}^{3} i^k \langle P_\alpha(v + i^k w), v + i^k w \rangle$$

exists for all v and w. This implies there is a positive linear operator P with $\|P\| \leq 1$ and $\langle Pv, w \rangle = \lim_\alpha \langle P_\alpha v, w \rangle$ for all v and w. Hence the net P_α converges weakly to P. We claim P_α converges strongly to P. Indeed, note

$$\begin{aligned}
\|(P - P_\alpha)v\|^2 &= \langle (P - P_\alpha)v, (P - P_\alpha)v \rangle \\
&= \langle (P - P_\alpha)^{1/2}v, (P - P_\alpha)^{3/2}v \rangle \\
&\leq \langle (P - P_\alpha)^{1/2}v, (P - P_\alpha)^{1/2}v \rangle^{1/2} \|(P - P_\alpha)^{3/2}(v)\| \\
&\leq \langle (P - P_\alpha)v, v \rangle^{1/2} \|v\| \to 0 \text{ as } \alpha \text{ increases.}
\end{aligned}$$

Now by Lemma II.5,

$$\begin{aligned}
|f_P(T) - f_{P_\alpha}(T)|^2 &= |f(T(P - P_\alpha))|^2 \\
&= |f(T(P - P_\alpha)^{1/2}(P - P_\alpha)^{1/2})|^2 \\
&\leq f(P - P_\alpha)f((P - P_\alpha)^{1/2}T^*T(P - P_\alpha)^{1/2}) \\
&\leq f(P - P_\alpha)\|f\|\,\|T\|^2 \to 0
\end{aligned}$$

since f is normal. Moreover, this limit is uniform for T in the unit ball of $\mathcal{A}$. Hence f_P is σ-weakly continuous on the unit ball. By Theorem VII.13 (ii), f_P is σ-weakly continuous. The Hausdorff maximality principle implies there is a maximal element P in $\mathcal{C}$.

We claim $P = I$. Suppose not. Set $Q = I - P$. Then $Q \neq 0$, and there is a vector v with $f(Q) < \omega_{v,v}(Q)$. Suppose T_i is an increasing net of positive operators in $\mathcal{A}$ satisfying $T_i \leq Q$ and $f(T_i) \geq \omega_{v,v}(T_i)$. Then $f(\sup T_i) \geq \omega_{v,v}(\sup T_i)$. Again by the Hausdorff maximality principle, there is a maximal $Q' \leq Q$ with $f(Q') \geq \omega_{v,v}(Q')$. Set $P' = Q - Q'$. Note $P' \neq 0$. Then because Q' is maximal, we see $f(A) < \omega_{v,v}(A)$ for all $A \in \mathcal{A}$ with $0 \leq A \leq P'$.

By Lemma II.3, $P'^*T^*TP' \leq \|T\|^2 P'^*P' \leq \|T\|^2\|P'\|P' \leq \|P'\|$. This implies

$$\begin{aligned}
|f_{P'}(T)|^2 &= |f(TP')|^2 \\
&\leq f(I)f(P'^*T^*TP) \\
&\leq f(I)\omega_{v,v}(P'^*T^*TP').
\end{aligned}$$

Hence $|f_{P'}(T)|^2 \leq f(I)\langle TP'v, TP'v\rangle$. One concludes $f_{P'}$ is strongly continuous and thus σ-strongly $*$ continuous. By Theorem VII.13 (i), $f_{P'}$ is σ-weakly continuous on $\mathcal{A}$. Hence $P + P' \leq I$ and $f_{P+P'}$ is σ-weakly continuous. This contradicts the maximality of P. Hence $P = I$, and $f = f_I$ is σ-weakly continuous. $\square$

LEMMA VII.21. *If Φ is positive, then $\Phi(T^*) = \Phi(T)^*$.*

PROOF. Let $\mathbb{H}$ be the Hilbert space for $\mathcal{B}$. Let $v \in \mathbb{H}$. Then $\omega_{v,v}|_{\mathcal{B}}$ is a positive linear functional on $\mathcal{B}$. Thus $\omega_{v,v} \circ \Phi$ is a positive linear functional on $\mathcal{A}$. By Corollary II.8, $\omega_{v,v}(\Phi(T^*)) = \overline{\omega_{v,v}(\Phi(T))} = \omega_{v,v}(\Phi(T)^*)$ for all v. Using the polarization formula $\omega_{v,w}(\Phi(T)) = \frac{1}{4}\sum_{k=0}^3 i^k \omega_{v+i^k w, v+i^k w}(\Phi(T))$, we obtain $\omega_{v,w}(\Phi(T^*)) = \omega_{v,w}(\Phi(T)^*)$ for all v, w. This gives $\Phi(T^*) = \Phi(T)^*$. $\square$

THEOREM VII.14. *Let $\mathcal{A}$, $\mathcal{B}$ be von Neumann algebras, and let Φ be a positive normal linear mapping from $\mathcal{A}$ to $\mathcal{B}$. Then Φ is σ-weakly continuous. Moreover, if there is a constant K with $\Phi(T^*)\Phi(T) \leq K\Phi(T^*T)$, then Φ is σ-strongly continuous. Conversely, if Φ is a σ-weakly continuous positive linear mapping from $\mathcal{A}$ to $\mathcal{B}$, then Φ is normal.*

PROOF. Let ϕ be a positive linear normal functional on $\mathcal{B}$. Then $\phi \circ \Phi$ is a positive normal functional on $\mathcal{A}$. By Proposition VII.13, $\phi \circ \Phi$ is σ-weakly continuous. But as seen in Section 2 of Chapter III, the σ-weak topology on $B(\mathbb{H})$ where $\mathbb{H}$ is the Hilbert space for $\mathcal{B}$ is defined by the seminorms $T \mapsto |\phi(T)|$ where $\phi(T) = \sum \langle Tx_i, x_i\rangle$ and $\sum \|x_i\|^2 < \infty$. Thus Φ is continuous from the σ-weak topology on $\mathcal{A}$ into the σ-weak topology on $\mathcal{B}$.

Now assume there is a $K > 0$ with $\Phi(T)^*\Phi(T) \leq K\Phi(T^*T)$. Let $T_\alpha \to 0$ σ-strongly. Then $T_\alpha^* T_\alpha$ converges σ-weakly to 0. Thus $\Phi(T_\alpha^* T_\alpha)$ converges σ-weakly to 0. Since $\Phi(T_\alpha)^*\Phi(T_\alpha) \leq K\Phi(T_\alpha^* T_\alpha)$, $\Phi(T_\alpha)^*\Phi(T_\alpha)$ converges σ-weakly to 0. (Again use the seminorms $T \mapsto |\sum\langle Tx_i, x_i\rangle|$ where $\sum \|x_i\|^2 < \infty$.) But then $\Phi(T_\alpha) \to 0$ σ-strongly.

Suppose Φ is σ-weakly continuous. Let $\mathcal{F}$ be a directed upper bounded set of positive elements in $\mathcal{A}$. Let $T = \sup \mathcal{F}$. Then $\lim_{S \in \mathcal{F}} S = T$ strongly and since $\mathcal{F}$ is norm bounded, $\mathcal{F}$ converges to T σ-strongly and thus σ-weakly. Thus $\lim_{S \in \mathcal{F}} \Phi(S) = \Phi(T)$ σ-weakly. Hence $\sup \Phi(\mathcal{F}) = \Phi(\sup \mathcal{F})$. $\square$

REMARK. If $\mathcal{B}$ is commutative, there is a K with $\Phi(T)^*\Phi(T) \leq K\Phi(T^*T)$.

PROOF. Let χ be a character of $\mathcal{B}$, i.e., $\chi \in \hat{\mathcal{B}}$, and suppose Φ is positive. Then $\chi \circ \Phi$ is positive on $\mathcal{A}$. Since $\chi(b^*) = \overline{\chi(b)}$, we see by the Cauchy-Schwarz inequality that

$$\chi(\Phi(T^*S))\chi(\Phi(T^*S)^*) \leq \chi(\Phi(T^*T))\chi(\Phi(S^*S)).$$

Since this is true for all χ, one has by the Gelfand transform that

$$\Phi(T^*S)\Phi(T^*S)^* \leq \Phi(T^*T)\Phi(S^*S).$$

Now take $S = I$ and $K = \|\Phi(I)\|$. □

COROLLARY VII.16. *Let $\mathcal{A}$ and $\mathcal{B}$ be von Neumann algebras. Let $\Phi : \mathcal{A} \to \mathcal{B}$ be a $*$ isomorphism of $\mathcal{A}$ onto $\mathcal{B}$. Then Φ is σ-weakly continuous.*

PROOF. $\Phi(\mathcal{A}^+) = \mathcal{B}^+$ since the positive elements have form X^*X. Thus

$$\Phi(\sup \mathcal{F}) = \sup \Phi(\mathcal{F})$$

for positive directed bounded sets $\mathcal{F}$. □

LEMMA VII.22. *Let $\mathcal{A}$ and $\mathcal{B}$ be von Neumann algebras, and suppose $\Phi : \mathcal{A} \to \mathcal{B}$ is a positive linear mapping. Then Φ is norm continuous.*

PROOF. Note if $0 \leq A \leq I$, then $0 \leq \Phi(A) \leq \Phi(I)$. But if $\|A\| \leq 1$, then $A = (P_1 - P_2) + i(P_3 - P_4)$ where $0 \leq P_j \leq I$. Thus $\|\Phi(A)\| \leq 4\|\Phi(I)\|$. □

PROPOSITION VII.14. *Let $\mathcal{A}$ be a von Neumann algebra on a Hilbert space $\mathbb{H}$. If M is a σ-weakly closed left sided ideal in $\mathcal{A}$, then there is a unique orthogonal projection $E \in \mathcal{A}$ with $M = \mathcal{A}E$. Moreover, if M is a two sided ideal, then $E \in \mathcal{A} \cap \mathcal{A}' = \mathcal{Z}(\mathcal{A})$; and M is a $*$ ideal.*

PROOF. Let $N = M \cap M^*$. Then N is a σ-weakly closed $*$ algebra and hence is a von Neumann algebra on the range of its identity projection E.

Set $\widetilde{M} = \{T \in \mathcal{A} : TE = T\}$. Since $E \in M$ and M is a left ideal, $\widetilde{M} \subset M$. Next suppose $T \in M$. Then $T = W|T|$ where $|T| = \sqrt{T^*T}$ and W is a partial isometry. Take $p_n(\lambda)$ to be a sequence of polynomials without constant terms converging uniformly to $\sqrt{\lambda}$ on $[0, \|T\|^2]$. Note since $T^*T \in N$, that $|T| = \sqrt{T^*T} = \lim p_n(T^*T) \in N$. Hence $|T|E = |T|$, and since $T = W|T|$, we see $TE = T$. Thus $\widetilde{M} = M$. This gives $M = \mathcal{A}E$. To see E is unique, suppose $M = \mathcal{A}F$

where F is an orthogonal projection in $\mathcal{A}$. Then F is also an identity for N, and hence $F = E$.

If M is two sided, then $M = \mathcal{A}E = \mathcal{A}E\mathcal{A}$; and hence M is a $*$ ideal and E is the unit in M. Hence $AE = (AE)E = E(AE) = (EA)E = EA$ for all $A \in \mathcal{A}$. Thus $E \in \mathcal{A} \cap \mathcal{A}'$. $\square$

THEOREM VII.15. *Let* $\Phi : \mathcal{A} \to \mathcal{B}$ *be a normal homomorphism between von Neumann algebras* $\mathcal{A}$ *and* $\mathcal{B}$. *Then* $\Phi(\mathcal{A})$ *is a strongly closed* $*$ *subalgebra of* $\mathcal{B}$. *Hence, if* $\Phi(\mathcal{A})$ *contains the identity operator,* $\Phi(\mathcal{A})$ *is a von Neumann algebra.*

PROOF. Since Φ is a positive linear transformation, $\Phi(I)$ is an orthogonal projection in the Hilbert space for $\mathcal{B}$. Moreover, $\Phi(A)(I - \Phi(I)) = (I - \Phi(I))\Phi(A)$ for all A in $\mathcal{A}$. Thus by replacing $\mathcal{B}$ by $\mathcal{B}_E$ where $E = \Phi(I)$ and where the Hilbert space for $\mathcal{B}_E$ is the range of E, we may assume $\Phi(I) = I$.

We claim it suffices to show $\Phi(\mathcal{A})_1$ is weakly (equivalently strongly) closed. Indeed, if this is shown, then $\Phi(\mathcal{A})_1$ is strongly closed in the von Neumann algebra $\overline{\Phi(\mathcal{A})}^w$. Since $\overline{\Phi(\mathcal{A})}^w = \overline{\Phi(\mathcal{A})}^s$, the Kaplansky density theorem would then imply $\Phi(\mathcal{A})_1$ is $(\overline{\Phi(\mathcal{A})}^w)_1$. Using Theorem VII.13 (iv), we would see $\Phi(\mathcal{A})$ is σ-strongly closed. By Corollary III.10, $\Phi(\mathcal{A})$ is strongly closed; and $\Phi(\mathcal{A})$ is a von Neumann algebra.

We now show $\Phi(\mathcal{A})_1$ is weakly closed. Note $\mathcal{A}_1$ is weakly (equivalently σ-weakly) compact. By Theorem VII.14, Φ is σ-weakly continuous. Thus $\Phi(\mathcal{A}_1)$ is σ-weakly compact. Lemma VII.22 implies $\Phi(\mathcal{A}_1)$ is bounded; and since it is σ-weakly closed, it is weakly closed. Set $\mathfrak{J} = \{A \in \mathcal{A} : \Phi(A) = 0\}$. Then $\mathfrak{J}$ is a σ-weakly closed $*$ ideal in $\mathcal{A}$. Proposition VII.14 implies $\mathfrak{J}$ has an identity orthogonal projection F in $\mathcal{A} \cap \mathcal{A}'$.

Thus $\mathcal{R} = \mathcal{A}(I - F)$ is a von Neumann algebra on the range of $I - F$, and Φ is one-to-one on $\mathcal{R}$. By Proposition II.4, Φ is an isometry on $\mathcal{R}$. Thus $\Phi(\mathcal{A}_1) = \Phi(\mathcal{R}_1) = \Phi(\mathcal{R})_1 = \Phi(\mathcal{A})_1$, and $\Phi(\mathcal{A})_1$ is weakly closed. $\square$

17. Weights on a von Neumann Algebra

DEFINITION. A weight on a von Neumann algebra $\mathcal{M}$ is a function $\phi : \mathcal{M}^+ \to [0, \infty]$ such that

(a) $\phi(A + B) = \phi(A) + \phi(B)$
(b) $\phi(\lambda A) = \lambda \phi(A)$ for $\lambda \geq 0$ where $0 \cdot \infty = 0$.

The weight ϕ is said to be **normal** if there is a family ϕ_i of positive normal linear functionals ϕ_i such that

$$\phi = \sum \phi_i.$$

ϕ is said to be **finite** if $\phi(A) < \infty$ for all $A > 0$ and **faithful** if $\phi(A) = 0$ implies $A = 0$. It is **semifinite** if the linear span of all $A > 0$ with $\phi(A) < \infty$ is weakly dense in $\mathcal{M}$.

For a weight ϕ on a von Neumann algebra $\mathcal{M}$, we set

$$\mathcal{N}_\phi = \{T : \phi(T^*T) < \infty, T \in \mathcal{M}\}$$
$$\mathcal{M}_\phi = \langle \{T \in \mathcal{M}^+ : \phi(T) < \infty\} \rangle$$
$$N_\phi = \{T \in \mathcal{M} : \phi(T^*T) = 0\}$$
$$M_\phi = \langle \{T \in \mathcal{M}^+ : \phi(T) = 0\} \rangle$$

where symbol $\langle \cdot \rangle$ means the linear span. In particular, ϕ is finite if $\mathcal{M}_\phi = \mathcal{M}$; it is semifinite if $\mathcal{M}_\phi$ is weakly dense in $\mathcal{M}$; and it is faithful if $M_\phi = \{0\}$.

REMARK. Let ϕ be a normal weight on a von Neumann algebra $\mathcal{M}$. Then $\phi = \sum \phi_i$ where ϕ_i are normal positive linear functionals on $\mathcal{M}$. By Propositions VII.13 and III.5, there are sequences of vectors $\{\xi_{i,j}\}_{j=1}^\infty$ with $\phi_i(T) = \sum \langle T\xi_{i,j}, \xi_{i,j} \rangle$. Combining these, we see ϕ is normal iff there is a set of vectors ξ_k such that $\phi(T) = \sum_k \langle T\xi_k, \xi_k \rangle$.

PROPOSITION VII.15. *Let ϕ be a weight on a von Neumann algebra $\mathcal{M}$. Suppose ϕ is lower σ-weakly continuous. Then for any bounded increasing net A_j in $\mathcal{M}^+$, $\lim_j \phi(A_j) = \phi(\lim_j A_j)$.*

PROOF. Let ϕ be lower σ-weakly continuous. Let $A_j \nearrow A$ in $\mathcal{M}^+$. By Theorem III.10, A_j converges strongly and hence weakly to A. Since this net is uniformly bounded, it converges σ-weakly. Thus $\phi(A) \leq \liminf_j \phi(A_j) \leq \limsup_j \phi(A_j) \leq \phi(A)$. □

PROPOSITION VII.16. *Let ϕ be a normal weight on a von Neumann algebra $\mathcal{M}$. Then ϕ is lower σ-weakly continuous; i.e., if $T_j \geq 0$ is a net converging σ-weakly to T, then $\phi(T) \leq \liminf \phi(T_j)$.*

PROOF. Since ϕ is normal, there is a family of vectors ξ_i in $\mathbb{H}$ such that $\phi(A) = \sum \langle A\xi_i, \xi_i \rangle$. Let F be a finite subset of the i's. Clearly, since $\langle T_j \xi_i, \xi_i \rangle \to \langle T\xi_i, \xi_i \rangle$ for each i, $\liminf_j \sum_i \langle T_j \xi_i, \xi_i \rangle \geq \sum_{i \in F} \langle T\xi_i, \xi_i \rangle$. Since this is true for all F, we have $\sum_i \langle T\xi_i, \xi_i \rangle \leq \liminf_j \sum_i \langle T_j \xi_i, \xi_i \rangle = \liminf_j \phi(T_j)$. □

LEMMA VII.23. *Let $\mathcal{A}$ be a C^* algebra with identity. Let $P \subseteq \mathcal{A}^+$. Assume $P + P \subseteq P$ and P is hereditary; i.e., $a \in P$, $0 \le b \le a \implies b \in P$. Then*

(i) $\mathcal{N} = \{x : x^*x \in P\}$ *is a left ideal;*

(ii) $\mathcal{M} = \mathcal{N}^*\mathcal{N} = \{\sum x_i^* y_i : x_i, y_i \in \mathcal{N}\}$ *is a $*$ subalgebra of $\mathcal{A}$, $\mathcal{M} \cap \mathcal{A}^+ = P$, and $\mathcal{M}$ is the linear span of P;*

(iii) *the norm closure of $\mathcal{M}$ is the norm closure of $\mathcal{N} \cap \mathcal{N}^*$;*

(iv) $\mathcal{M}$ *is an ideal iff $u^*\mathcal{M}u \subseteq \mathcal{M}$ for every unitary u in $\mathcal{A}$ iff $u^*xu \in P$ for all $x \in P$ and all unitary u in $\mathcal{A}$ iff $\mathcal{N}$ is an ideal;*

(v) $\mathcal{M}$ *is an ideal iff $x^*x \in P \implies xx^* \in P$.*

PROOF. Note $nP \subseteq P$ for $n = 1, 2, \cdots$ and the hereditary property implies $rP \subseteq P$ for $r \in \mathbb{R}^+$. Suppose x is in $\mathcal{N}$ and y is in $\mathcal{A}$. Lemma II.3 implies $(yx)^*(yx) = x^*y^*yx \le ||y||^2 x^*x \in P$. Thus $\mathcal{N}$ is a left ideal if it is a linear subspace. Clearly $rP \subseteq P$ for $r \in \mathbb{R}^+$ implies $\mathcal{N}$ is closed under scalar multiplication. To see $\mathcal{N}$ is closed under addition, note

$$0 \le (x - y)^*(x - y)$$
$$= 2(x^*x + y^*y) - (x + y)^*(x + y)$$
$$\le 2(x^*x + y^*y) \in P.$$

Thus $\mathcal{N}$ is a left ideal; consequentially $\mathcal{M}$ is a $*$ subalgebra of $\mathcal{A}$.

Let $\mathcal{M}^+ = \mathcal{M} \cap \mathcal{A}^+$. We show $\mathcal{M}^+ = P$. Let $a \in \mathcal{M}^+$. Then $a = \sum x_k^* y_k$ where x_k and y_k are in $\mathcal{N}$. Since $a = a^*$,

$$a = \frac{1}{4} \sum_k ((x_k + y_k)^*(x_k + y_k) - (x_k - y_k)^*(x_k - y_k))$$

$$\le \frac{1}{4} \sum_k (x_k + y_k)^*(x_k + y_k).$$

Thus a is in P by hereditarity. Conversely, let $a \in P$. Then $a^{1/2} \in \mathcal{N}$. So $a = a^{1/2}a^{1/2} \in \mathcal{M}$. To see the linear span of P is $\mathcal{M}$, note a can be written in form $\frac{1}{4} \sum_k \sum_{j=0}^{3} i^j (x_k + i^j y_k)^*(x_k + i^j y_k)$.

We now show the norm closure of $\mathcal{M}$ is the norm closure of $\mathcal{N} \cap \mathcal{N}^*$. Note $\overline{\mathcal{N} \cap \mathcal{N}^*}$ is a C^* subalgebra of $\mathcal{A}$. Let $x \ge 0$ be in $\overline{\mathcal{N} \cap \mathcal{N}^*}$. Then $x^{1/2} \in \overline{\mathcal{N} \cap \mathcal{N}^*}$. Hence $x^{1/2} = \lim y_n$ where $y_n \in \mathcal{N} \cap \mathcal{N}^*$. Thus $x = \lim y_n^2 \in \overline{\mathcal{M}}$. This implies $\overline{\mathcal{M}} \supseteq \overline{\mathcal{N} \cap \mathcal{N}^*}$. But $\mathcal{M} \subseteq \mathcal{N} \cap \mathcal{N}^*$. Hence $\overline{\mathcal{M}} = \overline{\mathcal{N} \cap \mathcal{N}^*}$.

Now suppose $u^*\mathcal{M}u \subseteq \mathcal{M}$ for all unitary u in $\mathcal{A}$. Since $\mathcal{M}$ is the linear span of P, this is equivalent to $u^*Pu = P$ for all

u. Thus both imply $u^*\mathcal{N}u \subseteq \mathcal{N}$ for all u; for if $x \in \mathcal{N}$, then $(u^*xu)^*(u^*xu) = u^*x^*xu \in P$. Conversely, suppose $u^*\mathcal{N}u \subseteq \mathcal{N}$ for all u; and let a be in P. Then $a^{1/2} \in \mathcal{N}$ and $u^*a^{1/2}u \in \mathcal{N}$. Hence $u^*au = (u^*a^{1/2}u)(u^*a^{1/2}u) \in \mathcal{M}$.

Since $\mathcal{N}$ is a left ideal, $u^*\mathcal{N}u = \mathcal{N}$ for all u is equivalent to $\mathcal{N}u = \mathcal{N}$ for all u. This is equivalent to $\mathcal{N}A = \mathcal{N}$ for every element of $\mathcal{A}$ is a finite linear span of unitaries. To see this, let $a \in \mathcal{A}$ be self adjoint. By scaling, we may assume $\|a\| \leq 1$. Then $u = a + i\sqrt{1 - a^2}$ is unitary and $a = \frac{1}{2}(u + u^*)$. Thus $u^*\mathcal{M}u = \mathcal{M}$ for all u iff $\mathcal{N}A = \mathcal{N}$. This implies $\mathcal{M}$ is a two sided ideal.

Finally, if $\mathcal{M}$ is an ideal and $x^*x \in P$, then $x = u^*y^{1/2}$ where $x^*x = y$. Thus $xx^* = u^*yu \in u^*\mathcal{M}^+u = \mathcal{M}^+$. Hence if $\mathcal{M}$ is an ideal, then $x^*x \in P$ implies $xx^* \in P$. Conversely, suppose $x^*x \in P$ implies $xx^* \in P$. For $y \in P$ and unitary $u \in \mathcal{A}$, set $x = u^*y^{1/2}$. Then $x^*x = y$ and $u^*yu = xx^* \in P$. Thus $u^*Pu = P$ for all u, and $\mathcal{N}$ is an ideal. $\square$

COROLLARY VII.17. *Let ϕ be a weight on a von Neumann algebra $\mathcal{M}$. Then $\mathcal{N}_\phi$ is a left ideal in $\mathcal{M}$ and $\mathcal{M}_\phi = \mathcal{N}_\phi^*\mathcal{N}_\phi$ is a $*$ subalgebra. Similarly, N_ϕ is a left ideal and $M_\phi = N_\phi^*N_\phi$ is a $*$ subalgebra. Moreover,*

$$(\mathcal{M}_\phi)^+ = \{T \geq 0 : \phi(T) < \infty\} \text{ and}$$
$$(M_\phi)^+ = \{T \geq 0 : \phi(T) = 0\}.$$

Furthermore, there exist a linear functional ϕ_f on $\mathcal{M}_\phi$ such that $\phi_f = \phi$ on $\mathcal{M}_\phi^+ = \mathcal{M}_\phi \cap \mathcal{M}^+$. ϕ_f satisfies $\phi_f(A^*) = \overline{\phi_f(A)}$.

PROOF. Define ϕ_f on $\mathcal{M}_\phi$ by $\phi_f(A - B + i(C - D)) = \phi(A) - \phi(B) + i\phi(C) - i\phi(D)$ if $A, B, C, D \in \mathcal{M}^+ \cap \mathcal{M}_\phi$. We need only show ϕ_f is well defined and has the stated properties. To see ϕ_f is well defined, note if $A - B + i(C - D) = 0$, then $A = B$ and $C = D$; and thus $\phi(A) - \phi(B) + i\phi(C) - i\phi(D) = 0$. ϕ_f is clearly linear and $\phi_f = \phi$ on $\mathcal{M}^+ \cap \mathcal{M}_\phi$. Moreover, $\phi_f((A - B + i(C - D))^*) = \phi_f(A - B + i(D - C)) = \overline{\phi_f(A - B + i(C - D))}$. $\square$

LEMMA VII.24. *Let ϕ be a weight on a von Neumann algebra $\mathcal{M}$. Then ϕ is semifinite if there is an increasing net E_λ of orthogonal projections converging strongly to I with $\phi(E_\lambda) < \infty$.*

PROOF. Note $E_\lambda \in \mathcal{N}_\phi \cap \mathcal{N}_\phi^*$. Hence if $A > 0$, $\sqrt{A}E_\lambda \in \mathcal{N}_\phi$ and $E_\lambda\sqrt{A} \in \mathcal{N}_\phi^*$. Thus $E_\lambda AE_\lambda \in \mathcal{M}_\phi$. Moreover $E_\lambda AE_\lambda \to A$ strongly.

Thus $\mathcal{M}_\phi^+$ is strongly dense in $\mathcal{M}^+$. Thus $\mathcal{M}_\phi$ is weakly dense in $\mathcal{M}$. $\square$

COROLLARY VII.18. *Let ϕ be a weight on a von Neumann algebra $\mathcal{M}$. Then if for each $B > 0$ there exists an $A > 0$ with $A < B$ and $\phi(A) < \infty$, then ϕ is semifinite.*

PROOF. Choose a maximal orthogonal family of projections E_α with $\phi(E_\alpha) < \infty$. We claim $\sum E_\alpha = I$. Indeed, if not, let $E = I - \sum E_\alpha$. If $E \neq 0$, then there is a nonzero $A \in \mathcal{M}^+$ with $\phi(A) < \infty$ and $A < E$. By the spectral theorem, there is a nonzero projection $F \in \mathcal{M}$ and a $\lambda > 0$ such that $\lambda F < A$. Thus $\lambda F < E$ and $\phi(F) < \infty$. This implies $F \leq E$, and thus the family $\{E_\alpha\}$ was not maximal. Hence there is a net I_η (finite sums of projections in the family $\{E_\alpha\}$) of orthogonal projections with $\phi(I_\eta) < \infty$ and $I_\eta \to I$ strongly. $\square$

PROPOSITION VII.17. *Let $\mathcal{M}$ be a von Neumann algebra. Then there exists a faithful normal semifinite weight on $\mathcal{M}$.*

PROOF. We claim there is a maximal family of nonzero normal linear functionals of the form $\omega_\xi(X) = \langle X\xi, \xi \rangle$ so that the subspaces $\overline{\mathcal{M}'\xi}$ are orthogonal. Let $\mathcal{C}$ be a chain of such families ordered by inclusion. Then $\cup \mathcal{C}$ is an upperbound. Hence by the Hausdorff maximality principle, a maximal family exists. Let $\{\omega_{\xi_i} : i \in I\}$ be such a family. Set E_i to be the orthogonal projection with range $\overline{\mathcal{M}'\xi_i}$. If $\oplus E_i < I$, then $\oplus \overline{\mathcal{M}'\xi_i} \neq \mathbb{H}$. Choose a nonzero vector $\xi \in \oplus \overline{\mathcal{M}'\xi_i}^\perp$. Then $\mathcal{M}'\xi \subseteq \overline{\mathcal{M}'\xi_i}^\perp$ for $\langle A'\xi, B'\xi_i \rangle = \langle \xi, A'^* B'\xi_i \rangle = 0$. Thus ω_ξ could be added to the family $\mathcal{C}$, and $\mathcal{C}$ would not be maximal. Hence $\oplus E_i = I$.

Set $\phi = \sum \omega_{\xi_i}$. Then ϕ is normal. Moreover, if $A > 0$, then $A^{1/2}E_i \neq 0$ for some i, and thus $A^{1/2}\xi_i \neq 0$. Therefore $\omega_{\xi_i}(A) = \langle A^{1/2}\xi_i, A^{1/2}\xi_i \rangle \neq 0$, and ϕ is faithful.

To see that ϕ is semifinite, let $\mathcal{F}$ be the set of all finite subsets F of I ordered by inclusion. For $\lambda \in \mathcal{F}$, set $E_\lambda = \sum_{i \in F} E_i$. The E_λ are orthogonal projections converging strongly to I. Moreover, $\phi(E_\lambda) < \infty$ for $E_i\xi_j = E_iE_j\xi_j = 0$ for $i \neq j$. Lemma VII.24 implies ϕ is semifinite. $\square$

18. The GNS Theorem for Normal Semifinite Weights

Let ϕ be a weight on a von Neumann algebra $\mathcal{M}$. By Lemma VII.23 and Corollary VII.17, $\mathcal{M}_\phi = \mathcal{N}_\phi^* \mathcal{N}_\phi$ is a $*$ algebra satisfying

$\mathcal{M}_\phi = \langle\{T \geq 0 : \phi(T) < \infty\}\rangle = \langle(\mathcal{M}_\phi)^+\rangle$. Moreover, ϕ has a unique linear extension ϕ_f to $\mathcal{M}_\phi$. Now $\mathcal{M}_\phi \subseteq \mathcal{N}_\phi \cap \mathcal{N}_\phi^*$ and $M_\phi \subseteq N_\phi \cap N_\phi^*$. Define

$$\langle X + N_\phi, Y + N_\phi\rangle = \phi_f(Y^*X) \text{ where } X, Y \in \mathcal{N}_\phi.$$

Note if $\langle X + N_\phi, X + N_\phi\rangle = 0$, then $\phi(X^*X) = 0$ and $X \in N_\phi$. The Cauchy–Schwarz inequality for the positive definite form $\langle\cdot,\cdot\rangle$ on $\mathcal{N}_\phi/N_\phi \times \mathcal{N}_\phi/N_\phi$ implies $\mathcal{N}_\phi/N_\phi$ is a pre-Hilbert space. Let $\mathbb{H}_\phi$ be its completion. Set $\Lambda_\phi(X) = X + N_\phi$ for $X \in \mathcal{N}_\phi$. There is a representation π_ϕ of $\mathcal{M}$ on $\mathbb{H}_\phi$ satisfying $\pi_\phi(X)\Lambda_\phi(Y) = \Lambda_\phi(XY)$ for $X \in \mathcal{M}$ and $Y \in \mathcal{N}_\phi$. Indeed, since $Y^*X^*XY \leq ||X||^2Y^*Y$,

$$||\Lambda_\phi(XY)||^2 = \phi(Y^*X^*XY)$$
$$\leq ||X||^2\phi(Y^*Y)$$
$$\leq ||X||^2||\Lambda_\phi(Y)||_\phi^2;$$

and thus $\pi_\phi(X)$ is a bounded operator on $\mathcal{N}_\phi/N_\phi$. Clearly, one has $\pi_\phi(X_1)\pi_\phi(X_2) = \pi_\phi(X_1X_2)$ and from

$$\langle\pi_\phi(X)\Lambda_\phi(Y), \Lambda_\phi(Z)\rangle_\phi = \phi_f(Z^*XY)$$
$$= \phi_f((X^*Z)^*Y)$$
$$= \langle\Lambda_\phi(Y), \pi_\phi(X^*)\Lambda_\phi(Z)\rangle_\phi,$$

one sees $\pi_\phi(X^*) = \pi_\phi(X)^*$.

REMARK. If ϕ is faithful and semifinite, one obtains a faithful representation of $\mathcal{M}$ on $\mathbb{H}_\phi$. This representation will be used to form a left Hilbert algebra.

Let ϕ be a faithful normal semifinite weight on a von Neumann algebra $\mathcal{M}$. We shall show that there is a natural embedding of $\mathcal{N}_\phi \cap \mathcal{N}_\phi^*$ into $\mathbb{H}_\phi$ such that $\mathcal{N}_\phi \cap \mathcal{N}_\phi^*$ considered as a subset of $\mathbb{H}_\phi$ is an achieved left Hilbert algebra. First note $N_\phi = \{0\}$ for if $\phi(T^*T) = 0$, then $T = 0$. In particular, $\mathcal{N}_\phi$ is a dense subspace of $\mathbb{H}_\phi$.

LEMMA VII.25. *Let ϕ be a normal weight on a von Neumann algebra $\mathcal{M}$. Let $B \in \mathcal{N}_\phi$. Then ϕ_B defined by $\phi_B(X) = \phi_f(B^*XB)$ is a normal positive linear functional on $\mathcal{M}$ and $|\phi_B(X)| \leq ||X||\phi(B^*B)$.*

PROOF. $B^*XB \in \mathcal{N}_\phi^*\mathcal{N}_\phi = \mathcal{M}_\phi$, and thus ϕ_B is well defined. Moreover,

$$\phi_B(X^*X) \geq 0 \text{ for } B^*X^*BX \in \mathcal{M}_\phi^+ = \{T \geq 0 : \phi(T) < \infty\}.$$

By Corollary II.8, ϕ_B is continuous and has norm $\phi_B(I) = \phi(B^*B)$. Finally, ϕ_B is normal; for if $\phi = \sum \omega_i$ where ω_i are positive normal linear functionals on $\mathcal{M}$, then $\phi_B = \sum \omega_{i_B}$, and the ω_{i_B} are positive normal functionals on $\mathcal{M}$. □

LEMMA VII.26. *A representation π of a von Neumann algebra $\mathcal{A}$ is normal iff for each ξ in a dense subspace of the Hilbert space of π, the linear functional $\omega_{\xi,\xi} \circ \pi$ is weakly continuous on the unit ball of $\mathcal{A}$. In particular, $\pi(\mathcal{A})$ is a von Neumann algebra.*

PROOF. Suppose π is normal. Now $\omega_{\xi,\xi}$ is a positive normal linear functional on $B(\mathbb{H})$. Thus $\omega_{\xi,\xi} \circ \pi$ is a normal linear functional on $\mathcal{A}$. By Proposition VII.13, $\omega_{\xi,\xi} \circ \pi$ is σ-weakly continuous. Thus $\omega_{\xi,\xi} \circ \pi$ is weakly continuous on $\mathcal{A}_1$, the unit ball of $\mathcal{A}$.

Conversely, suppose $\omega_{\xi,\xi} \circ \pi$ are weakly continuous on the unit ball $\mathcal{A}_1$ for each ξ in a dense subspace $\mathcal{D}$ of $\mathbb{H}_\pi$. Then, by Theorem VII.13, these functionals are strongly continuous on the unit ball of $\mathcal{A}$. But any $\omega_{\xi',\xi'} \circ \pi$ on the unit ball of $\mathcal{A}$ is a uniform limit of $\omega_{\xi,\xi} \circ \pi$ where $\xi \in \mathcal{D}$. Hence $\omega_{\xi,\xi} \circ \pi$ are strongly continuous on the unit ball of $\mathcal{A}$ for all $\xi \in \mathbb{H}_\pi$. Since $|\langle \pi(A)\xi, \eta \rangle| \leq \langle \pi(A)\xi, \xi \rangle^{1/2} \langle \pi(A)\eta, \eta \rangle^{1/2}$, it follows that $\omega_{\xi,\eta} \circ \pi$ are weakly continuous on the unit ball $\mathcal{A}_1$ for all $\xi, \eta \in \mathbb{H}_\pi$. This implies $A \mapsto \pi(A)$ is continuous on the unit ball $\mathcal{A}_1$ with the strong operator topology into $B(\mathbb{H}_\pi)$ with the strong operator topology. Indeed, $\|\pi(A)\xi\|^2 = \langle \pi(A^*A)\xi, \xi \rangle \to 0$ as $A \to 0$ strongly; for A^*A converges weakly to 0 as A converges strongly to 0 since $\langle A^*A\eta, \eta \rangle = \|A\eta\|^2 \to 0$. By Theorem VII.13(ii) and Theorem VII.14, strong continuity on the unit ball of $\mathcal{A}$ implies π is normal.

The last statement follows from Theorem VII.15. □

COROLLARY VII.19. *Suppose π is a normal representation of a von Neumann algebra $\mathcal{M}$. Then π is weakly continuous on the unit ball of $\mathcal{M}$.*

PROPOSITION VII.18. *Suppose ϕ is a normal weight on a von Neumann algebra $\mathcal{M}$. Then the representation π_ϕ of $\mathcal{M}$ is normal, and thus $\pi_\phi(\mathcal{M})$ is a von Neumann algebra on $\mathbb{H}_\phi$. Moreover, if ϕ is faithful and semifinite, then π_ϕ is faithful.*

PROOF. The positive linear functionals ϕ_B, $B \in \mathcal{N}_\phi$, are normal. But $\phi_B(X) = \omega_{\Lambda_\phi(B),\Lambda_\phi(B)}(X)$ for $X \in \mathcal{M}$. In particular, $\omega_{\xi,\xi} \circ \pi_\phi$ are normal positive linear functionals for all $\xi \in \mathcal{N}_\phi/N_\phi$. By Proposition VII.13 and Theorem VII.13, these functionals are weakly continuous on the unit ball of $\mathcal{M}$. Lemma VII.26 implies π_ϕ is normal, and $\pi_\phi(\mathcal{M})$ is a von Neumann algebra.

Suppose ϕ is faithful and semifinite. Then $N_\phi = \{0\}$, and $\mathcal{M}_\phi$ is weakly dense in $\mathcal{M}$. Hence $\mathcal{M}_\phi$ is strongly dense in $\mathcal{M}$. Thus if $A \in \mathcal{M}$ and $\pi_\phi(A) = 0$, then $AB = 0$ for all $B \in \mathcal{M}_\phi$ for $\mathcal{M}_\phi \subseteq N_\phi$. Hence $AB = 0$ for all $B \in \mathcal{M}$. Thus $A = 0$, and π_ϕ is faithful. $\square$

Define $\mathcal{U} = N_\phi \cap N_\phi^*$. Then $\mathcal{U}$ is a star algebra contained in $\mathcal{M}$.

PROPOSITION VII.19. *Let ϕ be a faithful normal semifinite weight on a von Neumann algebra $\mathcal{M}$. Then $\mathcal{U}$ is weakly dense in $\mathcal{M}$, and $\pi_\phi(\mathcal{U})$ is weakly dense in the Von Neumann algebra $\pi_\phi(\mathcal{M})$.*

PROOF. We know $\mathcal{M}_\phi$ is weakly dense in $\mathcal{M}$; and thus, since $\mathcal{M}_\phi \subseteq \mathcal{U}$, $\mathcal{U}$ is weakly dense in $\mathcal{M}$. The Kaplansky density theorem implies $\mathcal{U}_1$ is weakly dense in $\mathcal{M}_1$. By Corollary VII.19, π_ϕ is weakly continuous on the unit ball $\mathcal{M}_1$. Thus $\pi_\phi(\mathcal{U}_1)$ is weakly dense in $\pi_\phi(\mathcal{M}_1)$. Since π_ϕ is faithful, Proposition II.4 implies it is an isometry; and thus $\pi_\phi(\mathcal{M}_1) = \pi_\phi(\mathcal{M})_1$. Hence $\pi_\phi(\mathcal{U})$ is weakly dense in $\pi_\phi(\mathcal{M})$. $\square$

PROPOSITION VII.20. *Let ϕ be a faithful normal semifinite weight on von Neumann algebra $\mathcal{M}$. Then $\mathcal{U} = N_\phi \cap N_\phi^* \subseteq N_\phi + N_\phi \subseteq \mathbb{H}_\phi$ is a left Hilbert algebra where $(A + N_\phi)(B + N_\phi) = AB = AB + N_\phi$ and $(A + N_\phi)^\# = A^* + N_\phi = A^*$.*

PROOF. First note $\|AB\|_\phi = \|\pi_\phi(A)(B+N_\phi)\| \le \|A\|\,\|B+N_\phi\|_\phi$, and thus $B \mapsto AB$ is bounded on $\mathbb{H}_\phi$ for each $A \in \mathcal{U}$. Also note

$$\langle AB, C \rangle_\phi = \langle \pi_\phi(A)(B + N_\phi), C + N_\phi \rangle_\phi$$
$$= \langle B + N_\phi, \pi_\phi(A^*)(C + N_\phi) \rangle_\phi$$
$$= \langle B, A^*C \rangle_\phi$$

for A, B and C in $\mathcal{U}$.

To see $\mathcal{U}^2$ is $\mathbb{H}_\phi$ dense in $\mathcal{U}$, we know by Propositions VII.18 and VII.19 that π_ϕ is an isometry of $\mathcal{U}$ onto a weakly dense $*$ subalgebra of $\pi_\phi(\mathcal{M}F)$. By the Kaplansky Density Theorem, $\pi_\phi(\mathcal{U}_1)$ is strongly dense in $\pi_\phi(\mathcal{M})_1$. Hence there is a net I_η in $\mathcal{U}_1$ such that $\pi_\phi(I_\eta)$ converges strongly in $\pi_\phi(\mathcal{M})_1$ to the identity on $\mathbb{H}_\phi$. Thus $\pi_\phi(I_\eta)^2$ converges strongly. Thus if $A \in N_\phi$, $I_\eta^2 A \in \mathcal{U}(N_\phi \cap N_\phi^*)N_\phi \subseteq \mathcal{U}(N_\phi \cap N_\phi^*) = \mathcal{U}^2$. Moreover, $\|I_\eta^2 A - A\|_\phi = \|\pi(I_\eta)^2 A - A\|_\phi \to 0$ as η increases. Thus $\mathcal{U}^2$ is $\mathbb{H}_\phi$ dense in N_ϕ. Since N_ϕ is dense in $\mathbb{H}_\phi$, we see $\mathcal{U}^2$ is $\mathbb{H}_\phi$ dense in $\mathcal{U}$.

Finally assume $A_n \in \mathcal{U}$, $A_n \to 0$ in $\mathbb{H}_\phi$ and A_n^* is Cauchy in $\mathbb{H}_\phi$. To show $(A + N_\phi) \mapsto (A^* + N_\phi)$ extends to a closed operator, we show $A_n^* \to 0$ in $\mathbb{H}_\phi$.

Since ϕ is normal, there is a collection $\{v_j\}$ of vectors in $\mathbb{H}$ such that $\phi = \sum_j \omega_{v_j, v_j}$. Since $\phi(A_n^* A_n) \to 0$, $\lim_n \sum_j \langle A_n v_j, A_n v_j \rangle = 0$. Now $||A_m^* - A_n^*||_\phi^2 \to 0$ for large m and n means for any $\epsilon > 0$, one has a natural number $N(\epsilon)$ such that $\sum_j \langle (A_m^* - A_n^*)v_j, (A_m^* - A_n^*)v_j \rangle < \epsilon^2$ for $m, n > N(\epsilon)$. Thus for each j, $\lim_n A_n v_j = 0$, and $A_n^* v_j$ is Cauchy in $\mathbb{H}$. Let $w_j = \lim_n A_n^* v_j$.

Next note, since ϕ is faithful, that the vectors v_j are separating for the von Neumann algebra $\mathcal{M}$. Indeed, if $Av_j = 0$ for all j, then $\phi(A^* A) = 0$; and thus $A = 0$. This implies the collection $\{v_j\}$ is generating for the commutant $\mathcal{M}'$ (i.e., the linear span of the vectors Bv_j for $B \in \mathcal{M}'$ is dense in $\mathbb{H}$). But if $B \in \mathcal{M}'$,

$$\langle w_{j'}, Bv_j \rangle = \lim_{n \to \infty} \langle A_n^* v_{j'}, Bv_j \rangle$$
$$= \lim_{n \to \infty} \langle v_{j'}, BA_n v_j \rangle$$
$$= 0.$$

Hence $w_{j'} = 0$ for all j'. Thus $A_n^* v_j \to 0$ as $n \to \infty$ for each j. Now for each finite set F of j's, one has

$$\sum_{j \in F} ||A_m^* v_j - A_n^* v_j||^2 < \epsilon^2 \text{ if } m, n > N(\epsilon).$$

Letting $m \to \infty$, we see $\sum_{j \in F} ||A_n^* v_j||^2 \leq \epsilon^2$ for $n > N(\epsilon)$. Since this is true for all finite subsets F, $\phi(A_n A_n^*) \leq \epsilon^2$ for $n > N(\epsilon)$. Consequently $A_n^* \to 0$ in $\mathbb{H}_\phi$ as $n \to \infty$, and the operator $A \mapsto A^*$ on $\mathcal{U}$ is closeable. $\square$

Let $\mathcal{M}$ be a von Neumann algebra of operators on a Hilbert space $\mathbb{H}$. Let ϕ be a faithful normal semifinite weight on $\mathcal{M}$. Then by Proposition VII.20, $\mathcal{U}$ is a left Hilbert algebra on the Hilbert space $\mathbb{H}_\phi$. Moreover, by Proposition VII.19, $\pi_\phi : \mathcal{M} \to B(\mathbb{H}_\phi)$ is a normal isomorphism of $\mathcal{M}$ into $B(\mathbb{H}_\phi)$ and $\pi_\phi(\mathcal{M})$ is the left ring $\mathcal{L}(\mathcal{U})$ of the left Hilbert algebra $\mathcal{U}$.

We use the following two lemmas to show $\mathcal{U}$ is an achieved left Hilbert algebra.

LEMMA VII.27. *Let ξ' be a vector in $\mathbb{H}_\phi$ satisfying $\langle \pi_\phi(A)\xi', \xi' \rangle_\phi \leq \phi(A)$ for $A \in \mathcal{M}^+$. Then there is a $\xi \in \mathcal{U}'$ such that $F\xi = \xi$ and $\langle \pi_\phi(A)\xi', \xi' \rangle_\phi = \langle \pi_\phi(A)\xi, \xi \rangle_\phi$ for all $A \in \mathcal{M}$.*

PROOF. The operator $A \mapsto \pi_\phi(A)\xi'$ is bounded on $\mathcal{N}_\phi$ for

$$||\pi_\phi(A)\xi'||_\phi^2 = \langle \pi_\phi(A^*A)\xi', \xi'\rangle_\phi \le \phi(A^*A) = ||A||_\phi^2$$

and thus extends to a bounded operator $\pi_\phi'(\xi')$ on $\mathbb{H}_\phi$. Moreover, on $\mathcal{N}_\phi$, one has

$$\begin{aligned}
\pi_\phi'(\xi')\pi_\phi(A)B &= \pi_\phi'(\xi')AB \\
&= \pi_\phi(AB)\xi' \\
&= \pi_\phi(A)\pi_\phi(B)\xi' \\
&= \pi_\phi(A)\pi_\phi'(\xi')B.
\end{aligned}$$

Since $\mathcal{U}$ is dense in $\mathbb{H}_\phi$ and $\pi_\phi(\mathcal{U})$ is strongly dense in $\pi_\phi(\mathcal{M})$, we see $\pi_\phi'(\xi') \in \pi_\phi(\mathcal{M})'$. Now $\pi_\phi'(\xi')$ has a polar decomposition UP where U and P are in $\pi_\phi(\mathcal{M})'$ and U is a partial isometry from the closure of the range of $\pi_\phi(\xi')^*$ onto the closure of the range of $\pi_\phi(\xi')$. Now the linear subspaces $\pi_\phi(\xi')\mathbb{H}_\phi$, $\pi_\phi(\xi')\mathcal{U} = \pi_\phi(\mathcal{U})\xi'$, and $\langle \pi_\phi(\mathcal{M})\xi'\rangle$ have the same closures in $\mathbb{H}_\phi$. Hence the projection UU^* contains the vector ξ' in its range. Thus

$$\langle \pi_\phi(A)\xi', \xi'\rangle_\phi = \langle \pi_\phi(A)\xi', UU^*\xi'\rangle_\phi = \langle \pi_\phi(A)U^*\xi', U^*\xi'\rangle_\phi$$

for $A \in \mathcal{M}$. Set $\xi = U^*\xi'$. Since $\pi_\phi(A)U^*\xi' = U^*\pi_\phi(A)\xi'$ and U^* is bounded on $\mathbb{H}_\phi$, $A \mapsto \pi_\phi(A)\xi$ is bounded on $\mathcal{N}_\phi$ into $\mathbb{H}_\phi$. Moreover, $\xi \in \text{Dom}(F)$ and $F\xi = \xi$ for if $A, B \in \mathcal{U}$, then

$$\begin{aligned}
\langle S_{00}(AB), \xi\rangle_\phi &= \langle B^*A^*, \xi\rangle_\phi \\
&= \langle A^*, \pi_\phi(B)\xi\rangle_\phi \\
&= \langle A^*, \pi_\phi(B)U^*\xi'\rangle_\phi \\
&= \langle A^*, U^*\pi_\phi(B)\xi'\rangle_\phi \\
&= \langle A^*, U^*\pi_\phi'(\xi')B\rangle_\phi \\
&= \langle A^*, U^*UPB\rangle_\phi \\
&= \langle A^*, PB\rangle_\phi \\
&= \langle PA^*, B\rangle_\phi \\
&= \langle U^*UPA^*, B\rangle_\phi \\
&= \langle U^*\pi_\phi'(\xi')A^*, B\rangle_\phi \\
&= \langle U^*\pi_\phi(A^*)\xi', B\rangle_\phi \\
&= \langle \pi_\phi(A)^*U^*\xi', B\rangle_\phi \\
&= \langle \xi, AB\rangle_\phi.
\end{aligned}$$

Since $F = S_{00}^*$ and $\mathrm{Dom}(S_{00}) = \mathcal{U}^2$, we see $\xi \in \mathrm{Dom}\, F$ and $F\xi = \xi$. Since $A \mapsto \pi_\phi(A)\xi$ is bounded, $\xi \in \mathcal{U}'$. $\square$

LEMMA VII.28. *There is a family of vectors* $\{\xi_j : j \in J\}$ *in* $\mathcal{U}'$ *such that* $F\xi_j = \xi_j$ *and*

$$\phi(S) = \sum_j \langle \pi_\phi(S)\xi_j, \xi_j \rangle_\phi \ \textit{for } S \in \mathcal{M}^+.$$

PROOF. $\phi \circ \pi_\phi^{-1}$ is a faithful normal semifinite weight on $\pi_\phi(\mathcal{M})$. As seen in the remark before the proof of Proposition VII.15, there is a family of vectors ξ_j' in $\mathbb{H}_\phi$ such that

$$\phi \circ \pi_\phi^{-1}(S') = \sum_j \langle S'\xi_j', \xi_j' \rangle_\phi.$$

Thus $\phi(S) = \sum_j \langle \pi_\phi(S)\xi_j', \xi_j' \rangle_\phi$. By Lemma VII.27, for each j there is a vector $\xi_j \in \mathcal{U}'$ such that $F\xi_j = \xi_j$ and

$$\langle \pi_\phi(S)\xi_j', \xi_j' \rangle_\phi = \langle \pi_\phi(S)\xi_j, \xi_j \rangle_\phi \ \text{for all } S \in \mathcal{M}.$$

The result now follows. $\square$

We conclude with the following theorem.

THEOREM VII.16. *Let* ϕ *be a faithful normal semifinite weight on a von Neumann algebra* $\mathcal{M}$. *Then* π_ϕ *is a normal isomorphism of* $\mathcal{M}$ *into* $B(\mathbb{H}_\phi)$. *Moreover, if* $\mathcal{U} = \mathcal{N}_\phi^* \cap \mathcal{N}_\phi$, *then* $\mathcal{U}$ *is an achieved left Hilbert algebra in* $\mathcal{M} \cap \mathbb{H}_\phi$.

PROOF. We need to show $\mathcal{U}$ is an achieved left Hilbert algebra. This would follow if $J\mathcal{U} = \mathcal{U}'$. Indeed, by Theorem VII.8, $\mathcal{U}'$ is an achieved left Hilbert algebra; and thus by Theorem VII.9, $J\mathcal{U}' = \mathcal{U}''$.

We have $J\mathcal{U} \subseteq \mathcal{U}'$, and thus we must show $\mathcal{U}' \subseteq J\mathcal{U}$. Let $\xi' \in \mathcal{U}'$. Since $\xi' = \frac{\xi' + F\xi'}{2} + i\frac{\xi' - F\xi'}{2i}$, we may assume $F\xi' = \xi'$. By Corollary VII.5, $\pi_\phi'(\xi')$ is self adjoint. Thus $J\pi_\phi'(\xi')J$ is a self adjoint operator in $\pi_\phi(\mathcal{M})$ and so must have form $\pi_\phi(S)$ for some self adjoint $S \in \mathcal{M}$. By Lemma VII.8, $\pi_\phi(S)\eta' = J\pi_\phi'(\xi')J\eta' = \pi_\phi'(\eta')J\xi'$ for all $\eta' \in \mathcal{U}'$.

We show $S \in \mathcal{U}$. By Lemma VII.28, there is a family $\{\xi_j\}_{j \in J}$ of vectors in $\mathcal{U}'$ with $F\xi_j = \xi_j$ and

$$\phi(A) = \sum_j \langle \pi_\phi(A)\xi_j, \xi_j \rangle_\phi$$

for $A \in \mathcal{M}^+$. Thus for each finite subset $F \subseteq J$ and each $A \in \mathcal{U}$, one has

$$\|A\|_\phi^2 = \phi(A^*A)$$
$$\geq \sum_{j \in F} \langle \pi_\phi(A)\xi_j, \pi_\phi(A)\xi_j \rangle_\phi$$
$$= \sum_{j \in F} \|\pi'_\phi(\xi_j)A\|_\phi^2.$$

Since $\mathcal{U}$ is dense in $\mathbb{H}_\phi$, we have

$$\|\eta\|_\phi^2 \geq \sum_{j \in F} \|\pi'_\phi(\xi_j)\eta\|_\phi^2$$

for any $\eta \in \mathbb{H}_\phi$ and any finite subset F of J. Consequently one has $\sum_j \|\pi'_\phi(\xi_j)\eta\|_\phi^2 \leq \|\eta\|_\phi^2$ for all $\eta \in \mathbb{H}_\phi$. Hence

$$\phi(S^2) = \sum_j \langle \pi_\phi(S)\xi_j, \pi_\phi(S)\xi_j \rangle_\phi$$
$$= \sum_j \langle \pi'_\phi(\xi_j)J\xi', \pi'_\phi(\xi_j)J\xi' \rangle_\phi$$
$$= \sum_j \|\pi'_\phi(\xi_j)J\xi'\|_\phi^2$$
$$\leq \|J\xi'\|_\phi^2 < \infty.$$

Since S is self adjoint, $S \in \mathcal{N}_\phi^* \cap \mathcal{N}_\phi = \mathcal{U}$. Moreover, as seen earlier $\pi_\phi(S)\eta' = J\pi'_\phi(\xi')J\eta' = \pi'_\phi(\eta')J\xi'$ for $\eta' \in \mathcal{U}'$, and thus $\pi'_\phi(\eta')S = \pi'_\phi(\eta')J\xi'$. This implies $J\xi' = S$, and finally $\xi' = JS \in J\mathcal{U}$. $\square$

19. The Natural Weight on $\mathcal{L}(\mathcal{A})$

LEMMA VII.29. *Let $\mathcal{A}$ be an achieved left Hilbert algebra in a Hilbert space $\mathbb{H}$. Let $a \in \mathcal{A}$ and suppose $A \in \mathcal{L}(\mathcal{A})$ satisfies $A\pi(a) = \pi(a)A$. Then $Aa \in \mathcal{A}$, $\pi(Aa) = \pi(a)A$ and $S(Aa) = A^*Sa$. Furthermore, if P is an orthogonal projection in $\mathcal{L}(\mathcal{A})$ and $P\pi(a) = \pi(a)$, then $Pa = a$.*

PROOF. Let $\eta \in \mathcal{A}'$. Then since $\pi'(\mathcal{A}') \subseteq \mathcal{L}(\mathcal{A})'$,

$$\pi'(\eta)Aa = A\pi'(\eta)a$$
$$= A\pi(a)\eta$$
$$= \pi(a)A\eta.$$

Furthermore

$$(\pi(a)A)^*\eta = A^*\pi(a)^*\eta$$
$$= A^*\pi(Sa)\eta$$
$$= A^*\pi'(\eta)Sa$$
$$= \pi'(\eta)A^*Sa.$$

By Proposition VII.5, $Aa \in \mathcal{A}'' = \mathcal{A}$ and $SAa = A^*Sa$.

In the case when $P\pi(a) = \pi(a)$, one has

$$\pi'(\eta)Pa = P\pi'(\eta)a = P\pi(a)\eta = \pi(a)\eta$$

and

$$\pi(a)^*\eta = \pi'(\eta)Sa$$

for $\eta \in \mathcal{A}'$. By Proposition VII.5, $Pa \in \mathcal{A}$ and $\pi(Pa) = \pi(a)$. Thus $\pi(Pa)\eta = \pi(a)\eta$ for all $\eta \in \mathcal{A}'$. Equivalently, $\pi'(\eta)Pa = \pi'(\eta)a$ for all η, and thus since $\pi'(\mathcal{A}')$ generates $\mathcal{L}(\mathcal{A})'$, $Pa = a$. □

LEMMA VII.30. *Suppose R and S are positive operators in a von Neumann algebra $\mathcal{M}$ on a Hilbert space $\mathbb{H}$. If $R \leq S$, then there is a unique operator A such that $R^{1/2} = AS^{1/2}$ and $A = 0$ on $\ker(S)$. Furthermore, $A \in \mathcal{M}$ and $||A|| \leq 1$.*

PROOF. Consider the mapping on the range of $S^{1/2}$ defined by $S^{1/2}\xi \mapsto R^{1/2}\xi$. Note $||R^{1/2}\xi||^2 = \langle R\xi, \xi \rangle \leq \langle S\xi, \xi \rangle = ||S^{1/2}\xi||^2$, and thus this mapping is well defined and bounded. Use A to denote the linear operator on $\ker(S) \oplus \overline{\mathrm{Rang}(S^{1/2})}$ which is 0 on $\ker(S)$ and which on $\overline{\mathrm{Rang}(S^{1/2})}$ is the continuous extension of the operator $S^{1/2}\xi \mapsto R^{1/2}\xi$. Since $\ker S = \ker S^{1/2}$ and $S^{1/2}$ is self adjoint, A is defined on all of $\mathbb{H}$ for $\ker(S^{1/2})^* \oplus \overline{\mathrm{Rang}(S^{1/2})} = \mathbb{H}$. Clearly $R^{1/2} = AS^{1/2}$ and A is unique.

Let $T \in \mathcal{M}'$, the commutant to $\mathcal{M}$. Then $ATS^{1/2}\xi = AS^{1/2}T\xi = R^{1/2}T\xi = TR^{1/2}\xi = TAS^{1/2}\xi$. Thus $AT = TA$ on the $\overline{\mathrm{Rang}(S^{1/2})}$. For $\xi \in \ker S$, $TA\xi = 0 = AT\xi$ for $ST\xi = TS\xi = 0$. Hence $AT = TA$ on $\ker S$. Thus A is in $\mathcal{M}''$. □

DEFINITION. Let $\mathcal{A}$ be an achieved left Hilbert algebra in a Hilbert space $\mathbb{H}$. For $A \in \mathcal{L}(\mathcal{A})^+$ define

$$\phi_\mathcal{A}(A) = \begin{cases} ||a||^2 & \text{if } A^{1/2} = \pi(a), \quad a \in \mathcal{A} \\ \infty & \text{otherwise.} \end{cases}$$

We shall show $\phi_\mathcal{A}$ is a faithful normal semifinite weight on $\mathcal{L}(\mathcal{A})$. For convenience, we shall write $\phi = \phi_\mathcal{A}$ when the meaning is unambiguous.

LEMMA VII.31. *For $0 \leq A \leq B$ in $\mathcal{L}(\mathcal{A})$, $\phi(A) \leq \phi(B)$.*

PROOF. We may assume $B^{1/2} = \phi(b)$ for some $b \in \mathcal{A}$. By Lemma VII.30, there is an operator $R \in \mathcal{L}(\mathcal{A})$ with $||R|| \leq 1$ and $A^{1/2} = RB^{1/2}$. Set $a = Rb$. Then for $\eta \in \mathcal{A}'$,

$$\begin{aligned} \pi'(\eta)a &= \pi'(\eta)Rb \\ &= R\pi'(\eta)b \\ &= R\pi(b)\eta \\ &= RB^{1/2}\eta \\ &= A^{1/2}\eta. \end{aligned}$$

By Proposition VII.5, since $A^{1/2}$ is self adjoint, $a \in \mathcal{A}'' = \mathcal{A}$, $Sa = a$, and $\pi(a) = A^{1/2}$. Hence $\phi(A) = ||a||^2 = ||Rb||^2 \leq ||b||^2 = \phi(B)$. $\square$

LEMMA VII.32. *ϕ is additive, and $\phi(\lambda A) = \lambda\phi(A)$ if $\lambda \geq 0$ and $A \in \mathcal{L}(\mathcal{A})^+$.*

PROOF. Since ϕ is increasing, $\phi(A) = \infty$ or $\phi(B) = \infty$ implies $\phi(A + B) = \infty$. Hence we may suppose $\phi(A) < \infty$ and $\phi(B) < \infty$. Choose $a, b \in \mathcal{A}$ such that $\phi(a) = A^{1/2}$ and $\phi(b) = B^{1/2}$. By Lemma VII.30, there are $R, S \in \mathcal{L}(\mathcal{A})$ with $||R|| \leq 1$, $||S|| \leq 1$ and $R(A + B)^{1/2} = A^{1/2}$, $S(A + B)^{1/2} = B^{1/2}$ and $R = S = 0$ on $\ker(A + B)^{1/2}$. Set $P = R^*R + S^*S$. Then

$$\begin{aligned} A + B &= A^{1/2}A^{1/2} + B^{1/2}B^{1/2} \\ &= (A + B)^{1/2}R^*R(A + B)^{1/2} + (A + B)^{1/2}S^*S(A + B)^{1/2} \\ &= (A + B)^{1/2}P(A + B)^{1/2}. \end{aligned}$$

Hence

$$\langle P(A + B)^{1/2}\xi, (A + B)^{1/2}\xi' \rangle = \langle (A + B)^{1/2}\xi, (A + B)^{1/2}\xi' \rangle$$

for all ξ and ξ'. Since P vanishes on the orthogonal complement of the range of $(A + B)^{1/2}$, P is the orthogonal projection of $\mathbb{H}$ onto the closure of the range of $(A + B)^{1/2}$. Hence

$$(A + B)^{1/2} = P(A + B)^{1/2}$$
$$= R^*R(A + B)^{1/2} + S^*S(A + B)^{1/2}$$
$$= R^*A^{1/2} + S^*B^{1/2}.$$

Thus if $\xi = R^*a + S^*b$ and $\eta \in \mathcal{A}'$, one has

$$\pi'(\eta)\xi = \pi'(\eta)(R^*a + S^*b)$$
$$= R^*\pi'(\eta)a + S^*\pi'(\eta)b$$
$$= R^*\pi(a)\eta + S^*\pi(b)\eta$$
$$= (R^*A^{1/2} + S^*B^{1/2})\eta$$
$$= (A + B)^{1/2}\eta.$$

Proposition VII.5 gives $\xi \in \mathcal{A}'' = \mathcal{A}$ and $\pi(\xi) = (A + B)^{1/2}$. Moreover, since $P\pi(\xi) = \pi(\xi) = \pi(\xi)P$, Lemma VII.29 implies $P\xi = \xi$. Furthermore,

$$\pi'(\eta)R\xi = R\pi'(\eta)\xi$$
$$= R(A + B)^{1/2}\eta$$
$$= A^{1/2}\eta \text{ for } \eta \in \mathcal{A}'.$$

Thus again by Proposition VII.5, $R\xi \in \mathcal{A}'' = \mathcal{A}$ and $\pi(R\xi) = A^{1/2}$. Since $\pi(a) = A^{1/2}$, $a = R\xi$. Similarly, $b = S\xi$. Now

$$\phi(A + B) = \langle \xi, \xi \rangle$$
$$= \langle P\xi, \xi \rangle$$
$$= \langle (R^*R + S^*S)\xi, \xi \rangle$$
$$= \langle R\xi, R\xi \rangle + \langle S\xi, S\xi \rangle$$
$$= \langle a, a \rangle + \langle b, b \rangle$$
$$= \phi(A) + \phi(B).$$

The last statement is obvious. □

LEMMA VII.33. ϕ is faithful.

PROOF. Suppose $\phi(A) = 0$. Thus $A^{1/2} = \pi(a)$ and $\phi(A) = ||a||^2$. Hence $a = 0$, and thus $A = \pi(a)^2 = 0$. □

PROPOSITION VII.21. The weight ϕ is lower σ-weakly continuous.

PROOF. Let $\alpha \geq 0$. We show $\{A \in \mathcal{L}(\mathcal{A})^+ : \phi(A) \leq \alpha\}$ is σ-weakly closed. Since this is a convex set in $\mathcal{L}(\mathcal{A})$ and $\mathcal{L}(\mathcal{A})_*^* = \mathcal{L}(\mathcal{A})$ under the natural pairing $(f, A) \mapsto f(A)$, it suffices by Theorem VII.13 to show $C_{r,\alpha} = \{A \in \mathcal{L}(\mathcal{A})^+ : ||A|| \leq r, \phi(A) \leq \alpha\}$ is σ-weakly closed for all $r, \alpha \geq 0$. But since this a bounded convex set, it is equivalent to show that $C_{r,\alpha}$ is strongly closed. Let A_i be a net in $C_{r,\alpha}$ converging strongly to A. Clearly $A \geq 0$ and $||A|| \leq r$. Since $\phi(A_i) \leq \alpha$, there are $a_i \in \mathcal{A}$ with $\pi(a_i) \geq 0$ and $\pi(a_i)^2 = A_i$. Moreover, $||a_i||^2 = \phi(A_i) \leq \alpha$. Since $\pi(a_i) = \pi(a_i)^* = \pi(a_i^\#)$ and π is faithful, $a_i^\# = a_i$. Now note if $A_i \to A$ strongly, then $A_i^{1/2} \to A^{1/2}$ strongly. (Approximate $\sqrt{x}$ uniformly on $[0, \sup_i ||A_i||]$ by polynomials P(x) and use fact $P(A_i) \to P(A)$ strongly as i increases.) This implies $\pi(a_i)$ converges strongly to $A^{1/2}$. Moreover, since a_i is bounded and $a_i^\# = a_i$, Corollary VII.4 implies there is a $a \in \mathcal{A}''$ with $Sa = a$ such that $\pi(a) = A^{1/2}$ and $||a|| \leq \alpha^{1/2}$. Hence $\phi(A) = ||a||^2 \leq \alpha$. $\square$

PROPOSITION VII.22. *Let $A \in \mathcal{L}(\mathcal{A})$. Then there exists an $a \in \mathcal{A}$ with $A = \pi(a)$ iff $A \in \mathcal{N}_\phi \cap \mathcal{N}_\phi^*$. Moreover, if $a, b \in \mathcal{A}$, then $\pi(b)^*\pi(a) \in \mathcal{M}_\phi$ and $\phi_f(\pi(b)^*\pi(a)) = \langle a, b \rangle$.*

PROOF. Let $A \in \mathcal{L}(\mathcal{A})$. Let $A = U|A|$ be its polar decomposition. Then $A^* = U^*|A^*|$ is the polar decomposition of A^*. By Proposition VII.6, U, $|A|$, and $|A^*|$ are in $\mathcal{L}(\mathcal{A})$.

Assume $A = \pi(a)$ for some $a \in \mathcal{A}$. Then for $\eta \in \mathcal{A}'$, we have

$$\pi'(\eta)(U^*a) = U^*\pi'(\eta)a$$
$$= U^*\pi(a)\eta$$
$$= U^*A\eta$$
$$= |A|\eta.$$

By Proposition VII.5, $U^*a \in \mathcal{A}'' = \mathcal{A}$, $\pi(U^*a) = |A|$ and $SU^*a = U^*a$. Similarly

$$\pi'(\eta)(USa) = U\pi'(\eta)Sa$$
$$= U\pi(Sa)\eta$$
$$= U\pi(a)^*\eta$$
$$= UA^*\eta$$
$$= |A^*|\eta,$$

and thus again $USa \in \mathcal{A}$, $\pi(USa) = |A^*|$ and $SUSa = USa$. Hence $\phi(A^*A) = \phi(|A|^2) = \langle U^*a, U^*a \rangle = \langle UU^*a, a \rangle = \langle a, a \rangle$, for $UU^*\pi(a) = UU^*A = UU^*U|A| = U|A| = A = \pi(a)$; and thus by

Lemma VII.29, $UU^*a = a$. Similarly, one has $\phi(AA^*) = \phi(|A^*|^2) = \langle USa, USa \rangle = \langle U^*USa, Sa \rangle = \langle Sa, Sa \rangle$, for

$$
\begin{aligned}
U^*U\pi(Sa) &= U^*U\pi(a)^* \\
&= U^*UA^* \\
&= U^*UU^*|A^*| \\
&= U^*|A^*| \\
&= A^* \\
&= \pi(Sa).
\end{aligned}
$$

Thus $A = \pi(a) \in \mathcal{N}_\phi \cap \mathcal{N}_\phi^*$ and $\phi(A^*A) = \langle a, a \rangle$.

Conversely, suppose $A \in \mathcal{N}_\phi \cap \mathcal{N}_\phi^*$. Thus $\phi(A^*A) < \infty$ and $\phi(AA^*) < \infty$. Hence there are a and b in $\mathcal{A}$ such that $\pi(a) = |A|$ and $\pi(b) = |A^*|$. Hence

$$
\begin{aligned}
\pi'(\eta)(Ua) &= U\pi'(\eta)a \\
&= U\pi(a)\eta \\
&= U|A|\eta \\
&= A\eta
\end{aligned}
$$

and

$$
\begin{aligned}
\pi'(\eta)(U^*b) &= U^*\pi'(\eta)b \\
&= U^*\pi(b)\eta \\
&= U^*|A^*|\eta \\
&= A^*\eta
\end{aligned}
$$

for $\eta \in \mathcal{A}'$. By Proposition VII.5, $Ua \in \mathcal{A}$ and $\pi(Ua) = A$.

Finally if $a, b \in \mathcal{A}$, then $\pi(a), \pi(b) \in \mathcal{N}_\phi \cap \mathcal{N}_\phi^*$. Hence

$$
\begin{aligned}
\phi_f(\pi(b)^*\pi(a)) &= \frac{1}{4}\sum_{j=0}^{3}\phi(\pi(a+i^j b)^*\pi(a+i^j b)) \\
&= \frac{1}{4}\sum_{j=0}^{3}\langle a+i^j b, a+i^j b \rangle \\
&= \langle a, b \rangle.
\end{aligned}
$$

□

COROLLARY VII.20. *The weight $\phi = \phi_A$ is semifinite.*

PROOF. By Proposition VII.22, $\pi(b^\#)\pi(a) \in \mathcal{M}_\phi$ for $a, b \in \mathcal{A}$. Thus $\mathcal{M}_\phi \supseteq \pi(\mathcal{A})^2$. Since $\pi(\mathcal{A})$ is strongly dense in $\mathcal{L}(\mathcal{A})$, there is a net in $\pi(\mathcal{A})$ converging strongly to I. This implies the strong closure of $\pi(\mathcal{A})^2$ equals the strong closure of $\pi(\mathcal{A})$. Thus the strong closure of $\mathcal{M}_\phi$ equals $\mathcal{L}(\mathcal{A})$. Hence $\mathcal{M}_\phi$ is weakly dense in $\mathcal{L}(\mathcal{A})$. $\square$

20. Normality of the Natural Weight

PROPOSITION VII.23. *Let A_n be an increasing sequence of positive operators in $B(\mathbb{H})$. Set $T_n = (\frac{1}{n} + A_n)^{-1}A_n$. Then T_n is an increasing sequence of bounded operators which converges strongly to the orthogonal projection E whose range is the closure of the linear span of the union of the ranges of A_n.*

PROOF. Clearly $T_n \leq I$ for all n. We claim T_n is increasing. Indeed, note if $n < m$, then $\frac{1}{n} + A_n \leq \frac{1}{n} + A_m$. Hence by Lemma II.4, $(\frac{1}{n} + A_m)^{-1} \leq (\frac{1}{n} + A_n)^{-1}$. Now the functional calculus implies

$$\frac{1}{m}\left(\frac{1}{m} + A_m\right)^{-1} \leq \frac{1}{n}\left(\frac{1}{n} + A_m\right)^{-1}$$

for $n < m$. Hence $\frac{1}{m}\left(\frac{1}{m} + A_m\right)^{-1} \leq \frac{1}{n}\left(\frac{1}{n} + A_n\right)^{-1}$ for $n < m$. Thus

$$I - \frac{1}{n}\left(\frac{1}{n} + A_n\right)^{-1} \leq I - \frac{1}{m}\left(\frac{1}{m} + A_m\right)^{-1}.$$

Factoring yields

$$\left(\frac{1}{n} + A_n\right)^{-1}A_n \leq \left(\frac{1}{m} + A_m\right)^{-1}A_m.$$

Hence $T_n \leq T_m \leq I$ for $n < m$. By Theorem III.10, T_n converges strongly to an operator E. E is self adjoint for each T_n is self adjoint. Moreover, if $v \in (\text{Rang}A_n)^\perp$ for all n, then $A_n v = 0$ for all n; and thus $Ev = 0$.

Next note $\|(1 - T_m)A_m(1 - T_m)\| \leq \frac{1}{m}$ follows from the functional calculus. Indeed, if $A_m = \int \lambda\, dP(\lambda)$, then

$$(1 - T_m)A_m(1 - T_m) = \int \left(1 - (\frac{1}{m} + \lambda)^{-1}\lambda\right)^2 \lambda\, dP(\lambda)$$

$$= \int \left(\frac{\sqrt{m\lambda}}{1 + m\lambda}\right)^2 \frac{1}{m}\, dP(\lambda).$$

This gives $||(1 - T_m)A_m(1 - T_m)|| \leq \frac{1}{m}$. By Lemma II.3, for $n < m$ one has

$$||(1 - T_m)A_n(1 - T_m)|| \leq ||(1 - T_m)A_m(1 - T_m)|| \leq \frac{1}{m}.$$

Thus $\lim_m ||(1 - T_m)\sqrt{A_n}|| = 0$. Hence for each v, we see

$$\lim_{m \to \infty} \left(\sqrt{A_n} - T_m \sqrt{A_n}\right) \sqrt{A_n}v = 0.$$

Consequently, $T_m A_n v \to A_n v$ for each v as $m \to \infty$. Thus E is the projection onto the closure of the linear span of the union of the ranges of A_n. $\square$

LEMMA VII.34. *Let $\mathcal{A}$ be an achieved left Hilbert algebra, and let ϕ be the natural weight on $\mathcal{L}(\mathcal{A})$. Then*

$$\mathcal{L}(\mathcal{A})_0 \mathcal{M}_\phi \mathcal{L}(\mathcal{A})_0 = \mathcal{M}_\phi.$$

PROOF. It suffices to show $\mathcal{L}(\mathcal{A})_0 \mathcal{M}_\phi^+ \subseteq \mathcal{M}_\phi$, for $\mathcal{M}_\phi$ is a $*$ algebra and $\mathcal{M}_\phi$ is the linear span of $\mathcal{M}_\phi^+$. Suppose $T \in \mathcal{M}_\phi^+$. Then $\phi(T) < \infty$. Hence $T = \pi(a^2)$ where $\pi(a) \geq 0$ and $\phi(T) = ||a||^2$. According to Proposition VII.9, if $A \in \mathcal{L}(\mathcal{A})_0$, then $Aa \in \mathcal{A}$ and $A\pi(a) = \pi(Aa)$. By Proposition VII.22, $A\pi(a) \in \mathcal{N}_\phi^* \cap \mathcal{N}_\phi$. Thus $AT = A\pi(a)\pi(a) \in \mathcal{N}_\phi^* \mathcal{N}_\phi = \mathcal{M}_\phi$. $\square$

LEMMA VII.35. *Let $\mathcal{A}$ be a left Hilbert algebra. Then there is a family of pairwise orthogonal projections E_j in $\mathcal{L}(\mathcal{A}')$ such that $\sum E_j = I$, $\Delta(it)E_j\Delta(-it) = E_j$ for all $t \in \mathbb{R}$, and for each j there is a sequence $a_{j,n}$ in $\mathcal{A}'$ with $\pi'(a_{j,n}) \in \mathcal{M}_{\phi'}^+$ such that*

(a) $\pi'(a_{j,n}) \geq 0$ *for all n*
(b) $\pi'(a_{j,n}) \nearrow E_j$ *strongly for each j*
(c) *for each rational number r and each n there is an m such that*

$$\Delta(ir)\pi'(a_{j,n})\Delta(-ir) \leq \pi'(a_{j,m}).$$

PROOF. Let ϕ' be the natural weight on the achieved left Hilbert algebra $\mathcal{A}'$. Choose a maximal family E_j of pairwise orthogonal projections satisfying

$$\Delta(it)E_j\Delta(-it) = E_j$$

for all t and for which there is a sequence $a_{j,n}$ in $\mathcal{A}'$ with $\pi'(a_{j,n}) \in \mathcal{M}_{\phi'}^+$ and (a), (b) and (c) hold. We suppose $F = I - \sum E_j \neq 0$. We note $\Delta(it)F\Delta(-it) = F$ for all t. Hence $F \in \mathcal{L}(\mathcal{A}')_0$. There is a $c \in \mathcal{A}'$ such that $F\pi'(c^b) \neq 0$. Hence $F\pi'(c^b c)F \neq 0$. Set

$c' = c^b c$. By Proposition VII.22, $\pi'(c') \in \mathcal{M}_{\phi'}^+$. Lemma VII.34 implies $F\pi'(c')F \in \mathcal{M}_{\phi'}^+$. Hence, there is an $a \in \mathcal{A}'$ with $\pi'(a) = F\pi'(c')F$.

Let $\{r_j\}_{j=1}^\infty$ enumerate the rational numbers. For each n, set

$$A_n = \left(\frac{1}{n} + \sum_{j=1}^n \Delta(ir_j)\pi'(a)\Delta(-ir_j) \right)^{-1} \sum_{j=1}^n \Delta(ir_j)\pi'(a)\Delta(-ir_j).$$

We note since $\pi'(a) \in \mathcal{M}_{\phi'}^+$ that there is a b in $\mathcal{A}'$ with $\pi'(b) = \sqrt{\pi'(a)}$. Hence

$$\sum_{j=1}^n \Delta(ir_j)\pi'(a)\Delta(-ir_j) \in \mathcal{M}_{\phi'}^+$$

for $\Delta(ir_j)\pi'(a)\Delta(-ir_j) = \pi'(\Delta(-ir_j)b)^2$. But

$$A_n \le n \sum_{j=1}^n \Delta(ir_j)\pi'(a)\Delta(-ir_j).$$

Hence each $A_n \in \mathcal{M}_{\phi'}^+$. Moreover, Proposition VII.23 implies A_n is an increasing sequence of positive operators which converges strongly to the orthogonal projection E of $\mathbb{H}$ onto the closure of the linear span of the union of the ranges of $\Delta(ir_j)\pi'(a)\Delta(-ir_j)$. Hence $\Delta(it)E\Delta(-it) = E$ for all t.

Since $A_n \in \mathcal{M}_{\phi'}$, there is an $a_n \in \mathcal{A}'$ with $\pi'(a_n) = A_n$. Hence $\pi'(a_n) \ge 0$ and $\pi'(a_n) \nearrow E$ strongly. Clearly $E \le F$.

Finally, let r be a rational number and $n \ge 1$. Choose m so that $r + r_j \in \{r_1, r_2, \dots, r_m\}$ for $j = 1, 2, \dots, n$. Now

$$\Delta(ir)A_n\Delta(-ir) =$$

$$\Delta(ir)\left(\frac{1}{n} + \sum_{j=1}^n \Delta(ir_j)\pi'(a)\Delta(-ir_j) \right)^{-1} \sum_{j=1}^n \Delta(ir_j)\pi'(a)\Delta(-ir_j)\Delta(-ir)$$

$$= \left(\frac{1}{n} + \sum_{j=1}^n \Delta(i(r+r_j))\pi'(a)\Delta(-i(r+r_j)) \right)^{-1} \sum_{j=1}^n \Delta(i(r+r_j))\pi'(a)\Delta(-i(r+r_j))$$

$$\le \left(\frac{1}{m} + \sum_{j=1}^m \Delta(ir_j)\pi'(a)\Delta(-ir_j) \right)^{-1} \sum_{j=1}^m \Delta(ir_j)\pi'(a)\Delta(-ir_j) = A_m;$$

for Proposition VII.23 implies if $0 \le A \le B$ and $n \le m$, then

$$\left(\frac{1}{n} + A \right)^{-1} A \le \left(\frac{1}{m} + B \right)^{-1} B.$$

Hence adding E to the family E_j contradicts maximality. This gives $\sum E_j = I$. $\square$

COROLLARY VII.21. *Let $\mathcal{A}$ be an achieved left Hilbert algebra with natural weight ϕ. Suppose F is a nonzero orthogonal projection in $\mathcal{L}(\mathcal{A})$ with*

$$\Delta(it)F\Delta(-it) = F$$

for all t. Then there is a nonzero $A \in \mathcal{M}_\phi^+$ with $A \leq F$.

Let E_j be the family of projections given in Lemma VII.35. For each j, let $a_{j,n}$ be the sequence in $\mathcal{A}'$ with properties (a), (b) and (c). Since $\pi'(a_{j,n}) \in \mathcal{M}_{\phi'}^+$ and $\pi'(a_{j,n-1}) \leq \pi'(a_{j,n})$, $\pi'(a_{j,n}) - \pi'(a_{j,n-1}) \in \mathcal{M}_{\phi'}^+$. Thus for each n and j, there is a $\xi_{j,n} \in \mathcal{A}'$ such that $\pi'(\xi_{j,n}) = \sqrt{\pi'(a_{j,n}) - \pi'(a_{j,n-1})}$. Hence for each j, $\sum_n \pi'(\xi_{j,n}^2) = E_j$ in the strong operator topology. In particular

$$\sum_{j,n} \pi'(\xi_{j,n}^2) = I.$$

Define a normal weight ψ on $\mathcal{L}(\mathcal{A})^+$ by

$$\psi(T) = \sum_{j,n} \langle T\xi_{j,n}, \xi_{j,n} \rangle.$$

PROPOSITION VII.24. *$\psi(T) \leq \phi(T)$ for all $T \geq 0$, and $\psi(T) = \phi(T)$ for all $T \in \mathcal{M}_\phi^+$. Moreover, $\psi(\Delta(it)T\Delta(-it)) = \psi(T)$ for all $T \geq 0$ and $t \in \mathbb{R}$.*

PROOF. Let $T \in \mathcal{M}_\phi^+$. Hence $T^{1/2} = \pi(a)$ for some $a \in \mathcal{A}$ and

$$\psi(T) = \sum_{j,n} \langle T\xi_{j,n}, \xi_{j,n} \rangle = \sum_{j,n} \langle \pi(a)^2 \xi_{j,n}, \xi_{j,n} \rangle = \sum_{j,n} \langle \pi(a)\xi_{j,n}, \pi(a)\xi_{j,n} \rangle$$

$$= \sum_{j,n} \langle \pi'(\xi_{j,n})a, \pi'(\xi_{j,n})a \rangle = \sum_{j,n} \langle \pi'(\xi_{j,n})^2 a, a \rangle$$

$$= \sum_{j,n} \langle \pi'(\xi_{j,n}^2)a, a \rangle = \langle a, a \rangle = \phi(T).$$

Since $\phi(T) = \infty$ if $T \geq 0$ and $T \notin \mathcal{M}_\phi^+$, we have $\psi(T) \leq \phi(T)$ for all $T \in \mathcal{L}(\mathcal{A})^+$.

Let $r \in \mathbb{Q}$. We first show $\psi(\Delta(ir)T\Delta(-ir)) \leq \psi(T)$. Let $N > 0$ and let F be a finite subset of the j's. Lemma VII.35 implies there is an integer $M > 0$ such that for $j \in F$ and $n \leq N$ then $\Delta(-ir)\pi'(a_{j,n})\Delta(ir) \leq \pi'(a_{j,m})$ for some $m \leq M$.

Set $\psi_{F,N}(T) = \sum_{j \in F, n \leq N} \langle T\xi_{j,n}, \xi_{j,n} \rangle$. If $T = \pi(a)^2$ where $\pi(a) \geq 0$, then by Theorem VII.9, we see

$$\psi_{F,N}(\Delta(ir)T\Delta(-ir)) = \sum_{j \in F, n \leq N} \langle \Delta(ir)T\Delta(-ir)\xi_{j,n}, \xi_{j,n} \rangle$$

$$= \sum_{j \in F, n \leq N} \langle \Delta(ir)\pi(a)\Delta(-ir)\xi_{j,n}, \Delta(ir)\pi(a)\Delta(-ir)\xi_{j,n} \rangle$$

$$= \sum_{j \in F, n \leq N} \langle \pi(\Delta(ir)a)\xi_{j,n}, \pi(\Delta(ir)a)\xi_{j,n} \rangle$$

$$= \sum_{j \in F, n \leq N} \langle \pi'(\xi_{j,n})\Delta(ir)a, \pi'(\xi_{j,n})\Delta(ir)a \rangle$$

$$= \sum_{j \in F, n \leq N} \langle \pi'(\xi_{j,n})^2 \Delta(ir)a, \Delta(ir)a \rangle$$

$$= \sum_{j \in F, n \leq N} \langle \pi'(a_{j,n} - a_{j,n-1})\Delta(ir)a, \Delta(ir)a \rangle$$

$$= \sum_{j \in F} \langle \pi'(a_{j,N})\Delta(ir)a, \Delta(ir)a \rangle$$

$$= \sum_{j \in F} \langle \Delta(-ir)\pi'(a_{j,N})\Delta(ir)a, a \rangle$$

Since the $a_{j,N}$ have the properties listed in Lemma VII.35, one has

$$\Delta(ir)\pi'(a_{j,N})\Delta(-ir) \leq \pi'(a_{j,M}).$$

Thus

$$\psi_{F,N}(\Delta(ir)T\Delta(-ir)) \leq \sum_{j \in F} \langle \pi'(a_{j,M})a, a \rangle$$

$$= \sum_{j \in F, m \leq M} \langle \pi'(a_{j,m} - a_{j,m-1})a, a \rangle$$

$$= \sum_{j \in F, m \leq M} \langle \pi'(\xi_{j,m})^2 a, a \rangle$$

$$= \sum_{j \in F, m \leq M} \langle \pi'(\xi_{j,m})a, \pi'(\xi_{j,m})a \rangle$$

$$= \sum_{j \in F, m \leq M} \langle \pi(a)\xi_{j,m}, \pi(a)\xi_{j,m} \rangle$$

$$= \sum_{j \in F, m \leq M} \langle T\xi_{j,m}, \xi_{j,m} \rangle$$

$$= \psi_{F,M}(T).$$

Thus $\psi_{F,N}(\Delta(ir)T\Delta(-ir)) \leq \psi_{F,M}(T)$ for $T \in \mathcal{M}_\phi^+$.

Since $\mathcal{M}_\phi$ is weakly dense in $\mathcal{L}(\mathcal{A})$ and $\mathcal{M}_\phi \subseteq \mathcal{N}_\phi$, by Proposition III.11, there is a net V_λ in $\mathcal{N}_\phi^+$ such that $V_\lambda \nearrow I$ strongly. Set $U_\lambda = V_\lambda^2$. Then $0 \leq U_\lambda \leq I$ for each λ, $U_\lambda \in \mathcal{M}_\phi^+$, and U_λ converges strongly to I. Hence $U_\lambda = \pi(a_\lambda)^2$ where $a_\lambda \in \mathcal{A}$ and $\pi(a_\lambda) \geq 0$. Now by the note in the proof of Proposition VII.21, we have $\pi(a_\lambda) \to I$ strongly. Thus $\pi(a_\lambda)T\pi(a_\lambda) \to T$ strongly for any $T \in \mathcal{L}(\mathcal{A})^+$. Moreover, $\pi(a_\lambda)T\pi(a_\lambda) \leq \|T\| \pi(a_\lambda)^2 \in \mathcal{M}_\phi^+$. This gives

$$\psi_{F,N}(\Delta(ir)\pi(a_\lambda)T\pi(a_\lambda)\Delta(-ir)) \leq \psi_{F,M}(\pi(a_\lambda)T\pi(a_\lambda)).$$

Taking limits one obtains

$$\psi_{F,N}(\Delta(ir)T\Delta(-ir)) \leq \psi_{F,M}(T)$$

for all $T \in \mathcal{L}(\mathcal{A})^+$. Hence

$$\psi(\Delta(ir)T\Delta(-ir)) \leq \psi(T)$$

for $T \geq 0$ and all rational r.

By Proposition VII.16, ψ is lower σ-weakly continuous. Thus

$$\psi(\Delta(ir)T\Delta(-ir)) \leq \psi(T)$$

for all $r \in \mathbb{R}$ and all $T \in \mathcal{L}(\mathcal{A})^+$. This yields the invariance of ψ. $\quad\square$

LEMMA VII.36. *ψ is a faithful normal semifinite weight on the von Neumann algebra $\mathcal{L}(\mathcal{A})$.*

PROOF. Clearly ψ is positive and normal. We claim it is semifinite. Note $\mathcal{M}_\psi \supseteq \mathcal{M}_\phi$ for $\psi \leq \phi$. Also by Proposition VII.22, $\pi(\mathcal{A}^2) \subseteq \mathcal{M}_\phi$. But the strong closure of $\pi(\mathcal{A})\pi(\mathcal{A})$ contains $\pi(\mathcal{A})$ and thus is $\mathcal{L}(\mathcal{A})$. Hence $\mathcal{M}_\psi$ is strongly and hence weakly dense in $\mathcal{L}(\mathcal{A})$.

Now suppose $\psi(T) = 0$ and $T \neq 0$. Again let r_j enumerate the rationals. Proposition VII.23 implies the operators

$$T_n = \left(\frac{1}{n} + \sum_{j=1}^{n} \Delta(ir_j)T\Delta(-ir_j)\right)^{-1} \sum_{j=1}^{n} \Delta(ir_j)T\Delta(-ir_j)$$

are in $\mathcal{L}(\mathcal{A})$ and converge strongly upward to a nonzero projection E satisfying $\Delta(it)E\Delta(-it) = E$ for all t. Also

$$0 \leq \psi(T_n) \leq \sum_{j=1}^{n} \psi(\Delta(ir_j)T\Delta(-ir_j)) = 0 \text{ for all } n.$$

By Proposition VII.16, ψ is lower σ-weakly continuous and thus strongly continuous on bounded sets. Consequently $\psi(E) = 0$. But, by Corollary VII.21, if $E \neq 0$, there is a nonzero $A \in \mathcal{M}_\phi^+$ with $A \leq E$. By Proposition VII.24, $\phi(A) = \psi(A) = 0$. Finally, Lemma VII.33 implies $A = 0$, a contradiction. Thus ψ is faithful. $\square$

Using the faithful normal semifinite weight ψ on $\mathcal{L}(\mathcal{A})$ and the GNS construction, we may form $\mathcal{N}_\psi$, $\mathcal{M}_\psi$, $\mathbb{H}_\psi$ and a normal faithful representation Π_ψ on $\mathbb{H}_\psi$.

LEMMA VII.37. *Let $A \geq 0$ be in $\mathcal{L}(\mathcal{A})$. For each n, define*

$$A_n = \sqrt{\frac{n}{\pi}} \int_{-\infty}^\infty e^{-nt^2} \Delta(it) A \Delta(-it)\, dt.$$

Then each $A_n \in \mathcal{L}(\mathcal{A})_0$, and $A_n \to A$ strongly.

PROOF. We first show the mapping $is \mapsto \Delta(is) A_n \Delta(-is)$ has an entire extension which is uniformly bounded on vertical strips $-\delta \leq \mathrm{Re} z \leq \delta$. Indeed, set $F(z) = \sqrt{\frac{n}{\pi}} \int_{-\infty}^\infty e^{-n(iz+t)^2} \Delta(it) A \Delta(-it)\, dt$. It is easy to see $F(z)$ is holomorphic. Also

$$\|F(r+is)\| = \sqrt{\frac{n}{\pi}} \| \int_{-\infty}^\infty e^{-n(t-s+ir)^2} \Delta(it) A \Delta(-it)\, dt\|$$

$$= \sqrt{\frac{n}{\pi}} \| \int_{-\infty}^\infty e^{-n(t+ir)^2} \Delta(it+is) A \Delta(-it-is)\, dt\|$$

$$\leq \sqrt{\frac{n}{\pi}} \|A\| \int_{-\infty}^\infty e^{-n(t^2-r^2)}\, dt$$

$$= \|A\| e^{nr^2}$$

is bounded on the strip $-\delta \leq r \leq \delta$.
Moreover,

$$F(is) = \sqrt{\frac{n}{\pi}} \int_{-\infty}^\infty e^{-n(t-s)^2} \Delta(it) A \Delta(-it)\, dt$$

$$= \Delta(is) \left(\sqrt{\frac{n}{\pi}} \int_{-\infty}^\infty e^{-nt^2} \Delta(it) A \Delta(-it)\, dt \right) \Delta(-is)$$

$$= \Delta(is) A_n \Delta(-is).$$

To see A_n converges strongly to A, note

$$\|A_n v - Av\| = \|\sqrt{\frac{n}{\pi}} \int_{-\infty}^{\infty} e^{-nt^2} (\Delta(it) A \Delta(-it) v - Av)\, dt\|$$

$$\leq \sqrt{\frac{n}{\pi}} \int_{-\infty}^{\infty} e^{-nt^2} \|\Delta(it) A \Delta(-it) v - Av\|\, dt$$

$$\leq \sqrt{\frac{n}{\pi}} \int_{-\delta}^{\delta} e^{-nt^2} \|\Delta(it) A \Delta(-it) v - Av\|\, dt + 2\sqrt{\frac{n}{\pi}} \|A\|\, \|v\| \int_{|t| \geq \delta} e^{-nt^2}\, dt$$

$$\leq \max_{|t| \leq \delta} \|\Delta(it) A \Delta(-it) v - Av\| + 2\sqrt{\frac{n}{\pi}} \|A\|\, \|v\| \int_{|t| \geq \delta} e^{-nt^2}\, dt;$$

for $\sqrt{\frac{n}{\pi}} \int_{-\infty}^{\infty} e^{-nt^2}\, dt = 1$. Since $\sqrt{\frac{n}{\pi}} \int_{|t| \geq \delta} e^{-nt^2}\, dt \to 0$ as $n \to \infty$ for any $\delta > 0$, we see $A_n v \to Av$ as $n \to \infty$. $\quad\square$

THEOREM VII.17. *ϕ is normal.*

PROOF. Now let $A \in \mathcal{M}_\psi^+$. As seen in the proof of Proposition VII.24, there is a net U_λ in $\mathcal{M}_\phi^+$ satisfying $0 \leq U_\lambda \leq 1$ and $U_\lambda \to I$ strongly. Set

$$A_n = \sqrt{\frac{n}{\pi}} \int_{-\infty}^{\infty} e^{-nt^2} \Delta(it) A \Delta(-it)\, dt.$$

Lemma VII.37 gives $A_n \in \mathcal{L}(A)_0$ and $A_n \to A$ strongly. By Lemma VII.34, $A_n U_\lambda A_n \in \mathcal{M}_\phi^+$. Thus

$$\phi(A_n U_\lambda A_n) = \psi(A_n U_\lambda A_n) \leq \psi(A_n^2) \leq \|A_n\|\, \psi(A_n) \leq \|A\|\, \psi(A);$$

for $A_n U_\lambda A_n \leq \|U_\lambda\| A_n^2 \leq A_n^2$, $\|A_n\| \leq \|A\|$, and by lower σ-weak continuity,

$$\psi(A_n) \leq \sqrt{\frac{n}{\pi}} \int_{-\infty}^{\infty} e^{-nt^2} \psi(\Delta(it) A \Delta(-it))\, dt = \sqrt{\frac{n}{\pi}} \int_{-\infty}^{\infty} e^{-nt^2} \psi(A)\, dt = \psi(A).$$

Now $A_n U_\lambda A_n$ converges strongly to A_n^2 as λ increases. Since this is a bounded net, we have σ-weak convergence. Hence Proposition VII.21 implies $\phi(A_n^2) \leq \|A\|\, \psi(A)$. Again using the lower σ-weak continuity of ϕ, the boundedness of the sequence A_n, and the strong convergence of A_n to A, we see that

$$\phi(A^2) \leq \|A\|\psi(A).$$

We show $\phi = \psi$. To do this it suffices by Proposition VII.24 to show $\mathcal{M}_\psi^+ \subseteq \mathcal{M}_\phi^+$. We have shown if $A \in \mathcal{M}_\psi^+$, then $A^2 \in \mathcal{M}_\phi^+$. Thus every projection in $\mathcal{M}_\psi^+$ is in $\mathcal{M}_\phi^+$. Now let $A \in \mathcal{M}_\psi^+$, and let P be the resolution of the identity for A. Note if $c > 0$, then $cP[c, d] \leq A$

for any $d > c$. Thus $\psi(P[c, d]) \leq \frac{1}{c}\psi(A)$, and we see $P[c, d] \in \mathcal{M}_\phi^+$. Choose a sequence $\mathcal{P}_n$ of partitions where $\mathcal{P}_n$ is a partition of $[\frac{1}{n}, \|A\|]$ containing $[\frac{1}{n}, \frac{1}{n-1})$, every element in $\mathcal{P}_n$ except $[\frac{1}{n}, \frac{1}{n-1})$ is a subset of an element in $\mathcal{P}_{n-1}$, and each subset of $\mathcal{P}_n$ is an interval of length at most $\frac{1}{n}$. Set $A_n = \sum_{S \in \mathcal{P}_n} \inf S \cdot P(S)$. Note each $A_n \in \mathcal{M}_\phi^+$, for all $P(S) \in \mathcal{M}_\phi^+$. Also $A_n \nearrow A$ strongly. Since $\phi = \psi$ on $\mathcal{M}_\phi^+$, we see by Propositions VII.15, VII.16 and VII.21 that

$$\psi(A) = \lim_n \psi(A_n) = \lim_n \phi(A_n) = \phi(A).$$

Thus $\mathcal{M}_\psi^+ = \mathcal{M}_\phi^+$, and $\phi = \psi$. $\square$

Lemmas VII.32 and VII.33, Corollary VII.20, and Theorem VII.17 establish that $\phi_\mathcal{A}$ is a normal faithful semifinite weight on the von Neumann algebra $\mathcal{L}(\mathcal{A})$. It is called the natural weight on the left ring of the achieved left Hilbert algebra $\mathcal{A}$.

Combining these with Proposition VII.22 and Theorem VII.16, we obtain the following theorem.

THEOREM VII.18. *Let $\mathcal{A}$ be an achieved left Hilbert algebra in a Hilbert space $\mathbb{H}$. Let $\phi = \phi_\mathcal{A}$ be the natural weight on $\mathcal{L}(\mathcal{A})$. Then ϕ is faithful, normal, and semifinite. Moreover, the linear mapping U from $\mathcal{A}$ to $\mathcal{N}_\phi \cap \mathcal{N}_\phi^*$ defined by $Ua = \pi(a)$ is one-to-one and onto and is a left Hilbert algebra isometry from the achieved left Hilbert algebra $\mathcal{A}$ onto the achieved left Hilbert algebra $\mathcal{N}_\phi \cap \mathcal{N}_\phi^*$.*

21. Traces on a C* Algebra

DEFINITION. Let $\mathcal{A}$ be a C^* algebra. A trace on $\mathcal{A}$ is a function $\omega : \mathcal{A}^+ \to [0, \infty]$ satisfying

(a) $\omega(x + y) = \omega(x) + \omega(y)$
(b) $\omega(\lambda x) = \lambda\omega(x)$ if $\lambda \geq 0$ where $0 \cdot \infty = 0$
(c) $\omega(y^*y) = \omega(yy^*)$ for $y \in \mathcal{A}$.

The trace ω is said to be finite if $\omega(x) < \infty$ for $x \in \mathcal{A}^+$ and semifinite if $\omega(x) = \sup\{\omega(y) : y \leq x, \omega(y) < \infty\}$.

PROPOSITION VII.25. *Let ω be a trace on C^* algebra $\mathcal{A}$. Then the following hold:*

(a) *If $\mathcal{N}$ is the set of all $x \in \mathcal{A}$ with $\omega(x^*x) < \infty$, then $\mathcal{N}$ is a self adjoint ideal in $\mathcal{A}$; and if $\mathcal{M} = \langle \mathcal{N}^2 \rangle$, the linear span of products of two elements of $\mathcal{N}$, then $\mathcal{M}$ is the linear span of $\mathcal{M}^+ = \mathcal{A}^+ \cap \mathcal{M}$ and $\mathcal{M}^+$ is the set of all $x \in \mathcal{A}^+$ where*

$\omega(x) < \infty$. *Moreover, the closures of $\mathcal{M}$ and $\mathcal{N}$ in $\mathcal{A}$ are the same.*

(b) *There exists a unique linear form ω_f on $\mathcal{M}$ such that $\omega_f = \omega$ on $\mathcal{M}^+$.*

(c) *One has $\omega_f(x^*) = \overline{\omega_f(x)}$ for $x \in \mathcal{M}$, $\omega_f(zx) = \omega_f(xz)$ for $x \in \mathcal{M}$ and $z \in \mathcal{A}$, and $\omega_f(uv) = \omega_f(vu)$ for $u, v \in \mathcal{N}$.*

PROOF. Lemma VII.23 gives (a) while (b) and $\omega_f(x^*) = \overline{\omega_f(x)}$ follow from the argument given in Corollary VII.17.

Let x be in $\mathcal{N}$. Then $\omega_f(x^*x) = \omega(x^*x) = \omega(xx^*) = \omega_f(xx^*)$. Now if $u, v \in \mathcal{N}$, then $uv = \frac{1}{4}\sum_{j=0}^{3} i^j (u + i^j v^*)(u + i^j v^*)^*$ and $vu = \frac{1}{4}\sum_{j=0}^{3} i^j (u + i^j v^*)^*(u + i^j v^*)$. Thus $\omega_f(uv) = \omega_f(vu)$. Next note if $z \in \mathcal{A}$ and $u, v \in \mathcal{N}$, then $\omega_f(zuv) = \omega_f(v(zu)) = \omega_f((vz)u) = \omega_f(uvz)$ since $\mathcal{N}$ is an ideal. This implies $\omega_f(zx) = \omega_f(xz)$ for $x \in \mathcal{M}$ and $z \in \mathcal{A}$. □

We shall denote the ideal $\mathcal{M}$ by $\mathcal{M}_\omega$. Note if ω is finite, then $\mathcal{M}_\omega = \mathcal{A}$.

PROPOSITION VII.26. *Let $\mathcal{A}$ be a C^* algebra, and let π be a nondegenerate representation of $\mathcal{A}$. Let $\mathcal{U}$ be the von Neumann algebra generated by $\pi(\mathcal{A})$, and let t be a normal trace on $\mathcal{U}^+$. Let $\mathcal{M}$ be a self adjoint ideal in $\mathcal{A}$, and suppose $\pi(\mathcal{M}) \subseteq \mathcal{M}_t$ and is strongly dense in $\mathcal{U}$. Then $f = t \circ \pi|_{\mathcal{A}^+}$ is semifinite, lower semicontinuous; and if u_λ is an increasing approximate unit for $\overline{\mathcal{M}}$, one has $f(x) = \lim f(x^{1/2}u_\lambda x^{1/2})$ for all $x \in \mathcal{A}^+$.*

PROOF. Since $\pi(\mathcal{M})$ is strongly dense in $\mathcal{U}$, $\pi|_{\overline{\mathcal{M}}}$ is nondegenerate. Therefore $\pi(u_\lambda)$ converges strongly to I on $\mathbb{H}$. Let $x \in \mathcal{A}^+$. Then the increasing net $\pi(x^{1/2})\pi(u_\lambda)\pi(x^{1/2})$ in $\mathcal{U}^+$ converges strongly to $\pi(x)$. Since t is normal, Propositions VII.15 and VII.16 imply $t(\pi(x^{\frac{1}{2}}u_\lambda x^{\frac{1}{2}})) \to t(\pi(x)) = f(x)$; i.e., $f(x^{\frac{1}{2}}u_\lambda x^{\frac{1}{2}}) \to f(x)$. To see f is semifinite, we may by Proposition II.7 take $u_\lambda \in \mathcal{M}$. Thus $\pi(x^{\frac{1}{2}}u_\lambda x^{\frac{1}{2}}) \in \pi(\mathcal{M}) \subseteq \mathcal{M}_t$ and $f(x^{\frac{1}{2}}u_\lambda x^{\frac{1}{2}}) < \infty$. Hence f is semifinite. By Proposition VII.16, t is lower σ-weakly continuous. Thus t is norm lower semicontinuous. Hence f is lower semicontinuous. □

22. Unimodular Left Hilbert Algebras

Suppose $\mathcal{A}$ is a left Hilbert algebra in a Hilbert space $\mathbb{H}$. Then $\mathcal{A}$ is **unimodular** when the modular operator Δ is the identity. In this case, by Theorem VII.4, $J = S = F$. We shall use $*$ for the isometric conjugate linear involution obtained by closing the operator $a \mapsto a^{\#}$.

Next note $\mathcal{A} \subseteq \mathcal{A}'$. Indeed, $\xi a = (a^* \xi^*)^* = J\pi(a^*)J\xi$ for ξ and a in $\mathcal{A}$, and thus $\xi \mapsto \pi(\xi)a$ is a bounded operator. This gives $a \in \mathcal{A}'$ for $a \in \mathcal{A}$ and $\pi'(a) = J\pi(a^*)J$. In particular $\mathcal{A}' \subseteq \mathcal{A}''$, and thus $\mathcal{A} = \mathcal{A}'$ if $\mathcal{A}$ is an achieved left Hilbert algebra. Moreover, when $\mathcal{A}$ is achieved, $\pi(\mathcal{A})$ is an ideal in $\mathcal{L}(\mathcal{A})$. Indeed, let $T \in \mathcal{L}(\mathcal{A})$ and $a \in \mathcal{A}$. Then $b \mapsto \pi(b)a = \pi'(a)b$ is bounded. Thus $b \mapsto T\pi(b)a = T\pi'(a)b = \pi'(a)Tb$ is bounded. Thus $Tb \in \mathcal{A}''$. Hence $Tb \in \mathcal{A}$, and $\pi(Tb) = T\pi(b)$. Note $\pi(b)T = (T^*\pi(b^*))^* = \pi(T^*b^*)^* = \pi((T^*b^*)^*)$.

THEOREM VII.19. *Let $\mathcal{A}$ be an achieved left unimodular Hilbert algebra in a Hilbert space $\mathbb{H}$. Define ϕ on $\mathcal{L}(\mathcal{A})^+$ by*

$$\phi(T) = \begin{cases} \langle a, a \rangle & \text{if } \pi(a) = T^{1/2}, \, a \in \mathcal{A} \\ 0 & \text{otherwise.} \end{cases}$$

Then ϕ is a faithful, normal, semifinite trace on $\mathcal{L}(\mathcal{A})$; and

$$\{T \in \mathcal{L}(\mathcal{A}) : \phi(T^*T) < \infty\} = \pi(\mathcal{A}).$$

Moreover, $\langle a, b \rangle = \phi_f(\pi(b)^* \pi(a))$ for $a, b \in \mathcal{A}$.

PROOF. By Theorem VII.18, the natural weight on the von Neumann algebra $\mathcal{L}(\mathcal{A})$ is faithful, normal and semifinite; and the mapping U defined by $Ua = \pi(a)$ is an isometry of $\mathcal{A}$ onto $\mathcal{N} \cap \mathcal{N}^* = \{A \in \mathcal{L}(\mathcal{A}) : \phi(A^*A) < \infty \text{ and } \phi(AA^*) < \infty\}$.

We now show ϕ is a trace. Suppose $T \in \mathcal{L}(\mathcal{A})$ and $\phi(T^*T) < \infty$. Let $T = U|T|$ be its polar decomposition. Recall U is a partial isometry of the closure of the range of T^* (which is also the closure of the range of $|T|$) onto the closure of the range of T and $|T| = \sqrt{T^*T}$. By Proposition VII.6, U is in $\mathcal{L}(\mathcal{A})$. Choose $a \in \mathcal{A}$ with $\pi(a) = |T|$. Then $\phi(T^*T) = \langle a, a \rangle$. Next note $\pi(U^*Ua) = U^*U\pi(a) = |T| = \pi(a)$. This implies $U^*Ua = a$; for by Proposition VII.4, π is faithful. Since $|T^*| = U|T|U^*$, $|T^*| = U\pi(a)U^* = U\pi(a^*)U^* = U\pi((Ua)^*) = \pi(UJUa)$. Furthermore, $\pi(JUa) = \pi(Ua)^* = \pi(a)^*U^* = |T|U^* = U^*U|T|U^* = U^*U\pi(a)^*U^* = U^*U\pi(Ua)^* = U^*U\pi(JUa) = \pi(U^*UJUa)$, and thus $JUa = U^*UJUa$. Hence

$$\begin{aligned} \phi(TT^*) &= \phi(|T^*|^2) \\ &= \langle UJUa, UJUa \rangle \\ &= \langle U^*UJUa, JUa \rangle \\ &= \langle JUa, JUa \rangle \end{aligned}$$

$$= \langle Ua, Ua \rangle$$
$$= \langle U^*Ua, a \rangle$$
$$= \langle a, a \rangle$$
$$= \phi(T^*T).$$

Replacing T by T^*, we see $\phi(T^*T) = \phi(TT^*)$ if $\phi(TT^*) < \infty$. Thus ϕ is a trace. In particular $\mathcal{N}^* = \mathcal{N}$. This implies $\pi(\mathcal{A}) = \{T \in \mathcal{L}(\mathcal{A}) : \phi(T^*T) < \infty\}$.

The last statement follows by Proposition VII.22. □

REMARK. The semifiniteness of ϕ is a consequence of Corollary VII.20. A slightly stronger condition holds; namely for each $T > 0$ in $\mathcal{L}(\mathcal{A})$, there is an $S > 0$ of form $\pi(b)^*\pi(b)$, $b \in \mathcal{A}$ with $S < T$ holds. Indeed, since $\pi(\mathcal{A})$ is strongly dense in $\mathcal{L}(\mathcal{A})$, there is an $a \in \mathcal{A}$ with $\pi(a)T^{1/2} \neq 0$. Since $\pi(\mathcal{A})$ is a $*$ ideal in $\mathcal{L}(\mathcal{A})$, $\pi(a)T^{1/2} \in \pi(\mathcal{A})$. Moreover, $T^{1/2}\pi(a)^*\pi(a)T^{1/2} \leq ||\pi(a)||^2T$. Thus if $\pi(b) = \frac{\pi(a)T^{1/2}}{||\pi(a)||}$, then $\pi(b)^*\pi(b) \leq T$ and $\phi(\pi(b)^*\pi(b)) = \langle b, b \rangle$. Semifiniteness then follows from Corollary VII.18.

23. Bitraces

DEFINITION. Let $\mathcal{A}$ be a C^* algebra. A bitrace on $\mathcal{A}$ is a function $s : \mathcal{N} \times \mathcal{N} \to \mathbb{C}$ where $\mathcal{N}$ is a self adjoint ideal in $\mathcal{A}$ satisfying

   (a) s is a complex Hermitian nonnegative form
   (b) $s(y, x) = s(x^*, y^*)$
   (c) $s(zx, y) = s(x, z^*y)$ for $x, y \in \mathcal{N}$, $z \in \mathcal{A}$
   (d) for each $z \in \mathcal{A}$, $x \mapsto zx$ is continuous in the pre-Hilbert structure defined by s, and
   (e) the linear span of all products xy, $x, y \in \mathcal{N}$, is dense in $\mathcal{N}$ under the pre-Hilbert structure defined by s.

The ideal $\mathcal{N}$ is called the ideal of definition of s and is denoted by $\mathcal{N}_s$. The Cauchy-Schwarz inequality along with (b) and (c) imply the set N_s of all $x \in \mathcal{N}$ with $s(x, x) = 0$ is a self adjoint ideal in $\mathcal{A}$. Hence $\mathcal{N}_s/N_s$ is a pre-Hilbert space whose completion is denoted by $\mathbb{H}_s$. $\mathcal{N}_s/N_s$ is also a $\#$ algebra with operations $(x+N_s)(y+N_s) = xy+N_s$ and $(x + N_s)^\# = x^* + N_s$.

PROPOSITION VII.27. $\mathcal{N}_s/N_s$ is a left unimodular Hilbert algebra in $\mathbb{H}_s$. Moreover, for $a \in \mathcal{A}$, $\lambda(a)(x + N_s) = ax + N_s$ and $\rho(a)(x + N_s) = xa + N_s$ extend to bounded operators on $\mathbb{H}_s$ making λ into a $*$ representation of $\mathcal{A}$ and ρ into a $*$ anti-representation of $\mathcal{A}$.

PROOF. Clearly $\mathcal{N}_s/N_s$ is dense in $\mathbb{H}_s$ and is a $\#$ algebra. Moreover, $b \mapsto ab + N_s$ is bounded by (d); and thus $\lambda(a)$ is a bounded operator on $\mathcal{N}_s/N_s$ having a unique extension to $\mathbb{H}_s$. By (e), the linear span of $(\mathcal{N}_s/N_s)^2$ is dense in $\mathcal{N}_s/N_s$. Next note

$$
\begin{aligned}
\langle a(b + N_s), c + N_s \rangle &= s(ab, c) \\
&= s(b, a^*c) \\
&= \langle b + N_s, (a + N_s)^{\#}(c + N_s) \rangle_s.
\end{aligned}
$$

Also $a + N_s \mapsto a^* + N_s$ is an isometry of $\mathcal{N}_s/N_s$ and thus extends to a conjugate linear isometry S of $\mathbb{H}_s$. In particular, $\#$ is a closeable operator. We next note $F = S^* = S$. Indeed,

$$
\begin{aligned}
\langle b + N_s, S(a + N_s) \rangle_s &= s(b, a^*) \\
&= s(a, b^*) \\
&= \langle a + N_s, b^* + N_s \rangle
\end{aligned}
$$

for all $a \in \mathcal{N}_s$. Hence $F(b + N_s) = b^* + N_s$. This gives $F = S^* = S$. Since $\Delta = FS$, we see $\Delta = I$; and thus $\mathcal{N}_s/N_s$ is a left unimodular Hilbert algebra. Finally $J\lambda(a^*)J = \rho(a)$ implies $\rho(a)$ is a bounded operator for all $a \in \mathcal{A}$. That λ is a $*$ representation (equivalently ρ is a $*$ anti-representation) follows easily. $\square$

LEMMA VII.38. *Let $\mathcal{U}$ be a left unimodular Hilbert algebra in a Hilbert space $\mathbb{H}$. Then $\pi'(\mathcal{U})' = \pi'(\mathcal{U}')'$.*

PROOF. We know $\mathcal{U} \subseteq \mathcal{U}'$ and $J = S = F$. By Theorem VII.7, $J\mathcal{L}(\mathcal{U})J = \mathcal{L}(\mathcal{U}')$. Hence $J\pi(\mathcal{U})''J = \pi'(\mathcal{U}')''$. From this, one obtains $(J\pi(\mathcal{U})J)'' = \pi'(\mathcal{U}')''$. But by Corollary VII.7, we have $J\pi(\mathcal{U})J = \pi'(J\mathcal{U}) = \pi'(\mathcal{U}^{\#}) = \pi'(\mathcal{U})$. This gives $\pi'(\mathcal{U})'' = \pi'(\mathcal{U}')''$. Taking commutants gives $\pi'(\mathcal{U})' = \pi'(\mathcal{U}')'$. $\square$

PROPOSITION VII.28. *Let s be a bitrace on a C^* algebra $\mathcal{A}$. Let $\mathcal{U}_s$ be the von Neumann algebra generated by $\lambda(\mathcal{A})$ and $\mathcal{V}_s$ be the von Neumann algebra generated by $\rho(\mathcal{A})$. Then $\mathcal{U}_s = \mathcal{L}(\mathcal{N}_s/N_s)$ and $\mathcal{V}_s = \mathcal{L}((\mathcal{N}_s/N_s)')$. In particular, $\mathcal{U}_s$ and $\mathcal{V}_s$ are commutants.*

PROOF. Let π and π' be the natural representations of the unimodular Hilbert algebras $\mathcal{N}_s/N_s$ and $(\mathcal{N}_s/N_s)'$ on $\mathbb{H}_s$. Let a be in $\mathcal{N}_s$. Clearly, $\pi(a + N_s) = \lambda(a)$. Furthermore, since $\mathcal{N}_s/N_s \subseteq (\mathcal{N}_s/N_s)'$, we have $\pi'(a + N_s)(b + N_s) = (b + N_s)(a + N_s) = ba + N_s = \rho(a)(b + N_s)$. In particular, $\pi(\mathcal{N}_s/N_s) \subseteq \lambda(\mathcal{A})$ and $\pi'(\mathcal{N}_s/N_s) \subseteq \rho(\mathcal{A})$. By Lemma VII.38, $\pi'(\mathcal{N}_s/N_s)' = \pi'((\mathcal{N}_s/N_s)')'$. Thus

$$
\pi(\mathcal{N}_s/N_s) \subseteq \lambda(\mathcal{A}) \subseteq \rho(\mathcal{A})' \subseteq \pi'(\mathcal{N}_s/N_s)' = \pi'((\mathcal{N}_s/N_s)')' = \mathcal{L}((\mathcal{N}_s/N_s)')'.
$$

Taking commutants we obtain

$$\mathcal{L}((\mathcal{N}_s/N_s)') \subseteq \rho(A)'' \subseteq \lambda(A)' \subseteq \pi(\mathcal{N}_s/N_s)' = \mathcal{L}(\mathcal{N}_s/N_s)'.$$

But by Theorem VII.7, $\mathcal{L}(\mathcal{N}_s/N_s)' = \mathcal{L}((\mathcal{N}_s/N_s)')$. Hence

$$\mathcal{L}(\mathcal{N}_s/N_s)' = \rho(A)'' = \lambda(A)'.$$

Thus $\mathcal{U}_s' = \mathcal{V}_s = \mathcal{L}(\mathcal{N}_s/N_s)'$, and the result follows. $\square$

By Theorem VII.19, there is a natural faithful normal semifinite trace t_s on $\mathcal{L}(\mathcal{N}_s/N_s)$ such that $\mathcal{N}_{t_s} \supseteq \lambda(\mathcal{N}_s)$. By Proposition VII.20, Theorem VII.16, and the fact that $\mathcal{N}_{t_s} = \mathcal{N}_{t_s}^*$ since t_s is a trace, $\mathcal{N}_{t_s}$ is an achieved unimodular Hilbert algebra in $\mathbb{H}_{t_s}$. We recall the trace t_s on $\mathcal{L}(\mathcal{N}_s/N_s)$ satisfies

$$t_s(\lambda(a)^*\lambda(a)) = s(a,a) \text{ for } a \in \mathcal{N}_s.$$

LEMMA VII.39. *Let A be a C^* algebra, and let s be a bitrace on A. Form the corresponding unimodular left Hilbert algebra $A_0 = \mathcal{N}_s/N_s$ in $\mathbb{H}_s$, and let t_s be the natural normal semifinite trace on $\mathcal{L}(A_0)$. Then there is a unitary mapping U from $\mathbb{H}_s$ onto $\mathbb{H}_{t_s}$ defined on the dense subspace A_0 of $\mathbb{H}_s$ by $U(a + N_s) = \lambda(a) + N_{t_s}$. Moreover, U satisfies $U\lambda(a)U^{-1} = \pi(\lambda(a))$ for $a \in A$ where π is the representation of the von Neumann algebra $\mathcal{L}(A_0)$ obtained by the GNS construction on the trace t_s.*

PROOF. Let $\tilde{A}_0$ be the achieved left Hilbert algebra in $\mathbb{H}_s$ consisting of the elements $a \in \mathbb{H}_s$ where $a_0 \mapsto \rho(a_0)a$ is bounded on $\mathbb{H}_s$. We call the extension of this operator $\lambda(a)$. By Theorem VII.19, $\mathcal{N}_{t_s}$ consists of all operators $T \in \mathcal{L}(A_0)$ such that $T = \lambda(a)$ for some $a \in \tilde{A}_0$. Moreover, $\langle a, b \rangle_{\mathbb{H}_s} = \langle \lambda(a) + N_{t_s}, \lambda(b) + N_{t_s} \rangle_{t_s}$. Thus if $A, B \in \mathcal{N}_s$, $A + N_s, B + N_s \in \tilde{A}_0$ and $\langle A + N_s, B + N_s \rangle_{\mathbb{H}_s} = \langle \lambda(A) + N_{t_s}, \lambda(B) + N_{t_s} \rangle_{t_s}$. Hence U is an isometry from $\mathcal{N}_s/N_s$ into $\mathbb{H}_{t_s}$ which extends to an isometry from $\mathbb{H}_s$ into $\mathbb{H}_{t_s}$. To see U is onto, it suffices to note that the range of U contains $\mathcal{N}_{t_s}/N_{t_s}$. Let $T \in \mathcal{N}_{t_s}$. Then $T = \lambda(a)$ for some $a \in \tilde{A}_0$. Choose a sequence $a_i + N_s$ in $\mathcal{N}_s/N_s$ which converges to a in $\mathbb{H}_s$. Then $\lambda(a - a_i) + N_{t_s}$ converges to 0 in $\mathbb{H}_{t_s}$. Indeed, $\langle \lambda(a - a_i) + N_{t_s}, \lambda(a - a_i) + N_{t_s} \rangle_{t_s} = \|a - (a_i + N_s)\|_{\mathbb{H}_s}^2 \to 0$ as $i \to \infty$. Hence $U(\mathbb{H}_s) \supseteq \mathcal{N}_{t_s}/N_{t_s}$, and thus $U(\mathbb{H}_s) = \mathbb{H}_{t_s}$.

Finally, let $a \in \mathcal{A}$ and $b \in \mathcal{N}_s$. Then

$$
\begin{aligned}
U\lambda(a)(b + N_s) &= U(ab + N_s) \\
&= \lambda(ab) + N_{t_s} \\
&= \lambda(a)(\lambda(b) + N_{t_s}) \\
&= \pi(\lambda(a))(\lambda(b) + N_{t_s}) \\
&= \pi(\lambda(a))U(b + N_s).
\end{aligned}
$$

Thus $U\lambda(a)U^{-1} = \pi(\lambda(a))$ for $a \in \mathcal{A}$. $\square$

Using the isomorphism U^* to carry all operators back to $\mathbb{H}_s$, we have the following theorem.

THEOREM VII.20. *Let $\mathcal{A}$ be a C^* algebra, and let s be a bitrace on a self adjoint ideal $\mathcal{N}_s$. Then there is an achieved left unimodular Hilbert algebra $\tilde{A}$ in $\mathbb{H}_s$ containing $\mathcal{N}_s/N_s$. It consists precisely of the elements $\xi \in \mathbb{H}_s$ such that $a + N_s \mapsto \lambda(a)\xi$ and $a + N_s \mapsto \rho(a)\xi$ are continuous on the pre-Hilbert space $\mathcal{N}_s/N_s$.*

LEMMA VII.40. *Suppose s and s' are bitraces on a C^* algebra $\mathcal{A}$ with $\mathcal{N}_s \subseteq \mathcal{N}_{s'}$ and $s'|_{\mathcal{N}_s \times \mathcal{N}_s} = s$. Let T be the isometry of $\mathbb{H}_s$ into $\mathbb{H}_{s'}$ satisfying $T(x + N_s) = x + N_{s'}$ for $x \in \mathcal{N}_s$ with range $\mathbb{H}'_s$. Then the orthogonal projection P onto $\mathbb{H}'_s$ is in the center of $\mathcal{U}_{s'} = \mathcal{L}(\mathcal{N}_{s'}/N_{s'})$; and if t_s and $t_{s'}$ are the natural traces on $\mathcal{U}_s$ and $\mathcal{U}_{s'}$, then $t_s(X) = t_{s'}(TXT^*)$.*

PROOF. Note T is well defined since $N_s \subseteq N_{s'}$. Moreover, T is an isometry since $s'|_{\mathcal{N}_s \times \mathcal{N}_s} = s$. Since $T\lambda(a)(x + N_s) = ax + N_{s'} = \lambda'(a)T(x + N_s)$, we see $T\lambda(a) = \lambda'(a)T$ on $\mathbb{H}_s$ for each $a \in \mathcal{A}$. Similarly, $T\rho(a) = \rho'(a)T$ for all a. Thus the range of T is invariant under all $\lambda'(a)$ and all $\rho'(a)$. This implies $P \in \lambda'(\mathcal{A})' \cap \rho'(\mathcal{A})'$, and thus P is in the center of $\mathcal{U}_{s'} = \lambda'(\mathcal{A})''$. Now t_s is defined on $\mathcal{U}_s$ by

$$
t_s(X) = \begin{cases} \langle \xi, \xi \rangle & \text{if } \rho(x)\xi = X^{1/2}(x + N_s) \text{ for } x \in \mathcal{N}_s \\ \infty & \text{otherwise.} \end{cases}
$$

A similar definition holds for $t_{s'}$. Now note $T^*\lambda'(a) = \lambda(a)T^*$ and $T^*\rho'(a) = \rho(a)T^*$ for all a. Thus if $X \in \mathcal{U}_s$, then $TXT^* \in \rho'(\mathcal{A})' = \mathcal{U}_{s'}$.

If $x + N_s \mapsto \rho(x)\xi$ is a bounded operator on $\mathcal{N}_s/N_s$, we shall use the notation $\lambda(\xi)$ for its bounded extension rather than the standard notation $\pi(\xi)$. Similarly we use $\lambda'(\xi')$ for $\mathcal{N}_{s'}/N_{s'}$.

Now suppose $\lambda(\xi)$ is bounded for a $\xi \in \mathbb{H}_s$. Then if $x \in \mathcal{N}_{s'}$, $\rho'(x)T\xi = \rho'(x)PT\xi = P\rho'(x)T\xi$, and thus $(I - P)\rho'(x)T\xi = 0$. This gives $\lambda'(T\xi)(I - P)(x + N_{s'}) = 0$ for all $x \in \mathcal{N}_{s'}$. But if $x \in \mathcal{N}_s$, we see $\lambda'(T\xi)(x + N_{s'}) = \rho'(x)T\xi = T\rho(x)\xi = T\lambda(\xi)(x + N_s) = T\lambda(\xi)T^*(x + N_{s'})$. Combining these, we see if $\lambda(\xi)$ is bounded, then $\lambda'(T\xi)$ is bounded and $\lambda'(T\xi) = T\lambda(\xi)T^*$. In particular, if $\lambda(\xi) = X^{1/2}$, then $\lambda'(T\xi) = TX^{1/2}T^*$. Hence $t_s(X) = \langle \xi, \xi \rangle_s$ and $t_{s'}(TXT^*) = \langle T\xi, T\xi \rangle_{s'} = \langle \xi, \xi \rangle_s$. This gives $t_s(X) = t_{s'}(TXT^*)$.

Now suppose ξ' is bounded in $\mathbb{H}_{s'}$. Thus $x + N_{s'} \mapsto \rho'(x)\xi'$ extends to a bounded operator $\lambda'(\xi')$. Hence $x + N_{s'} \mapsto T^*\rho'(x)\xi' = \rho(x)T^*\xi'$ is bounded on $\mathcal{N}_{s'}/N_{s'}$. Hence $x + N_s \mapsto \rho(x)T^*\xi'$ is bounded. Hence $\lambda(T^*\xi')$ exists and

$$\lambda(T^*\xi')(x + N_s) = T^*\lambda'(\xi')(x + N_{s'}) = T^*\lambda'(\xi')T(x + N_s)$$

for $x \in \mathcal{N}_s$. This gives $\lambda(T^*\xi') = T^*\lambda'(\xi)T$ for any bounded ξ' in $\mathbb{H}_{s'}$. In particular, if $TX^{1/2}T^* = \lambda'(\xi')$, then $\lambda(T^*\xi') = T^*TX^{1/2}T^*T = X^{1/2}$. Thus $t_s(X) < \infty$ iff $t_{s'}(TXT^*) < \infty$, and then $t_s(X) = t_{s'}(TXT^*)$. $\square$

COROLLARY VII.22. *Let s be a bitrace on a C^* algebra $\mathcal{A}$. Let t_s be the standard faithful normal semifinite trace on the von Neumann algebra $\mathcal{L}(\mathcal{N}_s/N_s)$ generated by the left unimodular Hilbert algebra $\mathcal{N}_s/N_s$ on $\mathbb{H}_s$. Then $t_s \circ \lambda$ is a semifinite, lower semicontinuous trace on the C^* algebra $\mathcal{A}$. Moreover, if s' is any bitrace with $\mathcal{N}_s \subseteq \mathcal{N}_{s'}$ and $s'|_{\mathcal{N}_s \times \mathcal{N}_s} = s$, then $\mathcal{N}_s \subseteq \mathcal{N}_{s'} \subseteq \mathcal{N}_{t_{s'}\circ\lambda'} \subseteq \mathcal{N}_{t_s\circ\lambda}$.*

PROOF. $\mathcal{N}_s$ is a self adjoint ideal in $\mathcal{A}$; and by Proposition VII.28, $\lambda(\mathcal{N}_s)$ generates the von Neumann algebra $\mathcal{L}(\mathcal{N}_s/N_s) = \lambda(\mathcal{A})''$. In particular, $\lambda(\mathcal{N}_s)$ is strongly dense in $\mathcal{L}(\mathcal{N}_s/N_s)$. Moreover, by Theorem VII.19, t_s is a faithful normal semifinite trace on $\mathcal{L}(\mathcal{N}_s/N_s)$. Hence Proposition VII.26 implies $t_s \circ \lambda|_{\mathcal{A}^+}$ is a semifinite, norm lower semicontinuous trace on $\mathcal{A}$.

Now note that $\mathcal{N}_s \subseteq \mathcal{N}_{t_s\circ\lambda}$ since $\lambda(\mathcal{N}_s) \subseteq \mathcal{N}_{t_s}$. Thus if s' extends s, we need only show $\mathcal{N}_{t_{s'}\circ\lambda'} \subseteq \mathcal{N}_{t_s\circ\lambda}$. Let $a \in \mathcal{N}_{t_{s'}\circ\lambda'}$. Then $\lambda'(a) \in \mathcal{N}_{t_{s'}}$. Hence $t_{s'}(\lambda'(a^*a)) < \infty$, and there is a $\xi' \in \mathbb{H}_{s'}$ such that $\lambda'(a^*a)^{1/2} = \lambda'(\xi')$. But then $T^*\lambda'(\xi)T = \lambda(T^*\xi') = T^*\lambda'(a^*a)^{1/2}T = \lambda(a^*a)^{1/2}$ for $T^*\lambda'(a)T = \lambda(a)$. Thus $t_s(\lambda(a^*)\lambda(a)) = \langle T^*\xi', T^*\xi' \rangle_s \leq \langle \xi', \xi' \rangle_{s'} = t_{s'}(\lambda'(a^*)\lambda'(a)) < \infty$. $\square$

PROPOSITION VII.29. *Let $\mathcal{A}_0$ be a self adjoint dense subalgebra of a C^* algebra $\mathcal{A}$, and let $s_0 : \mathcal{A}_0 \times \mathcal{A}_0 \to \mathbb{C}$ be a mapping satisfying*

(a) *s_0 is a complex Hermitian nonnegative form;*

(b) $s_0(y, x) = s_0(x^*, y^*)$;

(c) $s_0(zx, y) = s_0(x, z^*y)$ for $x, y, z \in \mathcal{A}_0$;

(d) for each $z \in \mathcal{A}_0$, $s_0(zx, zx) \leq ||z||^2 s_0(x, x)$ for all $x \in \mathcal{A}_0$; and

(e) the elements xy, $x, y \in \mathcal{A}_0$, span a dense subspace of $\mathcal{A}_0$ under the pre-Hilbert structure defined by s_0.

Set $N = \{a \in \mathcal{A}_0 : s_0(a, a) = 0\}$. Then N is a $*$ ideal in $\mathcal{A}_0$, and $\mathcal{A}_0/N$ with involution defined by $(a + N)^\# = a^* + N$ is a unimodular left Hilbert algebra in the completion $\mathbb{H}$ of $\mathcal{A}_0/N$ equipped with Hermitian inner product $\langle a + N, b + N \rangle = s_0(a, b)$. Moreover, the natural representation λ_0 of $\mathcal{A}_0$ on $\mathbb{H}$ extends to a $*$-representation λ of $\mathcal{A}$; and if t is the natural normal faithful semifinite trace on $\mathcal{L}(\mathcal{A}_0/N)$, then s defined on $\mathcal{N}_s \times \mathcal{N}_s$ by $s(a, b) = t(\lambda(a)\lambda(b^*))$ where $\mathcal{N}_s = \lambda^{-1}(\mathcal{N}_t)$ extends s_0 and is a maximal bitrace on $\mathcal{A}$.

PROOF. N is a $*$ closed since $s_0(a^*, a^*) = s_0(a, a)$. It is a left ideal in $\mathcal{A}_0$, for $0 \leq s(ab, ab) \leq ||a||^2 s(b, b) = 0$ if $a \in \mathcal{A}_0$ and $b \in N$. Thus $\mathcal{A}_0/N$ is a pre-Hilbert space and has involution $(a + N)^\# = a^* + N$. Moreover, by (d) and (c), $\lambda_0(a)(b + N) = ab + N$ has a natural extension to a $*$ representation of $\mathcal{A}_0$ on $\mathbb{H}$ satisfying $||\lambda_0(a)|| \leq ||a||$ for $a \in \mathcal{A}_0$. Since $\mathcal{A}_0$ is dense in $\mathcal{A}$, λ_0 has a unique continuous extension to a $*$-representation λ of $\mathcal{A}$ on $\mathbb{H}$.

Now $a + N \mapsto a^* + N$ is an isometry of $\mathcal{A}_0/N$. Thus it has a unique extension to a conjugate linear isometry S of $\mathbb{H}$. That $\mathcal{A}_0/N$ is a left Hilbert algebra in $\mathbb{H}$ now follows from (c) and (e). $\mathcal{A}_0/N$ is unimodular since $\Delta = S^*S = I$.

Let t be the natural normal semifinite faithful trace on $\mathcal{L}(\mathcal{A}_0/N)$. Then $\mathcal{N}_t$ is a $*$ ideal in $\mathcal{L}(\mathcal{A}_0/N)$, and $\mathcal{N}_s = \lambda^{-1}(\mathcal{N}_t)$ is a $*$ ideal in $\mathcal{A}$. Moreover, since $\lambda(a) = \pi(a + N)$ for $a \in \mathcal{A}_0$ and $t(\lambda(a)^*\lambda(a)) = \langle a + N, a + N \rangle$ for $a \in \mathcal{A}_0$, we see $t(\lambda(a)\lambda(b)^*) = s_0(a, b)$ for $a, b \in \mathcal{A}_0$. Hence s extends s_0. Next note the Hilbert space structure defined by s on $\mathcal{N}_s/N_s$ has the same completion as the Hilbert space structure obtained by completing $\mathcal{A}_0/N$. Indeed, the completion of $\mathcal{A}_0/N$ is $\mathbb{H}$; and by Theorem VII.18, the mapping $U : \mathcal{A}_0/N \mapsto \mathbb{H}_t$ defined by $U(a + N) = \lambda(a)$ extends to a unitary isometry of $\mathbb{H}$ onto $\mathbb{H}_t$ which carries the achieved left Hilbert algebra $(\mathcal{A}_0/N)''$ onto $\mathcal{N}_t$. Now the mapping $a + N \mapsto a + N_s$ is an isometry from $\mathcal{A}_0/N$ into $\mathcal{N}_s/N_s$, and the mapping $a + N_s \mapsto \lambda(a)$ is an isometry of $\mathcal{N}_s/N_s$ into $\mathbb{H}_t$, and these compose to give U. Hence $\mathcal{A}_0/N \to \mathcal{N}_s/N_s \to \mathcal{N}_t/N_t$ are isometries which extend to the completions. In particular, $\mathcal{N}_s^2$ contains $\mathcal{A}_0^2$; and thus $\mathcal{N}_s^2$ is dense in pre-Hilbert structure on $\mathcal{N}_s/N_s$.

Thus s is a bitrace.

We finally note that s is maximal. Indeed, suppose $\mathcal{N}_{s'}$ is a $*$ ideal in $\mathcal{A}$ with $\mathcal{N}_s \subseteq \mathcal{N}_{s'}$ and $s(a,b) = s'(a,b)$ for $a, b \in \mathcal{N}_s$. By Corollary VII.22, $\mathcal{N}_s \subseteq \mathcal{N}_{s'} \subseteq \mathcal{N}_{t_{s'} \circ \lambda'} \subseteq \mathcal{N}_{t_s \circ \lambda_s}$. But $\mathcal{N}_{t_s \circ \lambda_s} \subseteq \mathcal{N}_{t \circ \lambda}$, for the mapping $a + N_s \mapsto \lambda(a) + N_t$ from $\mathcal{N}_s/N_s$ to $\mathcal{N}_t/N_t$ extends to an isometry V from $\mathbb{H}_s$ onto $\mathbb{H}_t$ carrying $\mathcal{N}_s/N_s$ into the achieved left Hilbert algebra $\mathcal{N}_t$ satisfying $V\pi_s(\lambda_s(a))V^* = \pi(\lambda(a))$ for $a \in \mathcal{N}_s$. Thus $\mathcal{N}_s \subseteq \mathcal{N}_{s'} \subseteq \mathcal{N}_{t \circ \lambda} = \mathcal{N}_s$, and $s = s'$. $\square$

EXAMPLE. Let G be a unimodular locally compact second countable group. Let $\mathcal{A}$ be $C^*(G)$, and let $\mathcal{A}_0 = C_c(G)$. Then $\mathcal{A}_0$ is a dense $*$ subalgebra of $\mathcal{A}$. Define s on $\mathcal{A}_0 \times \mathcal{A}_0$ by

$$s(f,g) = \int f(x)\bar{g}(x)\, dx$$

where dx is a left Haar measure on G. Then s satisfies
(a) s is a positive sesquilinear Hermitian form;
(b) $s(f,g) = s(g^*, f^*)$;
(c) $s(f * g, h) = s(g, f^* * h)$;
(d) $s(f * g, f * g) \le \|f\|^2_{C^*(G)} s(g,g)$; and
(e) the elements $f * g$ span a dense linear subspace of $\mathcal{A}_0$ relative to the inner product s.

PROOF. We check the statements. (a) is clear. For (b), note

$$\int f(x)\bar{g}(x)\, dx = \int f(x^{-1})\bar{g}(x^{-1})\Delta(x^{-1})\, dx$$

$$= \int g^*(x)\bar{f}^*(x)\Delta(x)\, dx$$

$$= s(g^*, f^*);$$

this is where we must have unimodularity. For (c), we have

$$s(f * g, h) = \int f * g(x)\bar{h}(x)\, dx$$

$$= \int\int f(y)g(y^{-1}x)\bar{h}(x)\, dy\, dx$$

$$= \int\int f(xy)g(y^{-1})\bar{h}(x)\, dy\, dx$$

$$= \int\int g(y)f(xy^{-1})\Delta(y^{-1})\bar{h}(x)\, dy\, dx$$

$$= \int g(y) \int \overline{f^*(yx^{-1})h(x)} \Delta(x^{-1}) \, dx \, dy$$

$$= \int g(y) \int \overline{f^*(yx)h(x^{-1})} \, dx \, dy$$

$$= \int g(y) \int \overline{f^*(x)h(x^{-1}y)} \, dx \, dy$$

$$= s(g, f^* * h).$$

For (d), $\lambda(f)\xi = f * \xi$ defines a nondegenerate representation of $C_c(G)$ on $L^2(G)$. Thus $||\lambda(f)|| \leq ||f||_{C^*(G)}$. Hence

$$s(f * g, f * g) = ||\lambda(f)g||_2^2 \leq ||\lambda(f)||^2 ||g||_2^2 \leq ||f||_{C^*(G)}^2 s(g, g).$$

Thus (a), (b), (c) and (d) hold. The nondegeneracy of λ implies (e) holds. □

For this example $N = N_s = \{0\}$. Thus $\mathcal{A}_0 = C_c(G)$ is a unimodular left Hilbert algebra in $L^2(G)$, and its completion $\mathbb{H}$ under $\langle \cdot, \cdot \rangle_s$ is $L^2(G)$. Thus for each $f \in \mathcal{A}_0$, there is a unique bounded operator $\lambda(f)$ on $\mathbb{H}$ extending $g \mapsto f * g$. (Note $\lambda(f)\xi = f * \xi$.) Also $||\lambda(f)|| \leq ||f||_{C^*(G)}$. Thus λ extends to a representation of $C^*(G)$ on $\mathbb{H}$. Moreover, $\lambda(C^*(G))$ and $\lambda(\mathcal{A}_0)$ generate the same von Neumann algebra $\mathcal{U}$.

Furthermore, if t is the natural normal faithful semifinite trace on $\mathcal{U} = \mathcal{L}(\mathcal{A}_0)$, then s defined on $\lambda^{-1}(\mathcal{N}_t)$ by $s(a, b) = t(\lambda(a)\lambda(b^*))$ is a maximal bitrace on $C^*(G)$. Also, for each $\xi \in L^2(G)$, one may consider the mappings $f \mapsto f * \xi$ and $f \mapsto \xi * f$ for $f \in \mathcal{A}_0 = C_c(G)$. If the first is bounded, we call it $\rho(\xi)$ and if the second is bounded we call it $\lambda(\xi)$. Then $\rho(\xi)$ and $\lambda(\xi)$ are both bounded or both unbounded. If they are both bounded, then ξ is said to be a bounded element in the Hilbert space. The set $\mathcal{A}'$ of bounded elements form an achieved left unimodular Hilbert algebra which is isomorphic to the achieved left Hilbert algebra $\mathcal{N}_t$ in the completion of $(\mathcal{N}_t/N_t, \langle \cdot, \cdot \rangle_t)$.

24. Disintegration of Positive Linear Functionals

PROPOSITION VII.30. *Let $\mathcal{A}$ be a separable C^* algebra and π be a ∗–representation of $\mathcal{A}$ on a separable Hilbert space $\mathbb{H}$. Suppose π has cyclic vector ξ, and π decomposes into a direct integral $\int^\oplus \pi_s \, d\mu(s)$ of representations π_s relative to a direct integral decomposition $\mathbb{H} = \int^\oplus \mathbb{H}_s \, d\mu(s)$ of a standard Borel Hilbert bundle $\mathcal{H}$. Let $s \mapsto \xi(s)$ be a Borel section of the Hilbert bundle with $\xi = \int^\oplus \xi(s) \, d\mu(s)$. Then for* *a.e.* s, $\xi(s)$ *is a cyclic vector for the representation* π_s; *and for each* $a \in \mathcal{A}$,

$$\langle \pi(a)\xi, \xi \rangle = \int \langle \pi_s(a)\xi(s), \xi(s) \rangle_s \, d\mu(s).$$

PROOF. We may assume the Hilbert bundle is trivial. Thus $\mathbb{H}_s = \mathbb{H}_0$ for all s. Let $\mathcal{A}_0 = \{a_i\}_{i=1}^\infty$ be a countable dense complex rational $*$ subalgebra of $\mathcal{A}$. Using a Gram-Schmidt like process there exists a sequence of Borel functions f_i from S into $\mathcal{A}$ so that for each i, $s \mapsto \pi(f_i(s))$ is a Borel field of operators; and

(a) $\langle \pi_s(f_i(s))\xi(s), \pi_s(f_j(s))\xi(s) \rangle_s = 0$ for $i \neq j$;
(b) $\pi_s(f_i(s))\xi(s)$ is either 0 or a unit vector;
(c) the linear span of $\pi_s(f_i(s))\xi(s)$ is dense in $\pi_s(\mathcal{A})\xi(s)$.

We indicate the inductive construction. Set

$$f_1(s) = \begin{cases} \dfrac{a_1}{||\pi_s(a_1)\xi(s)||_s^{1/2}} & \text{if } \pi_s(a_1)\xi(s) \neq 0 \\ 0 & \text{otherwise.} \end{cases}$$

Suppose $f_1, f_2, \cdots, f_n$ have been defined. For each s, set

$$a(s) = a_{n+1} - \sum_{i=1}^n \langle \pi_s(a_{n+1})\xi(s), \pi_s(f_i(s))\xi(s) \rangle_s f_i(s).$$

Then define f_{n+1} by

$$f_{n+1}(s) = \begin{cases} \dfrac{a(s)}{||a(s)||} & \text{if } a(s) \neq 0 \\ 0 & \text{otherwise.} \end{cases}$$

Note the linear span of $\pi_s(f_i(s))\xi(s)$ contains $\pi_s(\mathcal{A}_0)\xi(s)$ and thus is dense in $\pi_s(\mathcal{A})\xi(s)$.

Let $P(s)$ be the orthogonal projection of $\mathbb{H}_s$ onto the closure of $\pi_s(\mathcal{A})\xi(s)$. We note $s \mapsto P(s)$ is a Borel field of operators for the Hilbert bundle $\mathcal{H}$. Indeed,

$$P(s)\xi'(s) = \sum_{i=1}^\infty \langle \xi'(s), \pi_s(f_i(s))\xi(s) \rangle_s \pi_s(f_i(s))\xi(s)$$

is a Borel section of the Hilbert bundle $\mathcal{H}$ for each Borel section $\xi'(s)$ of $\mathcal{H}$.

Let $S_0 = \{s : P(s) \neq I_s\}$. Then S_0 is Borel. Moreover, $\{(s, v) : s \in S_0, ||v||_0 = 1, P(s)v = 0\}$ is a Borel subset of $S_0 \times \mathbb{H}_0$. By the von Neumann Selection Theorem I.15, there is a Borel function ξ' on S such that $\xi'(s) = 0$ for $s \notin S_0$ and $P(s)\xi'(s) = 0$ and $||\xi'(s)||_0 = 1$ for μ a.e. $s \in S_0$. This implies $\langle \pi(a)\xi, \xi' \rangle = 0$ for all $a \in \mathcal{A}$. Since ξ

is cyclic, $\xi' = 0$; and S_0 has measure 0. Hence $\xi(s)$ is a cyclic vector for π_s for almost all s.

Finally, note

$$\langle \pi(a)\xi, \xi \rangle = \int_S \langle \pi_s(a)\xi(s), \xi(s) \rangle_s \, d\mu(s)$$

for all a follows from $\pi = \int^\oplus \pi_s \, d\mu(s)$. □

COROLLARY VII.23. *Let ω be a positive bounded linear functional on a separable C^* algebra $\mathcal{A}$. Let π_ω be the representation of $\mathcal{A}$ obtained on $\mathbb{H}_\omega$ by the GNS construction. Suppose $\pi_\omega = \int^\oplus \pi_s \, d\mu(s)$ is a direct integral decomposition of π_ω relative to a Hilbert bundle $\mathcal{H}$. If $\xi_\omega = \int^\oplus \xi_\omega(s) \, d\mu(s)$ is the canonical cyclic vector for π_ω, then for almost all s, $\xi_\omega(s)$ is the canonical cyclic vector for π_s relative to positive bounded linear functional $\omega_s(a) \equiv \langle \pi_s(a)\xi_\omega(s), \xi_\omega(s) \rangle_s$.*

PROOF. This follows easily from Theorem II.7. □

25. Traces on Type I von Neumann Algebras

PROPOSITION VII.31. *Let $\mathbb{H}$ be a Hilbert space, and let t be a normal semifinite trace on $B(\mathbb{H})$. Then $\mathcal{N}_t = \mathcal{HS}(\mathbb{H}, \mathbb{H})$, and $\mathcal{M}_t$ is the space of trace class operators on $\mathbb{H}$. Moreover, there is a $\lambda > 0$ such that $t(A) = \lambda Tr(A)$ for $A \geq 0$.*

PROOF. Let v be a unit vector in $\mathbb{H}$. Then $E_v(w) = \langle w, v \rangle v$ is a rank one projection in $B(\mathbb{H})$. Note the positive operators smaller than E_v have form rE_v where $0 \leq r \leq 1$. Hence $\infty > t(E_v) = \lambda > 0$ since t is faithful and semifinite. Moreover, if v' is any other unit vector, there is a unitary operator U with $Uv = v'$. Hence $UE_vU^* = E_{v'}$. Thus $t(E_{v'}) = t(UE_vU^*) = t(E_vU^*U) = t(E_v) = \lambda$. Hence $t(P) = \lambda$ for all rank one orthogonal projections. Thus $t(P) = \lambda Tr(P)$ for all finite rank orthogonal projections P. Moreover, any orthogonal projection P is the strong limit of an increasing net of finite rank projections. Indeed, let $\{e_i\}_{i \in I}$ be an complete orthonormal basis of $P(\mathbb{H})$. Order the finite subsets F of I by inclusion; and for each F, set P_F to be the orthogonal projection onto the linear span of F. Then for each $v \in \mathbb{H}$, $P_F v \to Pv$ strongly. Hence $P_F \nearrow P$ strongly as F increases. By normality, $t(P) = \lambda Tr(P)$ for any orthogonal projection P.

Now let A be any positive operator. By the spectral theorem, for any $\epsilon > 0$, there is a positive operator A_ϵ with finite spectrum satisfying $A \geq A_\epsilon$ and $||A - A_\epsilon|| < \epsilon$. Indeed, if $A = \int_{\sigma(A)} \lambda \, dP(\lambda)$

and $E_1, E_2, \cdots, E_n$ is a partition of $\sigma(A)$ into disjoint Borel subsets with $\sup_{\lambda \in E_i} \lambda - \inf_{\lambda \in E_i} \lambda < \epsilon$, then take $A_\epsilon = \sum_{i=1}^n \lambda_i P(E_i)$ where $\lambda_i = \inf_{\lambda \in E_i} \lambda$. By doing refinements, we may assume $A_{2^{-n}} \nearrow A$ as $n \to \infty$. Hence $t(A) = \lambda \mathrm{Tr}(A)$ for all $A \geq 0$.

Since $\mathcal{M}_t$ is the linear span of $A \geq 0$ with $t(A) < \infty$, we see $\mathcal{M}_t$ is the space of trace class operators on $\mathbb{H}$. Clearly, $\mathcal{N}_t$ is $\mathcal{HS}(\mathbb{H}, \mathbb{H})$. $\square$

REMARK. Tr is a faithful normal semifinite trace on the von Neumann algebra $B(\mathbb{H})$. Moreover, $\mathbb{H}_{\mathrm{Tr}} = \mathcal{HS}(\mathbb{H}, \mathbb{H})$; and $\mathcal{HS}(\mathbb{H}, \mathbb{H})$ is an achieved left unimodular Hilbert algebra in the Hilbert space $\mathcal{HS}(\mathbb{H}, \mathbb{H})$.

PROPOSITION VII.32. *Suppose ν is a σ-finite measure on a standard Borel space S and $\mathbb{H}$ is a separable Hilbert space. Let $\mathcal{A}$ be the type I von Neumann algebra $L^\infty(S, \nu, B(\mathbb{H}))$. If t is a normal, semifinite trace on $\mathcal{A}$, then there is a unique σ-finite measure $\mu \prec \nu$ satisfying*

$$t(A) = \int Tr(A(s)) \, d\mu(s)$$

for each positive operator $A \in L^\infty(S, \nu, B(\mathbb{H}))$. Conversely, if μ is a σ-finite measure with $\mu \prec \nu$, then

$$A \mapsto \int Tr(A(s)) \, d\mu(s)$$

is a normal semifinite trace. It is faithful iff $\mu \sim \nu$.

PROOF. Let E_v be the operator $w \mapsto \langle w, v \rangle v$. Then if v is a unit vector, E_v is a rank 1 orthogonal projection in $B(\mathbb{H})$. For each Borel subset F of S, the operator $(1_F E_v)(s) = 1_F(s) E_v$ is in $\mathcal{A}^+$. Moreover, if v and w are unit vectors in $\mathbb{H}$ and T is a unitary operator on $\mathbb{H}$ carrying v to w, then $1_S T$ is a unitary operator on $L^2(S, \mathbb{H})$ satisfying $(1_S T)(1_F E_v)(1_S T)^* = 1_F E_w$. In particular, since t is a trace, $t(1_F E_w) = t(1_F E_v)$. Hence $\mu(F) \equiv t(1_F E_v)$ where v is a unit vector in $\mathbb{H}$ is well defined. Moreover, using the normality of t and Propositions VII.16 and VII.15, one sees that μ is a measure on S. We further note this measure is absolutely continuous with respect to ν.

To see that μ is σ-finite, we exploit the semifiniteness of the trace t. Consider the operator $1_S E_v$ where v is a unit vector. It is in $\mathcal{A}^+$. By the semifiniteness of t and the strong separability of $\mathcal{A}$, there is a sequence of elements A_n in the linear span of all positive elements $P \in \mathcal{A}^+$ with $t(P) < \infty$ satisfying $A_n \to 1_S E_v$ in the strong operator topology. In particular, there is a countable family $\{P_i\}$ of positive projections in $\mathcal{A}$ with $\nu(S - \cup_i\{s : P_i(s) \neq 0\}) = 0$ and $t(P_i) < \infty$ for all i. Since $\mu \prec \nu$, we see $\cup_i\{s : P_i(s) \neq 0\}$ is μ-conull in S. Thus to see μ is σ-finite, it suffices to show $\{s : P(s) \neq 0\}$ has finite μ measure for each projection $P \in \mathcal{A}$ with $t(P) < \infty$. Since P is a projection in $L^\infty(S, \nu, B(\mathbb{H}))$, Proposition III.12 implies we may assume $P(s)$ is an orthogonal projection in $B(\mathbb{H})$ for each s.

Let $\{e_i\}_{i=1}^\infty$ be a complete orthonormal basis of $\mathbb{H}$. Thus for each i, $s \mapsto P(s)e_i$ is Borel. We do a Gram-Schmidt type orthonormalization process. Namely define f_1 by

$$f_1(s) = \begin{cases} \frac{P(s)e_1}{||P(s)e_1||} & \text{if } P(s)e_1 \neq 0 \\ 0 & \text{otherwise.} \end{cases}$$

Suppose $f_1, \dots, f_n$ have been defined. For each s, let

$$p(s) = P(s)e_{n+1} - \sum_{k=1}^n \langle P(s)e_{n+1}, f_k(s)\rangle f_k(s).$$

Define f_{n+1} by

$$f_{n+1}(s) = \begin{cases} \frac{p(s)}{||p(s)||} & \text{if } p(s) \neq 0 \\ 0 & \text{otherwise.} \end{cases}$$

Then for each s, the range of $P(s)$ is the closure of the linear span of the set of vectors $\{f_i(s) : i = 1, 2, \dots\}$. Define f by $f(s) = \sum_{n=1}^\infty 2^{-n} f_n(s)$. Set

$$v(s) = \begin{cases} \frac{f(s)}{||f(s)||} & \text{if } f(s) \neq 0 \\ 0 & \text{otherwise.} \end{cases}$$

Then $P(s) \neq 0$ iff $v(s) = P(s)v(s) \neq 0$.

Pick a unit vector v_0. The set

$$\{(s, T) : T \in \mathcal{U}(\mathbb{H}), Tv_0 = v(s) \text{ where } v(s) \neq 0\}$$

is a Borel subset of $S \times B(\mathbb{H})$ whose projection into S is the set of s's with $P(s) \neq 0$. By the von Neumann Selection Theorem (see Theorem I.14), there is a strongly Borel function $s \mapsto T(s)$ such that $T(s) \in \mathcal{U}(\mathbb{H})$ and $T(s)v_0 = v(s)$ for ν a.e. s where $v(s) \neq 0$. In particular, $T(s)^*P(s)T(s)v_0 = v(s)$ for ν a.e. s where $P(s) \neq 0$. This implies $T^*PT \geq 1_{\{s:P(s)\neq 0\}}E_{v_0}$. Thus

$$\mu(\{s : P(s) \neq 0\}) = t(1_{\{s:P(s)\neq 0\}}E_{v_0}) \leq t(T^*PT) = t(P) < \infty;$$

and we see μ is σ-finite.

Next note $A \mapsto \int \mathrm{Tr}(A(s)) \, d\mu(s)$ defines a normal trace. Indeed, we have $\int \mathrm{Tr}(A(s)) \, d\mu(s) = \sum_i \int \langle A(s)e_i, e_i \rangle \, d\nu(s) = \sum_i \omega_{e_i(\cdot), e_i(\cdot)}(A)$ where $e_i(\cdot)$ is the element in $L^2(S, \nu, \mathbb{H})$ with $e_i(s) = e_i$ for all s.

We show this trace is t. Let $A \in L^\infty(S, \nu, B(\mathbb{H}))^+$ have spectral resolution $A = \int_{[0,\infty]} \lambda \, dP(\lambda)$. By Theorem II.5, each projection $P(E)$ where E is a Borel subset of S commutes with all the operators in $L^\infty(S, \nu, B(\mathbb{H}))'$; and thus each $P(E)$ belongs to $L^\infty(S, \nu, B(\mathbb{H}))$. As seen in the proof of Proposition VII.31, there is a sequence A_n of positive operators such that $A_n \nearrow A$ strongly; and each A_n is a finite positive linear combination of projections of form $P(E_i)$. Thus by Propositions VII.16 and VII.15, it suffices to show

$$t(P) = \int \mathrm{Tr}(P(s)) \, d\mu(s)$$

for all orthogonal projections P in $L^\infty(S, \nu, B(\mathbb{H}))$. As seen earlier, there are Borel functions f_i on S such that each $f_i(s)$ is either a unit vector or 0, the $f_i(s)$ are orthogonal, and the range of $P(s)$ is the closure of the linear span of the $f_i(s)$. Set $P_n(s)v = \sum_{i=1}^n \langle v, f_i(s) \rangle f_i(s)$. Then $P_n \in L^\infty(S, \nu, B(\mathbb{H}))$, and $P_n \nearrow P$ strongly. Thus since both $t(P_n) \nearrow t(P)$ and $\int \mathrm{Tr}(P_n(s)) \, d\mu(s) \nearrow \int \mathrm{Tr}(P(s)) \, d\mu(s)$, we need only show

$$t(P_n) = \int \mathrm{Tr}(P_n(s)) \, d\mu(s).$$

But each $P_n(s)$ is a finite sum of the orthogonal projections $E_i(s)$ defined by $E_i(s)(v) = \langle v, f_i(s) \rangle f_i(s)$. Hence we need only check that $t(Q) = \int \mathrm{Tr}(Q(s)) \, d\mu(s)$ where Q satisfies $Q(s) = E_{v(s)}$ for some Borel function $s \mapsto v(s) \in \mathbb{H}$ with $||v(s)|| = 0$ or 1. The argument above for σ-finiteness shows there is a unitary T in $L^\infty(S, \nu, B(\mathbb{H}))$ such that $T(s)E_{v(s)}T(s)^* = 1_{\{s:v(s)\neq 0\}}(s)E_{v_0}$. Hence

$$\begin{aligned} t(Q) &= t(TQT^*) \\ &= t(1_{\{s:v(s)\neq 0\}}E_{v_0}) \\ &= \mu(\{s : v(s) \neq 0\}) \\ &= \int \mathrm{Tr}(Q(s)) \, d\mu(s). \end{aligned}$$

Uniqueness of the measure μ follows from the fact that if

$$t(A) = \int \mathrm{Tr}(A(s)) \, d\mu'(s)$$

for positive operators A, then

$$\mu'(F) = \int \mathrm{Tr}(1_F(s)E_v)\, d\mu'(s) = t(1_F E_v) = \mu(F)$$

for any unit vector $v \in \mathbb{H}$.

Conversely, suppose μ is a σ-finite measure with $\mu \prec \nu$. We have seen that $t(A) = \int \mathrm{Tr}(A(s))\, d\mu(s)$ is a normal trace. Moreover, if $A > 0$, we have seen there is a nonzero projection P of form $P(s) = E_{v(s)}$ and a $\lambda > 0$ such that $\lambda P \leq A$. Using the σ-finiteness of μ, we may assume $\mu\{s : v(s) \neq 0\} < \infty$. Hence $B = \lambda P$ satisfies $B > 0$, $B \leq A$, and $t(B) < \infty$. Semifiniteness now follows by Corollary VII.18.

Clearly, $\mu \sim \nu$ iff t is faithful. $\square$

THEOREM VII.21. *Let $\mathcal{H}$ be a Hilbert bundle over the standard Borel space S. Suppose $n(s)$ is a Borel function on S with values in $\{\infty, 1, 2, 3, \dots\}$ and ν is a σ-finite measure on S. Then if t is a normal semifinite trace on the separable type I von Neumann algebra $A = \int^{\oplus} n(s)B(\mathbb{H}_s)\, d\nu(s)$, then there is a unique σ-finite measure $\mu \prec \nu$ such that*

$$t\left(\int^{\oplus} n(s)A(s)\, d\nu(s)\right) = \int \mathrm{Tr}_s(A(s))\, d\mu(s)$$

for any Borel field of positive operators $s \mapsto A(s)$ relative to the Hilbert bundle $\mathcal{H}$. Conversely, if $\mu \prec \nu$ and μ is σ-finite, then $A \mapsto \int \mathrm{Tr}_s(A(s))\, d\mu(s)$ for $A = \int^{\oplus} n(s)A(s)\, d\mu(s) \in A^+$ defines a normal semifinite trace.

PROOF. By Theorem III.12, we may assume $\mathcal{H}$ is a direct sum of trivial Hilbert bundles $\mathcal{H}_k$ where $\mathcal{H}_k = S_k \times \mathbb{H}_k$ and $\dim \mathbb{H}_k = k$. For $l = \infty, 1, 2, \dots$, set $S_{k,l} = n^{-1}(l) \cap S_k$. Using this decomposition, one sees $A = \oplus A_{k,l}$ where $A_{k,l} = L^{\infty}(S_{k,l}, \nu, lB(\mathbb{H}_k))$. By normality, the result will follow if for any normal semifinite trace $t_{k,l}$ on $A_{k,l}$, there is a unique σ-finite measure μ on $S_{k,l}$ such that $t_{k,l}$ has form $t_{k,l}(\int_{S_k}^{\oplus} lA(s)\, d\nu(s)) = \int_{S_k} \mathrm{Tr}_k(A(s))\, d\mu(s)$ for any strongly Borel function $s \mapsto A(s) \in B(\mathbb{H}_k)^+$. But if $t_{k,l}$ is a normal semifinite trace on $L^{\infty}(S_k, \nu, lB(\mathbb{H}_k))$, $t'_{k,l}$ defined by $t'_{k,l}(A) = t_{k,l}(lA)$ for $A \in L^{\infty}(S_{k,l}, \nu, B(\mathbb{H}_k)^+)$ is a normal semifinite trace on $L^{\infty}(S_{k,l}, \nu, B(\mathbb{H}_k))$. The existence and uniqueness of μ now follows Proposition VII.32.

Conversely, let μ be a σ-finite measure on S absolutely continuous relative to ν. Again we may assume $\mathcal{H}$ is a direct sum of trivial Hilbert bundles. By Proposition VII.32, $t_{k,l}$ on $L^{\infty}(S_{k,l}, \nu, lB(\mathbb{H}_k))$

defined by $t_{k,l}(A) = \int_{S_{k,l}} \mathrm{Tr}_k(A(s))\, d\mu(s)$ is a normal semifinite trace. Thus $t = \oplus_{k,l} t_{k,l}$ is normal and semifinite. $\square$

26. Bitraces on Separable Type I C* Algebras

LEMMA VII.41. *Let s be a bitrace on a C* algebra $\mathcal{A}$. Suppose S is a norm dense subset in $\mathcal{A}$ and $S \subseteq \mathcal{N}_s$. Then the set $\{xy + N_s : x, y \in S\}$ spans a dense subspace in $\mathbb{H}_s$. In particular, if $\mathcal{A}$ is separable, then $\mathbb{H}_s$ is separable.*

PROOF. By the definition of a bitrace, the elements $xy + N_s$ where $x, y \in \mathcal{N}_s$ span a dense subpsace of $\mathbb{H}_s$. Now λ and ρ are continuous representations of $\mathcal{A}$ on $\mathbb{H}_s$. Thus if x_i and y_j are nets in S which converge in norm to x and y, then

$$\|x_i y_j - xy + N_s\|_s \leq \|x_i y_j - x_i y + N_s\|_s + \|x_i y - xy + N_s\|_s$$
$$\leq \|\rho(y_j - y)(x_i + N_s)\|_s + \|\lambda(x_i - x)(y + N_s)\|_s$$
$$\leq \|y_j - y\|_{\mathcal{A}} \|x_i + N_s\|_s + \|x_i - x\|_{\mathcal{A}} \|y + N_s\|_s$$

which can be made small by fixing x_i near x and then choosing y_j near y. $\square$

REMARK. Let π be a $*$-representation of a $*$-algebra $\mathcal{A}$ on a Hilbert space $\mathbb{H}$. Then $a \mapsto \pi(a^*)$ is a $*$ anti-representation of $\mathcal{A}$ on $\mathbb{H}$. We denote it by π^*.

THEOREM VII.22. *Let $\mathcal{A}$ be a separable type I C* algebra, and let s be a bitrace on $\mathcal{A}$. Suppose $\mathcal{N}_s$ is norm dense in $\mathcal{A}$. Let λ and ρ be the associated left and right representations of $\mathcal{A}$ on the separable Hilbert space $\mathbb{H}_s$. Let J_s be the natural conjugation on $\mathbb{H}_s$ obtained by extending the isometry $a + N_s \mapsto a^* + N_s$ on $\mathcal{N}_s/N_s$ to $\mathbb{H}_s$. Then there is a σ-finite measure μ on $\hat{\mathcal{A}}$ such that if $[\sigma] \mapsto \sigma$ is a Borel cross section of $\hat{\mathcal{A}}$ into the irreducible $*$-representations of $\mathcal{A}$ and $[\sigma] \mapsto \mathbb{H}_\sigma$ is the corresponding Hilbert bundle, then:*

(a) *$\mathbb{H}_s = \int^{\oplus} \mathbb{H}_\sigma \otimes \bar{\mathbb{H}}_\sigma \, d\mu([\sigma]) = \int^{\oplus} \mathcal{HS}(\mathbb{H}_\sigma)\, d\mu([\sigma])$;*

(b) *for each $a \in \mathcal{A}^+$, $[\sigma] \mapsto Tr(\sigma(a))$ is Borel; and for each $a \in \mathcal{A}$, $[\sigma] \mapsto \sigma(a) \otimes \bar{I}$ and $[\sigma] \mapsto I \otimes \bar{\sigma}(a)$ are Borel fields of bounded operators relative to the Hilbert bundle $[\sigma] \mapsto \mathbb{H}_\sigma$;*

(c) *for each σ, let J_σ denote the conjugate linear involutive isometry on $\mathbb{H}_\sigma \otimes \bar{\mathbb{H}}_\sigma$ satisfying $J_\sigma(v \otimes \bar{w}) = w \otimes \bar{v}$; then $[\sigma] \mapsto J_\sigma$ is Borel;*

(d) *$t_s(\lambda(a)) = \int Tr(\sigma(a))\, d\mu([\sigma])$ for $a \geq 0$, $\lambda = \int^{\oplus} \sigma \otimes \bar{I}\, d\mu([\sigma])$, $\rho = \int I \otimes \bar{\sigma}^*\, d\mu([\sigma])$, and $J_s = \int^{\oplus} J_\sigma\, d\mu([\sigma])$;*

(e) *Moreover, μ is uniquely determined by the condition*

$$s(a,b) = \int Tr(\sigma(ab^*)) \, d\mu([\sigma]) \text{ for } a,b \in \mathcal{N}_s.$$

Also if $\mathcal{U}$ and $\mathcal{V}$ are the von Neumann algebras generated by $\lambda(\mathcal{A})$ and $\rho(\mathcal{A})$, then $\mathcal{U} = \int^{\oplus} B(\mathbb{H}_\sigma) \otimes I \, d\mu([\sigma])$ and $\mathcal{V} = \int^{\oplus} I \otimes B(\bar{\mathbb{H}}_\sigma) \, d\mu([\sigma])$. In particular, the center of $\mathcal{U}$ is $\int^{\oplus} \mathbb{C}(I_\sigma \otimes \bar{I}_\sigma) \, d\mu(\sigma)$.

PROOF. Form the Hilbert space $\mathbb{H}_s$ and the representations $\lambda = \lambda_s$, $\rho = \rho_s$, and the operator $J = J_s$ on $\mathbb{H}_s$. Let $t = t_s$ be the natural normal semifinite trace defined on the von Neumann algebra $\mathcal{U} = \lambda(\mathcal{A})''$. Since $\mathcal{A}$ is type I and λ is a representation of $\mathcal{A}$, Theorem VI.14 implies there is a finite measure ν on the standard Borel space $\hat{\mathcal{A}}$ and a Borel function $n : \hat{\mathcal{A}} \to \{\infty, 1, 2, \dots\}$ such that $\lambda = \int^{\oplus} n([\sigma])\sigma \, d\nu(\sigma)$. In particular, the Hilbert space for λ is $\int^{\oplus} n([\sigma])\mathbb{H}_\sigma \, d\nu([\sigma])$; and

$$\mathcal{U} = \int^{\oplus} n([\sigma])B(\mathbb{H}_\sigma) \, d\nu([\sigma]).$$

By Theorem III.16, the commutant of $\mathcal{U}$ is given by

$$\mathcal{V} = \int^{\oplus} M_{n([\sigma])}(\mathbb{C}) \, d\nu([\sigma]).$$

Applying Theorem VII.21, we see there is a unique σ-finite measure $\mu \prec \nu$ such that if $A \geq 0$, then

$$t\left(\int^{\oplus} n([\sigma])A([\sigma]) \, d\nu([\sigma])\right) = \int Tr(A([\sigma])) \, d\mu([\sigma]).$$

For notational convenience, we write $n(\sigma)$, $A(\sigma)$, etc. for $n([\sigma])$, $A([\sigma])$, etc.

We now carefully construct the Hilbert space $\mathbb{H}_t$. $\mathcal{N}_t$ consists of all bounded Borel fields $n(\sigma)A(\sigma)$ such that $\int Tr(A(\sigma)^*A(\sigma)) \, d\mu(\sigma) < \infty$, and N_t consists of all such fields such that $Tr(A(\sigma)^*A(\sigma)) = 0$ for μ a.e. σ. For each σ, we consider the Hilbert space $\mathcal{HS}(\mathbb{H}_\sigma)$. We give a Borel structure that makes $\sigma \mapsto \mathcal{HS}(\mathbb{H}_\sigma)$ into a Hilbert bundle. Indeed, let $\mathcal{H} = \{(\sigma, A) : A \in \mathcal{HS}(\mathbb{H}_\sigma)\}$, and let f_n be a countable family of Borel sections of the Hilbert bundle $\sigma \mapsto \mathbb{H}_\sigma$ satisfying (1) and (2) from Section 9 of Chapter III in the definition of a Hilbert bundle. As seen in the argument of the proof of Theorem III.12, we may assume $\{f_1(\sigma), \dots, f_{N(\sigma)}(\sigma) : f_{N(\sigma)}(\sigma) \neq 0, f_{N(\sigma)+1}(\sigma) = 0\}$ is a complete orthonormal basis of $\mathbb{H}_\sigma$ for all σ. Define $E_{i,j}(\sigma) \in \mathcal{HS}(\mathbb{H}_\sigma)$ by $E_{i,j}(\sigma)(v) = \langle v, f_j(\sigma) \rangle f_i(\sigma)$. Then an easy argument shows $\{E_{i,j}(\sigma) : i, j \leq n(\sigma)\}$ is a complete orthonormal basis of $\mathcal{HS}(\mathbb{H}_\sigma)$. These sections define a standard Borel structure on $\mathcal{H}$. Indeed, take the smallest Borel structure on $\mathcal{H}$ such that $(\sigma, A) \mapsto \sigma$ and $(\sigma, A) \mapsto \langle A, E_{i,j}(\sigma)\rangle_{\mathcal{HS}(\mathbb{H}_\sigma)}$ is Borel for all i and j. To see this is standard, usual arguments using a Borel isomorphism of the Hilbert bundle $\sigma \mapsto \mathbb{H}_\sigma$ onto a Borel direct sum of trivial Hilbert bundles allow one to consider the special case where the bundle $\sigma \mapsto \mathbb{H}_\sigma$ is the trivial bundle $S \times \mathbb{H}$ where $S \subseteq \hat{A}$. In this case the bundle $\mathcal{H}$ is $S \times \mathcal{HS}(\mathbb{H})$, and the Borel structure is the smallest that makes $(s, A) \mapsto s$ and $(s, A) \mapsto \langle A, E_{i,j}\rangle$ Borel for all i and j. Since the $E_{i,j}$ form an orthonormal basis of $\mathcal{HS}(\mathbb{H})$, it follows that this Borel structure is the product Borel structure of S and the Borel structure from the Hilbert vector space topology on $\mathcal{HS}(\mathbb{H})$. Thus this structure is standard. Moreover, this implies a mapping f from a Borel space X into $\mathcal{HS}$ is Borel iff $x \mapsto p(f(x))$ where $p(\sigma, A) = \sigma$ is Borel and $x \mapsto \langle f(x), E_{i,j}(p(f(x)))\rangle_{p(f(x))}$ is Borel for all i and j. In particular, a mapping $\sigma \mapsto (\sigma, A(\sigma))$ where $A(\sigma) \in \mathcal{HS}(\mathbb{H}_\sigma)$ is Borel iff $\sigma \mapsto A(\sigma)$ is a Borel field of operators relative to the Hilbert bundle $\sigma \mapsto \mathbb{H}_\sigma$. Indeed, this follows from the fact that

$$\begin{aligned}
\langle A(\sigma), E_{i,j}(\sigma)\rangle_{\mathcal{HS}(\mathbb{H}_\sigma)} &= \mathrm{Tr}(A(\sigma)E_{i,j}(\sigma)^*) \\
&= \mathrm{Tr}(A(\sigma)E_{j,i}(\sigma)) \\
&= \sum_k \langle A(\sigma)E_{j,i}(\sigma)f_k(\sigma), f_k(\sigma)\rangle \\
&= \langle A(\sigma)f_j(\sigma), f_i(\sigma)\rangle_{\mathbb{H}_\sigma}.
\end{aligned}$$

In particular, the mapping $\mathcal{N}_t/N_t \to \int^\oplus \mathcal{HS}(\mathbb{H}_\sigma)\, d\mu(\sigma)$ given by $A \mapsto (\sigma \mapsto A(\sigma))$ where $A(\sigma)$ is set to 0 if $\mathrm{Tr}(A(\sigma)A(\sigma)^*) = \infty$ is an isometry. Moreover, its range is dense. Indeed, the range contains all sections $\sigma \mapsto A(\sigma)$ where $A(\sigma) \in \mathcal{HS}(\mathbb{H}_\sigma)$, $\int \|A(\sigma)\|^2_{\mathcal{HS}(\mathbb{H}_\sigma)}\, d\mu(\sigma) < \infty$, and ess $\sup_\sigma \|A(\sigma)\|_{B(\mathbb{H}_\sigma)} < \infty$, for $\mathcal{U} = \int^\oplus n(\sigma)B(\mathbb{H}_\sigma)\, d\nu(\sigma)$. In particular, the range contains all the sections $1_F E_{i,j}$ where F is a Borel subset of $\hat{A}$ and $E_{i,j}$ are the basis sections given earlier. Since the linear span of the $1_F E_{i,j}$ are dense in $\int^\oplus \mathcal{HS}(\mathbb{H}_\sigma)\, d\mu(\sigma)$, we see $A \mapsto (\sigma \mapsto A(\sigma))$ mapping $\mathcal{N}_t/N_t$ into $\int^\oplus \mathcal{HS}(\mathbb{H}_\sigma)\, d\mu(\sigma)$ extends to a unitary isomorphism U.

We now return to the proof. By Lemma VII.39, the mapping V defined on $\mathcal{N}_s/N_s$ by $V(a + N_s) = \lambda(a) + N_t$ extends to a unitary mapping of $\mathbb{H}_s$ onto $\mathbb{H}_t$ satisfying $V\lambda(a) = \pi(\lambda(a))V$. Hence UV is a unitary mapping of $\mathbb{H}_s$ onto $\int^\oplus \mathcal{HS}(\mathbb{H}_\sigma)\, d\mu(\sigma)$ satisfying $UV(a +$

$N_s) = U(\lambda(a) + N_t) = (\sigma \mapsto \sigma(a)) \in \int^\oplus \mathcal{HS}(\mathbb{H}_\sigma) \, d\mu(\sigma)$. Moreover, if $a \in \mathcal{A}^+$, then $\lambda(a) \in \mathcal{A}^+$; and $t(\lambda(a)) = \int \mathrm{Tr}(\sigma(a)) \, d\mu(\sigma)$.

We now recall that $\mathcal{HS}(\mathbb{H}) = \mathbb{H} \otimes \bar{\mathbb{H}}$; and under this identification, $A \otimes \bar{I}(T) = AT$ and $I \otimes \bar{A}(T) = TA^*$ where A is an operator in $B(\mathbb{H})$ and $v \otimes \bar{w} \mapsto w \otimes \bar{v}$ on $\mathbb{H} \otimes \bar{\mathbb{H}}$ gives the mapping $J(T) = T^*$ on $\mathcal{HS}(\mathbb{H})$. (See Section 2.)

Next note $\sigma \mapsto \sigma(a) \otimes \bar{I}$ is a bounded Borel field of operators on the Hilbert bundle $\sigma \mapsto \mathcal{HS}(\mathbb{H}_\sigma)$. Indeed, it suffices to note $\sigma \mapsto \langle \sigma(a) \otimes \bar{I}(E_{i,j}(\sigma)), E_{k,l}(\sigma) \rangle_{\mathcal{HS}(\mathbb{H}_\sigma)}$ is Borel for all i, j, k and l. But this mapping is

$$\sigma \mapsto \mathrm{Tr}(\sigma(a)E_{i,j}(\sigma)E_{k,l}(\sigma)^*) = \mathrm{Tr}(\sigma(a)E_{i,j}(\sigma)E_{l,k}(\sigma))$$
$$= \delta_{j,l}\mathrm{Tr}(\sigma(a)E_{i,k}(\sigma))$$
$$= \delta_{j,l}\langle \sigma(a)f_i(\sigma), f_k(\sigma) \rangle$$

which is Borel in σ since $\sigma \mapsto \sigma(a)$ is Borel relative to Hilbert bundle $\sigma \mapsto \mathbb{H}_\sigma$. Similarly $\sigma \mapsto I \otimes \bar{\sigma}(a)$ is Borel for each $a \in \mathcal{A}$. To see that $\sigma \mapsto J_\sigma$ is Borel, note

$$\langle J_\sigma E_{i,j}(\sigma), E_{k,l}(\sigma) \rangle_{\mathcal{HS}(\mathbb{H}_\sigma)} = \langle E_{i,j}(\sigma)^*, E_{k,l}(\sigma) \rangle$$
$$= \langle E_{j,i}(\sigma), E_{k,l}(\sigma) \rangle$$
$$= \mathrm{Tr}(E_{j,i}(\sigma)E_{k,l}(\sigma)^*)$$
$$= \mathrm{Tr}(E_{j,i}(\sigma)E_{l,k}(\sigma))$$
$$= \delta_{i,l}\mathrm{Tr}(E_{j,k}(\sigma))$$
$$= \delta_{i,l}\delta_{j,k}$$

which is constant in σ.

Now for $b \in \mathcal{N}_s$,

$$UV\lambda(a)(b + N_s) = UV(ab + N_s)$$
$$= U(\lambda(ab))$$
$$= \sigma \mapsto \sigma(ab)$$
$$= \sigma \mapsto \sigma(a)\sigma(b)$$
$$= \int^\oplus (\sigma(a) \otimes \bar{I})\sigma(b) \, d\mu(\sigma)$$
$$= \left(\int^\oplus \sigma(a) \otimes \bar{I} \, d\mu(\sigma) \right) UV(b + N_s).$$

Similarly,

$$UV\rho(a)(b + N_s) = (\sigma \mapsto \sigma(b)\sigma(a))$$
$$= \sigma \mapsto (I \otimes \overline{\sigma(a)^*})\sigma(b)$$
$$= \sigma \mapsto (I \otimes \bar{\sigma}^*(a))\sigma(b)$$
$$= \left(\int^\oplus I \otimes \bar{\sigma}^*(a)\, d\mu(\sigma)\right) UV(b + N_s)$$

for $\bar{\sigma}$ is the $*$-representation on $\bar{\mathbb{H}}$ defined by $\bar{\sigma}(a)\bar{v} = \overline{\sigma(a)v}$.
Next note

$$UVJ(b + N_s) = UV(b^* + N_s)$$
$$= \sigma \mapsto \sigma(b^*)$$
$$= \sigma \mapsto \sigma(b)^*$$
$$= \sigma \mapsto J_\sigma(\sigma(b))$$
$$= \left(\int^\oplus J_\sigma\, d\mu(\sigma)\right) UV(b + N_s).$$

We have just seen that UV is a unitary mapping carrying the representation λ and ρ to $\int_{\hat{\mathcal{A}}}^\oplus \sigma \otimes \bar{I}\, d\mu(\sigma)$ and $\int_{\hat{\mathcal{A}}}^\oplus I \otimes \bar{\sigma}^*\, d\mu(\sigma)$, respectively. Since $\rho(\mathcal{A})'' = \lambda(\mathcal{A})'$ and $\lambda(\mathcal{A})'' = \rho(\mathcal{A})'$, we see $UVUV^*U^* = \int_{\hat{\mathcal{A}}}^\oplus B(\mathbb{H}_\sigma) \otimes \bar{I}\, d\mu(\sigma)$ and $UVVV^*U^* = \int_{\hat{\mathcal{A}}}^\oplus I \otimes B(\bar{\mathbb{H}}_\sigma)\, d\mu(\sigma)$. In particular, the center of these two von Neumann algebras is $\int_{\hat{\mathcal{A}}}^\oplus \mathbb{C}I \otimes \bar{I}\, d\mu(\sigma)$.

Finally we show the uniqueness of the σ-finite measure μ. We have already seen that $t(\lambda(a)) = \int_{\hat{\mathcal{A}}} \mathrm{Tr}(\sigma(a))\, d\mu(\sigma)$ for $a \in \mathcal{A}^+$. Let $\tilde{\mu}$ be another measure satisfying

$$s(a, b) = \int \mathrm{Tr}(\sigma(ab^*))\, d\tilde{\mu}(\sigma) \text{ for } a, b \in \mathcal{N}_s.$$

First consider the von Neumann algebra $\tilde{\mathcal{U}} = \int_{\hat{\mathcal{A}}}^\oplus B(\mathbb{H}_\sigma)\, d\tilde{\mu}(\sigma)$ and the representation $\tilde{\lambda}$ of $\mathcal{A}$ on $\tilde{\mathbb{H}} = \int_{\hat{\mathcal{A}}}^\oplus \mathbb{H}_\sigma\, d\tilde{\mu}(\sigma)$ given by $\tilde{\lambda}(a) = \int^\oplus \sigma(a)\, d\tilde{\mu}(\sigma)$. By Theorem VII.21, $\tilde{t}(\tilde{A}) = \int_{\hat{\mathcal{A}}} \mathrm{Tr}(\tilde{A}(\sigma))\, d\tilde{\mu}(\sigma)$ for $\tilde{A} \geq 0$, defines a normal trace $\tilde{t}$ on $\tilde{\mathcal{U}}$. Moreover, by assumption, if $a, b \in \mathcal{N}_s$, $\tilde{\lambda}(a)$ and $\tilde{\lambda}(b)$ belong to $\tilde{\mathcal{U}}_t$ and $\tilde{t}(\tilde{\lambda}(ab^*)) = s(a, b)$.

Now, since $\mathcal{N}_s$ is a dense $*$ ideal in $\mathcal{A}$, the proof of Proposition II.7 shows there is an increasing sequence $0 < u_1 < u_2 < \cdots$ in the linear span of $\mathcal{N}_s^2$ such that the u_i form an approximate unit for $\mathcal{A}$. By Proposition VII.26, $t \circ \lambda|_{\mathcal{A}^+}$ and $\tilde{t} \circ \tilde{\lambda}|_{\mathcal{A}^+}$ are lower semicontinuous semifinite traces on $\mathcal{A}$ satisfying $\tilde{t}(\tilde{\lambda}(a)) = \lim \tilde{t}(\tilde{\lambda}(a^{1/2}u_i a^{1/2}))$ and $t(\lambda(a)) = \lim t(\lambda(a^{1/2}u_i a^{1/2}))$ for $a \in \mathcal{A}^+$.

But $\tilde{t}(\tilde{\lambda}(a^{1/2}u_i a^{1/2})) = t(\lambda(a^{1/2}u_i a^{1/2}))$ if u_i is in the linear span of $\mathcal{N}_s^2$. Indeed, suppose $u = \sum_{k=1}^n a_k b_k$ where $a_k, b_k \in \mathcal{N}_s$. Then

$$\tilde{t}(\tilde{\lambda}(a^{1/2}u a^{1/2})) = \sum_k \tilde{t}(\tilde{\lambda}(a^{1/2}a_k b_k a^{1/2}))$$

$$= \sum_k s(a^{1/2}a_k, a^{1/2}b_k^*)$$

$$= \sum_k t(\lambda(a^{1/2}a_k b_k a^{1/2}))$$

$$= t(\lambda(a^{1/2}u_i a^{1/2})).$$

Thus $t(\lambda(a)) = \tilde{t}(\tilde{\lambda}(a))$, and we see

$$\int_{\hat{\mathcal{A}}} \mathrm{Tr}(\sigma(a)) \, d\mu(\sigma) = \int_{\hat{\mathcal{A}}} \mathrm{Tr}(\sigma(a)) \, d\tilde{\mu}(\sigma)$$

for all $a \in \mathcal{A}^+$.

Next suppose E is a closed subset of $\hat{\mathcal{A}}$ in the hull–kernel topology. Then $E = \mathrm{Hull}(I)$ where I is a closed two sided ideal in $\mathcal{A}$. Thus $\sigma(x) = 0$ if $\sigma \in E$ and $x \in I$, and $\sigma(I) \neq 0$ if $\sigma \notin E$.

Now let u_i be an increasing sequence in I^+ that is an approximate unit for the C^* algebra I. Then $\sigma(a^{1/2}u_i a^{1/2}) = 0$ if $\sigma \in E$. If $\sigma \notin E$, the subspace $\sigma(I)\mathbb{H}_\sigma$ is nonzero and invariant under the irreducible representation σ. Hence it is dense in $\mathbb{H}_\sigma$. But $\sigma(a^{1/2}u_i a^{1/2})\sigma(b)v$ converges to $\sigma(ab)v = \sigma(a)\sigma(b)v$ for $a \in \mathcal{A}^+$, $b \in I$ and $v \in \mathbb{H}_\sigma$. Thus $\sigma(a^{1/2}u_i a^{1/2})$ increases strongly to $\sigma(a)$ for $\sigma \notin E$ and is 0 for $\sigma \in E$. Hence if a is a positive element in $\mathcal{A}$, we see $\mathrm{Tr}(\sigma(a^{1/2}u_i a^{1/2}))$ increases monotonically to $1_{\hat{\mathcal{A}}-E}(\sigma)\mathrm{Tr}(\sigma(a))$; and thus

$$\int_{\hat{\mathcal{A}}-E} \mathrm{Tr}(\sigma(a)) \, d\mu(\sigma) = \int_{\hat{\mathcal{A}}-E} \mathrm{Tr}(\sigma(a)) \, d\tilde{\mu}(\sigma)$$

for all closed subsets E of $\hat{\mathcal{A}}$.

Consider $\mathcal{N}_{to\lambda}$. As seen by Proposition VII.25, the linear span $\mathcal{M}$ of $(\mathcal{N}_{to\lambda})^2$ is norm dense in $\mathcal{N}_{to\lambda}$ and since $\mathcal{N}_{to\lambda} \supseteq \mathcal{N}_s$, we see $\mathcal{M}$ is norm dense in $\mathcal{A}$. Moreover, $\mathcal{M}^+ = \{a : a \geq 0 \text{ and } t(\lambda(a)) < \infty\}$; and the linear span of $\mathcal{M}^+$ is $\mathcal{M}$. Now if $a \in \mathcal{M}^+$, $W \mapsto \int_W \mathrm{Tr}(\sigma(a)) \, d\mu(\sigma)$ and $W \mapsto \int_W \mathrm{Tr}(\sigma(a)) \, d\tilde{\mu}(\sigma)$ are finite measures on the standard Borel space $\hat{\mathcal{A}}$ that agree on the hull-kernel open subsets of $\hat{\mathcal{A}}$. But by Corollary VI.6, these generate the Borel structure. Hence these measures are equal for $a \in \mathcal{M}^+$. In particular, since

$\mathrm{Tr}(\sigma(a)) > 0$ if $\sigma(a) > 0$, the measures $\mu|_{\{\sigma:\sigma(a)\neq 0\}}$ and $\tilde{\mu}|_{\{\sigma:\sigma(a)\neq 0\}}$ are equal.

But since $\mathcal{M}$ is dense in $\mathcal{A}$, the linear span of $\mathcal{M}^+$ is $\mathcal{M}$, and $\mathcal{A}$ is separable, there is a sequence a_i in $\mathcal{M}^+$ such that if $\sigma \in \hat{\mathcal{A}}$, then $\sigma(a_i) > 0$ for some i. Thus $\cup_i \{\sigma : \sigma(a_i) > 0\} = \hat{\mathcal{A}}$, and $\mu = \tilde{\mu}$. $\quad\square$

27. The Plancherel Theorem

THEOREM VII.23. *Let G be a second countable, unimodular, locally compact, type I group; and let m be a Haar measure on G. Then there is a unique σ-finite measure μ on $\hat{G}$ such that for $f \in L^2(G) \cap L^1(G)$,*

$$\int_G |f(x)|^2 \, dm(x) = \int_{\hat{G}} \mathrm{Tr}(\sigma(f * f^*)) \, d\mu(\sigma).$$

Moreover, if λ and η are the left and right regular representations of G on $L^2(G)$ and J is the conjugate linear isometry $J(f) = f^*$ of $L^2(G)$, then there is a unitary isomorphism W of $L^2(G)$ onto $\int^\oplus \mathbb{H}_\sigma \otimes \bar{\mathbb{H}}_\sigma \, d\mu([\sigma])$ where $[\sigma] \mapsto \sigma$ is a Borel cross section of $\hat{G}$ and $[\sigma] \mapsto \mathbb{H}_\sigma$ is the corresponding Hilbert bundle with

   (a) $W\lambda(g)W^* = \int^\oplus \sigma(g) \otimes \bar{I} \, d\mu([\sigma])$ *and*
       $W\eta(g)W^{-1} = \int^\oplus I \otimes \bar{\sigma}(g) \, d\mu([\sigma])$ *for all $g \in G$;*
   (b) $WJW^* = \int^\oplus J_\sigma \, d\mu([\sigma])$ *where $J_\sigma(v \otimes \bar{w}) = w \otimes \bar{v}$ on $\mathbb{H}_\sigma \otimes \bar{\mathbb{H}}_\sigma$;*
   (c) *if $\mathcal{U} = \lambda(G)''$ and $\mathcal{V} = \rho(G)''$, then $W\mathcal{U}W^* = \int^\oplus B(\mathbb{H}_\sigma) \otimes \bar{I} \, d\mu([\sigma])$ and $W\mathcal{V}W^* = \int^\oplus I \otimes B(\bar{\mathbb{H}}_\sigma) \, d\mu([\sigma])$; moreover, $W(\mathcal{U} \cap \mathcal{V})W^* = \int^\oplus \mathbb{C}(I \otimes \bar{I}) \, d\mu([\sigma])$.*

PROOF. By Corollary IV.9, there is a natural correspondence $\pi \mapsto \pi' \mapsto \pi''$ between the unitary representations π of G, the nondegenerate $*$ representations π' of $L^1(G)$, and the nondegenerate $*$ representations π'' of $C(G)^*$. Moreover, the von Neumann algebras generated by the corresponding representations are all equal. We shall use π and context to denote π, π', or π''.

Now $\lambda(y)f(x) = f(y^{-1}x)$ and $\eta(y)f(x) = f(xy)$ for $f \in L^2(G)$ and $y \in G$. Hence if f and g are in $C_c(G)$, $\lambda(f)g = f * g$ and $\eta(\check{f})(g) = g * f = \rho(f)g$ where $\check{f}(x) = f(x^{-1})$.

Now by the example in Section 23 and Proposition VII.29, there is a maximal bitrace s on $C(G)^*$ extending $(f, g) \mapsto \int f(x)\bar{g}(x) \, dm(x)$ on $C_c(G) \times C_c(G) \subseteq C(G)^* \times C(G)^*$. Moreover, $\mathcal{N}_s \supseteq C_c(G)$ and by Proposition VII.29, $\mathcal{N}_s/N_s$ is a left unimodular Hilbert algebra in $\mathbb{H}_s = L^2(G)$ and $\lambda(f)\xi = f * \xi$ and $\rho(f)\xi = \xi * f$ for $f \in C_c(G)$ and $\xi \in L^2(G)$. Using the natural Borel isomorphism from $\hat{G}$ onto

$\hat{C}^*(G)$, Theorem VII.22 implies the existence of a σ-finite measure μ on $\hat{G}$ and a unitary isomorphism W for $L^2(G)$ onto $\int^\oplus \mathbb{H}_\sigma \otimes \bar{\mathbb{H}}_\sigma \, d\mu([\sigma])$ such that $W\lambda(a)W^{-1} = \int_{\hat{G}}^\oplus \sigma(a) \otimes \bar{I} \, d\mu([\sigma])$ and $W\rho(a)W^{-1} = \int_{\hat{G}}^\oplus I \otimes \bar{\sigma}^*(a) \, d\mu([\sigma])$ for $a \in C^*(G)$. Moreover, $t_s(\lambda(a)) = \int \mathrm{Tr}(\sigma(a)) \, d\mu([\sigma])$ for $a \in C^*(G)^+$, $J = \int_{\hat{G}}^\oplus J_\sigma \, d\mu([\sigma])$, $WUW^{-1} = \int^\oplus B(\mathbb{H}_\sigma) \otimes \bar{I} \, d\mu([\sigma])$, $WVW^{-1} = \int^\oplus I \otimes B(\bar{\mathbb{H}}_\sigma) \, d\mu([\sigma])$, and $W(U \cap V)W^{-1} = \int^\oplus CI \otimes \bar{I} \, d\mu([\sigma])$.

In particular, if $g \in G$, then $W\lambda(g)W^{-1} = \int \sigma(g) \otimes \bar{I} \, d\mu([\sigma])$, and since $\eta(f) = \rho(\check{f}) = \int^\oplus I \otimes \bar{\sigma}^*(\check{f}) \, d\mu([\sigma])$, then $\eta(g) = \int^\oplus I \otimes \bar{\sigma}(g) \, d\mu([\sigma])$, for the representation $\bar{\sigma}$ on G integrated to $L^1(G)$ is $\bar{\sigma}^*(\check{f})$. Indeed,

$$\bar{\sigma}^*(\check{f})\bar{v} = \bar{\sigma}(\check{f}^*)\bar{v} = \overline{\sigma(\check{f}^*)\bar{v}} = \overline{\sigma(\check{f}^*)v} = \overline{\sigma(\check{f})v}$$

$$= \overline{\int \bar{f}(g)\sigma(g)v \, dm(g)} = \int f(g)\overline{\sigma(g)v} \, dm(g)$$

$$= \int f(g)\bar{\sigma}(g)\bar{v} \, dm(g) = \bar{\sigma}(f)\bar{v}$$

for $f \in L^1(G)$.

We now note that if $f \in L^2(G) \cap L^1(G)$, then $f \in \mathcal{N}_s$. Indeed, by Proposition VII.29, $f \in \mathcal{N}_s$ iff $\lambda(f) \in \mathcal{N}_{t_s}$; and since t_s is the natural normal, faithful, semifinite trace on $\mathcal{U} = \mathcal{L}(\mathbb{H}_{t_s})$, $\mathcal{N}_{t_s}$ is an achieved left von Neumann algebra. In particular, an element $f \in \mathbb{H}_{t_s}$ is in $\mathcal{N}_{t_s}$ iff $\lambda(f)$ is bounded. But $\lambda(f)$ is bounded for $f \in L^2(G) \cap L^1(G)$. Now $s(a, b) = \int \mathrm{Tr}(\sigma(a)\sigma(b)^*) \, d\mu([\sigma])$ for a and b in $\mathcal{N}_s$; and if $f, g \in L^2(G) \cap L^1(G)$, then $f, g \in \mathcal{N}_s$. Thus

$$\int_G f(x)\bar{g}(x) \, dm(x) = \int_{\hat{G}} \mathrm{Tr}(\sigma(f)\sigma(g)^*) \, d\mu([\sigma]).$$

Finally we show the uniqueness of μ. Suppose $\tilde{\mu}$ is another σ-finite measure such that

$$\int_G f(x)\bar{g}(x) \, dm(x) = \int_{\hat{G}} \mathrm{Tr}(\sigma(f)\sigma(g)^*) \, d\tilde{\mu}([\sigma])$$

for $f, g \in L^2(G) \cap L^1(G)$. Let $\mathcal{N}$ be the set of all $a \in C^*(G)$ such that

$$\int_{\hat{G}} \mathrm{Tr}(\sigma(a)\sigma(a)^*) \, d\tilde{\mu}([\sigma]) = \int_{\hat{G}} \mathrm{Tr}(\sigma(a)\sigma(a)^*) \, d\mu([\sigma]) < \infty.$$

Then $\mathcal{N}$ is a two sided ideal in $C^*(G)$ containing $C_c(G)$. Moreover, as seen in the proof of Theorem VII.22, there is a nondegenerate representation $\tilde{\lambda}$ and a normal trace $\tilde{t}$ such that $\tilde{t}$ is defined on $C^*(G)^+$

by $\tilde{t}(a) = \int \mathrm{Tr}(\sigma(a)) \, d\tilde{\mu}([\sigma])$. Hence Proposition VII.26 applies to both λ and $\tilde{\lambda}$ and if u_i used in the proof of Theorem VII.22 are taken to be in the linear span of $\mathcal{N}^2$ instead of $\mathcal{N}_s^2$, we obtain

$$\tilde{t}(\tilde{\lambda}(a)) = t_s(\lambda(a))$$

for all $a \in C^*(G)^+$. This gives

$$\int_{\hat{G}} \mathrm{Tr}(\sigma(a)) \, d\mu([\sigma]) = \int_{\hat{G}} \mathrm{Tr}(\sigma(a)) \, d\tilde{\mu}([\sigma])$$

for $a > 0$; and again following the proof of Theorem VII.22, we have $\mu = \tilde{\mu}$. $\square$

CHAPTER VIII

THE FOURIER–STIELTJES ALGEBRA

1. Introduction

In this chapter we shall use several of the topologies on $\mathbf{B}(\mathbb{H})$ to put a Banach algebra structure on the space of matrix coefficients of unitary representations of G.

Namely, let G be a locally compact group. Then $L^1(G)$ is a Banach $*$-algebra with an approximate unit. The universal enveloping C^* algebra $C^*(G)$ of $L^1(G)$ is known as the C^* algebra of the group G. Let τ be the natural mapping of $L^1(G)$ into $C^*(G)$ defined in Section 7 of Chapter II. Lemma IV.5 shows τ is an inclusion of $L^1(G)$ into $C^*(G)$. Let R be the collection of unitary representations of G. Then Corollary IV.9 implies this may be thought of both as the collection of nondegenerate $*$-representations of the $*$ algebra $L^1(G)$ and as the collection of nondegenerate $*$ representations of $C^*(G)$. By Theorem IV.13, the continuous positive linear functionals ω on $L^1(G)$ are in one-to-one correspondence with the positive definite functions ϕ on G. This correspondence is given by $\omega_\phi(f) = \int f(x)\phi(x)\,dx$ for $f \in L^1(G)$. Moreover, the positive definite functions on G are defined in terms of matrix coefficients having form $\phi(g) = \langle \pi(g)v, v \rangle$ for some $\pi \in R$. Using Corollary II.12, we summarize the situation in the following remark.

REMARK. If ω is a continuous positive linear functional on $C^*(G)$, then ω is the unique extension of a continuous positive linear functional ω on $L^1(G)$, and thus $\omega = \omega_\phi$ for some positive definite function ϕ on G. Also, if π is a nondegenerate $*$-representation of $C^*(G)$, then π is the unique extension of a nondegenerate $*$-representation of $L^1(G)$ and thus satisfies $\pi(\tau(f)) = \int f(x)\pi(x)\,dx$ for some unique unitary representation π of G.

Thus the positive continuous linear functionals of $C^*(G)$ are identifiable with the positive definite matrix coefficients $\phi(g) = \langle \pi(g)v, v \rangle$ where $\pi \in R$.

DEFINITION. The Fourier-Stieltjes algebra $B(G)$ of G is defined to be the space $\{\langle \pi(g)v, w \rangle : v, w \in \mathbb{H}(\pi) \text{ where } \pi \in R\}$.

We shall see that $B(G)$ is the dual space of $C^*(G)$ under the identification

$$\omega_b(\tau(f)) = \int f(g)b(g)\, dg$$

where $\tau : L^1(G) \to C^*(G)$ is the natural inclusion. Using the norm from the dual space, $B(G)$ is a Banach $*$ algebra which will be shown to be a predual of the universal enveloping von Neumann algebra $W^*(G)$ of G. This will take most of this chapter.

In particular, using this norm, one has $||b|| = ||\omega_b||$ if $b \in B(G)$. Thus one would have $|\int f(g)b(g)\, dg| \leq ||\tau(f)||\, ||b|| \leq ||f||_1 ||b||$ for all $f \in L^1(G)$. This implies $||b||_\infty \leq ||b|| = ||\omega_b||$, and thus convergence in $B(G)$ will force uniform convergence.

Multiplication and an adjoint operation on $B(G)$ are defined by

$$(b_1 b_2)(g) = b_1(g)b_2(g) \text{ and } b^*(g) = \overline{b(g)}.$$

To see $B(G)$ is closed under multiplication and adjoints, one uses tensor products and conjugate representations of unitary representations in R.

PROPOSITION VIII.1. $B(G)$ is a $*$ algebra with identity.

PROOF. We first note it is closed under summation. Indeed, let $b_1(g) = \langle \pi_1(g)v, v' \rangle$ and $b_2(g) = \langle \pi_2(g)w, w' \rangle$ where π_1 is a unitary representation on $\mathbb{H}$ and π_2 is a unitary representation on $\mathbb{K}$. Then define $\pi_1 \oplus \pi_2$ on $\mathbb{H} \oplus \mathbb{K}$ by $\pi_1 \oplus \pi_2(g) = \pi_1(g) \oplus \pi_2(g)$. Then $b_1(g) + b_2(g) = \langle (\pi_1 \oplus \pi_2)(g)(v, w), (v', w') \rangle$. Also note $b_1(g)b_2(g) = \langle (\pi_1 \otimes \pi_2)(g)v \otimes w, v' \otimes w' \rangle$. Since $cb_1(g) = \langle \pi_1(g)cv, v' \rangle$, we see $B(G)$ is an algebra. Moreover, $\bar{b}_1(g) = \overline{\langle \pi_1(g)v, v' \rangle} = \langle \bar{\pi}_1(g)v, v' \rangle_{\bar{\mathbb{H}}}$. $\square$

2. The Hahn Decomposition

DEFINITION. A linear functional ω on a $*$ algebra $\mathcal{A}$ is said to be Hermitian if $\omega(x^*) = \overline{\omega(x)}$ for all $x \in \mathcal{A}$.

REMARK. Every linear function ω on a $*$ algebra $\mathcal{A}$ is a unique linear combination $\omega_1 + i\omega_2$ where ω_1 and ω_2 are Hermitian.

PROOF. $\omega_1(x) = \frac{\omega(x) + \overline{\omega(x^*)}}{2}$ and $\omega_2(x) = \frac{\omega(x) - \overline{\omega(x^*)}}{2i}$ are the unique Hermitian linear functionals satisfying $\omega = \omega_1 + i\omega_2$. $\square$

THEOREM VIII.1 (THE HAHN DECOMPOSITION). *Suppose ϕ is a continuous Hermitian linear functional on a C^* algebra $\mathcal{A}$. Then there exist continuous positive linear functionals p_+ and p_- such that $\phi = p_+ - p_-$ and $||\phi|| = ||p_+|| + ||p_-||$.*

PROOF. Note if $x \in \mathcal{A}$, then $x = x_1 + ix_2$ where $x_i^* = x_i$ for $i = 1, 2$. Hence any Hermitian ϕ is determined by its restriction to $\mathcal{A}_h = \{x : x^* = x\}$. But ϕ on $\mathcal{A}_h$ is real valued.

Let $\Omega = \mathcal{A}_{+,\le 1}^*$ be the set of positive linear functionals on $\mathcal{A}$ with norm bounded above by 1. Then Ω is compact and Hausdorff in the weak $*$ topology.

Each $x \in \mathcal{A}_h$ defines a continuous real valued function f_x on Ω given by $f_x(\omega) = \langle x, \omega \rangle = \omega(x)$. Note $||f_x||_\infty = ||x||$. Indeed, by Corollary II.14,

$$||x|| = \sup_{\pi \in R, ||u|| \le 1} |\langle \pi(x)u, u \rangle|$$
$$= \sup_{\omega \in \Omega} |\omega(x)|$$
$$= \sup_{\omega \in \Omega} |f_x(\omega)| = ||f_x||_\infty.$$

The collection $V = \{f_x : x \in \mathcal{A}_h\}$ forms a vector subspace of $C(\Omega)$. Define F on this subspace by $F(f_x) = \phi(x)$. F is linear. Moreover, $|F(f_x)| = |\phi(x)|$; and thus the norm of F on V is $||F|| = \sup_{||f_x||_\infty = 1} |\phi(x)| = \sup_{||x||=1, x \in \mathcal{A}_h} |\phi(x)| \le ||\phi||$. By the Hahn–Banach Theorem, F has a continuous extension G on $C(\Omega)$ with $||G|| = ||F||$. The Riesz representation theorem implies there is a real finite measure μ with $|\mu| = ||F|| \le ||\phi||$ satisfying $\phi(x) = \int f_x(\omega) \, d\mu(\omega)$. The measure μ on Ω has Hahn decomposition $\mu = \mu_+ - \mu_-$ where $\mu_\pm$ are positive finite measures on Ω and $\mu_+(\Omega) + \mu_-(\Omega) = |\mu|$. Define $p_+(x) = \int f_x \, d\mu_+$ and $p_-(x) = \int f_x \, d\mu_-$. Extend $p_\pm$ to all of $\mathcal{A}$ by $p_\pm(x_1 + ix_2) = p_\pm(x_1) - ip_\pm(x_2)$. Note $p_\pm(x) \ge 0$ if $x \ge 0$ and $\phi = p_+ - p_-$. Note also that $|p_\pm(x)| \le ||f_x||_\infty |\mu_\pm| = ||x|| \, |\mu_\pm|$. Thus using $||p|| = \sup_{||x|| \le 1} p(x^*x)$ from Corollary II.9, we see $||p_\pm|| \le |\mu_\pm|$. Note $||\phi|| = ||p_+ - p_-|| \le ||p_+|| + ||p_-|| \le |\mu_+| + |\mu_-| = |\mu| = ||G|| \le ||\phi||$, and so $||\phi|| = ||p_+|| + ||p_-||$. $\square$

REMARK. This decomposition of Hermitian continuous linear functionals on a C^* algebra can be shown to be unique. In the case when

$\mathcal{A} = C(X)$, the decomposition of a real linear functional of $\mathcal{A}^*$ corresponds to the Hahn decomposition of the corresponding measure μ. This uniqueness argument is measure theoretic and not topological. The study of von Neumann algebras is the analog of the measure theory on X. The uniqueness of the decomposition will be shown using arguments concerning Hermitian linear functionals on von Neumann algebras.

3. A Predual of $B(G)$

LEMMA VIII.1. *Suppose $f \in L^1(G)$ and p is a positive definite function on G. Then $||pf||_{C^*(G)} \leq ||p||_\infty ||f||_{C^*(G)}$.*

PROOF. Let $p(x) = \langle \pi(x)v, v \rangle$ where π is a unitary representation. Note $||p||_\infty = ||v||^2$, and thus pf is in L^1. Now if q is a positive definite function with $q(e) \leq 1$, then $q(x) = \langle \rho(x)w, w \rangle$ where ρ is a unitary representation and $||w|| \leq 1$. Hence

$$q((pf)^* * (pf)) = \int q(x) \int (pf)^*(y)(pf)(y^{-1}x)\, dy\, dx$$

$$= \iint q(x)\Delta(y^{-1})\overline{p(y^{-1})f(y^{-1})}\, p(y^{-1}x)f(y^{-1}x)\, dy\, dx$$

$$= \iint q(x)\bar{p}(y)\bar{f}(y)p(yx)f(yx)\, dy\, dx$$

$$= \iint q(y^{-1}x)\bar{p}(y)\bar{f}(y)p(x)f(x)\, dy\, dx$$

$$= \iint \langle \rho(x)w, \rho(y)w \rangle \langle v, \pi(y)v \rangle \langle \pi(x)v, v \rangle f(x)\bar{f}(y)\, dx\, dy$$

$$= \iint \langle \rho(x)w \otimes v \otimes \pi(x)v, \rho(y)w \otimes \pi(y)v \otimes v \rangle f(x)\bar{f}(y)\, dx\, dy$$

$$= \langle (\rho \otimes I \otimes \pi)(f)w \otimes v \otimes v, (\rho \otimes \pi \otimes I)(f)w \otimes v \otimes v \rangle\, dx\, dy$$

$$\leq ||(\rho \otimes I \otimes \pi)(f)||\, ||(\rho \otimes \omega \otimes I)(f)||\, ||w||^2 ||v||^2 ||v||^2$$

$$\leq ||f||^2_{C^*(G)} ||v||^4.$$

Using Theorem II.8, Corollary II.12, and Theorem IV.13, we see

$$||pf||^2_{C^*(G)} \leq ||f||^2_{C^*(G)} ||p||^2_\infty.$$

□

THEOREM VIII.2. *Let $B(G)$ be the Fourier-Stieltjes algebra of the group G. Then $B(G)$ is the dual space of $C^*(G)$.*

PROOF. Note $L^1(G) \subset C^*(G)$, and $C^*(G)$ is the completion of $L^1(G)$ when $L^1(G)$ is equipped with the norm

$$||f||_{C^*(G)} = \sup_{w \in P} w(f^*f)^{1/2} = \sup_{\pi \in R} ||\pi(f)||.$$

By Corollary II.12, we know that every continuous positive linear functional on $C^*(G)$ is the unique extension of a continuous positive linear functional w on $L^1(G)$. But by Theorem IV.13, we have $w = w_\phi$ for some positive definite function ϕ on G. Thus if w is a continuous positive linear functional on $C^*(G)$, there is a positive definite function $\phi \in B(G)$ such that

$$w(f) = \int f(x)\phi(x) \, dx$$

for $f \in L^1(G) \subset C^*(G)$. Now by the Hahn decomposition theorem, we know if $w \in C^*(G)^*$, then $w = w^1 + iw^2 = w_+^1 - w_-^1 + i(w_+^2 - w_-^2)$ where $w_j^\pm$ are positive linear functionals on $C^*(G)$. Thus w is defined by the element $b(g) = \phi_+^1(g) - \phi_+^1(g) + i(\phi_+^2(g) - \phi_-^2(g))$ where the $\phi_\pm^j$ are the positive definite functions defining the continuous positive linear functionals $w_\pm^j$. $\square$

REMARK. The space $B(G)$ has a natural norm; namely $||b|| = ||w_b||$ as a continuous linear functional on $C^*(G)$. With this norm $B(G)$ is a Banach space. One can show this makes $B(G)$ into a Banach $*$ algebra. However to do this, we will need a finer analysis of the linear functionals on a C^* algebra. This analysis involves an analysis of the σ weakly continuous linear functionals on a von Neumann algebra.

4. The Dual of the C* Algebra of Compact Operators

REMARK (DISCRETE NONCOMMUTATIVE MEASURE THEORY). Let X be a set with the discrete topology. Let $C_0(X)$ be the commutative C^* algebra of all continuous functions on X which vanish at ∞. Thus $f \in C_0(X)$ iff $\lim_{x \to \infty} f(x) = 0$. More specifically,

$$C_0(X) = \left\{ \sum_{n=1}^{\infty} \lambda_n 1_{\{x_n\}} : \lim_{n \to \infty} \lambda_n = 0 \right\}.$$

Let $\mathbb{H}$ be a Hilbert space with the same cardinality as X. Set $\mathcal{K}(\mathbb{H})$ to be the noncommutative C^* algebra consisting of all compact operators on $\mathbb{H}$.

There is an analogy between the dual space of the commutative C^* algebra $C_0(X)$ and the dual space of $\mathcal{K}(\mathbb{H})$. In fact, by

Corollary II.7, $\mathcal{K}(\mathbb{H})$ consists of all operators $\sum \lambda_n \xi_n \otimes \bar{\eta}'_n$, where ξ_n and η_n are orthonormal sequences of vectors, $\lim_n \lambda_n = 0$, and $\xi \otimes \bar{\eta}$ is the operator defined by $\xi \otimes \bar{\eta}(v) = \langle v, \eta \rangle \xi$ for $v \in \mathbb{H}$. Note the dual of $C_0(X)$ is $L_1(X)$, and the double dual $C_0(X)^{**}$ is given by $C(X)^{**}_0 = L_1(X)^* = L_\infty(X)$. We shall see shortly that $\mathcal{K}(\mathbb{H})^* = B_1(\mathbb{H})$, the space of trace class operators on $\mathbb{H}$; and thus by Theorem III.5, $\mathcal{K}(\mathbb{H})^{**} = \mathbb{B}(\mathbb{H})$. Furthermore, by Theorem III.5, the natural identification between the σ weakly continuous linear functionals on $B(\mathbb{H})$ and the space $B_1(\mathbb{H})$ of trace class operators is the mapping $\sum \lambda_i \omega_{e_i, e'_i} \mapsto \sum \lambda_i e_i \otimes \bar{e}'_i$ where $\{e_i\}$ and $\{e'_i\}$ are orthonormal sequences and $\{\lambda_i\}$ is a summable sequence of positive real numbers. One has $||f|| = \sum \lambda_i = \mathrm{Tr}(|\sum \lambda_i e_i \otimes \bar{e}'_i|)$.

Let $\omega \in \mathcal{K}(\mathbb{H})^*$. Thus ω is a norm continuous linear functional on the space of compact operators. In particular, the mapping

$$(\xi, \eta) \mapsto \omega(\xi \otimes \bar{\eta})$$

is a bounded sesquilinear form. Hence there is a bounded operator $t(\omega)$ on $\mathbb{H}$ such that

$$\omega(\xi \otimes \bar{\eta}) = \langle t(\omega)\xi, \eta \rangle.$$

We note $t(\omega_{\xi,\eta}) = \xi \otimes \bar{\eta}$. Indeed

$$\begin{aligned}
\omega_{\xi,\eta}(\xi' \otimes \bar{\eta}') &= \langle \xi' \otimes \bar{\eta}'(\xi), \eta \rangle \\
&= \langle \langle \xi, \eta' \rangle \xi', \eta \rangle \\
&= \langle \langle \xi', \eta \rangle \xi, \eta' \rangle \\
&= \langle \xi \otimes \bar{\eta}(\xi'), \eta' \rangle.
\end{aligned}$$

The mapping $\omega \mapsto t(\omega)$ will give the natural isometric linear isomorphism of $B(\mathbb{H})_*$ with $B_1(\mathbb{H})$.

THEOREM VIII.3. *The mapping $\omega \mapsto T$ where T is the unique bounded operator on $\mathbb{H}$ satisfying*

$$\omega(\xi \otimes \bar{\eta}) = \langle T\xi, \eta \rangle$$

is a linear isometry of $\mathcal{K}(\mathbb{H})^*$ onto $B_1(\mathbb{H})$.

PROOF. Note if $\omega \in \mathcal{K}(\mathbb{H})^*$, then $T = t(\omega)$. We claim T is trace class. Indeed, if $\{e_\alpha\}$ and $\{e'_\alpha\}$ are complete orthonormal bases of

$\mathbb{H}$, choose for each α a c_α in $\mathbb{C}$ with $|c_\alpha| = 1$ and $c_\alpha \langle t(\omega)e_\alpha, e'_\alpha \rangle \geq 0$. Then if F is a finite set of α's,

$$\sum_{\alpha \in F} |\langle t(\omega)e_\alpha, e'_\alpha \rangle| = \sum_F \langle t(\omega)c_\alpha e_\alpha, e'_\alpha \rangle$$
$$= \sum_F \omega(c_\alpha e_\alpha \otimes \bar{e}'_\alpha)$$
$$= \omega(\sum_F c_\alpha e_\alpha \otimes \bar{e}'_\alpha)$$
$$\leq ||\omega||$$

for $|| \sum c_\alpha e_\alpha \otimes \bar{e}'_\alpha || = 1$. Thus $\sum_\alpha |\langle Te_\alpha, e'_\alpha \rangle| \leq ||\omega||$. Corollary III.8 implies T is trace class, and $||T||_1 \leq ||\omega||$.

Conversely, suppose T is trace class. Define a linear functional ω on $\mathcal{K}(\mathbb{H})$ by $\omega(A) = \text{Tr}(AT)$. Note Theorem III.5 implies ω is bounded on $\mathcal{K}(\mathbb{H})$, and $||\omega|| \leq ||T||_1$. Also $\omega(\xi \otimes \bar{\eta}) = \text{Tr}((\xi \otimes \bar{\eta})T) = \text{Tr}(\xi \otimes \overline{T^*\eta}) = \langle \xi, T^*\eta \rangle = \langle T\xi, \eta \rangle$. Thus $t(\omega) = T$, and $||\omega|| \leq ||T||_1$. $\square$

5. Polar Decompositions

In Theorem VIII.1, we showed every Hermitian continuous linear functional on a C^* algebra $\mathcal{A}$ has a decomposition of form $f = p_1 - p_2$ where $p_j \geq 0$ and $||f|| = ||p_1|| + ||p_2||$. Our intent is to show the uniqueness of this decomposition. The major tool is a polar decomposition theorem for continuous linear functionals analogous to polar decompositions of measures in the abelian case.

More specifically we will show that if $\mathcal{A}$ is a von Neumann algebra of operators on $\mathbb{H}$ and $f \in \mathcal{A}_*$, that is f is σ weakly continuous and linear, then there is a partial isometry $U \in \mathcal{A}$ and an $|f| \in (\mathcal{A}_*)_+$ with

$$f(x) = \langle x, f \rangle = \langle Ux, |f| \rangle = |f|(Ux).$$

DEFINITION. Let f be a linear functional on an algebra $\mathcal{A}$. For $a \in \mathcal{A}$, define $f \cdot a$ and $a \cdot f$ by

$$\langle f \cdot a, x \rangle = f \cdot a(x) = f(ax) = \langle f, a \cdot x \rangle$$
$$\langle a \cdot f, x \rangle = a \cdot f(x) = f(xa) = \langle xa, f \rangle.$$

Note if $\mathcal{A}$ is a Banach algebra, then $||f \cdot a|| \leq ||f|| \, ||a||$ and $||a \cdot f|| \leq ||a|| \, ||f||$.

LEMMA VIII.2. *Let A be a C^* algebra with identity. If T is an extreme point in the unit ball A_1 of A, then T^*T and TT^* are projections.*

PROOF. Note $0 \leq T^*T \leq I$. Also $||T^*T|| = ||T||^2 = 1$. Let B be the commutative C^* algebra generated by T^*T and I. Let Δ be its compact spectrum. Set f to be the Gelfand transform of T^*T. Then $0 \leq f \leq 1$, and f has value 1 somewhere on Δ. Suppose there is an $x \in \Delta$ with $0 < f(x) < 1$. Choose a continuous function g on Δ with $1 \geq g \geq 0$, $g \neq 0$ near x, $f(1 + g)^2 \leq 1$, and $g = 0$ wherever $f = 1$. Then $\sup f(1 + g)^2 = \sup f(1 - g)^2 = 1$. Let $g = \hat{U}$ where $U \in B$. Then $U \geq 0$ and

$$||T^*T(I + U)^2|| = ||T^*T(I - U)^2|| = 1.$$

Since $(I + U)$ and T^*T commute, we see $T(I + U)$ and $T(I - U)$ are in A_1. But $T = \frac{1}{2}(T(I + U) + T(I - U))$. Clearly $T(I + U) \neq T$. This is a contradiction to extremeness. Finally note if T is extreme, then T^* is also extreme. $\square$

REMARK. The following are equivalent:

(a) T is a partial isometry
(b) T^*T is a projection
(c) $TT^*T = T$
(d) TT^* is a projection
(e) $T^*TT^* = T^*$.

PROOF. To see (b) implies (c), note

$$||TT^*T - T||^2 = ||(TT^*T - T)^*(TT^*T - T)||$$
$$= ||T^*TT^*TT^*T - T^*TT^*T - T^*TT^*T + T^*T||$$
$$= 0.$$

Clearly (c) implies (d), for $TT^*T = T$ implies $TT^*TT^* = TT^*$. The rest follow by similar arguments. $\square$

LEMMA VIII.3. *Let A be a von Neumann algebra on a Hilbert space $\mathbb{H}$. Let $f \in A_*$. Let E be an orthogonal projection in A. Then*

(i) $||f||^2 \geq ||f \cdot E||^2 + ||f \cdot (I - E)||^2$
(ii) *If $||f|| = ||f \cdot E||$, then $f = f \cdot E$.*

PROOF. Clearly (1) implies (2). For (1), we first note we may assume $\mathcal{A} = B(\mathbb{H})$. Indeed by Lemma VII.20, $\mathcal{A}_*$ is isometricly isomorphic to $B(\mathbb{H})_*/\mathcal{A}^\perp$. Hence given an $\epsilon > 0$, there is a $g \in B(\mathbb{H})_*$ with $g|_\mathcal{A} = f$ and $||f|| \leq ||g|| \leq ||f|| + \epsilon$. If we show $||g||^2 \geq ||g \cdot E||^2 + ||g \cdot (I - E)||^2$, we would have $(||f|| + \epsilon)^2 \geq ||f \cdot E||^2 + ||f \cdot (I - E)||^2$ for all $\epsilon > 0$. This would give (1). Note the norms for f are supremums over the unit ball of $\mathcal{A}$ while those for g are supremums on the unit ball of $B(\mathbb{H})$.

Hence we may assume $\mathcal{A} = B(\mathbb{H})$. Now by Theorem III.3, $\overline{B(\mathbb{H})_\sim} = B(\mathbb{H})_*$; and since $||g \cdot a|| \leq ||g|| \, ||a||$, we may by taking limits assume $f \in B(\mathbb{H})_\sim$. By Lemma III.2, $f = \sum_{i=1}^n \lambda_i \omega_{e_i, e_i'}$ where the e_i, e_i' are orthonormal sequences, $\lambda_i > 0$, and $||f|| = \sum \lambda_i$. Now

$$\begin{aligned}
||f \cdot E||^2 &= \sup_{||X|| \leq 1} |f(EX)|^2 \\
&= \sup_{||X|| \leq 1} \left| \sum \lambda_i \langle EXe_i, e_i' \rangle \right|^2 \\
&\leq \sup_{||X|| \leq 1} \left(\sum \lambda_i |\langle Xe_i, Ee_i' \rangle| \right)^2 \\
&\leq \left(\sum \lambda_i ||Ee_i'|| \right)^2.
\end{aligned}$$

Using a similar inequality for $f \cdot (I - E)$, we see

$$\begin{aligned}
||f \cdot E||^2 + ||f \cdot (I - E)||^2 &\leq \left(\sum \lambda_i ||Ee_i'|| \right)^2 + \left(\sum \lambda_i ||(I - E)e_i'|| \right)^2 \\
&= \sum \lambda_i^2 ||Ee_i'||^2 + \sum \lambda_i^2 ||(I - E)e_i'||^2 \\
&\quad + 2 \sum_{1 \leq i < j \leq n} \lambda_i \lambda_j \left(||Ee_i'|| \, ||Ee_j'|| + ||(I - E)e_i'|| \, ||(I - E)e_j'|| \right) \\
&\leq \sum \lambda_i^2 ||Ee_i'||^2 + \sum \lambda_i^2 ||(I - E)e_i'||^2 \\
&\quad + 2 \sum_{i < j} \lambda_i \lambda_j \left(||Ee_i'||^2 + ||(I - E)e_i'||^2 \right)^{1/2} \left(||Ee_j'||^2 + ||(I - E)e_j'||^2 \right)^{1/2} \\
&= \sum \lambda_i^2 ||e_i'||^2 + 2 \sum_{i < j} \lambda_i \lambda_j ||e_i'|| \, ||e_j'|| \\
&= \left(\sum \lambda_i \right)^2 \\
&= ||f||^2.
\end{aligned}$$

$\square$

LEMMA VIII.4. *Let $\mathcal{A}$ be a C^* algebra of operators on $\mathbb{H}$. Suppose $I \in \mathcal{A}$ and $f \in \mathcal{A}_*$. Let $\phi \in \mathcal{A}_+^*$ with $||f|| = ||\phi||$. Suppose $f = \phi \cdot s$*

for some $s \in \mathcal{A}$ with $||s|| \leq 1$ and $||f|| = ||f \cdot t||$ for some $t \in \mathcal{A}$ where $||t|| \leq 1$ and $f \cdot t \geq 0$. Then $f \cdot t = \phi$.

PROOF. Do the GNS construction for $(\mathcal{A}, \phi)$ letting π_ϕ be the corresponding $*$ representation of $\mathcal{A}$ on $\mathbb{H}_\phi$ with cyclic vector ξ_ϕ satisfying $\phi(a) = \langle \pi_\phi(a)\xi_\phi, \xi_\phi \rangle$. Using Corollary II.8, one sees $||f \cdot t|| = \langle f \cdot t, I \rangle$. Hence

$$
\begin{aligned}
\langle \xi_\phi, \pi_\phi(t^*s^*)\xi_\phi \rangle &= \phi(st) \\
&= \langle \phi \cdot s, t \rangle \\
&= f(t) \\
&= \langle f \cdot t, I \rangle \\
&= ||f \cdot t|| \\
&= ||f|| \\
&= ||\phi|| \\
&= ||\xi_\phi||^2.
\end{aligned}
$$

But $|\langle \xi_1, \xi_2 \rangle| = ||\xi_1|| \, ||\xi_2||$ iff ξ_1 and ξ_2 are linearly dependent. Thus $\pi(t^*s^*)\xi_\phi = \xi_\phi$. Hence for $r \in \mathcal{A}$, we have

$$
\begin{aligned}
f \cdot t(r) &= \langle f, tr \rangle \\
&= \langle \phi \cdot s, tr \rangle \\
&= \langle \phi, str \rangle \\
&= \langle \pi_\phi(str)\xi_\phi, \xi_\phi \rangle \\
&= \langle \pi_\phi(r)\xi_\phi, \pi_\phi(t^*s^*)\xi_\phi \rangle \\
&= \langle \pi_\phi(r)\xi_\phi, \xi_\phi \rangle \\
&= \phi(r).
\end{aligned}
$$

□

LEMMA VIII.5. *Let $\mathcal{A}$ be a von Neumann algebra. Let $\phi \in \mathcal{A}_{*+}$. Then there exists a largest orthogonal projection $E \in \mathcal{A}$ with $\phi(E) = 0$. $I - E$ is called the **support** of ϕ.*

PROOF. Let $\mathcal{N} = \{T : \phi(T^*T) = 0\}$. By Lemma II.5, $\mathcal{N}$ is a left ideal in $\mathcal{A}$. Note $\mathcal{N} = \{T \in \mathcal{A} : \phi(ST) = 0 \text{ for } S \in \mathcal{A}\}$. Indeed, if $T \in \mathcal{N}$, then $|\phi(ST)| \leq \phi(SS^*)^{1/2}\phi(T^*T)^{1/2} = 0$. $\mathcal{N}$ is σ weakly closed; for if T_α converges to T in the σ weak topology, then ST_α converges to ST in this topology; and since ϕ is σ weakly continuous, we see if $T_\alpha \in \mathcal{N}$ for all α, then $\phi(ST) = 0$ for all $S \in \mathcal{A}$. By Proposition III.11, $\mathcal{N}$ contains an orthogonal projection

E with $\mathcal{N} = \mathcal{A}E$. Clearly E is the largest orthogonal projection with $\phi(E) = 0$. E is unique; for if F is an orthogonal projection and $\phi(F) = 0$, then $F \in \mathcal{N}$ for $F^*F = F$; and thus $F = FE$ and $F \leq E$. □

PROPOSITION VIII.2. *Let ϕ be a positive σ weakly continuous linear functional on a von Neumann algebra $\mathcal{A}$ having support E. Then $\phi|_{EAE}$ is faithful; i.e., if $X \in E\mathcal{A}E$ and $X \geq 0$, then $\phi(X) = 0$ implies $X = 0$.*

PROOF. Suppose $X \in E\mathcal{A}E$ and $\phi(X) = 0$. Then $EXE = X$ implies $EX^{1/2}E = X^{1/2}$. Hence $\phi(X^{1/2*}X^{1/2}) = \phi(X) = 0$. Thus $X^{1/2} \in \mathcal{N}_\phi$. Hence $X^{1/2} = X^{1/2}(I - E)$ since $\mathcal{N}_\phi = \mathcal{A}(I - E)$. Thus $(I - E)X^{1/2} = X^{1/2}$. Hence $EX^{1/2}E = 0$. Thus $X^{1/2} = 0$, and consequentially $X = 0$. □

Recall if $\mathcal{A}$ is a von Neumann algebra on the Hilbert space $\mathbb{H}$ and E is an orthogonal projection in $\mathcal{A}$, then $E\mathcal{A}E$ is a von Neumann algebra on $E\mathbb{H}$ called $\mathcal{A}_E$.

LEMMA VIII.6. *Let $\mathcal{A}$ be a von Neumann algebra with identity I. Let $\phi \in \mathcal{A}_{*+}$. Let E be the support of ϕ. Then $\phi = E \cdot \phi = \phi \cdot E$, and $\|\phi\| = \phi(E)$.*

PROOF.
$$|(I - E) \cdot \phi(Y)| = |\phi(Y(I - E))|$$
$$\leq \phi(Y^*Y)^{1/2}\phi(I - E)^{1/2} = 0$$

and
$$|\phi \cdot (I - E)(Y)| = |\phi((I - E)Y)|$$
$$\leq \phi(I - E)^{1/2}\phi(Y^*Y) = 0.$$
Thus $\phi = E \cdot \phi = \phi \cdot E$ and $\|\phi\| = \phi(I) = \phi(E)$. □

THEOREM VIII.4 (POLAR DECOMPOSITION). *Suppose $\mathcal{A}$ is a von Neumann algebra on a Hilbert space $\mathbb{H}$. Let $f \in \mathcal{A}_*$.*

(i) *There is a pair (ϕ, U) where $\phi \in (\mathcal{A}_*)_+$ and U is a partial isometry in $\mathcal{A}$ such that $f = \phi \cdot U$ and $UU^* = \operatorname{supp}\phi$. Also $f \cdot U^* = \phi$.*

(ii) *Suppose (ϕ', U') is another pair with $\phi' \in (\mathcal{A}_*)_+$, $U'U'^* \leq \operatorname{supp}\phi'$, $f = \phi' \cdot U'$, and $f \cdot U'^* = \phi'$. Then $\phi = \phi'$ and $U = U'$.*

PROOF. (Existence) We start by finding a U such that $f \cdot U^*(I) = ||f||$. Note we may assume $||f|| = 1$. Let $\mathcal{A}_2 = \{X \in \mathcal{A}_1 : f(X) = 1\}$. Now the unit ball $\mathcal{A}_1$ of $\mathcal{A}$ is weakly compact, and $X \mapsto \langle X, f \rangle$ is weakly continuous. Hence there is an $X_0 \in \mathcal{A}_1$ with $|f(X_0)| = 1$. Thus $\mathcal{A}_2 \neq \emptyset$. $\mathcal{A}_2$ is a weakly compact convex subset of $\mathcal{A}_1$ and so must have an extreme point V. An easy argument shows $\text{ext}\mathcal{A}_2 \subset \text{ext}\mathcal{A}_1$. Lemma VIII.2 implies V^*V is a projection. Thus V is a partial isometry. Now $f \cdot V(I) = 1 = ||f \cdot V||$ for $||f \cdot V|| \leq 1$. By Corollary II.8, we see $f \cdot V \geq 0$. Set $\phi = f \cdot V$, and let E be the support of ϕ. Define U by $U = EV^*$. Note $V^*V \geq E$. Indeed, $\phi(V^*V) = f(VV^*V) = f(V) = 1$ since V is a partial isometry. Thus by Lemma VIII.5, $I - V^*V \leq I - E$, and we see $V^*V \geq E$. Hence $E = EV^*VE = UU^*$. Moreover, $f \cdot U^* = \phi$. Indeed,

$$\begin{aligned}
\langle f \cdot U^*, X \rangle &= \langle f, U^*X \rangle \\
&= \langle f, VEX \rangle \\
&= \langle f \cdot V, EX \rangle \\
&= \langle \phi, EX \rangle \\
&= \langle \phi, X \rangle.
\end{aligned}$$

Next note $\phi \cdot U = f$. Indeed, set $F = U^*U$. Then $\phi \cdot U = f \cdot U^* \cdot U = f \cdot F$. Clearly $||f \cdot F|| \leq 1$. Note also that

$$\begin{aligned}
f \cdot F(U^*) &= \langle f \cdot U^*U, U^* \rangle \\
&= \langle f, U^*UU^* \rangle \\
&= \langle f, U^* \rangle \\
&= \langle f \cdot U^*, I \rangle \\
&= \langle \phi, I \rangle \\
&= 1.
\end{aligned}$$

Thus $||f \cdot F|| = 1 = ||f||$, and (ii) of Lemma VIII.3 implies $f = f \cdot F = \phi \cdot U$.

(Uniqueness) Suppose $f \cdot U'^* = \phi'$, $\phi' \cdot U' = f$, and $U'U'^* \leq \text{supp} \, \phi'$. First note $||f|| = ||\phi'||$. Indeed,

$$||\phi'|| = ||f \cdot U'^*|| \leq ||f|| = ||\phi' \cdot U'|| \leq ||\phi'||.$$

Thus $\phi = \phi'$. Indeed, $f \cdot U^* = \phi \geq 0$, $||\phi|| = ||f||$, $f \cdot U'^* \geq 0$, and $||f \cdot U'^*|| = ||f||$. Hence Lemma VIII.4 gives $\phi' = f \cdot U'^* = \phi$.

We show $U = U'$. Set $X = UU'^*$. Note $\phi(X) = 1$. Indeed,

$$\begin{aligned}
\phi(X) &= \langle \phi, UU'^* \rangle \\
&= \langle \phi \cdot U, U'^* \rangle \\
&= \langle f, U'^* \rangle \\
&= \langle f \cdot U'^*, 1 \rangle \\
&= \langle \phi', 1 \rangle \\
&= \|f\| = 1.
\end{aligned}$$

Also $\|\phi\| = \phi(I) = \phi \cdot E(I) = \phi(E)$, for E is the support of ϕ. Since $\phi = \phi'$, $E = \operatorname{supp}(\phi')$. Hence $E \geq U'U'^*$ by hypothesis. Since $UU^* = E$, we have $EXE = EUU'^*E = UU'^* = X$. Hence $EX = X = XE$, and $EX^* = X^* = X^*E$.

Now note

$$\begin{aligned}
0 &\leq \phi((E - X)(E - X)^*) \\
&= \phi(E) - \phi(X) - \phi(X^*) + \phi(XX^*) \\
&= \phi(X^*X) - \phi(X^*) \\
&\leq \|\phi\| - \overline{\phi(X)} = 0,
\end{aligned}$$

for $\phi(E) = 1$ and $\phi(X) = 1$.

But $\phi|_{EAE}$ is faithful by Proposition VIII.2. Hence $(E - X)(E - X)^* = 0$, and thus $E - X = 0$. Hence $E = X$. Now U is a partial isometry. Thus $UU^* = E = UU'^*$.

Suppose $\xi \in E\mathbb{H}$. Then

$$\|UU'^*\xi\| = \|E\xi\| = \|\xi\|.$$

Since $\|U'^*\xi\| \leq \|\xi\|$, we see $\|U'^*\xi\| = \|\xi\|$; and $U'^*\xi$ belongs to the initial domain of U. Thus $U^*UU'^*\xi = U'^*\xi$, and consequently $U^*X\xi = U'^*\xi$ for $\xi \in E\mathbb{H}$. Since $X = E$, we have $U^* = U'^*$ on the range of E. But $U^*(I - E) = U^*(I - UU^*) = 0$. Since $U'U'^* \leq E$, $Ev = 0$ implies $U'^*v = 0$. Thus on the range of $I - E$, we see both U'^* and U^* vanish. Thus $U'^* = U^*$. This gives $U = U'$. $\square$

DEFINITION. Let $\mathcal{A}$ be a von Neumann algebra on $\mathbb{H}$. Let $f \in \mathcal{A}_*$. Define $|f| = \phi$ if $\phi \geq 0$ and there is a partial isometry U such that $f = \phi \cdot U$, $\phi = f \cdot U^*$ and $UU^* \leq \operatorname{supp} \phi$.

PROPOSITION VIII.3. *Let $f \in \mathcal{A}_*$ where $\mathcal{A}$ is a von Neumann algebra. Then $|f^*| = U^* \cdot |f| \cdot U = U^* \cdot f$ and $\operatorname{supp}|f^*| = U^*U$.*

PROOF. We have $f = |f| \cdot U$, $|f| = f \cdot U^*$, and $UU^* = \operatorname{supp}|f|$. Now $f^*(X) = \overline{f(X^*)} = \overline{|f| \cdot U(X^*)} = \overline{|f|(UX^*)} = \overline{|f|(XU^*)|} = U^* \cdot |f|(X)$ for $X \in \mathcal{A}$. Thus $f^* = U^* \cdot |f| = U^* \cdot |f| \cdot U \cdot U^*$ since $UU^* = \operatorname{supp}|f|$. This gives $U^* \cdot |f| \cdot U = f^* \cdot U$. Furthermore, note $U^*|f|U \geq 0$.

We need to show $U^*U \leq \operatorname{supp} U^*|f|U$. First note if E is a projection in $\mathcal{A}$, then $\langle U^*|f|U, E \rangle = \langle |f|, UEU^* \rangle = 0$ implies $UEU^* \leq I - \operatorname{supp}|f|$. This implies $U(I - \operatorname{supp} U^*|f|U)U^* \leq I - \operatorname{supp}|f|$. Thus if $G = \operatorname{supp} U^*|f|U$, we see $U(I - G)U^* = 0$, for $UU^* = \operatorname{supp}|f|$. This gives $(I - G)U^* = 0$, and thus $(I - G)U^*U = 0$. Hence $U^*U \leq G = \operatorname{supp} U^* \cdot |f| \cdot U$, and by Theorem VIII.4 (ii), we see $|f^*| = U^* \cdot |f| \cdot U$ and $\operatorname{supp}|f^*| = U^*U$. $\square$

Compare the following theorem with Theorem VIII.1 for C^* algebras.

THEOREM VIII.5. *Suppose f is an Hermitian σ weakly continuous linear functional on a von Neumann algebra $\mathcal{A}$. There exist positive σ weakly continuous linear functionals ϕ^+ and ϕ^- with disjoint supports so that $f = \phi^+ - \phi^-$. Moreover, $|f| = \phi^+ + \phi^-$; and this decomposition is unique.*

PROOF. Since $f = f^*$, Theorem VIII.4 and Proposition VIII.3 imply $U = U^*$ and $U^*|f|U = |f|$. Since U^2 is a projection, one has $\sigma(U) \subset \{-1, 0, 1\}$. Hence $U = E_+ - E_-$ where $E_\pm$ are orthogonal projections and $E_+E_- = 0$. Also $U^2 = \operatorname{supp}|f| = E_+ + E_-$. Thus $|f| = (E_+ - E_-)|f|(E_+ - E_-)$. Also by Lemma VIII.6, $|f| = (E_+ + E_-)|f|(E_+ + E_-)$, for $(E_+ + E_-)$ is the support of $|f|$. Combining these give

$$|f| = E_+ \cdot |f| \cdot E_+ + E_- \cdot |f| \cdot E_-.$$

This implies $|f| \cdot E_\pm = E_\pm|f|E_\pm = E_\pm \cdot |f|$, and hence $|f| = |f| \cdot E_+ + |f| \cdot E_-$ and $f = |f| \cdot U = |f| \cdot E_+ - |f| \cdot E_-$. Define $\phi^\pm = E_\pm|f|E_\pm$. These are clearly positive and σ weakly continuous on $\mathcal{A}$. We have $f = \phi^+ - \phi^-$ and $|f| = \phi^+ + \phi^-$. Clearly the $\operatorname{supp} \phi^\pm = (\operatorname{supp}|f|)E_\pm$.

To see uniqueness, suppose $f = \phi_1 - \phi_2$ where ϕ_j, $j = 1, 2$, are positive σ weakly continuous and have disjoint supports. We claim $f = (\phi_1 + \phi_2) \cdot (E_1 - E_2)$ is the polar decomposition of f. Here $E_j = \operatorname{supp} \phi_j$. Clearly $\phi_1 + \phi_2$ is positive, and the support of $\phi_1 + \phi_2$ is $E_1 + E_2$. Thus if $V = E_1 - E_2$, we have $VV^* = \operatorname{supp}|f|$ and $f = |f| \cdot V$ and $|f| = f \cdot V^*$. Thus, using the uniqueness of the polar decomposition, we see $V = U$ and $\phi_1 + \phi_2 = |f| = \phi^+ + \phi^-$. Also $E_1 + E_2 = \operatorname{supp}|f| = E_+ + E_-$. Since $E_1 - E_2 = V = U = E_+ - E_-$, we see $E_1 = E_+$ and $E_2 = E_-$. Thus $\phi_1 = f \cdot E_1 = f \cdot E_+ = \phi^+$. Similarly, $\phi_2 = \phi^-$. $\square$

REMARK. Let ϕ_1 and ϕ_2 be positive σ weakly continuous linear functionals. Then $||\phi_1 - \phi_2|| = ||\phi_1|| + ||\phi_2||$ iff the supports of ϕ_1 and ϕ_2 are disjoint.

PROOF. Suppose the supports are disjoint. Let E_j be the support of ϕ_j, $j = 1, 2$. By Lemma VIII.6, $||\phi_j|| = \phi_i(E_j)$. Set $U = E_1 - E_2 \in \mathcal{A}$. Then $||U|| \leq 1$ and $|(\phi_1 - \phi_2)(U)| = \phi_1(E_1) + \phi_2(E_2) = ||\phi_1|| + ||\phi_2||$. Thus $||\phi_1 - \phi_2|| \geq ||\phi_1|| + ||\phi_2|| \geq ||\phi_1 - \phi_2||$.

Conversely, suppose $||\phi_1 - \phi_2|| = ||\phi_1|| + ||\phi_2||$. Let $E_j = \operatorname{supp} \phi_j$. Set $f = \phi_1 - \phi_2$. Now $f = \phi^+ - \phi^-$ where $\phi^\pm$ have disjoints supports $E_\pm$, and $||f|| = f(E_+ - E_-) = \phi_1(E_+) + \phi_2(E_-) - \phi_1(E_-) - \phi_2(E_+) = ||\phi_1|| + ||\phi_2||$. Hence $\phi_1(E_-) = 0$ and $\phi_2(E_+) = 0$. Hence $||\phi_1|| + ||\phi_2|| = f(E_+ - E_-) = (\phi_1 - \phi_2)(E_+ - E_-)$. Thus

$$\phi_1(E_+) + \phi_2(E_-) = ||\phi_1|| + ||\phi_2||.$$

Hence $\phi_1(E_+) = ||\phi_1|| = \phi_1(I)$ and $\phi_2(E_-) = ||\phi_2|| = \phi_2(I)$. This gives $E_+ \geq \operatorname{supp} \phi_1 = E_1$ and $E_- \geq \operatorname{supp} \phi_2 = E_2$. Thus $E_1 E_2 \leq E_+ E_- = 0$. $\square$

6. The Arens' Construction

Let $\mathcal{A}$ be a normed $*$ algebra. Let $\mathcal{A}^*$ and $\mathcal{A}^{**}$ be the first and second dual spaces. For $x \in \mathcal{A}^{**}$ and $f \in \mathcal{A}^*$, define $x \cdot f \in \mathcal{A}^*$ by

$$x \cdot f(a) = x(f \cdot a).$$

Recall $f \cdot a(b) = \langle f \cdot a, b \rangle = \langle f, ab \rangle = f(ab)$. The Arens' multiplication is defined on $\mathcal{A}^{**}$ by

$$\langle xy, f \rangle = \langle x, y \cdot f \rangle.$$

Note

$$|\langle xy, f \rangle| \leq ||x|| \, ||y \cdot f|| \leq ||x|| \, ||y|| \, ||f||,$$

and thus $||xy|| \leq ||x|| \, ||y||$. Define $x^* \in \mathcal{A}^{**}$ for $x \in \mathcal{A}^{**}$ by

$$\langle x^*, f \rangle = \overline{\langle x, f^* \rangle}.$$

We will show later $\mathcal{A}^{**}$ with these operations is isomorphic to a von Neumann algebra.

7. Bidual of a C* Algebra

PROPOSITION VIII.4. *Let $\mathcal{A}$ be a C^* subalgebra of $B(\mathbb{H})$. Let $\mathcal{B}$ be the weak closure of $\mathcal{A}$, and suppose $I \in \mathcal{B}$. Furthermore, assume each norm continuous linear functional f on $\mathcal{A}$ is σ strongly continuous and has σ-strongly continuous extension $\tilde{f}$ to $\mathcal{B}$. Then*

(i) *$f \xrightarrow{i} \tilde{f}$ is an isometric isomorphism of $\mathcal{A}^*$ onto the predual $\mathcal{B}_*$ of $\mathcal{B}$.*

(ii) *For each $y \in \mathcal{B}$, define $\hat{y} \in \mathcal{A}^{**}$ by $\hat{y}(f) = \tilde{f}(y)$ for $f \in \mathcal{A}^*$. Then $y \mapsto \hat{y}$ is an isometric isomorphism of $\mathcal{B}$ onto $\mathcal{A}^{**}$. This map is the transpose $i^t : \mathcal{B}^*_* = \mathcal{B} \to \mathcal{A}^{**}$. Its restriction to $\mathcal{A}$ is the natural inclusion of $\mathcal{A}$ into $\mathcal{A}^{**}$.*

PROOF. By the Kaplansky density theorem, $\mathcal{A}_1$ is σ strongly dense in $\mathcal{B}_1$. Hence each extension $\tilde{f}$, if it exists, is unique. Note if $\tilde{f}$ is σ strongly continuous, then by Theorem VII.13, it is σ weakly continuous. Hence $\tilde{f} \in \mathcal{B}_*$. Also $||\tilde{f}|| = \sup_{\mathcal{B}_1} |\tilde{f}(x)| = \sup_{\mathcal{A}_1} |f(x)|$ since $\tilde{f}$ is σ strongly continuous and $\mathcal{A}_1$ is σ strongly dense in $\mathcal{B}_1$. Thus $i : \mathcal{A}^* \to \mathcal{B}_*$ is an isometry. It is onto since $i(f|_\mathcal{A}) = f$. This concludes (i).

Note $i^t(y)$ is defined by $i^t(y)(f) = y(i(f)) = y(\tilde{f}) = \tilde{f}(y) = \hat{y}(f)$ for $y \in \mathcal{B}$, $f \in \mathcal{A}^*$. In particular, if $y \in \mathcal{A} \subset \mathcal{B}$, then $\hat{y}(f) = \tilde{f}(y) = f(y)$; and i^t is the natural imbedding of $\mathcal{A}$ into $\mathcal{A}^{**}$. We claim i^t is an isometry. Note

$$||i^t(y)|| = \sup_{f \in \mathcal{A}^*_1} |\tilde{f}(y)|$$
$$= \sup_{f \in \mathcal{B}_{*,1}} |f(y)|$$
$$= ||y||,$$

since $\mathcal{B}_*$ contains the functionals $\omega_{\xi,\eta}$. Clearly i^t is onto, for i is one-to-one and $\overline{\text{Rang}\,(i^t)} = (\text{Ker}\,i)^\perp$ by the Hahn–Banach Theorem. $\square$

COROLLARY VIII.1. *Let $\mathcal{A}$ be the C^* algebra of compact operators on a Hilbert space $\mathbb{H}$. Then*

(i) *Each $f \in \mathcal{A}^*$ is σ strongly continuous on $\mathcal{A}$ and has a unique σ strongly continuous extension $\tilde{f}$ to $\mathcal{B} = \bar{\mathcal{A}}^w = B(\mathbb{H})$.*

(ii) *The map $f \mapsto i(f) = \tilde{f}$ is an isometric isomorphism of $\mathcal{A}^*$ onto the predual $B(\mathbb{H})_*$ of $B(\mathbb{H})$.*

(iii) *The map from $B(\mathbb{H})$ to $\mathcal{A}^{**}$ given by $y \mapsto \hat{y}$ where $\hat{y}(f) = \tilde{f}(y)$ is an isometric isomorphism extending the canonical mapping of $\mathcal{A}$ into $\mathcal{A}^{**}$.*

PROOF. By Lemma III.1, the finite rank operators are weakly dense in $B(\mathbb{H})$. Since they are in $\mathcal{A}$, we see $\bar{\mathcal{A}}^w = B(\mathbb{H})$. Now let $f \in \mathcal{A}^*$. Then, by Theorem VIII.3, f is given by a trace class operator, and $f(C) = \sum_i \lambda_i \omega_{\xi_i, \eta_i}(C)$ where the $\{\xi_i\}$ and $\{\eta_i\}$ are orthonormal sequences and $\lambda_i \geq 0$ with $\sum \lambda_i < \infty$. Thus $\tilde{f} = \sum \omega_{\sqrt{\lambda_i}\xi_i, \sqrt{\lambda_i}\eta_i}$ is a σ weakly and hence σ strongly continuous linear functional on $B(\mathbb{H})$ extending f. $\square$

COROLLARY VIII.2. *Let $\mathcal{A}$ be a C^* algebra. Let π be the universal representation $\oplus_{f \in \mathcal{A}_+^*} \pi_f$ of $\mathcal{A}$. Let $\mathcal{B} = \overline{\pi(\mathcal{A})}^w$. Then $I \in \mathcal{B}$. Moreover,*

(i) *Any positive normal linear form f on $\mathcal{B}$ is of form $f = \omega_{\xi,\xi}$. Each σ strongly continuous linear functional on $\mathcal{B}$ is weakly continuous, and the σ weak topology on $\mathcal{B}$ is the weak topology on $\mathcal{B}$.*

(ii) *If $f \in \mathcal{A}^*$, then there exists a σ weakly continuous linear functional $\tilde{f}$ on $\mathcal{B}$ such that $\tilde{f}(\pi(x)) = f(x)$ for $x \in \mathcal{A}$.*

(iii) *The mapping $f \xrightarrow{i} \tilde{f}$ is an isometric isomorphism of $\mathcal{A}^*$ onto the predual $\mathcal{B}_*$ of $\mathcal{B}$ which takes $\mathcal{A}_+^*$ onto $(\mathcal{B}_*)_+$; also $(\tilde{f})^* = (\widetilde{f^*})$.*

(iv) *For each $y \in \mathcal{B}$, define $\hat{y} \in \mathcal{A}^{**}$ by $\hat{y}(f) = \tilde{f}(y)$. Then $\hat{y} = i^t(y)$ and $i^t : \mathcal{B} \to \mathcal{A}^{**}$ is an isometric isomorphism. Moreover, if $y \in \mathcal{A}$, then $i^t(\pi(y))(f) = \tilde{f}(\pi(y)) = f(y)$, and thus $i^t \circ \pi$ is the natural inclusion of $\mathcal{A}$ into $\mathcal{A}^{**}$. This isomorphism makes $\mathcal{A}^{**}$ isometricly isomorphic to the von Neumann algebra $\mathcal{B}$. With this algebraic structure, $\mathcal{A}^{**}$ is called the universal enveloping von Neumann algebra of the C^* algebra $\mathcal{A}$.*

(v) *The isomorphism $i^t : \mathcal{B} \to \mathcal{A}^{**}$ is bicontinuous between the weak topology on $\mathcal{B}$ and the $\sigma(\mathcal{A}^{**}, \mathcal{A}^*)$ topology on $\mathcal{A}^{**}$.*

PROOF. We first show $I \in \mathcal{B}$. By Theorem III.9, we need only show π is nondegenerate. Suppose $v \perp \operatorname{Rang} \pi(a)$ for all $a \in \mathcal{A}$. But $v = \sum v_f$ where $v_f \in \mathbb{H}_{\pi_f}$. Since $\langle v, \pi(a^*a)v \rangle = 0$, we see $\langle \pi_f(a)v_f, \pi_f(a)v_f \rangle = 0$ for all f and a. Thus $\pi_f(a)v_f = 0$ for all f and a. Thus $v_f \perp \operatorname{Rang} \pi_f(a)$ for all f. Since the π_f are nondegenerate, we see $v_f = 0$ for all f and so $v = 0$. Thus $I \in \mathcal{B}$.

(i) Let f be a positive normal form on $\mathcal{B}$. By Proposition VII.13, f is σ strongly and thus norm continuous. Hence the norm continuity of the representation π implies $f \circ \pi$ is a continuous positive linear functional on $\mathcal{A}$. Thus there is a ξ_f with $f(\pi(a)) = \langle \pi(a)\xi_f, \xi_f \rangle$. Therefore f is weakly continuous on $\pi(\mathcal{A})$, and since $\overline{\pi(\mathcal{A})}^w = \mathcal{B}$, we see $f(b) = \langle b\xi_f, \xi_f \rangle = \omega_{\xi_f, \xi_f}(b)$ for all b in $\mathcal{B}$. Now let f be a σ strongly continuous linear functional on $\mathcal{B}$. Then f is a linear combination of four positive normal forms. Indeed, f is σ weakly continuous and can be written as a linear combination of two self adjoint σ weakly continuous linear functionals and these using the Hahn decomposition may be written as the sum of two elements in $(\mathcal{B}_*)_+$. But, by Proposition VII.13, σ weakly continuous positive linear functionals are normal. Hence each σ strongly continuous linear functional on $\mathcal{B}$ is weakly continuous. Now the σ weak topology on $\mathcal{B}$ contains the weak topology, and since every σ weakly continuous linear functional is weakly continuous and the σ weakly continuous linear functionals determine the σ weak topology (recall the semi-norms for the σ weak topology have form $\|T\| = |\sum \omega_{\xi_n, \eta_n}(T)|$ where $\sum \|\xi_n\|^2 < \infty$ and $\sum \|\eta_n\|^2 < \infty$), we see the σ weak topology on $\mathcal{B}$ is the weak topology on $\mathcal{B}$. This shows (i).

(ii) Let $f \in \mathcal{A}^*$. We may suppose that $f \geq 0$ by Theorem VIII.1 (the Hahn decomposition). But then $f(a) = \langle \pi(a)\xi_f, \xi_f \rangle$ for some ξ_f. Define $\tilde{f}(b) = \langle b\xi_f, \xi_f \rangle$. Then $\tilde{f}$ is weakly continuous and $\tilde{f} \circ \pi = f$.

(iii) All of (iii) follows from Proposition VIII.4 except that we must check the image of $\mathcal{A}^*_+$ is $(\mathcal{B}_*)_+$ and $(\tilde{f})^* = (f^*\tilde{\,})$. Let $f \geq 0$ in $\mathcal{A}^*$. Then $\tilde{f}(b) = \langle b\xi_f, \xi_f \rangle$. Therefore $\tilde{f} \geq 0$. Suppose $\tilde{f} \geq 0$. Then $f = \tilde{f} \circ \pi \geq 0$. Finally if $f \in \mathcal{A}^*$, then $f = f_1 - f_2 + i(f_3 - f_4)$ where $f_j \geq 0$ and $(\tilde{f}_i)^*(b) = \overline{\langle b^*\xi_{f_i}, \xi_{f_i} \rangle} = \overline{\langle b^*\xi_{f_i^*}, \xi_{f_i^*} \rangle} = \langle b\xi_{f_i^*}, \xi_{f_i^*} \rangle = (f_i^*\tilde{\,})(b)$.

(iv) follows from Proposition VIII.4.

(v) By (i) the weak topology on $\mathcal{B}$ is the σ weak topology. Thus $x_\alpha \to x$ σ weakly in $\mathcal{B}$ iff $\langle x_\alpha, \tilde{f} \rangle \to \langle x, \tilde{f} \rangle$ for all $\tilde{f} \in \mathcal{B}_*$ iff $\langle x_\alpha, i(f) \rangle \to \langle x, i(f) \rangle$ for all $f \in \mathcal{A}^*$ iff $\langle i^t(x_\alpha), f \rangle \to \langle i^t(x), f \rangle$ for all $f \in \mathcal{A}^*$ iff $i^t(x_\alpha) \to i^t(x)$ in the $\sigma(\mathcal{A}^{**}, \mathcal{A}^*)$ topology. $\square$

PROPOSITION VIII.5. *The Arens' multiplication and adjoint on $\mathcal{A}^{**}$ are precisely the universal enveloping von Neumann algebra multiplication and adjoint in $\mathcal{A}^{**}$. Hence $\mathcal{A}^{**}$ with the Arens' structure is $*$ isometricly isomorphic to the universal enveloping von Neumann algebra of $\mathcal{A}$.*

PROOF. Let $x, y \in \mathcal{A}^{**}$. We need to show if $i^t X = x$ and $i^t Y = y$, then $i^t(XY) = xy$ and $i^t(X^*) = x^*$. Since $i^t X = x$ and $i^t Y = y$, we have $x(f) = \tilde{f}(X)$ and $y(f) = \tilde{f}(Y)$. We first calculate $(y \cdot \tilde{f})$. Indeed, since $(f \cdot \tilde{a})(\pi(a')) = f \cdot a(a') = f(aa') = \tilde{f}(\pi(aa')) = \tilde{f} \cdot \pi(a)(\pi(a'))$

and $\pi(\mathcal{A})$ is weakly dense in $\mathcal{B}$, we have $(f \cdot \tilde{a}) = \tilde{f} \cdot \pi(a)$. This gives

$$
\begin{aligned}
(y \cdot \tilde{f})(\pi(a)) &= y \cdot f(a) \\
&= \langle y, f \cdot a \rangle \\
&= \langle (f \cdot \tilde{a}), Y \rangle \\
&= \langle \tilde{f} \cdot \pi(a), Y \rangle \\
&= \langle \tilde{f}, \pi(a)Y \rangle \\
&= \langle Y \cdot \tilde{f}, \pi(a) \rangle,
\end{aligned}
$$

and since both sides are strongly continuous and $\pi(\mathcal{A})$ is weakly and hence strongly dense in $\mathcal{B}$, we see $(y \cdot \tilde{f}) = Y \cdot \tilde{f}$. Thus

$$
\begin{aligned}
xy(f) &= x(y \cdot f) \\
&= (y \cdot \tilde{f})(X) \\
&= Y \cdot \tilde{f}(X) \\
&= \tilde{f}(XY).
\end{aligned}
$$

This shows $i^t(XY) = xy$, and i^t preserves multiplication.

For adjoints, using (iii) of Corollary VIII.2, note

$$
\begin{aligned}
i^t(X^*)(f) &= \langle i^t(X^*), f \rangle \\
&= \langle \tilde{f}, X^* \rangle \\
&= \overline{(\tilde{f})^*(X)} \\
&= \overline{(f^*)\tilde{\,}(X)} \\
&= \overline{\langle x, f^* \rangle} \\
&= \langle x^*, f \rangle.
\end{aligned}
$$

□

8. Some Properties of the Universal Enveloping Representation

PROPOSITION VIII.6. *Let $\mathcal{A}$ be a C^* algebra. Let ω be the universal representation of $\mathcal{A}$ on $\mathbb{H}_\omega$. Let ρ be any representation of $\mathcal{A}$ on $\mathbb{H}_\rho$. Let $\mathcal{B} = \overline{\omega(\mathcal{A})}^w$ be the universal enveloping von Neumann algebra. Then there is a unique normal representation $\tilde{\rho}$ of $\mathcal{B}$ on $\mathbb{H}_\rho$ such that $\tilde{\rho}(\mathcal{B}) = \overline{\rho(\mathcal{A})}^w$ in $B(\mathbb{H}_\rho)$ and $\tilde{\rho}(\omega(a)) = \rho(a)$ for $a \in \mathcal{A}$. (Recall $\mathcal{A}^{**} \cong \mathcal{B}$ when $\mathcal{A}^{**}$ has the Arens' structure.)*

PROOF. $\omega(\mathcal{A})$ is σ weakly dense in $\mathcal{B}$. Thus if $\tilde{\rho}_1$ and $\tilde{\rho}_2$ are normal representations, Theorem VII.14 implies they are σ weakly continuous; and hence if they agree on $\omega(\mathcal{A})$, they agree on $\mathcal{B}$. Hence one has uniqueness. Let ρ be a representation of $\mathcal{A}$. Then $\rho = \oplus_{f \in S} \pi_f$ where $S \subset \mathcal{A}_+^*$ (the space on which $\rho = 0$ can be excluded). Hence ρ is a subrepresentation of ω. Define K_ρ to be the projection of $\mathbb{H}_\omega$ onto $\mathbb{H}_\rho = \oplus_{f \in S}\mathbb{H}_{\pi_f}$. Set $\tilde{\rho}(b) = K_\rho b K_\rho$. Then $\overline{\rho(\mathcal{A})}^w = \tilde{\rho}(B)$ and $\tilde{\rho}(\omega(a)) = \omega(a)|_{K_\rho(\mathbb{H}_\omega)} = \rho(a)$. $\square$

We are now able to show the uniqueness in the Hahn decomposition for self adjoint continuous linear functionals on a C^* algebra.

THEOREM VIII.6 (UNIQUENESS-HAHN DECOMPOSITION). *Let $\mathcal{A}$ be a C^* algebra, and let ϕ be a continuous Hermitian linear functional on $\mathcal{A}$. Then there exist unique p_+ and p_- in $(\mathcal{A}^*)_+$ such that $\phi = p_+ - p_-$ and $\|\phi\| = \|p_+\| + \|p_-\|$.*

PROOF. Set $\mathcal{B} = \overline{\omega(\mathcal{A})}^w$. By Corollary VIII.2, there is a unique $\tilde{\phi} \in \mathcal{B}_*$ such that $\phi(a) = \tilde{\phi}(\omega(a))$, where ω is the universal representation of $\mathcal{A}$. Since the correspondence $\phi \mapsto \tilde{\phi}$ is an isometry, we have $\|\phi\| = \|\tilde{\phi}\|$. Moreover, by Corollary VIII.2, $\tilde{\phi}^* = \tilde{\phi}$. By Theorem VIII.5, there is a unique decomposition $\tilde{\phi} = \tilde{p}_+ - \tilde{p}_-$ where $p_\pm$ are σ weakly positive continuous linear functionals on $\mathcal{B}$ with $\|\tilde{\phi}\| = \|\tilde{p}_+\| + \|\tilde{p}_-\|$. The isometric $*$ isomorphism $f \mapsto \tilde{f}$ of Corollary VIII.2 from $\mathcal{A}^*$ onto $\mathcal{B}_*$ gives the result. $\square$

REMARK. Suppose $\mathcal{A}$ is a C^* algebra, and $\omega : \mathcal{A} \to \mathcal{B}$ is a representation of $\mathcal{A}$ into a von Neumann algebra such that for every representation ρ of $\mathcal{A}$, there is a unique normal representation $\tilde{\rho}$ of $\mathcal{B}$ with $\tilde{\rho} \circ \omega = \rho$. This is a universal property that uniquely determines the universal enveloping von Neumann algebra of $\mathcal{A}$ up to a σ weakly continuous isomorphism. To argue this, one can invoke the uniqueness of the Hahn decomposition and the GNS construction to show for each linear functional f in $\mathcal{A}^*$, there is a unique σ-weakly continuous linear function $\tilde{f} \in \mathcal{B}_*$ such that $\tilde{f} \circ \omega = f$. Indeed, it suffices, by the uniqueness of the Hahn decomposition, to show each positive linear functional p has a unique extension $\tilde{p}$ where $\tilde{p}$ is σ-weakly continuous. The GNS construction and the universal property of ω can be used to give an extension; and uniqueness follows from Proposition VII.18, for σ-weakly continuous positive functionals are normal and give normal representations via the GNS construction. Using the Hahn–Banach Theorem, one then sees $\omega(\mathcal{A})$ is σ-weakly dense in

$\mathcal{B}$. Theorem VII.14 would then imply any two von Neumann algebras with the universal property of Proposition VIII.6 are σ-weakly isomorphic.

Let $W^*(G)$ be the universal enveloping von Neumann algebra of the universal enveloping C^* algebra $C^*(G)$ of $L^1(G)$, where G is a locally compact group.

Let π be a unitary representation of G on $\mathbb{H}_\pi$. Then, by Theorem IV.11, π lifts to a representation of the space $M(G)$ of regular Borel measures on G by

$$\pi'(\mu) = \int \pi(x)\,d\mu(x).$$

The representation π' has nondegenerate restriction to $L^1(G)$. Moreover, $\pi'(\mu) \in \pi(G)''$, for by Corollary I.10, discrete measures are weakly dense in $M(G)$. By Corollary IV.9, we know π' lifts from a nondegenerate representation of $L^1(G)$ to a representation π'' of $C^*(G)$; i.e., $\pi''(\tau(f)) = \pi'(f)$, where $\tau : L^1(G) \to C^*(G)$ is the natural injection. By Proposition VIII.6, there is a unique normal representation π''' on $W^*(G)$ with $\pi''' \circ i = \pi''$, where $i : C^*(G) \to W^*(G)$ is the natural injection of $C^*(G)$ into $W^*(G) = C^*(G)^{**}$ and $C^*(G)^{**}$ has the Arens' structure. We no longer use the primes; we use π to denote all three of these representations, identifying them by context.

Let ω be the natural representation of $C^*(G)$ given by

$$\omega = \oplus_{f \in C^*(G)_+^*}\, \pi_f.$$

Then $W^*(G) = \omega(C^*(G))^{**}$, and the natural extension of ω to $W^*(G)$ is the identity representation; i.e., $\tilde\omega = Id$.

Note ω is nondegenerate on $C^*(G)$, for each representation π_f is nondegenerate on $\mathbb{H}_f$. Hence the corresponding representation of $L^1(G)$ is nondegenerate. This implies ω comes from a nondegenerate representation of $L^1(G)$. Thus by Theorem IV.11, there is a unitary representation ω of G which corresponds to the representation ω on $L^1(G)$, ω on $C^*(G)$, and Id on $W^*(G)$. Namely $\omega = \oplus_{\phi \in P(G)}\pi_\phi$, for the positive linear functionals on $C^*(G)$ correspond to the positive definite functions on G. This can be summarized using the identifications given by the correspondences in Corollary II.12 and Theorem

IV.13 by

$$\overline{\omega(C^*(G))}^w = W^*(G)$$
$$\overline{\omega(L^1(G))}^{\text{norm}} = \omega(C^*(G))$$
$$C^*(G)^*_+ = (L^1(G)^*)_+$$
$$L^1(G)^*_+ = P(G).$$

Using the proper context for π and ω, we have the following proposition.

PROPOSITION VIII.7. *Let π be a unitary representation of G on a Hilbert space $\mathbb{H}$. Then for $g \in G$, $\pi(g) = \pi(\omega(g))$.*

PROOF. Let $f \in L^1(G)$ and let $g \in G$. Let ϵ_g be the point mass measure at g. Note $\epsilon_g * f = \lambda(g)f$ where $\lambda(g)$ is defined by $\lambda(g)f(x) = f(g^{-1}x)$. (We are using a left invariant Haar measure on G.) Hence

$$\begin{aligned}
\pi(g)\pi(f) &= \pi(\epsilon_g)\pi(f) \\
&= \pi(\epsilon_g * f) \\
&= \pi(\tau(\epsilon_g * f)) \\
&= \pi(\omega(\tau(\epsilon_g * f))) \\
&= \pi(\omega(\epsilon_g * f)) \\
&= \pi(\omega(\epsilon_g)\omega(f)) \\
&= \pi(\omega(\epsilon_g))\pi(\omega(f)) \\
&= \pi(\omega(\epsilon_g))\pi(\omega(\tau f)) \\
&= \pi(\omega(\epsilon_g))\pi(f).
\end{aligned}$$

In the argument above, we are using the fact that π and ω in each context is a representation of G, of $L^1(G)$, of $C^*(G)$, and $W^*(G)$ along with

$$\pi(\omega(\tau(f))) = \pi(\tau(f)) = \pi(f) = \int f(x)\pi(x)\,dx \text{ and}$$
$$\omega(\tau(f)) = \omega(\omega(\tau(f))) = \omega(f) = \int f(x)\omega(x)\,dx.$$

Since the representation π on $L^1(G)$ is nondegenerate, we see $\pi(g) = \pi(\omega(\epsilon_g))$. $\square$

By definition, $B(G)$ is the linear span of $P(G)$, and by Theorem VIII.2 can be identified with $C^*(G)^*$. Now $B(G)$ has a natural pointwise $*$ algebra structure. We shall show $B(G)$ with this $*$ algebra structure and with norm from $C^*(G)^*$ is a Banach $*$ algebra.

Let $CRB(G)$ denote the space of bounded uniformly right continuous functions on G. Thus $f \in CRB(G)$ iff f is bounded and for each $\epsilon > 0$, there is a neighborhood U of e such that $|f(xy) - f(x)| < \epsilon$ if $y \in U$.

LEMMA VIII.7. *Let $u_\alpha \in C_c(G)$ be an approximate unit in $L^1(G)$ satisfying $u_\alpha(x^{-1}) = u_\alpha(x)$. Let $\mu \in M(G)$. Then*

(i) $\int (\mu * u_\alpha)(x)k(x)dx \to \int k(x)d\mu(x)$ *for all $k \in CRB(G)$.*
(ii) $w(\mu * u_\alpha)$ *converges strongly to $w(\mu)$ where w is the universal representation of G.*

PROOF. (i) Let $k \in CRB(G)$. Then

$$\langle \mu * u_\alpha, k \rangle = \int \int k(xy)u_\alpha(y)\,d\mu(x)\,dy$$
$$= \int \int k(xy)u_\alpha(y^{-1})dy\,d\mu(x)$$
$$= \int k * u_\alpha(x)\,d\mu(x)$$
$$\to \int k(x)\,d\mu(x),$$

for

$$|k * u_\alpha(x) - k(x)| = |\int (k(xy) - k(x))u_\alpha(y^{-1})\,dy| \to 0$$

uniformly since k is right uniformly continuous and $\operatorname{supp} u_\alpha \to \{e\}$.

(ii) Note $w(u_\alpha)w(f) \to w(f)$ for each f in $L^1(G)$, for $u_\alpha * f \to f$. Since w is a nondegenerate representation of $L^1(G)$, the span of the ranges of all the $w(f)$ is dense. Thus $\|w(u_\alpha)\| \leq 1$ for all α implies $w(u_\alpha) \to I$ strongly. Hence $w(\mu * u_\alpha) = w(\mu)w(u_\alpha) \to w(\mu)$ strongly. $\square$

PROPOSITION VIII.8. *Let w be the universal unitary representation of G. Then for $\mu \in M(G)$, $b \in B(G) \cong C^*(G)^*$, we have $\langle \tilde{b}, w(\mu) \rangle = \int b(x)\,d\mu(x)$. (Recall each $b \in C^*(G)^*$ has a natural σ weakly continuous extension $\tilde{b}$ on $C^*(G)^{**} = w(G)'' = W^*(G)$.)*

PROOF. We may assume $b \in P(G)$. We first take $\mu = f(x)\,dx$. Then

$$\langle b, \omega(f) \rangle = \langle b, f \rangle = \int f(x)b(x)\,dx.$$

Next note, by Corollary IV.10, that b is in $CRB(G)$. Now let u_α be as in Lemma VIII.7. Then

$$\langle b, \omega(\mu * u_\alpha) \rangle = \langle \tilde{b}, \omega(\mu * u_\alpha) \rangle$$
$$\rightarrow \langle \tilde{b}, \omega(\mu) \rangle,$$

since $\tilde{b}$ is σ weakly and hence strongly continuous on bounded subsets of $W^*(G)$. But

$$\langle b, \omega(\mu * u_\alpha) \rangle = \int \mu * u_\alpha(x)b(x)\,dx \rightarrow \int b(x)\,d\mu(x)$$

by (i) of Lemma VIII.7. $\square$

9. Imbedding G into $W^*(G)$

Note the σ weak topology on $W^*(G)$ is precisely the $\sigma(W^*(G), B(G))$ topology. Indeed, the correspondence $b \rightarrow \tilde{b}$ is one-to-one and onto between $B(G)$ and $W^*(G)_*$, the space of σ weakly continuous linear functionals, with the latter defining the σ weak topology on $W^*(G)$.

THEOREM VIII.7. *The mapping $\omega : G \rightarrow W^*(G)$ is a topological isomorphism of G into $W^*(G)$, where $W^*(G)$ has the σ weak (equivalently the $\sigma(W^*(G), B(G))$) topology.*

PROOF. ω is continuous for $\langle \tilde{b}, \omega(g) \rangle = \langle \tilde{b}, \omega(\epsilon_g) \rangle = \int b(x)\,d\epsilon_g(x) = b(g)$. Now let U be a symmetric compact neighborhood of e. Then $1_U * 1_U \in P(G)$. Indeed, an easy calculation shows

$$1_U * 1_U(x) = \langle \lambda(x)1_U, 1_U \rangle$$

where λ is the left regular representation of G. Hence suppose $\omega(g_\alpha) \rightarrow \omega(e)$ in $\sigma(W^*(G), B(G))$. Then if g_α doesn't converge to e, we may assume we have a neighborhood V of e such that $g_\alpha \notin V$ for all α. Choose U as above with $U^2 \subset V$. Then if $b = 1_U * 1_U$, we have

$$\langle \tilde{b}, \omega(g_\alpha) \rangle \rightarrow \langle \tilde{b}, \omega(e) \rangle = b(e) = m(U) > 0,$$

while $\langle \tilde{b}, \omega(g_\alpha) \rangle = b(g_\alpha) = 0$, since $b = 1_U * 1_U$ vanishes off U^2. This is a contradiction. $\square$

10. Invariant Sets in $B(G)$

Each $b \in B(G)$ is identified with an element $\tilde{b}$ of $W^*(G)_*$. Hence if $X \in W^*(G)$, $X \cdot b$ and $b \cdot X$ are defined.

LEMMA VIII.8. *Let* $X_\alpha \to X$ *strongly in* $W^*(G)$. *Let* $b \in B(G)$. *Then* $X_\alpha \cdot b \to X \cdot b$ *in* $W^*(G)^*$ *norm. If* $X_\alpha^* \to X_\alpha^*$, *then* $b \cdot X_\alpha \to b \cdot X$ *in* $W^*(G)^*$ *norm.*

PROOF. By Theorem VIII.4, $b = p \cdot V$ where $p \in (W^*(G)_*)_+$ and $VV^* = \operatorname{supp} \phi$. Now using Lemma II.5,

$$
\begin{aligned}
\|X_\alpha \cdot b - X \cdot b\| &= \sup_{y \in W^*(G), \|y\| \le 1} |\langle X_\alpha \cdot p \cdot V - X \cdot p \cdot V, y \rangle| \\
&= \sup_{\|y\| \le 1} |\langle p, Vy(X_\alpha - X) \rangle| \\
&\le \sup_{\|y\| \le 1} p((Vy)(Vy)^*)^{1/2} p((X_\alpha - X)^*(X_\alpha - X))^{1/2} \\
&\le \|p\|^{1/2} p((X_\alpha - X)^*(X_\alpha - X))^{1/2} \to 0;
\end{aligned}
$$

for if $X_\alpha \to X$ strongly, then $(X_\alpha - X)^*(X_\alpha - X) \to 0$ weakly, and by (i) of Corollary VIII.2, p is weakly continuous, for the σ-weak topology is the same as the weak topology. The argument for the other case is similar. $\square$

Using Theorem VIII.2 and (iii) of Corollary VIII.2, we isometrically identify $B(G)$, $C^*(G)^*$ and $W^*(G)_*$. For $b \in B(G)$, we use b and if necessary for clarity $\tilde{b}$ to denote the corresponding element in $W^*(G)_*$.

PROPOSITION VIII.9. *Let* S *be a normed closed subspace of* $B(G)$. *The following are equivalent:*

(i) $S \cdot X \subset SX$ *for all* $X \in W^*(G)$

(ii) $S \cdot X \subset S$ *for all* $X \in C^*(G)$

(iii) $S \cdot \omega(g) \subset S$ *for all* $g \in G$

(iv) S *is left invariant in* $B(G)$ *under the left regular representation*

(v) $S = P \cdot B(G)$ *where* P *is a projection in* $W^*(G)$.

PROOF. (i) implies (iii) is clear. (i) implies (ii) for $b \cdot X = \tilde{b} \cdot \omega(X)$ for $X \in C^*(G)$. (ii) implies (i) follows from Lemma VIII.8 and the fact that $\omega(C^*(G))$ is strongly dense in $W^*(G)$. Similarly (iii) implies (i) since the strong closure of the linear span of $\omega(G)$ is $W^*(G)$. To see (v) implies (i), note $B(G) \cdot X \subset B(G)$; for $B(G) = W^*(G)_*$ and $Y \mapsto XY$ is σ weakly continuous. To show (i) implies (v), we use the notion of polar sets. Since $B(G) = W^*(G)_*$, we have $B(G)^* = W^*(G)$. We note $S^0 = S^\perp$, since S is a subspace. We note S^0 is σ weakly closed, for $S^0 = \cap_{b \in S} \ker b$. Also if $X \in S^\perp$ and $Y \in W^*(G)$, then $b(YX) = b \cdot Y(X) = 0$, since $S \cdot W^*(G) \subset S$. By Proposition VII.14, there is a unique projection $Q \in S^0$ with $S^0 = W^*(G)Q$. Now $(W^*(G)_*, W^*(G))$ is a dual pair; and thus by Proposition VI.10, $(S^0)^0$ is the $\sigma(B(G), W^*(G))$ closed convex hull of S. But S is $\sigma(B(G), W^*(G))$ closed; for by the Hahn–Banach Theorem, if b is not in S, then there is an $X \in B(G)^*$ such that $X = 0$ on S and $\langle X, b \rangle = 1$. Hence, since S is convex, we have $(S^0)^0 = S$. But then

$$
\begin{aligned}
S = (S^0)^0 &= \{b : \langle b, X \rangle = 0 \text{ for all } X \in S^0\} \\
&= \{b : \langle b, XQ \rangle = 0 \text{ for all } X \in W^*(G)\} \\
&= \{b : \langle b, X(I - Q) \rangle = \langle b, X \rangle \text{ for all } X \in W^*(G)\} \\
&= \{b : (I - Q)b = b\} \\
&= (I - Q)B(G).
\end{aligned}
$$

Finally we show (iii) and (iv) are equivalent. For $f \in L^1(G)$, one has $\lambda(g)f = \epsilon_g * f$; and thus

$$
\begin{aligned}
\langle b \cdot \omega(g), f \rangle &= \langle b, \omega(g)\omega(f) \rangle \\
&= \langle b, \omega(\epsilon_g * f) \rangle \\
&= \langle b, \omega(\lambda(g)f) \rangle \\
&= \int b(x)f(g^{-1}x)\,dx \\
&= \int b(gx)f(x)\,dx.
\end{aligned}
$$

This gives $b \cdot \omega(g) = \lambda(g^{-1})b$ which implies the equivalence. $\square$

REMARK. Let Z be a central projection in $W^*(G)$. Then $B(G) \cdot Z = Z \cdot B(G)$ is a two sided invariant subspace of $B(G)$. By Proposition VII.14 and the above proof, each norm closed two sided invariant subspace is of this form.

11. The Support of a Representation

Let π be a unitary representation of G on $\mathbb{H}_\pi$. Then π integrates to a representation $L^1(G)$ which extends to $C^*(G)$. The resulting π factors through $C^*(G)$; and by Proposition VIII.6, this representation extends to a normal representation π of $W^*(G)$. Let

$N = \{X \in W^*(G) : \pi(X) = 0\}$. Then N is a two sided $*$ ideal. By Proposition VII.14, there is a central projection $Z(\pi)$ in $W^*(G)$ such that $N = W^*(G)(I - Z(\pi))$. $Z(\pi)$ is called the (central) **support** of π. Again using Proposition VIII.6, $\pi(W^*(G))$ is a von Neumann algebra on $\mathbb{H}_\pi$; and π factors uniquely through $W^*(G)/N \cong Z[\pi]W^*(G)$. The resulting homomorphism from $W^*(G)Z(\pi)$ to $\pi(W^*(G))$ is normal and hence a σ weak isometry.

LEMMA VIII.9. *Let $b \in B(G)$. Then there is a $p \in P(G)$ and a partial isometry $V \in W^*(G)$ such that $b = p \cdot V$ where $VV^* = supp\,(p) = E$. Moreover, if π_p is the cyclic unitary representation of G with cyclic vector ξ_p satisfying $p(g) = \langle \pi_p(g)\xi_p, \xi_p \rangle$ for $g \in G$ and π is the corresponding normal representation of $W^*(G)$, then $p(X) = \langle \pi(X)\xi_p, \xi_p \rangle$ for $X \in W^*(G)$. Also $E \leq Z(\pi_p)$.*

PROOF. By Proposition VIII.8,

$$p(\omega(\epsilon_g)) = p(g) = \langle \pi_p(g)\xi_p, \xi_p \rangle.$$

Thus $\langle \pi_p(\sum \lambda_i g_i)\xi_p, \xi_p \rangle = p(\sum \lambda_i \omega(\epsilon_{g_i}))$. But both sides are normal and hence σ-weakly continuous on $\omega(G)'' = W^*(G)$, and thus they are equal on $W^*(G)$. This gives $\langle \pi(X)\xi_p, \xi_p \rangle = p(X)$ for $X \in W^*(G)$.

Let E be the support of p. Note

$$p(I - Z(\pi_p)) = \langle \pi(I - Z(\pi_p))\xi_p, \xi_p \rangle = 0.$$

Thus $I - Z(\pi_p) \leq I - E$, and hence $E \leq Z(\pi_p)$. $\square$

Let Z be a central projection in $W^*(G) = \omega(W^*(G))$ where ω is the universal representation. Then $ZW^*(G)$ is a von Neumann algebra in $B(Z\mathbb{H}_\omega)$.

LEMMA VIII.10. *If Z is a central projection in $W^*(G)$, then $(ZW^*(G))_* = Z \cdot W^*(G)_*$.*

PROOF. Each $Z \cdot b \in (ZW^*(G))_*$ if $b \in W^*(G)_*$. By the Hahn–Banach Theorem, each $b \in (ZW^*(G))_*$ has an extension to a $b' \in W^*(G)_*$, for $ZW^*(G)$ is σ weakly closed in $W^*(G)$. But then $b = Z \cdot b'$. $\square$

DEFINITION. Let $\mathcal{M}$ be a von Neumann algebra, and let $E \in \mathcal{M}$ be an orthogonal projection. Then the mapping $M \mapsto EME \in B(E\mathbb{H})$ is called an **induction**. If $E \in \mathcal{M}'$, it is called a **reduction**.

Let Z be a central projection in $W^*(G)$. Then $\omega_Z(X) = XZ$ is a representation of $W^*(G)$ on $Z\mathbb{H}_\omega$ with support Z. We shall show all normal representations of $W^*(G)$ are 'quasi-equivalent' to one of this form.

LEMMA VIII.11. *Let* $b \in B(G)$. *Suppose* $b = p \cdot V$ *where* $p \in P(G)$ *and* V *is a partial isometry in* $W^*(G)$ *with* $VV^* = \operatorname{supp} p$. *Then* $b \in Z(\pi_p) \cdot B(G)$.

PROOF. Since $p(X) = \langle \pi_p(X)\xi_p, \xi_p \rangle$, we have

$$
\begin{aligned}
b(X) &= \langle p \cdot V, X \rangle \\
&= \langle p, VX \rangle \\
&= \langle \pi_p(VX)\xi_p, \xi_p \rangle \\
&= \langle \pi_p(X)\xi_p, \pi_p(V)^*\xi_p \rangle \\
&= \langle \pi_p(X)\xi_p, \eta_p \rangle
\end{aligned}
$$

where $\eta_p = \pi_p(V)^*\xi_p$. Now π_p is normal, and

$$
\pi_p : W^*(G) \to \pi_p(W^*(G)) \subset B(\mathbb{H}_{\pi_p}).
$$

Thus the pretranspose ${}^t\pi_p : \pi_p(W^*(G))_* \to W^*(G)_*$ is given by

$$
({}^t\pi_p f)(X) = f(\pi_p(X)).
$$

Note Theorem VII.14 implies ${}^t\pi_p(f)$ is σ weakly continuous, for f is σ-weakly continuous and π_p is normal. But $\pi_p : Z(\pi_p)W^*(G) \to \pi_p(W^*(G))$ is a σ weak topological isomorphism. Therefore, using Lemma VIII.10, ${}^t\pi_p : \pi_p(W^*(G))_* \to (Z(\pi_p)W^*(G))_* = Z(\pi_p) \cdot W^*(G)_* = Z(\pi_p) \cdot B(G)$ is an isomorphism; i.e., ${}^t\pi_p$ has range the set of all $b \in B(G)$ such that $b = 0$ on $(I - Z(\pi_p))W^*(G)$. But if $b = p \cdot V$ where $VV^* = \operatorname{supp} p$, then by Lemma VIII.9, one has $\operatorname{supp} p \le Z(\pi_p)$. Thus $b \in Z(\pi_p) \cdot B(G)$. $\quad\square$

12. The Norm on $B(G)$

LEMMA VIII.12. *Let* $b \in B(G) = W^*(G)_* = C^*(G)^*$, *and suppose* $b = p \cdot V$ *where* $VV^* = \operatorname{supp} p$ *and* $p \ge 0$. *Then*

$$
\begin{aligned}
\|b\| &= \sup_{f \in L^1(G), \|\omega(f)\| \le 1} \left| \int b(x)f(x)\, dx \right| \\
&= \sup_{f \in L^1(G), \|\tau(f)\|_{C^*(G)} \le 1} \left| \int b(x)f(x)\, dx \right|
\end{aligned}
$$

$$= \sup_{|| \sum \lambda_i \omega(g_i)|| \leq 1} |\langle b, \sum \lambda_i \omega(g_i) \rangle|$$

$$= ||b \cdot Z(\pi_p)||$$

$$= \sup_{|| \sum \lambda_i Z(\pi_p) \omega(g_i)|| \leq 1} |\langle b, \sum \lambda_i \omega(g_i) \rangle|.$$

PROOF. Let $\langle \omega(G) \rangle$ denote the linear span of $\omega(G)$. Then

$$\overline{\langle \omega(G) \rangle}^w = \overline{\omega(L^1(G))}^w = \overline{\omega(C^*(G))}^w = W^*(G).$$

Also ω is an isometry of $C^*(G)$ into $W^*(G)$, and $\overline{L^1(G)}^{\text{norm}} = C^*(G)$. By the Kaplansky density theorem, the unit balls of $\langle \omega(G) \rangle$, $\omega(L^1(G))$, and $\omega(C^*(G))$ are dense in the σ weak topology in the unit ball of $W^*(G)$. These facts together with Lemma VIII.11 and Proposition VIII.8 give the result. $\square$

COROLLARY VIII.3. Let $b = \langle \pi(\cdot)\xi, \eta \rangle \in B(G)$. Then $||b|| \leq ||\xi|| \, ||\eta||$.

PROOF. Let $\tau : L^1(G) \to C^*(G)$ be the inclusion mapping. Since $||\omega(f)|| = ||\tau(f)|| = \sup_{\pi \in R} ||\pi(f)||$, one has

$$||b|| = \sup_{f \in L^1(G), ||\omega(f)|| \leq 1} \left| \int f(x)\langle \pi(g)\xi, \eta \rangle \, dx \right|$$

$$= \sup_{||\omega(f)|| \leq 1} |\langle \pi(f)\xi, \eta \rangle|$$

$$\leq \sup_{||\omega(f)|| \leq 1} ||\pi(f)|| \, ||\xi|| \, ||\eta||$$

$$\leq ||\xi|| \, ||\eta||.$$

$\square$

In Section 1, we noted $||b||_\infty \leq ||b||$. The following is another argument for this result.

COROLLARY VIII.4. $||b||_{B(G)} \geq ||b||_\infty$.

PROOF. By Proposition VIII.8, $b(g) = \langle b, \omega(g)\rangle$. Now, by Lemma VIII.12,

$$
\begin{aligned}
||b|| &= \sup_{||\sum \lambda_i \omega(g_i)|| \leq 1} |\langle b, \sum \lambda_i \omega(g_i)\rangle| \\
&\geq \sup_g |\langle b, \omega(g)\rangle| \\
&= \sup_G |b(g)| \\
&= ||b||_\infty.
\end{aligned}
$$

□

LEMMA VIII.13. *If* $p(x) = \langle \pi(x)\xi, \xi\rangle$ *where* $\pi \in R$, *then* $||p||_{B(G)} = p(e) = ||p||_\infty$.

PROOF. By Corollary VIII.4, we have $||p||_\infty \leq ||p||$; and Proposition IV.18 gives $p(e) = ||p||_\infty$. Also, by Lemma VIII.12,

$$
\begin{aligned}
||p|| &= \sup_{f \in L^1(G), ||\tau(f)|| \leq 1} \int p(x) f(x)\, dx \\
&= \sup_{f \in L^1(G), ||\tau(f)|| \leq 1} \int f(x)\langle \pi(x)\xi, \xi\rangle\, dx \\
&= \sup_{f \in L^1(G), ||\tau(f)|| \leq 1} |\langle \pi(f)\xi, \xi\rangle| \\
&\leq \sup_{f \in L^1(G), ||\tau(f)|| \leq 1} ||\pi(f)||\, ||\xi||^2 \\
&\leq ||\xi||^2 \leq ||p||_\infty,
\end{aligned}
$$

for $||\tau(f)|| = \sup_{\pi' \in R} ||\pi'(f)|| \geq ||\pi(f)||$.   □

LEMMA VIII.14. *Let* $b = p \cdot V$ *where* $p \geq 0$, V *is a partial isometry in* $W^*(G)$, *and* $VV^* = \mathrm{supp}\, p$. *Let* ξ_p *be a cyclic vector satisfying* $p(x) = \langle \pi_p(x)\xi_p, \xi_p\rangle$. *Set* $\eta_p = \pi_p(V^*)\xi_p$. *Then* $b(x) = \langle \pi_p(x)\xi_p, \eta_p\rangle$ *and* $||b|| = ||\xi_p||\, ||\eta_p||$.

PROOF. By Lemma VIII.9, $\langle \pi_p(X)\xi_p, \xi_p\rangle = \langle p, X\rangle$ for all $X \in W^*(G)$; and by Proposition VIII.7, $\pi_p(g) = \pi_p(\omega(g))$ for all $g \in G$.

Hence, we see

$$
\begin{aligned}
b(g) &= \langle b, \omega(\epsilon_g) \rangle \\
&= \langle b, \omega(g) \rangle \\
&= \langle p \cdot V, \omega(g) \rangle \\
&= \langle p, V\omega(g) \rangle \\
&= \langle \pi_p(V\omega(g))\xi_p, \xi_p \rangle \\
&= \langle \pi_p(\omega(g))\xi_p, \pi_p(V)^*\xi_p \rangle \\
&= \langle \pi_p(g)\xi_p, \eta_p \rangle
\end{aligned}
$$

By Corollary VIII.3, $||b|| \le ||\xi_p|| \, ||\eta_p||$; and by Theorem VIII.4, $||b|| = ||p||$. But Lemma VIII.13 gives $||b|| = ||p|| = ||\xi_p||^2 \ge ||\xi_p|| \, ||\pi_p(V^*)\xi_p|| = ||\xi_p|| \, ||\eta_p|| \ge ||b||$ since V is a partial isometry and $||\pi_p(V)|| \le ||V|| \le 1$. $\square$

LEMMA VIII.15. *Let $b, b_1, b_2 \in B(G)$. Then $||b_1 b_2|| \le ||b_1|| \, ||b_2||$. If $\check{b}(g) = b(g^{-1})$, $\bar{b}(g) = \overline{b(g)}$, and $b \cdot x(g) = b(xg)$ for $x \in G$, then $b \mapsto \check{b}$, $b \mapsto \bar{b}$, and $b \mapsto b \cdot x$ are isometric linear or antilinear isometric automorphisms of $B(G)$.*

PROOF. By Lemma VIII.14, we can select ξ_j, η_j, $j = 1, 2$ so that

$$
b_j(g) = \langle \pi_j(g)\xi_j, \eta_j \rangle \text{ and } ||b_j|| = ||\xi_j|| \, ||\eta_j||.
$$

Now $b_1(g)b_2(g) = \langle \pi_1 \otimes \pi_2(g)\xi_1 \otimes \xi_2, \eta_1 \otimes \eta_2 \rangle$. Thus by Corollary VIII.3, $||b_1 b_2|| \le ||\xi_1|| \, ||\xi_2|| \, ||\eta_1|| \, ||\eta_2|| = ||b_1|| \, ||b_2||$.

Note if $b = \langle \pi(\cdot)\xi, \eta \rangle$ where $||b|| = ||\xi|| \, ||\eta||$, then $\check{b}(g) = \langle \xi, \pi(g)\eta \rangle = \langle \bar{\pi}(g)\eta, \xi \rangle$ and $\bar{b}(g) = \langle \bar{\pi}(g)\xi, \eta \rangle$. Thus $||\check{b}|| \le ||\xi|| \, ||\eta|| = ||b||$ and $||\bar{b}|| \le ||\xi|| \, ||\eta|| \le ||b||$. But $\check{\check{b}} = b$ and $\bar{\bar{b}} = b$. Hence $||b|| = ||\bar{\bar{b}}|| \le ||\bar{b}|| \le ||b||$ and $||b|| = ||\check{\check{b}}|| \le ||\check{b}|| \le ||b||$.

Finally $b \cdot x(g) = b(xg) = \langle \pi(g)\xi, \pi(x^{-1})\eta \rangle$. Hence

$$
||b \cdot x|| \le ||\xi|| \, ||\pi(x^{-1})\eta|| = ||b||.
$$

But then $||b|| = ||b \cdot x \cdot x^{-1}|| \le ||b \cdot x|| \le ||b||$. $\square$

THEOREM VIII.8. *$B(G)$ equipped with norm from $C^*(G)^*$ is a Banach $*$ algebra.*

PROOF. This is a combination of Proposition VIII.1, Theorem VIII.2, and Lemma VIII.15. $\square$

13. Quasi-equivalence of Representations

DEFINITION. Let $\mathcal{A}$ be a $*$ algebra, and let π_1 and π_2 be $*$ representations of $\mathcal{A}$. Then π_1 is quasi-equivalent to π_2 (written $\pi_1 \approx \pi_2$) if there is an isomorphism $\Phi : \pi_1(\mathcal{A})'' \to \pi_2(\mathcal{A})''$ satisfying $\Phi(\pi_1(a)) = \pi_2(a)$ for all $a \in \mathcal{A}$. If π_1 and π_2 are unitary representations of a locally compact group, they are quasi-equivalent if there is an isomorphism $\Phi : \pi_1(G)'' \to \pi_2(G)''$ satisfying $\Phi(\pi_1(g)) = \pi_2(g)$ for $g \in G$.

PROPOSITION VIII.10. *If π_1 and π_2 are unitary representations of G, then $\pi_1 \approx \pi_2$, $\pi_1' \approx \pi_2'$, $\pi_1'' \approx \pi_2''$, and $\pi_1''' \approx \pi_2'''$ are equivalent.*

PROOF. First, recall by Theorem VII.14 that any $*$ representation of a von Neumann algebra is normal iff it is σ-weakly continuous. In particular, by Corollary VII.16, any isomorphism between two von Neumann algebras is both normal and bi-σ weakly continuous. We also recall if $\langle \pi_j(G) \rangle$ denotes the linear span of the elements $\pi_j(g)$ for $g \in G$, then all three of the $*$ subalgebras $\langle \pi_j(G) \rangle$, $\pi_j'(L^1(G))$, and $\pi_j''(C^*(G))$ are σ-weakly dense in $\pi_j''(W^*(G)) = \pi_j(G)'' = \pi_j'(L^1(G))'' = \pi_j''(C^*(G))''$.

First suppose $\pi_1''' \approx \pi_2'''$, and Φ is the implementing isomorphism. Then $\pi_2'''(\omega(a)) = \Phi(\pi_1'''(\omega(a)))$ for $a \in C^*(G)$. But $\pi_j'''(\omega(a)) = \pi_j''(a)$ when $a \in C^*(G)$. Thus $\pi_1'' \approx \pi_2''$. Similarly, if $\pi_1'' \approx \pi_2''$ on $C^*(G)$ and Φ satisfies $\pi_2''(a) = \Phi(\pi_1''(a))$ for $a \in C^*(G)$, then if $f \in L^1(G)$ and τ is the inclusion of $L^1(G)$ into $C^*(G)$, one has $\pi_2''(\tau(f)) = \Phi(\pi_1''(\tau(f)))$ for $f \in L^1$. Since $\pi_j'' \circ \tau = \pi_j'$, we see $\pi_1' \approx \pi_2'$. Also, if $\pi_1' \approx \pi_2'$, one would have $\Phi(\pi_1'(f)) = \pi_2'(f)$ for $f \in L^1(G)$. But then $\Phi(\pi_1(g)\pi_1'(f)) = \Phi(\pi_1'(\lambda(g)f)) = \pi_2'(\lambda(g)f) = \pi_2(g)\pi_2'(f)$ where $\lambda(g)f(x) = f(g^{-1}x)$. Thus $\Phi(\pi_1(g))\Phi(\pi_1'(f)) = \pi_2(g)\Phi(\pi_1'(f))$ for all $f \in L^1(G)$. Now since Φ is a bi-σ weakly continuous homeomorphism, $\Phi(\pi_1'(L^1(G)))$ is σ-weakly dense. Since $I \in \pi_2'(L^1(G))''$, the σ-weak closure of $\Phi(\pi_1'(L^1(G)))$, one would then have $\Phi(\pi_1(g)) = \pi_2(g)$ for $g \in G$.

Suppose $\pi_1 \approx \pi_2$. We show $\pi_1''' \approx \pi_2'''$. Note we have an isomorphism Φ satisfying $\Phi(\pi_1(g)) = \pi_2(g)$ for $g \in G$. Let ω be the universal representation of G. Now $\omega(G)'' = W^*(G)$, and hence $\langle \omega(G) \rangle$ is σ-weakly dense in $W^*(G)$. Thus if $X \in W^*(G)$, there is a net $X_i \in \langle \omega(G) \rangle$ such that X_i converges σ-weakly to X. Since Φ is σ-weakly continuous, one has $\Phi(\pi_1'''(X_i)) \to \Phi(\pi_1'''(X))$. But by Proposition VIII.7, $\pi_j'''(\omega(g)) = \pi_j(g)$ for $g \in G$. Thus $\Phi(\pi_1'''(X_i)) = \pi_2'''(X_i)$ for each i. Hence $\pi_2'''(X_i) \to \Phi(\pi_1'''(X))$. But since π_2''' is

σ-weakly continuous, $\pi_2'''(X_i) \to \pi_2'''(X)$. Thus $\Phi(\pi_1'''(X)) = \pi_2'''(X)$ for all $X \in W^*(G)$. This gives $\pi_1''' \approx \pi_2'''$. $\square$

THEOREM VIII.9. *There is a bijection between quasi-equivalence classes of unitary representations on G and central projections in $W^*(G)$ given by $\pi \leftrightarrow Z(\pi)$ where $Z(\pi)$ is the support of π in $W^*(G)$.*

PROOF. Suppose $\pi_1 \approx \pi_2$. Using Proposition VIII.10, we see one has $\Phi(\pi_1'''(X)) = \pi_2'''(X)$ for some $*$ isomorphism Φ from $\pi_1(G)''$ onto $\pi_2(G)''$. In particular $X \in \ker \pi_1'''$ iff $X \in \ker \pi_2'''$. Thus $Z(\pi_1) = Z(\pi_2)$.

Suppose Z is a central projection in $W^*(G)$. Let π''' be the representation of $W^*(G)$ given by $\pi'''(X) = X|_{Z\mathbb{H}}$ where $\mathbb{H}$ is the Hilbert space for $W^*(G)$. Note π''' is normal and nondegenerate, and $\ker(\pi''') = (I - Z)W^*(G)$. Set $\pi(g) = \pi'''(\omega(g))$ for $g \in G$. Then $Z(\pi) = Z$.

Now assume $Z(\pi_1) = Z(\pi_2)$. (We again discontinue the use of the primes.) Then π_1 is an isomorphism of $Z(\pi_1)W^*(G)$ onto $\pi_1(W^*(G))$, and π_2 is an isomorphism of $Z(\pi_2)W^*(G)$ onto $\pi_2(W^*(G))$. Since $Z(\pi_1) = Z(\pi_2)$, $\Phi = \pi_2 \circ \pi_1^{-1}$ is an isomorphism of $\pi_1(W^*(G))$ onto $\pi_2(W^*(G))$. By Proposition VIII.6, $\pi_j(W^*(G)) = \pi_j(G)''$. Moreover, if $X \in W^*(G)$, then $\Phi(\pi_1(X)) = \pi_2(X)$. Indeed, $\pi_1(X) = \pi_1(Z(\pi_1)X)$, and thus $\pi_1^{-1}(\pi_1(X)) = Z(\pi_1)X$. Hence $\Phi(\pi_1(X)) = \pi_2(Z(\pi_1)X) = \pi_2(Z(\pi_2)X) = \pi_2(X)$. Thus $\pi_1 \approx \pi_2$. $\square$

REMARK. Let π_1 and π_2 be unitary representations of G, and π_1''' and π_2''' be their normal extensions to $W^*(G)$. Set $\pi_1 \otimes \pi_2$ to be the tensor product unitary representation of G. Let π''' be its normal extension to $W^*(G)$. Then $\pi''' \neq \pi_1''' \otimes \pi_2'''$, for $X \mapsto \pi_1'''(X) \otimes \pi_2'''(X)$ is not linear. Set $\mathcal{S} = \{S \in W^*(G) : \pi'''(S) = \pi_1'''(S) \otimes \pi_2'''(S) \text{ for all } \pi_1 \text{ and } \pi_2\}$. Then $\mathcal{S} \subset \sigma(B(G))$, the spectrum of $B(G)$. Indeed, if $S \in \mathcal{S}$, define h_S on $B(G)$ by $h_S(b) = \langle b, S \rangle$. Then if $b_j(g) = \langle \pi_j(g)\xi_j, \eta_j \rangle$ for $j = 1$ and $j = 2$, we have $b_1b_2(g) = \langle \pi_1 \otimes \pi_2(g)\xi_1 \otimes \xi_2, \eta_1 \otimes \eta_2 \rangle$, and hence $h_S(b_1b_2) = \langle \pi'''(S)\xi_1 \otimes \xi_2, \eta_1 \otimes \eta_2 \rangle = \langle \pi_1'''(S)\xi_1, \eta_1 \rangle \langle \pi_2'''(S)\xi_2, \eta_2 \rangle = h_S(b_1)h_S(b_2)$. In particular, $\omega(G) \subset \sigma(B(G))$.

14. The Fourier Algebra

PROPOSITION VIII.11. *Let π be a unitary representation of G. Then $Z[\pi] \cdot B(G)$ is the $B(G)$ norm closure of the linear span of all matrix coefficients $x \mapsto \langle \pi(x)\xi, \eta \rangle$ where ξ and η are in $\mathbb{H}_\pi$.*

PROOF. First note $Z[\pi] \cdot B(G)$ is closed in $B(G)$. Indeed, Lemma VIII.12 shows norm convergence in $B(G)$ with $C^*(G)^*$ norm is the same as convergence in norm in $W^*(G)_*$. Thus if $Z[\pi]b_n \to b$, we have $b(Z[\pi]X) = \lim_n Z[\pi] \cdot b_n(Z[\pi]X) = \lim_n Z[\pi] \cdot b_n(X) = b(X)$. Thus $b \in Z[\pi] \cdot B(G)$.

Set A to be the $B(G)$ closure of the linear span of all matrix coefficients $x \mapsto \langle \pi(x)\xi, \eta \rangle$, and let $b(x) = \langle \pi(x)\xi, \eta \rangle$ for some ξ and η in $\mathbb{H}_\pi$. We claim $b(x) \in Z[\pi] \cdot B(G)$. Indeed,

$$\begin{aligned}
Z[\pi] \cdot b(Y) &= b(Z[\pi]Y) \\
&= \langle \pi(Z[\pi]Y)\xi, \eta \rangle \\
&= \langle \pi((I - Z[\pi] + Z[\pi])Y)\xi, \eta \rangle \\
&= \langle \pi(Y)\xi, \eta \rangle \\
&= b(Y),
\end{aligned}$$

for $\ker \pi = W^*(G)(I - Z[\pi])$. This gives $A \subseteq Z(\pi) \cdot B(G)$.

But A is a closed two sided invariant subspace of $B(G)$. Indeed,

$$\begin{aligned}
\langle \pi(xy)\xi, \eta \rangle &= \langle \pi(y)\xi, \pi(x^{-1})\xi \rangle \\
&= \langle \pi(x)\pi(y)\xi, \eta \rangle,
\end{aligned}$$

and by Lemma VIII.15, both left and right translations are isometries of $B(G)$. By Proposition VIII.9 and its following remark, there is a central projection $Z \in W^*(G)$ such that $A = Z \cdot B(G)$. Hence, if $b \in B(G)$, we see $Z[\pi] \cdot Z \cdot b = Z \cdot b$. This gives $b(Z[\pi]ZX) = b(ZX)$ for all $X \in B(G)$ and $b \in B(G) = W^*(G)_*$. Hence $Z[\pi]ZX = ZX$ for all $X \in W^*(G)$. Taking $X = I$, we see $Z \leq Z[\pi]$.

Now note if c is a matrix coefficient of π, we have $Z \cdot c = c$. Thus $(Z[\pi] - Z) \cdot c = 0$. Hence $\langle \pi(Z[\pi] - Z)\xi, \eta \rangle = 0$ for all ξ and η. Thus $Z[\pi] - Z \in \ker \pi = (I - Z[\pi])W^*(G)$, and we see $Z[\pi] - Z \leq I - Z[\pi]$. This gives $Z = Z[\pi]$. $\square$

DEFINITION. The Fourier algebra, $A(G)$, where G is a locally compact group is defined to be

$$\overline{B(G) \cap C_c(G)}^{B(G)}.$$

LEMMA VIII.16. $A(G)$ is a closed translation invariant two sided ideal in $B(G)$.

PROOF. Clearly $B(G) \cap C_c(G)$ is a translation invariant two sided $*$ ideal. By Lemma VIII.15, translation, conjugation and multiplication are $B(G)$ norm continuous. $\square$

REMARK. There are several equivalent definitions for the Fourier algebra. Namely

(a) $A(G) = L^2(G) * L^2(\check{G})$, where $\check{f}(x) = f(x^{-1})$
(b) $A(G) = \overline{B(G) \cap C_c(G)}^{B(G)}$
(c) $A(G) = Z[\lambda] \cdot B(G)$ where λ is the left regular representation of G on $L^2(G)$.

REMARK. In the case when G is abelian, the spectrum of the commutative C^* algebra $C^*(G)$ is $\hat{G}$, the character group of G. Hence $C^*(G)$ is isometric to $C_0(\hat{G})$, and thus $B(G) = C^*(G)^*$ is isometricly isomorphic to $M(\hat{G}) = C_0(\hat{G})^*$. The isomorphism from $M(\hat{G})$ onto $B(G)$ is given by the Fourier transform. Moreover, the Fourier algebra is the Fourier transform of $L^1(\hat{G})$.

Symbols

Bibliography

[1] J. Aarnes and R. Kadison. Pure states and approximate identities. *Proc. Amer. Math. Soc.*, (21):749–752, 1969.

[2] C. A. Akemann and P. A. Ostrand. Computing norms in group C*-algebras. *Amer. J. Math.*, (98):1015–1048, 1976.

[3] N. I. Akhiezer and I. M. Glazman. *The Theory of Linear Operators in Hilbert Spaces*. Frederick Ungar Publishing Co., New York, 1961. (Translated from Russian).

[4] W. Ambrose. Spectral resolution of unitary operators. *Duke Math. J.*, (2):589–595, 1944.

[5] W. Ambrose. Structure theorem for a special class of Banach algebras. *Trans. Amer. Math. Soc.*, (57):364–386, 1945.

[6] W. Ambrose. The L^2-system of a unimodular group, I. *Trans. Amer. Math. Soc.*, (65):27–48, 1949.

[7] H. Araki and R. V. Kadison. *C* Algebras and Applications to Physics*. Lecture Notes in Mathematics 650. Springer-Verlag, New York, 1978.

[8] H. Araki and E. J. Woods. A classification of factors. *Publ. Res. Inst. Mathematical Sci. Society A*, (4):51–130, 1968/69.

[9] R. Arens. On a theorem of Gelfand and Neumark. *Proc. Nat. Acad. Sci. U.S.A.*, (32):237–239, 1946.

[10] R. Arens. Linear topological division algebras. *Bull. Amer. Math. Soc. (N.S.)*, (53):623–630, 1947.

[11] R. Arens. Representations of *-algebras. *Duke Math. J.*, (14):269–282, 1947.

[12] R. Arens. Approximation in, and representation of, certain Banach algebras. *Amer. J. Math.*, (71):783–790, 1949.

[13] R. Arens. A generalization of normed rings. *Pacific J. Math.*, (2):455–471, 1952.

[14] R. Arens and I. Kaplansky. Topological representation of algebras. *Trans. Amer. Math. Soc.*, (63):457–481, 1948.

[15] R. Arens and I. Kaplansky. The adjoint of a bilinear operation. *Proc. Amer. Math. Soc.*, (2):839–848, 1951.

[16] V. Y. Arsenin and A. A. Lyapunov. Theory of A-sets. *Uspekhi Mat. Nauk*, (5):45–108, 1950.

[17] W. Arveson. Subalgebras of C*-algebras. *Acta Math.*, (123):141–224, 1969.

[18] W. Arveson. *An Invitation to C* Algebras*. Springer-Verlag, New York, 1976.

[19] W. Arveson. *Ten Lectures on Operator Algebras*. CBMS 55. American Mathematical Society, Providence, Rhode Island, 1984.

[20] L. Auslander, L. Green, and F. Hahn. *Flows on Homogeneous Spaces*. Annals of Mathematical Studies 53. 1963.

[21] L. Auslander and L. W. Green. Flows on solvmanifolds. *Bull. Amer. Math. Soc. (N.S.)*, (69):745–746, 1963.

[22] L. Auslander and L. W. Green. G induced flows. *Amer. J. Math.*, (88):43–60, 1966.

[23] L. Auslander and B. Kostant. Quantization and representations of solvable Lie groups. *Bull. Amer. Math. Soc. (N.S.)*, (73):692–695, 1967.

[24] L. Auslander and B. Kostant. Polarization and unitary representations of solvable Lie groups. *Invent. Math.*, (14):255–354, 1971.

[25] L. Auslander and C. C. Moore. *Unitary Representations of Solvable Lie Groups*. Memoirs American Mathematical Society 62. American Mathematical Society, Providence, Rhode Island, 1966.

[26] G. Bachman. *Elements of Abstract Harmonic Analysis*. Academic Press, New York and London, 1964.

[27] G. Bachman and L. Narici. *Functional Analysis*. Academic Press, New York and London, 1966.

[28] W. G. Bade. On Boolean algebras of projections and algebras of operators. *Trans. Amer. Math. Soc.*, (86):345–360, 1955.

[29] V. Bargman. Irreducible unitary representations of the Lorentz group. *Ann. of Math. (2)*, (48):568–640, 1947.

[30] V. Bargman. On unitary ray representations of continuous groups. *Ann. of Math. (2)*, (59):1–46, 1954.

[31] H. Bauer. *Wahrscheinlichkeits-theorie und Grundzuge der Masstheorie*. Walter de Gruyter & Co., Berlin, 1968.

[32] S. K. Berberian. The regular ring of a finite AW^*-algebra. *Ann. of Math. (2)*, (65):224–240, 1957.

[33] S. K. Berberian. $N \times N$ matrices over AW^*-algebra. *Amer. J. Math.*, (80):37–44, 1958.

[34] F. A. Berezin, I. M. Gelfand, M. I. Graev, and M. A. Naimark. Representations of groups. *Uspekhi Mat. Nauk*, (11):13–40, 1958.

[35] P. Bernat. Sur les représentations unitaires des groupes de Lie résolubles. *Ann. Sci. École Norm. Sup. (4)*, (82):37–99, 1965.

[36] P. Bernat et al. *Représentations des Groupes de Lie Résolubles.* Dunod, Paris, 1972.

[37] I. N. Bernstein. All reductive algebraic groups are liminal. *Funktsional. Anal. i Prilozhen.*, (8: 2):3–6, 1974. Russian.

[38] I. N. Bernstein. All reductive p-adic groups are tame. *Functional Anal. Appl.*, (8), 1974. (In Russian).

[39] G. Birkoff. *Lattice Theory.* Colloquium Publications XXV. American Mathematical Society, Providence, Rhode Island, 1967.

[40] R. J. Blattner. Positive definite measures. *Proc. Amer. Math. Soc.*, (14):423–428, 1963.

[41] R. J. Blattner. Group extension representations and the structure space. *Pacific J. Math.*, (15):1101–1113, 1965.

[42] R. J. Blattner. Quantization and representation theory. *Proc. Sympos. Pure Math.*, (26):147–165, 1973.

[43] E. Blum. A theory of analytic functions in Banach algebras. *Trans. Amer. Math. Soc.*, (78):343–370, 1955.

[44] F. Bonsall. A minimal property of the norm in some Banach algebras. *J. London Math. Soc. (2)*, (29):156–164, 1954.

[45] Bourbaki. *Éléments de mathématique, Vol. XXIX, Intégration.* 1963.

[46] Bourbaki. *General Topology, Parts 1, 2.* Addison-Wesley, Massachusetts, 1966.

[47] A. M. Bouvier. La formule de Plancherel pour les groupes localement compacts séparables. *C. R. Acad. Sci. Paris Sér. I Math.*, (285):357–360, 1977.

[48] J. Brezin. *Unitary Representations for Solvable Lie Groups.* Memoirs American Mathematical Society 79. American Mathematical Society, Providence, Rhode Island, 1968.

[49] T. Bröcker and T. Dieck. *Representations of Compact Lie Groups.* Springer-Verlag, New York Berlin Heidelberg, 1985.

[50] M. Broise. Commutateurs dans le groupe unitaire d'un facteur. *J. Math. Pures Appl. (9)*, (46):299–312, 1967.

[51] M. Broise. Sur les vecteurs séparateurs des algèbres de von Neumann. *J. Funct. Anal.*, (1):281–289, 1967.

[52] D. Bures. *Abelian Subalgebras of Von Neumann Algebras.* Memoirs American Mathematical Society 110. American Mathematical Society, Providence, Rhode Island, 1971.

[53] J. W. Calkin. Two-sided ideals and convergence in the ring of bounded operators in Hilbert space. *Ann. of Math. (2)*, (42):839–873, 1941.

[54] J. Carmona and M. Vergne. *Non-Commutative Harmonic Analysis.* Lecture Notes in Mathematics 587. Springer-Verlag, Berlin Heidelberg New York, 1977.

[55] H. Cartan. Sur la mesure de Haar. *C. R. Acad. Sci. Paris Sér. I Math.*, (211):759–782, 1940.

[56] H. Cartan and R. Godement. Théorie de la dualité et analyse harmonique dans les groupes abéliens localement compacts. *Ann. Sci. École Norm. Sup. (4)*, (64):79–99, 1947.

[57] C. Chevalley. *Theory of Lie Groups*, volume I. Princeton University Press, Princeton, 1946.

[58] C. Chevalley. *Théory des Groupes de Lie*. Hermann, Paris, 1968.

[59] J. Coleman. Induced and subduced representations. In E. M. Loebl, editor, *Group Theory and Its Applications*, pages 57–118. Academic Press, New York, 1968.

[60] F. Combes. Relations entre formes positives sur une C*-algèbre. *C. R. Acad. Sci. Paris Sér. I Math.*, (260):5435–5438, 1965.

[61] F. Combes. Représentations d'une C*-algèbre et formes linéaires positives. *C. R. Acad. Sci. Paris Sér. I Math.*, (260):5993–5996, 1965.

[62] F. Combes. Etude des représentations tracées d'une C*-algèbre. *C. R. Acad. Sci. Paris Sér. I Math.*, (262):116–117, 1966.

[63] A. Connes. Une classification des facteurs de type III. *Ann. Sci. École Norm. Sup. (4)*, (6):133–252, 1973.

[64] A. Connes. Sur la classification des facteurs de type II. *C. R. Acad. Sci. Paris Sér. I Math.*, (281):13–15, 1975.

[65] C. W. Curtis and I. Reiner. *Representation Theory of Finite Groups and Associative Algebras*. Interscience, New York, 1966.

[66] A. V. Daele. *Continuous Crossed Products and Type III Algebras*. Lecture Series Notes 31. London Mathematical Society, Cambridge, 1978.

[67] W. Darsow. Positive definite functions and states. *Ann. of Math. (2)*, (60):447–453, 1954.

[68] E. B. Davies. On the Borel structure of C* algebras. *Comm. Math. Phys.*, (8):147–163, 1968.

[69] J. Dixmier. Les anneaux d'opérateurs de classe finie. *Ann. Sci. École Norm. Sup. (4)*, (66):209–261, 1949.

[70] J. Dixmier. Mesure de Haar et trace d'un opérateur. *C. R. Acad. Sci. Paris Sér. I Math.*, (228):152–154, 1949.

[71] J. Dixmier. Les fonctionnelles linéaires sur l'ensemble des opérateurs bornés d'un espace de Hilbert. *Ann. of Math. (2)*, (51):387–408, 1950.

[72] J. Dixmier. Sur la réduction des anneaux d'opérateurs. *Ann. Sci. École Norm. Sup. (4)*, (68):185–202, 1951.

[73] J. Dixmier. Formes linéaires sur un anneau d'opérateurs. *Bull. Soc. Math. France*, (81):9–39, 1953.

[74] J. Dixmier. Sur les anneaux d'opérateurs dans les espaces hilbertiens. *C. R. Acad. Sci. Paris Sér. I Math.*, (238):439–441, 1954.

[75] J. Dixmier. Sur les représentations unitaires des groupes de Lie nilpotents, II. *Bull. Soc. Math. France*, (85):325–328, 1957.

[76] J. Dixmier. *Les C*-Algébres et Leurs Représentationes.* Gauthier-Villars, Paris, 1964.

[77] J. Dixmier. *Les Algébres d'Opérateurs dans L'espace Hilbertien.* Gauthier-Villars, Paris, second edition, 1969.

[78] R. S. Doran and J. M. G. Fell. *Representations of Algebras, Locally Compact Groups, and Banach * Algebraic Bundles/Basic Theory of Groups and Algebras.* Academic Press, New York, 1988.

[79] R. Douglas. *C*-Algebra Extensions and K-Homology.* Annals of Mathematical Studies 95. Princeton University Press, Princeton, 1980.

[80] N. Dunford. Resolutions of the identity for commutative B* algebras of operators. *Acta Szeged.,* (12):51–56, 1950.

[81] N. Dunford and J. T. Schwartz. *Linear Operators, Parts I, II.* Interscience, New York-London-Sydney, 1957, 1963.

[82] H. A. Dye. On the geometry of projections in certain operator algebras. *Ann. of Math. (2),* (61):73–89, 1955.

[83] H. A. Dye. On groups of measure preserving transformations, I. *Amer. J. Math.,* (81):119–159, 1959.

[84] H. A. Dye. On groups of measure preserving transformations, II. *Amer. J. Math.,* (86):551–576, 1963.

[85] R. Edwards. Multiplicative norms on Banach algebras. *Math. Proc. Cambridge Philos. Soc.,* (47):473–474, 1951.

[86] E. G. Effros. A decomposition theory for representations of C*-algebras. *Trans. Amer. Math. Soc.,* (107):83–106, 1963.

[87] E. G. Effros. Order ideals in a C*-algebra and its dual. *Duke Math. J.,* (30):391–412, 1963.

[88] E. G. Effros. The Borel space of von Neumann algebras on a separable Hilbert space. *Pacific J. Math.,* (15):1153–1164, 1964.

[89] E. G. Effros. Convergence of closed subsets in a topological space. *Proc. Amer. Math. Soc.,* (16):929–931, 1965.

[90] E. G. Effros. Transformation groups and C*-algebras. *Ann. of Math. (2),* (81):38–55, 1965.

[91] E. G. Effros. The canonical measures for a separable C* algebra. *Amer. J. Math.,* (92):56–60, 1970.

[92] E. G. Effros and F. Hahn. *Locally Compact Transformation Groups and C*-Algebras.* Memoirs American Mathematical Society 75. American Mathematical Society, Providence, Rhode Island, 1967.

[93] L. Ehrenpreis and F. I. Mautner. Uniformly bounded representations of groups. *Proc. Nat. Acad. Sci. U.S.A.,* (41):231–233, 1955.

[94] J. A. Ernest. A decomposition theory for unitary representations of locally compact groups. *Trans. Amer. Math. Soc.,* (104):252–277, 1962.

[95] P. Eymard. L 'Algèbre de Fourier d'un groupe localement compact. *Bull. Soc. Math. France,* (92):181–236, 1964.

[96] R. Fabec. Normal ergodic actions and extensions. *Israel J. Math.*, pages 175–186, 1981.

[97] R. Fabec. Cocycles, extensions of group actions, and bundle representations. *J. Funct. Anal.*, (56):79–98, 1984.

[98] R. Fabec. Induced group actions, representations, and fibered skew product extensions. *Trans. Amer. Math. Soc.*, (301):489–513, 1987.

[99] J. Fell. *An Extension of Mackey's Method to Banach *-Algebraic Bundles*. Memoirs American Mathematical Society 90. American Mathematical Society, Providence, Rhode Island, 1969.

[100] J. M. G. Fell. Representations of weakly closed algebras. *Math. Ann.*, (133):118–126, 1957.

[101] J. M. G. Fell. C*-algebras with smooth dual. *Illinois J. Math.*, (4):221–230, 1960.

[102] J. M. G. Fell. The dual space of C*-algebras. *Trans. Amer. Math. Soc.*, (94):365–403, 1960.

[103] J. M. G. Fell. The structure of algebras of operator fields. *Acta Math.*, (106):233–280, 1961.

[104] J. M. G. Fell. A new proof that nilpotent groups are CCR. *Proc. Amer. Math. Soc.*, (13):93–99, 1962.

[105] J. M. G. Fell. Weak containment and induced representations of groups. *Canad. J. Math.*, (14):237–268, 1962.

[106] J. M. G. Fell. Weak containment and induced representations of groups II. *Trans. Amer. Math. Soc.*, (110):424–447, 1964.

[107] J. M. G. Fell. *Induced Representations and Banach*-Algebraic Bundles*. Lecture Notes in Mathematics 582. Springer-Verlag, Berlin Heidelberg New York, 1977.

[108] P. A. Fillmore. *Notes on Operator Theory*. Van Nostrand Reinhold, New York, 1968.

[109] G. B. Folland. *A Course in Abstract Harmonic Analysis*. Studies in advanced mathematics. CRC Press, Boca Raton, 1995.

[110] N. Friedman. *Introduction to Ergodic Theory*. Van Nostrand Reinhold Company, New York, 1970.

[111] B. Fuglede and R. Kadison. On determinants and a property of the trace in finite factors. *Proc. Nat. Acad. Sci. U.S.A.*, (37):425–431, 1951.

[112] B. Fuglede and R. Kadison. Determinant theory in finite factors. *Ann. of Math. (2)*, (55):520–530, 1952.

[113] M. Fukamiya. On a theorem of Gelfand and Neumark and the B*-algebra. *Kumamoto J. Math.*, (1):17–22, 1952.

[114] S. A. Gaal. *Linear Analysis and Representation Theory*. Springer-Verlag, New York Heidelberg Berlin, 1973.

[115] P. Gard. Symmetrized *n*-th powers of induced representations. *Journal Phys. A. Mathematical Nucl. Gen.*, (6):1807–1828, 1973.

[116] I. M. Gelfand. On normed rings. *Dokl. Akad. Nauk UzSSR*, (23):430–432, 1939.

[117] I. M. Gelfand. Ideale und primäre ideale in normierten ringen. *Matem. Sbornik, 9*, (51):41–48, 1941. (Russian summary).

[118] I. M. Gelfand. Normierte ringe. *Matem. Sbornik, 9*, (51):3–24, 1941.

[119] I. M. Gelfand. Spherical functions on symmetric Riemann spaces. *Dokl. Akad. Nauk UzSSR*, (70):5–8, 1950.

[120] I. M. Gelfand and M. Graev. Analogue to Plancherel's formula for classical groups. *Trudy Mat. Inst. Steklov.*, (4):375–404, 1955.

[121] I. M. Gelfand and M. I. Graev. Traces of unitary représentations of the real unimodular group. *Dokl. Akad. Nauk UzSSR*, (100):1037–1040, 1955.

[122] I. M. Gelfand, M. I. Graev, and N. Y. Vilenkin. *Generalized Functions, Integral Geometry and Representation Theory*. Academic Press, New York and London, 1966.

[123] I. M. Gelfand and A. Kolmogorov. On rings of continuous functions on topological spaces. *Dokl. Akad. Nauk UzSSR*, (22):11–15, 1939.

[124] I. M. Gelfand and M. A. Naimark. On the embedding of normed rings into the ring of operators in Hilbert space. *Mat. Sb.*, 12(54):197–213, 1943.

[125] I. M. Gelfand and M. A. Naimark. Unitary representations of the Lorentz group. *Izv. Ross. Akad. Nauk Ser. Mat.*, (11):411–504, 1947.

[126] I. M. Gelfand and M. A. Naimark. Analogue of the Plancherel formula for the complex unimodular group. *Dokl. Akad. Nauk UzSSR*, (63):609–612, 1948.

[127] I. M. Gelfand and M. A. Naimark. Rings with involution and their representations. *Izv. Ross. Akad. Nauk Ser. Mat.*, (12):445–480, 1948.

[128] I. M. Gelfand and M. A. Naimark. Unitary representations of the classical groups. *Trudy Mat. Inst. Steklov.*, (36):1–288, 1950.

[129] I. M. Gelfand and D. Raikov. On the theory of characters of commutative topological groups. *Dokl. Akad. Nauk UzSSR*, (28):195–198, 1940.

[130] I. M. Gelfand and D. Raikov. Continuous unitary representations of locally bicompact groups. *Matem. Sbornik 13*, (55):301–316, 1943.

[131] I. M. Gelfand and G. Shilov. Commutative normed rings. *Uspekhi Mat. Nauk*, (12):48–146, 1946.

[132] J. Glimm. Locally compact transformation groups. *Trans. Amer. Math. Soc.*, (101):124–138, 1961.

[133] J. Glimm. Type I C*-algebras. *Ann. of Math. (2)*, (73):572–612, 1961.

[134] R. Godement. Les fonctions de type positif et la theorie des groupes. *Trans. Amer. Math. Soc.*, (63):1–84, 1948.

[135] R. Goodman. *Nilpotent Lie Groups*. Lecture Notes in Mathematics 562. Springer-Verlag, Berlin Heidelberg New York, 1976.

[136] K. Gottfried. *Quantum Mechanics*. W. A. Benjamin, Inc., New York-Amsterdam, 1966.

[137] L. W. Green. Spectra of nilflows. *Bull. Amer. Math. Soc. (N.S.)*, (67):414–415, 1961.

[138] W. Greub. *Linear Algebra*. Springer-Verlag, New York, 1967.

[139] E. Griffin. Some contributions to the theory of rings of operators. *Trans. Amer. Math. Soc.*, (75):471–504, 1953.

[140] E. Griffin. Some contributions to the theory of rings of operators, II. *Trans. Amer. Math. Soc.*, (79):389–400, 1955.

[141] R. Gunning and H. Rossi. *Analytic Functions of Several Complex Variables*. Prentice-Hall, New Jersey, 1965.

[142] A. Gurevich. Unitary representation in Hilbert space of compact topological groups. *Matem. Sbornik, 13*, (55):79–86, 1943.

[143] U. Haagerup. Normal weights on W*-algebras. *J. Funct. Anal.*, (19):302–318, 1975.

[144] U. Haagerup. The standard form of von Neumann algebras. *Math. Scand.*, (37):271–283, 1975.

[145] A. Haar. Der massbegriff in der theorie der kontinuierlichen gruppen. *Ann. of Math. (2)*, (34):147–169, 1933.

[146] P. Halmos. *Lectures on Boolean Algebras*. Van Nostrand Reinhold, London, 1963.

[147] P. Halmos. *A Hilbert Space Problem Book*. D. Van Nostrand, 1967.

[148] P. Halmos. *Measure Theory*. D. Van Nostrand Company, Inc., Princeton, 1968.

[149] P. R. Halmos. *Ergodic Theory*. Chelsea, New York, 1956.

[150] P. R. Halmos. *Introduction to Hilbert Space*. Chelsea, New York, second edition, 1957.

[151] P. R. Halmos and J. Von Neumann. Operator algebras in classical mechanics. II. *Ann. of Math. (2)*, (43):332–350, 1942.

[152] Harish-Chandra. Plancherel formula for complex semisimple Lie groups. *Proc. Nat. Acad. Sci. U.S.A.*, (37):813–818, 1951.

[153] Harish-Chandra. Representations of semisimple Lie groups. I, II, III. *Trans. Amer. Math. Soc.*, (75):185–243, 1953.

[154] Harish-Chandra. On the Plancherel formula for the right-invariant functions on a semisimple Lie group. *Proc. Nat. Acad. Sci. U.S.A.*, (40):200–204, 1954.

[155] Harish-Chandra. The Plancherel formula for complex semisimple Lie groups. *Trans. Amer. Math. Soc.*, (76):485–528, 1954.

[156] Harish-Chandra. Harmonic analysis on semi simple Lie groups. *Bull. Amer. Math. Soc. (N.S.)*, (76):529–551, 1970.

[157] S. Helgason. *Differential Geometry and Symmetric Spaces*. Academic Press, New York and London, 1962.

[158] H. Helson. On the ideal structure of group algebras. *Ark. Mat.*, (2):83–86, 1952.

[159] H. Helson. *Lectures on Invariant Subspaces*. Academic Press, New York and London, 1964.

[160] R. Hermann. *Lectures in Mathematical Physics*. W. A. Benjamin, Inc., Reading, Massachusetts, 1972.

[161] I. N. Herstein. *Rings with Involution*. University of Chicago Press, Chicago, 1976.

[162] E. Hewitt. Rings of real-valued continuous functions, I. *Trans. Amer. Math. Soc.*, (64):45–99, 1948.

[163] E. Hewitt. A new proof of Plancherel's theorem for locally compact groups. *Acta Sci. Math. (Szeged)*, (24):219–227, 1963.

[164] E. Hewitt and K. A. Ross. *Abstract Harmonic Analysis*, volume I, II. Springer-Verlag, New York, 1963, 1970.

[165] E. Hewitt and K. Stromberg. *Real and Abstract Analysis*. Springer-Verlag, New York Heidelberg Berlin, 1965.

[166] E. Hille. *Analytic Function Theory*, volume I, II. Blaisdell Publishing Co., Waltham, Massachusetts, 1959, 1962.

[167] R. Howe. Quantum mechanics and partial differential equations. *J. Funct. Anal.*, (38):188–254, 1980.

[168] K. Iséki. On B*-algebras. *Nederl. Akad. Wetensch. Proc. Ser. A*, (15):12–14, 1953.

[169] K. Iwasawa. On group rings of topological groups. *Proc. Imp. Acad. Tokyo*, (20):67–70, 1944.

[170] N. Jacobson. The radical and semi-simplicity for arbitrary rings. *Amer. J. Math.*, (67):300–320, 1945.

[171] N. Jacobson. A topology for the set of primitive ideals in an arbitrary ring. *Proc. Nat. Acad. Sci. U.S.A.*, (31):333–338, 1945.

[172] N. Jacobson. On the theory of primitive rings. *Ann. of Math. (2)*, (48):8–21, 1947.

[173] N. Jacobson. *Structure of Rings*. Colloquium Publications 37. American Mathematical Society, Providence, Rhode Island, 1964.

[174] J. Jauch. *Foundations of Quantum Mechanics*. Addison-Wesley, Reading, Massachusetts, 1966.

[175] R. Kadison. Isometries of operator algebras. *Ann. of Math. (2)*, (54):325–338, 1951.

[176] R. Kadison. Infinite unitary groups. *Trans. Amer. Math. Soc.*, (72):386–399, 1952.

[177] R. Kadison. Infinite general linear groups. *Trans. Amer. Math. Soc.*, (76):68–91, 1954.

[178] R. Kadison. Multiplicity theory for operator algebras. *Proc. Nat. Acad. Sci. U.S.A.*, (41):169–173, 1955.

[179] R. Kadison. On the additivity of the trace in finite factors. *Proc. Nat. Acad. Sci. U.S.A.*, (41):385–387, 1955.

[180] R. Kadison. The trace in finite operator algebras. *Proc. Amer. Math. Soc.*, (12):973–977, 1961.

[181] R. V. Kadison and J. R. Ringrose. *Fundamentals of the Theory of Operator Algebras, Vols. I, II.* Academic Press, New York, 1986.

[182] I. Kaplansky. Topological rings. *Amer. J. Math.*, (69):153–193, 1947.

[183] I. Kaplansky. Dual rings. *Ann. of Math. (2)*, (49):689–701, 1948.

[184] I. Kaplansky. Locally compact rings. *Amer. J. Math.*, (70):447–459, 1948.

[185] I. Kaplansky. Regular Banach algebras. *J. Indian Math. Soc. (N.S.)*, (12):57–62, 1948.

[186] I. Kaplansky. Topological rings. *Bull. Amer. Math. Soc. (N.S.)*, (54):809–826, 1948.

[187] I. Kaplansky. Groups with representations of bounded degree. *Canad. J. Math.*, (1):105–112, 1949.

[188] I. Kaplansky. Normed algebras. *Duke Math. J.*, (16):399–418, 1949.

[189] I. Kaplansky. Primary ideals in group algebras. *Proc. Nat. Acad. Sci. U.S.A.*, (35):133–136, 1949.

[190] I. Kaplansky. Topological representation of algebras. II. *Trans. Amer. Math. Soc.*, (68):62–75, 1950.

[191] I. Kaplansky. The structure of certain operator algebras. *Trans. Amer. Math. Soc.*, (70):219–255, 1951.

[192] I. Kaplansky. A theorem on rings of operators. *Pacific J. Math.*, (1):227–232, 1951.

[193] I. Kaplansky. Algebras of type I. *Ann. of Math. (2)*, (56):460–472, 1952.

[194] I. Kaplansky. Ring isomorphisms of Banach algebras. *Canad. J. Math.*, (6):374–381, 1954.

[195] I. Kaplansky. Generalization of the group duality principle. *Dokl. Akad. Nauk UzSSR*, (138):275–278, 1961.

[196] I. Kaplansky. *Rings of Operators.* W. A. Benjamin, 1968.

[197] I. Kaplansky. *Lie Algebras and Locally Compact Groups.* The University of Chicago Press, Chicago and London, 1971.

[198] D. Kastler and D. Robinson. Invariant states in statistical mechanics. *Comm. Math. Phys.*, (3):151–180, 1966.

[199] J. Kelley. *General Topology.* Van Nostrand Reinhold Company, New York-Toronto-London-Melbourne, 1955.

[200] A. A. Kirillov. Unitary representations of nilpotent Lie groups. *Uspekhi Mat. Nauk*, (17):57–110, 1962. In Russian.

[201] A. A. Kirillov. Infinite dimensional unitary representations of a second order matrix group with elements in a locally compact field. *Dokl. Akad. Nauk UzSSR*, (150):740–743, 1963.

[202] A. A. Kirillov. Plancherel's formula for nilpotent groups. *Functional Anal. Appl.*, (1):84–85, 1967.

[203] A. A. Kirillov. *Elements of the Theory of Representations.* Springer-Verlag, Berlin Heidelberg New York, 1976.

[204] A. Kleppner. On the intertwining number theory. *Proc. Amer. Math. Soc.*, (12):731–733, 1961.

[205] A. Kleppner. Multipliers on abelian groups. *Math. Ann.*, (158):11–34, 1965.

[206] A. Kleppner. Representations induced from compact sub-groups. *Amer. J. Math.*, (88):544–552, 1966.

[207] A. Kleppner. The Plancherel formula for groups extensions II. *Ann. Sci. École Norm. Sup. (4)*, (6):103–132, 1973.

[208] A. Kleppner and R. Lipsman. The Plancherel formula for group extensions I. *Ann. Sci. École Norm. Sup. (4)*, (5):459–516, 1972.

[209] R. Korenblyubi. On some special commutative normed rings. *Dokl. Akad. Nauk UzSSR*, (64):281–284, 1949.

[210] K. Kuratowski. *Topology*, volume I. Academic Press, New York and London, 1966.

[211] G. Larman. Projecting and uniformizing Borel sets with K_σ sections. *Mathematika*, (19):231–244, 1972.

[212] H. Leptin. Darstellungen lokal kompakter gruppen II. *Invent. Math.*, (4):68–86, 1967.

[213] H. Leptin. Verallgemeinerte L^1-algebren und projektive darstellungen lokal kompakter gruppen I. *Invent. Math.*, (3):257–281, 1967.

[214] H. Leptin. Darstellungen verallgemeinerte L^1-algebren. *Invent. Math.*, (5):192–215, 1968.

[215] L. H. Loomis. *An Introduction to Abstract Harmonic Analysis*. D. Van Nostrand Company, Inc., Princeton, New Jersey, 1953.

[216] G. W. Mackey. Imprimitivity for representations of locally compact groups, I. *Proc. Nat. Acad. Sci. U.S.A.*, (35):537–545, 1949.

[217] G. W. Mackey. On a theorem of Stone and von Neumann. *Duke Math. J.*, (16):226–313, 1949.

[218] G. W. Mackey. On induced representations of groups. *Amer. J. Math.*, (73):576–592, 1951.

[219] G. W. Mackey. Induced representations of locally compact groups, I: The Frobenius reciprocity theorem. *Ann. of Math. (2)*, (55):101–139, 1952.

[220] G. W. Mackey. Induced representations of locally compact groups, II: The Frobenius reciprocity theorem. *Ann. of Math. (2)*, (58):193–231, 1953.

[221] G. W. Mackey. Borel structure in groups and their duals. *Trans. Amer. Math. Soc.*, (85):134–165, 1957.

[222] G. W. Mackey. Unitary representations of group extensions, I. *Acta Math.*, (99):265–311, 1958.

[223] G. W. Mackey. Point realizations of transformation groups. *Illinois J. Math.*, (6):327–335, 1962.

[224] G. W. Mackey. Ergodic theory, group theory and differential geometry. *Proc. Nat. Acad. Sci. U.S.A.*, (50):1184–1191, 1963.

[225] G. W. Mackey. Infinite dimensional group representations. *Bull. Amer. Math. Soc. (N.S.)*, (69):628–686, 1963.

[226] G. W. Mackey. *Mathematical Foundations of Quantum Mechanics.* W. A. Benjamin, Inc., New York, 1963.

[227] G. W. Mackey. Ergodic transformations with pure point spectrum. *Illinois J. Math.*, (8):593–600, 1964.

[228] G. W. Mackey. Ergodic theory and virtual groups. *Math. Ann.*, (166):187–207, 1966.

[229] G. W. Mackey. *Group Representations.* Lectures delivered at Oxford University, mimeographed. 1966/67.

[230] G. W. Mackey. *Induced Representations of Groups and Quantum Mechanics.* W. A. Benjamin, New York, 1968.

[231] G. W. Mackey. Virtual groups. In *Topological Dynamics*, pages 335–364. W. A. Benjamin, New York-Amsterdam, 1968.

[232] G. W. Mackey. Ergodicity in the theory of group representations. *Actes Cong. Int. Math. Nice*, (2):401–405, 1970.

[233] G. W. Mackey. Products of subgroups and projective multipliers. *Coll. Mathematical Societats, Journal Bolya*, (5):401–405, 1970. Hilbert space operators.

[234] G. W. Mackey. Ergodic theory and its significance for probability theory and statistical mechanics. *Adv. Math.*, (12):178–268, 1974.

[235] G. W. Mackey. *The Theory of Unitary Group Representations.* University of Chicago Press, Chicago and London, 1976.

[236] S. Matsushita. Plancherel's theorem on general locally compact groups. *Journ. Inst. Polytechn. Osaka City Univ., A*, (4):63–70, 1953.

[237] S. Matsushita. Positive linear functionals on self-adjoint B-algebras. *Proc. Japan Acad. Ser. A Math. Sci.*, (29):427–430, 1953.

[238] S. Matsushita. Sur le théorme de Plancherel. *Proc. Japan Acad. Ser. A Math. Sci.*, (30):557–561, 1954.

[239] S. Matsushita. Positive functionals and representation theory on Banach algebras, I. *Journ. Inst. Polytechn. Osaka City Univ., A*, (6):1–18, 1955.

[240] F. I. Mautner. The completeness of the irreducible unitary representations of a locally compact group. *Proc. Nat. Acad. Sci. U.S.A.*, (34):52–54, 1948.

[241] F. I. Mautner. Unitary representations of locally compact groups, I. *Ann. of Math. (2)*, (51):1–25, 1950.

[242] F. I. Mautner. Unitary representations of locally compact groups, II. *Ann. of Math. (2)*, (52):528–556, 1950.

[243] F. I. Mautner. A generalization of the Frobenius reciprocity theorem. *Proc. Nat. Acad. Sci. U.S.A.*, (37):431–435, 1951.

[244] F. I. Mautner. On the decomposition of unitary representations of Lie groups. *Proc. Amer. Math. Soc.*, (2):490–496, 1951.

[245] F. I. Mautner. Induced representations. *Amer. J. Math.*, (74):737–758, 1952.

[246] F. I. Mautner. Geodesic flows and unitary representations. *Proc. Nat. Acad. Sci. U.S.A.*, (40):33–36, 1954.

[247] F. I. Mautner. Note on the Fourier inversion formula for groups. *Trans. Amer. Math. Soc.*, (78):371–384, 1955.

[248] Y. Misonou. Operator algebras of type I. *Kadai Math. Sem. Reports*, (3):87–90, 1953.

[249] C. C. Moore. Extensions and low dimensional cohomology of locally compact groups I and II. *Trans. Amer. Math. Soc.*, (113):40–63, 64–86, 1964.

[250] C. C. Moore. Ergodicity of flows on homogeneous spaces. *Amer. J. Math.*, (88):154–178, 1966.

[251] C. C. Moore. Restrictions of unitary representations to subgroups and ergodic theory: Group extensions and group cohomology. Number 6 in Lecture Notes in Physics, pages 1–35. Springer-Verlag, Berlin-Heidelberg-New York, 1970.

[252] C. C. Moore. *Harmonic Analysis on Homogeneous Spaces*. Proc. Symp. Pure Math. 26. American Mathematical Society, Providence, Rhode Island, 1972.

[253] C. C. Moore. Extensions and cohomology for locally compact groups. III. *Trans. Amer. Math. Soc.*, (221):1–33, 1976.

[254] C. C. Moore. Ergodic theory and von Neumann algebras. Number 38 in Proc. Sympos. Pure Math., pages 179–226. American Math. Soc., Providence, R. I., 1982. (Part 21).

[255] C. C. Moore and J. Wolf. Square integrable representations of nilpotent Lie groups. *Trans. Amer. Math. Soc.*, (185):445–462, 1973.

[256] C. C. Moore and R. J. Zimmer. Groups admitting ergodic actions with generalized discrete spectrum. *Invent. Math.*, (51):171–188, 1979.

[257] R. T. Moore. *Measurable, Continuous and Smooth Vectors for Semigroups and Representations,AMS*. Memoirs American Mathematical Society 78. American Mathematical Society, Providence, Rhode Island, 1968.

[258] R. D. Mosak. *Banach Algebras*. The University of Chicago Press, Chicago and London, 1975.

[259] F. Murray and J. Von Neumann. On rings of operators I. *Ann. of Math. (2)*, (37):116–229, 1936.

[260] F. Murray and J. Von Neumann. On rings of operators II. *Trans. Amer. Math. Soc.*, (41):208–248, 1937.

[261] F. Murray and J. Von Neumann. On rings of operators IV. *Ann. of Math. (2)*, (44):716–808, 1943.

[262] M. A. Naimark. Positive definite operator functions on a commutative group. *Izv. Ross. Akad. Nauk Ser. Mat.*, (7):237–244, 1943.

[263] M. A. Naimark. Rings of operators in Hilbert space. *Uspekhi Mat. Nauk*, (4):84–147, 1949.

[264] M. A. Naimark. Description of all irreducible unitary representations of the classical groups. *Dokl. Akad. Nauk UzSSR*, (84):883–886, 1952.

[265] M. A. Naimark. On the description of all unitary representations of the complex classical groups, I. *Matem. Sbornik 35*, (77):317–356, 1954.

[266] M. A. Naimark. On the decomposition into factors of a representation of a unitary representation of a locally compact group. *Sibirsk. Mat. Zh.*, (2):89–99, 1961.

[267] M. A. Naimark. On the structure of factors of the representations of a locally compact group. *Dokl. Akad. Nauk UzSSR*, (148):775–778, 1963.

[268] M. A. Naimark. *Normed Rings*. Wolters-Noordhoff, The Netherlands, 1970.

[269] M. A. Naimark and S. V. Fomin. Continuous direct sums of Hilbert spaces and some applications. *Uspekhi Mat. Nauk*, (10):111–142, 1955. Russian.

[270] M. Nakamura. The two-sided representations of an operator algebra. *Proc. Japan Acad. Ser. A Math. Sci.*, (27):172–176, 1951.

[271] M. Nakamura and Z. Takeda. Normal states of commutative operator algebras. *Tôhoku Math. J.*, (5):109–121, 1953.

[272] H. Nakano. Hilbert algebras. *Tôhoku Math. J.*, (2):4–23, 1950.

[273] E. Nelson. Analytic vectors. *Ann. of Math. (2)*, (70):572–615, 1959.

[274] T. Ogasawara and K. Yoshinaga. A non-commutative theory of integration for operators. *Journ. Sci. Hiroshima Univ.*, A, (18):311–347, 1955.

[275] M. Orihara and T. Tsuda. The two sided regular representation of a locally compact group. *Mem. Fac. Sci. Kyushu Univ. Ser. A*, (6):21–29, 1951.

[276] R. Pallu de la Barriére. Algèbres auto-adjointes faiblement fermées et algèbres hilbertiennes de classe finie. *C. R. Acad. Sci. Paris Sér. I Math.*, (232):1994–1995, 1951.

[277] R. Pallu de la Barriére. Isomorphisme des *-algèbres faiblement fermées d'opérateurs. *C. R. Acad. Sci. Paris Sér. I Math.*, (234):795–797, 1952.

[278] R. Pallu de la Barriére. Algèbres unitaires et espaces d'Ambrose. *Ann. Sci. École Norm. Sup. (4)*, (70):381–401, 1953.

[279] R. Pallu de la Barriére. Sur les algèbres d'opérateurs dans les espaces hilbertiens. *Bull. Soc. Math. France*, (82):1–52, 1954.

[280] K. R. Parthasarathy. *Probability Measures on Metric Spaces*. Academic Press, New York, 1967.

[281] K. R. Parthasarathy. *Multipliers on Locally Compact Groups*. Lecture Notes in Mathematics 93. Springer-Verlag, Berlin Heidelberg New

York, 1969.

[282] K. R. Parthasarathy, R. Ranga Rao, and V. S. Varadarajan. Representations of complex semi simple Lie groups and Lie algebras. *Ann. of Math. (2)*, (85):383–429, 1967.

[283] A. L. Paterson. *Amenability*. American Mathematical Society, Providence, Rhode Island, 1988.

[284] L. S. Pontryagin. *Topological Groups*. Princeton University Press, Princeton, New Jersey, 1946. (Translated from Russian).

[285] A. Povzner. On positive functions on an abelian group. *Dokl. Akad. Nauk UzSSR*, (28):294–295, 1940.

[286] R. Powers. Representations of uniformly hyperfinite algebras and their associated von Neumann rings. *Ann. of Math. (2)*, (86):138–171, 1967.

[287] L. Pukansky. The theorem of Radon-Nikodym in operator rings. *Amer. J. Math.*, (15):149–156, 1954.

[288] L. Pukansky. *Leçons sur les Représentations des Groupes*. Dunod, Paris, 1967.

[289] L. Pukansky. On the characters and the Plancherel formula of nilpotent groups. *J. Funct. Anal.*, (1):255–280, 1967.

[290] L. Pukansky. On the theory of exponential groups. *Trans. Amer. Math. Soc.*, (126):487–507, 1967.

[291] L. Pukansky. On the unitary representations of exponential groups. *J. Funct. Anal.*, (2):73–113, 1968.

[292] L. Pukansky. Representations of solvable Lie groups. *Ann. Sci. École Norm. Sup. (4)*, (4):464–608, 1971.

[293] L. Pukansky. The primitive ideal space of solvable Lie groups. *Invent. Math.*, (22):75–118, 1973.

[294] L. Pukansky. Characters of connected Lie groups. *Acta Math.*, (133):81–137, 1974.

[295] L. Pukansky. Lie groups with completely continuous representations. *Bull. Amer. Math. Soc. (N.S.)*, (81):1061–1063, 1975.

[296] D. Raikov. On positive definite functions. *Dokl. Akad. Nauk UzSSR*, (36):857–862, 1940.

[297] D. Raikov. Positive definite functions on commutative groups with invariant measure. *Dokl. Akad. Nauk UzSSR*, (28):296–300, 1940.

[298] D. Raikov. Generalized duality theorem for commutative groups with an invariant measure. *Dokl. Akad. Nauk UzSSR*, (30):589–591, 1941.

[299] D. Raikov. Harmonic analysis on commutative groups with Haar measure and the theory of characters. *Trudy Mat. Inst. Steklov.*, (14):1–86, 1945.

[300] D. Raikov. On the theory of normed rings with involution. *Dokl. Akad. Nauk UzSSR*, (84):387–390, 1946.

[301] D. Raikov. On various types of convergence of positive definite functions. *Dokl. Akad. Nauk UzSSR*, (58):1279–1282, 1947.

[302] A. Ramsay. Virtual groups and group actions. *Adv. Math.*, (6):253–322, 1971.

[303] A. Ramsay. Boolean duals of virtual groups. *J. Funct. Anal.*, (15):56–101, 1974.

[304] A. Ramsay. Subobjects of virtual groups. *Pacific J. Math.*, (87):389–454, 1980.

[305] A. Ramsay. Topologies on measured groupoids. *J. Funct. Anal.*, (47):314–343, 1982.

[306] J. Renault. *A Groupoid Approach to C* Algebras*. Lecture Notes in Mathematics 793. Springer-Verlag, Berlin Heidelberg New York, 1980.

[307] C. Rickart. Banach algebras with an adjoint operation. *Ann. of Math. (2)*, (47):528–550, 1946.

[308] C. Rickart. *General Theory of Banach Algebras*. D. Van Nostrand, 1960.

[309] M. Rieffel. Induced representations of C* algebras. *Adv. Math.*, (13):176–257, 1974.

[310] F. Riesz and B. Sz.-Nagy. *Functional Analysis*. Frederick Ungar Publishing Co., New York, 2nd edition, 1955.

[311] R. Rigelof. Induced representations of locally compact groups. *Acta Math.*, (125):155–187, 1970.

[312] A. P. Robertson and W. J. Robertson. *Topological Vector Spaces*. Cambridge University Press, Great Britain, 1964.

[313] H. L. Royden. *Real Analysis*. 1988.

[314] W. Rudin. *Fourier Analysis on Groups*. Wiley (Interscience), New York, 1962.

[315] W. Rudin. *Real and Complex Analysis*. McGraw-Hill, New York, 2nd edition, 1966.

[316] W. Rudin. *Functional Analysis*. McGraw-Hill, New York, 1973.

[317] L. Rumshisky. On some classes of positive functions. *Dokl. Akad. Nauk UzSSR*, (33):105–108, 1941.

[318] K. Saitô. On the preduals of W*-algebras. *Tôhoku Math. J.*, (19):324–331, 1967.

[319] S. Sakai. A characterization of W*-algebras. *Pacific J. Math.*, (6):763–773, 1956.

[320] S. Sakai. On the σ-weak topology of W*-algebras. *Proc. Japan Acad. Ser. A Math. Sci.*, (62):329–332, 1956.

[321] S. Sakai. On topological properties of W*-algebras. *Proc. Japan Acad. Ser. A Math. Sci.*, (33):439–444, 1957.

[322] S. Sakai. On linear functionals of W*-algebras. *Proc. Japan Acad. Ser. A Math. Sci.*, (34):571–574, 1958.

[323] S. Sakai. *C*-Algebras and W*-Algebras*. Springer-Verlag, New York, 1971.

[324] S. Saks and A. Zygmund. *Analytic Functions.* Elsevier Publishing Company, Amsterdam-London-New York, third edition, 1971.

[325] I. E. Segal. The group ring of a locally compact group I. *Proc. Nat. Acad. Sci. U.S.A.*, (27):348–352, 1941.

[326] I. E. Segal. Representation of certain commutative Banach algebras. *Bull. Amer. Math. Soc. (N.S.)*, (52):421–422, 1946.

[327] I. E. Segal. The group algebra of a locally compact group. *Trans. Amer. Math. Soc.*, (61):69–105, 1947.

[328] I. E. Segal. Irreducible representations of operator algebras. *Bull. Amer. Math. Soc. (N.S.)*, (53):73–88, 1947.

[329] I. E. Segal. A kind of abstract integration pertinent to locally compact groups, I. *Bull. Amer. Math. Soc. (N.S.)*, (55):46, 1949.

[330] I. E. Segal. Two-sided ideals in operator algebras. *Ann. of Math. (2)*, (50):856–865, 1949.

[331] I. E. Segal. An extension of Plancherel's formula to separable unimodular locally compact groups. *Ann. of Math. (2)*, (52):272–292, 1950.

[332] I. E. Segal. The two-sided regular representation of a unimodular locally compact group. *Ann. of Math. (2)*, (51):293–298, 1950.

[333] C. Series. Ergodic actions of product groups. *Harvard Ph. D. thesis*, 1976.

[334] G. Shilov. Ideals and subrings of the ring of continuous functions. *Dokl. Akad. Nauk UzSSR*, (22):7–10, 1939.

[335] G. Shilov. On the extension of maximal ideals. *Dokl. Akad. Nauk UzSSR*, (29):83–85, 1940.

[336] G. Shilov. Sur la théorie des idéaux dans les anneaux normés de fonctions. *Dokl. Akad. Nauk UzSSR*, (27):900–903, 1940.

[337] Y. G. Sinai. *Introduction to Ergodic Theory.* Princeton University Press, Princeton, 1977.

[338] W. F. Stinespring. Positive functions on C*-algebras. *Proc. Amer. Math. Soc.*, (6):211–216, 1955.

[339] M. Stone. On one-parameter unitary groups in Hilbert space. *Ann. of Math. (2)*, (33):843–648, 1932.

[340] M. Stone. Applications of the theory of Boolean rings to general topology. *Trans. Amer. Math. Soc.*, (41):375–481, 1937.

[341] E. Störmer. Positive linear maps of operator algebras. *Acta Math.*, (110):233–278, 1963.

[342] S. Strătilă and L. Zsidó. *Lectures on von Neumann Algebras.* Abacus Press, England, 1979.

[343] O. Takenouchi. On the maximal Hilbert algebras. *Tôhoku Math. J.*, (3):123–131, 1951.

[344] M. Takesaki. Covariant representations of C* algebras and their locally compact automorphism groups. *Acta Math.*, (119):273–303, 1967.

[345] M. Takesaki. *Tomita's Theory of Modular Hilbert Algebras.* Lecture Notes in Mathematics 128. Springer-Verlag, New York, 1970.

[346] M. Takesaki. *Structure of Factors and Automorphism Groups.* Regional Conference Series in Mathematics 51. American Mathematical Society, Providence, Rhode Island, 1983.

[347] M. Tomita. On rings of operators in non-separable Hilbert spaces. *Mem. Fac. Sci. Kyushu Univ. Ser. A*, (7):129–168, 1953.

[348] M. Tomita. Representations of operator algebras. *Math. J. Okayama Univ.*, (3):147–173, 1954.

[349] P. Tondeur. *Introduction to Lie Groups and Transformation Groups.* Lecture Notes in Mathematics 7. Springer-Verlag, 1965.

[350] D. M. Topping. *Lectures on Von Neumann Algebras.* Van Nostrand, Great Britain, 1971.

[351] F. Treves. *Topological Vector Spaces, Distributions and Kernels.* Academic Press, New York, 1967.

[352] A. Van Daele. A new approach to the Tomita-Takesaki theory of generalized Hilbert algebras. *J. Funct. Anal.*, (15):378–393, 1974.

[353] A. Van Daele. The Tomita-Takesaki theory for von Neumann algebras with a separating and cyclic vector. In *C* Algebras and Their Applications to Statistical Mechanics and Quantum Field Theory*, number 60 in Rendiconti della Scuola internazionale di fisica, pages 19–28. North-Holland, Amsterdam-New York, 1976.

[354] V. S. Varadarajan. Groups of automorphisms of Borel spaces. *Trans. Amer. Math. Soc.*, (109):191–220, 1963.

[355] V. S. Varadarajan. *Geometry of Quantum Theory*, volume I, II. D. Van Nostrand Company, Inc., Princeton, 1968, 1970.

[356] V. S. Varadarajan. *Lie Groups, Lie Algebras and Their Representations.* Prentice-Hall, New Jersey, 1974.

[357] N. J. Vilenkin. *Special Functions and the Theory of Group Representations.* Translations of Mathematical Monographs 22. American Mathematical Society, Providence, Rhode Island, 1968.

[358] J. Voisin. On some unitary representations of the Galilei group; Irreducible representations of the Galilei group. *J. Math. Phys.*, (6):1519–1529, 1965.

[359] J. Voisin. On the unitary representations of the Galilei group II: Two particle systems. *J. Math. Phys.*, (6):1822–1832, 1965.

[360] J. Von Neumann. Zur algebra der funktionaloperatoren und theorie der normalen operatoren. *Math. Ann.*, (102):370–427, 1929.

[361] J. Von Neumann. Über funktionen von funktionaloperatoren. *Anal. Math.*, (32):191–226, 1931.

[362] J. Von Neumann. On a certain topology for rings of operators. *Ann. of Math. (2)*, (37):111–115, 1936.

[363] J. Von Neumann. On regular rings. *Proc. Nat. Acad. Sci. U.S.A.*, (22):707–713, 1936.

[364] J. Von Neumann. On rings of operators, III. *Ann. of Math. (2)*, (41):94–161, 1940.

[365] J. Von Neumann. On some algebraic properties of operator rings. *Ann. of Math. (2)*, (44):709–715, 1943.

[366] J. Von Neumann. On rings of operators, reduction theory. *Ann. of Math. (2)*, (50):401–485, 1949.

[367] N. Wallach. *Harmonic Analysis on Homogeneous Spaces*. Marcel Dekker, Inc., New York, 1973.

[368] G. Warner. *Harmonic Analysis on Semisimple Lie Groups*, volume I, II. Springer-Verlag, New York Heidelberg Berlin, 1972.

[369] A. Weil. *L'intégration dans les groupes topologiques et ses applications*. Hermann & Co., Paris, 1940.

[370] A. Weil. Sur certains groupes d'opérateurs unitaires. *Acta Math.*, (111):143–211, 1964.

[371] J. J. Westman. Harmonic analysis on groupoids. *Pacific J. Math.*, (27):621–632, 1968.

[372] J. J. Westman. Virtual group homomorphisms with dense range. *Illinois J. Math.*, (20):41–47, 1976.

[373] H. Weyl. *The Theory of Groups and Quantum Mechanics*. Dover Publications, Inc., New York, second edition, 1931.

[374] A. S. Wightman. On the localizability of quantum mechanical systems. *Rev. Mod. Phys.*, (34):845–872, 1962.

[375] F. Wright. A reduction for algebras of finite type. *Ann. of Math. (2)*, (60):560–570, 1954.

[376] H. Yoshizawa. Unitary representations of locally compact groups. *Osaka J. Math.*, (1):81–89, 1949.

[377] H. Yoshizawa. A proof of the Plancherel theorem. *Proc. Japan Acad. Ser. A Math. Sci.*, (30):276–281, 1954.

[378] K. Yosida. *Functional Analysis*. Springer-Verlag, New York, 2nd edition, 1968.

[379] G. Zeller-Meier. Products croisés d'une C* algèbre par un groupe d'automorphism. *J. Math. Pures Appl. (9)*, (47):101–239, 1968.

[380] R. J. Zimmer. Compact nilmanifold extensions of ergodic actions. *Trans. Amer. Math. Soc.*, (223):397–406, 1976.

[381] R. J. Zimmer. Ergodic actions with generalized discrete spectrum. *Illinois J. Math.*, (20):555–588, 1976.

[382] R. J. Zimmer. Extensions of ergodic group actions. *Illinois J. Math.*, (20):373–409, 1976.

[383] R. J. Zimmer. Cocycles and the structure of ergodic group actions. *Israel J. Math.*, (26):214–220, 1977.

[384] R. J. Zimmer. Normal ergodic actions. *J. Funct. Anal.*, (25):286–305, 1977.

[385] R. J. Zimmer. Orbit spaces of unitary representations, ergodic theory, and simple Lie groups. *Ann. of Math. (2)*, (106):573–588, 1977.

[386] R. J. Zimmer. Amenable ergodic group actions and an application to Poisson boundaries of random walks. *J. Funct. Anal.*, (27):350–372, 1978.

[387] R. J. Zimmer. Induced and amenable ergodic actions of Lie groups. *Ann. Sci. École Norm. Sup. (4)*, (11):407–428, 1978.

[388] R. J. Zimmer. *Essential Results of Functional Analysis.* The University of Chicago Press, Chicago and London, 1990.

Index

Z